T0137137

Studies in Computational Intelligence

Volume 1019

Series Editor

Janusz Kacprzyk, Polish Academy of Sciences, Warsaw, Poland

The series "Studies in Computational Intelligence" (SCI) publishes new developments and advances in the various areas of computational intelligence—quickly and with a high quality. The intent is to cover the theory, applications, and design methods of computational intelligence, as embedded in the fields of engineering, computer science, physics and life sciences, as well as the methodologies behind them. The series contains monographs, lecture notes and edited volumes in computational intelligence spanning the areas of neural networks, connectionist systems, genetic algorithms, evolutionary computation, artificial intelligence, cellular automata, self-organizing systems, soft computing, fuzzy systems, and hybrid intelligent systems. Of particular value to both the contributors and the readership are the short publication timeframe and the world-wide distribution, which enable both wide and rapid dissemination of research output.

Indexed by SCOPUS, DBLP, WTI Frankfurt eG, zbMATH, SCImago.

All books published in the series are submitted for consideration in Web of Science.

More information about this series at https://link.springer.com/bookseries/7092

Allam Hamdan · Aboul Ella Hassanien ·
Timothy Mescon · Bahaaeddin Alareeni
Editors

Technologies, Artificial Intelligence and the Future of Learning Post-COVID-19

The Crucial Role of International Accreditation

Springer

Editors
Allam Hamdan
College of Business and Finance
Ahlia University
Manama, Bahrain

Timothy Mescon
AACSB International
Tampa, FL, USA

Aboul Ella Hassanien
College of Computers and Artificial
Intelligence
Cairo University
Giza, Egypt

Bahaaeddin Alareeni
Majan University College
Muscat, Oman

ISSN 1860-949X   ISSN 1860-9503 (electronic)
Studies in Computational Intelligence
ISBN 978-3-030-93923-6   ISBN 978-3-030-93921-2 (eBook)
https://doi.org/10.1007/978-3-030-93921-2

This Springer imprint is published by the registered company Springer Nature Switzerland AG
The registered company address is: Gewerbestrasse 11, 6330 Cham, Switzerland

Foreword

I am pleased to write this foreword, I have passion for education transformation and deeply believe that sharing best practices from different levels such as faculty, curriculum designers and academic managers will provide an oversight to the readers on quality of teaching and learning and support excellence. This book is edited by a number of diverse academic leaders that will enrich the knowledge of learning in terms of future learning and the use of technology.

Learning and education is vital for society, economic growth and sustainability. Learning providers suffered a number of challenges during COVID-19 that was tackled by a number of technologies. A number of research papers prior to COVID-19 pandemic called for the need of digital transformation which could be used to support multiple dimensions including learning. During COVID-19 digital transformation and technology embarked and countries worldwide transformed, in particular, in the area of teaching and learning.

Low-income countries suffered from lack of infrastructure that enabled moving fast with digital technologies. There has been calls by United Nations to support countries to establish a quality education. Other countries with well-established ICT infrastructure were able to transform in a very short period. However, a number of challenges encountered dealing with emerging technologies and maintaining ethical standards was one of the serious challenges faced worldwide. Despite the academic misconduct procedures set, learning providers found a high number of unethical practices.

Gamification and digital applications were used as part of learning that is based on artificial intelligence. A number of practices shared a positive and interesting finding; however, the impact of such applications on learners' progression is yet ambiguous. This book will share with faculty and practitioners' best practices that could be carried away as well as shed the light on future research that could be taken away by scholars.

International Accreditation and Quality Assurance Standards act as drivers for education transformation. A number of accreditation and quality assurance standards were upgraded and developed to assess the effectiveness of online and blended learning. This book will add a value for accreditation and quality assurance executives and academic leaders to set new strategic dimension to measure the effectiveness of technology in learning.

I could assure the readers that this book will be the real guide for learning transformation with a number of best practices from a number of countries worldwide. This book will support adaptation of new technological methods which could help in designing new pedagogical methods that are sustainable for quality education.

Esra Saleh Al Dhaen, PFHEA
Assistant Professor
Executive Director Centre for
Accreditation and Quality Assurance
Ahlia University
Manama, Bahrain
esaldhaen@ahlia.edu.bh

Preface

Education delivery requires the support of new technologies such as Artificial Intelligence (AI), the Internet of Things (IoT), big data and machine learning to fight and aspire to new diseases. The academic community and those interested in education agree that education after the Corona pandemic will not be the same as before.

This book aims to assess the experience of education during COVID-19 pandemic and explore the future of application of Artificial Intelligence (AI), the Internet of Things (IoT), big data, machine learning and simulation in education. The book also questions the role of accreditation bodies (e.g. AACSB, etc.) to ensure the effectiveness and efficiency of technology tools in achieving distinguished education in times of crisis.

This book includes 36 chapters by authors from several countries, and all the chapters have been evaluated by the editorial board and reviewed based on double-blind peer-review system by at least two reviewers.

The chapters of the book are divided into five main parts:

Distance learning solutions: the role of artificial intelligence.
Online Examination Systems, preventing cheating and the ethical, social and professional view of them.
Experiences and practices of technology application in times of the COVID-19 pandemic.
Applying Simulation Tools in Education During the COVID-19 Pandemic.
The governmental support in distance education and Digital learning management systems.

The chapters of this book present a selection of high-quality research on the theoretical and practical levels, which ground the uses of technology and artificial intelligence in education. We hope that the contribution of this book will be at the academic level and decision-makers in the various education and executive levels.

Manama, Bahrain   Allam Hamdan
Giza, Egypt   Aboul Ella Hassanien
Tampa, USA   Timothy Mescon
Güzelyurt, Turkey   Bahaaeddin Alareeni

Acknowledgments The editors wish to dedicate this work to Prof. Abdulla Al Hawaj, the Founding President of Ahlia University and the President Prof. Mansoor Alaali for the unconditioned and continuous support of academic research.

Contents

Distance Learning Solutions: The Role of Artificial Intelligence

Using Student Opinions to Guide Investments in Assessment Methods

Abdullah Al-Bahrani, Hana Bawazir, Omar Al-Ubaydli, Allam Hamdan, and Mohammed Al Mannai

1 Introduction

Student assessment has been evolving organically for some time. The digitization of education has played an important role, both because of the challenges that easily-accessible digital resources present to assessors, and also because of the increasing demand for remote education. The Covid-19 pandemic has reinforced many of the pre-existing trends in educational assessment, while accentuating the importance of solutions that are compatible with distance learning.

The uncertainty regarding the ability to return to in-person learning during the 2021/22 academic year and the sharp increase in demand for remote learning together require academic administrations to carefully consider their educational investments for the coming period. These include decisions regarding student assessment methods, which must be flexible, inclusive, and fair.

A. Al-Bahrani
Department of Economics and Center for Economics Education, Northern Kentucky University, Highland Heights, USA

H. Bawazir
Department of Economics and Finance, University of Bahrain, Zallaq, Kingdom of Bahrain

O. Al-Ubaydli (✉)
Studies and Research Directorate, Derasat, Kingdom of Bahrain
e-mail: omar@omar.ec

Department of Economics and Mercatus Center, George Mason University, Fairfax, USA

A. Hamdan
College of Business and Finance, Ahlia University, Manama, Bahrain

M. Al Mannai
Department of Innovation and Technology Management, Arabian Gulf University, Manama, Kingdom of Bahrain

A. Hamdan et al. (eds.), *Technologies, Artificial Intelligence and the Future of Learning Post-COVID-19*, Studies in Computational Intelligence 1019,
https://doi.org/10.1007/978-3-030-93921-2_1

During the planning phase, administrators must pay attention to student opinions regarding assessment methods. It is tempting and convenient to fixate on technical solutions as the price of such technologies falls, and as educators and students become increasingly comfortable with digitized education. However, student perceptions are integral to the success of assessment methods, both because they are a central stakeholder who deserve to be heard, and also because students possess valuable insights regarding the effectiveness of the different options.

This chapter explores the options available to administrators in light of the pre-Covid-19 trends, and in the wake of the educational shock caused by the pandemic. The chapter also presents the results of a survey conducted on UK students regarding their opinions of different assessment methods. We synthesize these different perspectives and present educational administrators with a series of recommendations.

The rest of this chapter is structured as follows. Section 2 examines trends in student assessment. Section 3 presents and analyzes survey data on student opinions regarding assessment. Section 4 is a discussion of the findings. Section 5 concludes.

2 Trends in Student Assessment

2.1 The Goals of Student Assessment

There are two main reasons why instructors use assessment. The first, which is the main purpose of assessment in a traditional education environment, is to certify student learning and to graduate students. Effective assessment allows students to demonstrate what they know after completing the course. Instructors using assessment for these reasons are interested in Quality Assurance (QA) [24].

The second view on assessment is that assessment is part of the learning process. Assessing students allows them to engage with the content and improves learning [2]. This view of assessment makes it difficult to separate its contribution to learning from other learning activities.

The choice of assessment is a function of what it is expected to accomplish. In the QA approach, assessments provide information to internal or external stakeholders on the quality of learning, level of achievement attained by graduates, and a measure of academic standards [8]. In this model, assessment and instructional design is intended to collect data and to be self-evaluated or evaluated by peer reviewers.

As student populations become increasingly heterogeneous, reliability of quality assurance has become problematic [25]. Moreover, this assessment strategy does not consider the value of the assessment to the student. While some assessment at the course level is designed to report quality assurance data, instructors are more likely to employ assessment as a learning strategy and to develop intrinsic motivation through feedback and engagement [24]. This has created a shift from standardized assessment strategies towards more inclusive and diverse assessment tools [11].

Assessment choices increase when instructors perceive assessments as part of the learning process. Any activity that allows students to engage with the content and be an active participant in the learning process can be perceived as an assessment strategy. A student-centered assessment is designed to allow students to do something with the course content that allows them to develop their intrinsic motivation for learning [2, 21].

2.2 Ensuring Attribution

An explicit property that any assessment method seeks is attribution: "clear evidence that the work has been produced by the candidate," [4]. As any professor or student knows, breakdowns of attribution are not pure chance events. Instead, they frequently represent purposeful attempts at cheating [27].

Students engage in academic misconduct for several reasons, including [10]: the pursuit of the financial benefits of superior performance, mediated by the labor market benefits of good grades [5]; the pursuit of the recognition associated with being on honor rolls and other prizes for excellence; the desire to exert less effort in preparing for an assessment; and so on.

There are many methods of cheating, and they constantly evolve in response to technological progress [10]. They include perennially primitive techniques such as students covertly exchanging notes during exams or retrieving illicit information during trips to the bathroom, or more advanced methods such as concealed ear pieces.

Educators seek to ensure attribution because cheating undermines the pursuit of all of the traditional goals of assessment methods. In the case of assessment as a certifier, cheating invalidates the certification process, diminishing its value to prospective employers and educators. In light of the importance of educational credentials to earnings in modern labor markets [5], systemic cheating can be seen to have macroeconomic implications. Similarly, when it comes to assessment as a learning aid, cheating undermines the process as it breaks the links between teaching, learning, and engagement. Alongside these instrumental criticisms, people generally dislike cheating on purely ethical grounds [23], as it is inherently deceitful.

In all cases, it is notable that the costs of cheating are not just borne by the perpetrator (whether or not they are caught in the act); all stakeholders are impacted negatively, including those who choose to behave honestly.

Given the consensus regarding the unacceptability of cheating, when developing assessment methods, educators allocate considerable resources to preventing cheating, and these are reflected in a wide array of techniques deployed to that end [1]. Countermeasures include making it harder to cheat, such as giving students copies of the same exam but with the questions randomly reordered, or keeping desks far apart. They also involve increasing the likelihood that an educator detects cheating, such as by installing security cameras or proctoring exams while students are facing away from the proctor, allowing the educator to use the threat of punishment as a way of motivating honest behavior [10].

As the consequences of successfully and unsuccessfully cheating have evolved, and the technological possibilities have developed, the overall incidence of cheating has changed, as has the nature of the methods used. During the period 1960–1995, prior to the proliferation of the internet, the frequency of self-reported cheating increased substantially [20], possibly due to the rising labor market stakes associated with educational credentials. However it is worth noting that during this time period, the technology of assessment was generally constant.

During the last quarter century, the internet has opened up new opportunities for cheaters operating in traditional environments, while also changing the nature of education due to the advent of remote learning [28]. One of the most notable trends has been a sharp increase in the rate of plagiarism, which is the practice of taking someone else's work or ideas and passing them off as one's own. For example, whereas 3% of surveyed students downloaded internet papers in 2009, this figure rose to 10% by 2013 [25]. As a result, educators have experimented with a wide variety of methods for combating plagiarism [18].

One particularly concerning form of plagiarism has been ghost writing. Plagiarism usually involves the person whose ideas are being copied being an involuntary participant in the act, such as when a scholar has their arguments reproduced by a student without attribution. However, a narrow class of plagiarism involves the source voluntarily allowing the plagiarist to falsely claim their work, sometimes in exchange for financial compensation, as occurs when a student commissions a paper from a freelance writer [19].

The reason that this type of plagiarism is more concerning is that it is much harder to detect: as an accomplice in the act, the source may take steps to support that plagiarist's claim of innocence, including ensuring that the paper is made to order and therefore does not exist in the databases that educators use when trying to detect plagiarism.

Ultimately, when educators make decisions about assessment methods that include choices regarding the countermeasures to cheating, there are important consequences for the learning environment, with concomitant impacts on student behavior [4]. Well-designed assessment systems can incentivize students to increase effort and develop better learning habits, and so it is imperative that academic administrators carefully consider the options available to them.

In this chapter, we focus on analyzing the methods that educators use to combat cheating, with an emphasis on three techniques that have become particularly widespread during the Covid-19 pandemic (see below).

The first is the use of anti-plagiarism software such as TurnItIn, which is surely familiar to readers of this book. The large volume of accumulated experience with anti-plagiarism software has provided education scholars with a clear sense of its pros and cons [17]. Among its advantages are that it is fast and reasonably priced, and it has demonstrated effectiveness as a plagiarism deterrent. It also confers pedagogical benefits upon users by educating them regarding how to write ethically.

However, anti-plagiarism software remains imperfect. The databases are large and continue to grow, but they remain incomplete, and as mentioned above, their achilles heel is professional ghostwriting. Moreover, the software only detects the crudest forms of plagiarism, whereby text is almost exactly copied. Stealing ideas and repackaging them without indicating the original source is enough to avoid detection by anti-plagiarism software, despite representing a clear act of plagiarism. Similarly, some critics have expressed philosophical concerns with the software, regarding it as a heavy-handed and vastly inferior substitute to engaging students about ethical research practices.

The second method we examine is lockdown browsers. These are customized web browsers that are designed specifically for the act of remote assessment. They work by restricting what students can do on their computer during assessment. For example, students are denied access to other applications (including messaging software that can be used for synchronous communication); they cannot use functions such as copy, paste, and screenshots; and all actions are recorded in real time permitting proctoring. Modern versions of lockdown browsers work with a webcam that looks at the student, and in conjunction with artificial intelligence algorithms, they can alert proctors to suspicious activity such as looking away from the screen for extended periods, and can be used to confirm the student's identity.

Like anti-plagiarism software, lockdown browsers have their own advantages and disadvantages [16]. They are inexpensive and scalable, especially when used with artificial intelligence powered webcams and desktop monitors. However, since students use their own computers, they suffer from the possibility of ill-timed technical failures stemming from each student having their own computer/operating system/browser etc. Moreover, the commercial versions that involve proprietary closed-source software may have security flaws that can be exploited by tech-savvy students.

Beyond these technical deficiencies, lockdown browsers that are integrated with webcams have been accused of deploying racially discriminating algorithms. These include students with black or brown skin being asked to shine more light on themselves during identification, or simply failing to detect the face of non-white students at all [26].

The third and final anti-cheating method we will analyze in this paper is one of the oldest and most intuitive techniques: using a centralized test center. This is strongly associated with remote learning environments where students may never meet one another or the professor. As mentioned above, using a test center allows educators to use a wide range of traditional anti-cheating measures, such as strategic seating environments and bans on the use of mobile telephones.

Most of these advantages that test centers offer stem from the enhanced control that educators have over the testing environment. However, test centers present logistical challenges, especially when students have a great deal of heterogeneity in their access to a test center and their overall availability. Moreover, test centers are less scalable than virtual proctoring methods when students are geographically dispersed, especially in the case of online courses where students may be in different countries [22].

2.3   The Impact of Covid-19 on Student Assessment

As readers of this book will know from first-hand experience, the Covid-19 pandemic dramatically impacted the teaching and assessment methods available to educators [6], due to the switch to remote learning. A full description is beyond the scope of this chapter; instead, we will focus on the assessment-related changes witnessed.

At the outset, it is worth noting that the pandemic caught almost all educators by surprise, meaning that there was a lot of uncertainty and confusion among professors and students, resulting in a large amount of trial and error. The general state of chaos has two implications for educators going forward.

The first is the need to be careful and judicious in drawing conclusions about optimal assessment methods, as there is still much that we do not know. Second, educators need to engage their students systematically to infer what does and doesn't work well, through surveys and other tools [7, 12].

The pandemic presented several challenges to educators seeking to assess students [13]. First, infrastructure: educators and students vary in their access to equipment that can play a role in assessment. For example, while it may seem straightforward to replace a requirement for a face-to-face presentation with one conducted using video conferencing software, not all students have access to a computer with a webcam and microphone, and they may lack the financial resources to address this lack of equipment. Stable, high speed access to the internet was a significant problem for many students during the pandemic. Sometimes, the infrastructure problem manifests itself in simply not having a quiet room to work in when at home.

Second, educators and students vary in their familiarity with remote learning software. Thus, a student experiencing a technical difficulty during an online exam might not be able to resolve it.

Third, the pandemic caused significant disruption to daily work schedules for both educators and students. Previously agreed-upon class times may have become infeasible due to the need to arrange childcare, or due to a change in shift times for students who work while studying part time.

Despite these difficulties, educators and students deployed a variety of strategies to adapt to the pandemic [7, 12]. The simplest was to use lockdown browsers to produce a remote version of an in-class proctored assessment, i.e., each student would receive the same exam, which they would perform on their computer while having their access to various applications restricted, and while having their actions monitored via a webcam powered by artificial intelligence. In the case of in-class oral presentations, these were done via video conferencing software.

Educators sometimes introduced small modifications to this assessment strategy to decrease the likelihood of cheating in written exams, such as assigning each student a unique and random subset of questions from a larger common bank. Moreover, to overcome challenges associated with patchy internet access, educators gave students extra time than usual to perform written exams. In the context of assessments that were already asynchronous, such as homeworks and take-home exams, educators essentially made no alterations to the established methods, beyond giving students extra time due to the difficulties of balancing one's daily responsibilities during a pandemic.

Once the pandemic began to subside, and educational institutions partially reopened, some educators exercised the option of traditional, centralized exams while continuing remote teaching. Thus, a large hall with social distancing precautions would be prepared, and students would come to take the exam while being proctored face to face.

The other option that many educators used was partially or even fully shifting the weight of assessment from synchronous format, such as proctored exams, to asynchronous format, such as homeworks and take-home exams. This was sometimes caused by a belief that online exams in real time were a flawed substitute for face to face ones, due to the aforementioned challenges relating to infrastructure and equity/fairness. An essential element of this transition was a greater reliance on anti-plagiarism software to maintain attribution.

These changes occurred during the pandemic and are nominally transient, but there is an expectation that some will persist even after the pandemic ends. This is because some of the changes are chronically desirable, but were previously avoided due to the large investment required to implement them, e.g., buying webcams for all students, or ensuring that all professors (including senior ones) know how to use video conferencing software. Alternatively, it could be because the pandemic itself has caused fundamental changes in the market for education.

One such change is permanently offering students the option of attending a class in person or remotely. This is partially due to the fact that some students enjoy the convenience of remote learning [14]. Moreover, remote learning can increase access to groups that are traditionally underrepresented in classrooms, such as those with physical disabilities. For some universities, remote learning is also a way of increasing enrollment and revenues [29]. If these predictions are realized, then educators and academic administrators will need to make important decisions regarding their future assessment-related investments [3].

3 Student Opinions on Assessment Methods

3.1 Survey Goals

The academic literature on assessment methods that we cited above, along with our own experiences as educators, yield many insights on the prevailing state of assessment methods and on the options available for academic administrators. For administrators to make sound judgments regarding the future of assessment methods, it is important for them to complement these insights with information garnered from the students themselves, for two reasons.

First, students are central stakeholders, and so they have a right to contribute to decisions that impact them directly. If students feel that they have been marginalized by administrators making important choices regarding the future of assessment methods, their disenchantment may undermine the assessment methods' effectiveness due to non-compliance as a form of protest, and also due to a potential decrease in the demand for higher education.

Second, students are a valuable source of information regarding the pros and cons of individual assessment methods. For example, they are likely to have information about the latest security flaws that students exploit to gain an unfair advantage.

With this in mind, we decided to survey students to better understand their views regarding assessment methods.

3.2 Survey Questions

The questions relating to this chapter were included in a longer survey that addressed a variety of education-related issues. Beyond some standard demographic questions, the key question posed to participants was:

Online education creates a risk of students cheating. For each of the following statements, indicate how much you agree with the statement on a scale of 1-to-5, where 1 indicates "strongly disagree", 3 indicates "neutral", and 5 indicates "strongly agree".

- Tools designed to detect plagiarism, like Turnitin, are effective in deterring students from plagiarizing.
- Tools designed to detect plagiarism, like Turnitin, create a fair grading system.
- Software that stops students from browsing the internet on their computer while they are doing an exam (lockdown browsers) makes it more difficult for students to cheat during exams.
- Software that stops students from browsing the internet on their computer while they are doing an exam (lockdown browsers) creates a fair grading system.
- Administering an exam at a test center instead of an online exam makes it harder to cheat.
- Administering an exam at a test center instead of an online exam creates a fair grading system.

Ideally, we would have also included questions that would allow us to infer the reasons behind participants' choices. For example, in the event that a participant regarded a test center as being unfair, or plagiarism detection software as ineffective, we would have liked to understand the factors that explain that decision. However, due to our limited financial resources, making the survey longer meant gathering a small number of observations.

The whole survey took approximately six minutes to complete, though the elements relevant to this chapter required around one minute to complete.

3.3 Participants, Sampling and Survey Distribution

The target population was 1,000 adult bachelor's and master's students currently residing in the UK. We chose the UK because the distribution language was English, and we wanted to ensure that no students were prevented from participating due to language barriers. Moreover, the UK has a mature education system and large population, making it easier for us to secure participants. We excluded doctoral students because many do not take courses and would therefore not be exposed to the assessment methods we wished to inquire about.

A total of 966 UK students participated. The participants' key demographic traits were as follows:

- 68% were female, and 32% were male.
- 68% were aged 18–24, with the remaining participants being aged 25 or over.
- 72% were bachelor's students, and 28% were master's students.
- The participants' majors were extremely diverse, spanning all elements of the natural sciences, social sciences, humanities, and other disciplines.

The survey was distributed via the online platform "Prolific" during the first quarter of 2021. An online system was used to facilitate recruitment as none of the authors were residing in the UK. Participants are paid an hourly rate to voluntarily take part in surveys. They register their interest with Prolific, and when the survey is online, they are made aware of its availability should they wish to participate, on the condition that they satisfy the aforementioned participation criteria. They can participate via a personal computer, a tablet, or a mobile telephone.

3.4 Results

Figure 1 shows the sample mean response for each of the six main questions. Though the question asked participants to evaluate their agreement on a scale of 1 to 5, we have rescaled it to −2 to +2 so that 0 indicates neutrality. Note that the standard deviation for each of the responses is approximately 1, implying a considerable degree of variation in the responses.

Result 1: On average, participants agreed that all anti-cheating techniques were effective, but a significant percentage of participants were neutral or disagreed.

On a scale of −2 to +2, for perceived effectiveness, the sample means were 0.83 for plagiarism detection software, 0.52 for lockdown browsers, and 1.4 for test centers, i.e., they were all positive. Simple t-tests confirm that these averages are all statistically significantly greater than zero ($p < 1\%$). However, in each case, a considerable proportion of participants responded neutrally or negatively: 29% for plagiarism detection software, 39% for lockdown browsers, and 11% for test centers.

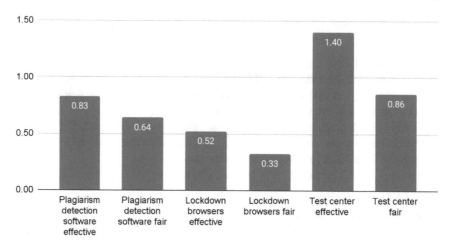

Fig. 1 Survey results

These data indicate that while UK students generally perceive these methods to be effective, this view is far from unanimous, and a considerable proportion are either ambivalent or hold negative attitudes.

Result 2: Participants perceived a centralized test center to be the most effective anti-cheating technique, followed by plagiarism detection software and lockdown browsers.

In terms of ranking, a test center is clearly deemed to be the most effective option, followed by anti-plagiarism software, and then lockdown browsers. All the bilateral comparisons differ in a statistically significant way according to a simple t-test ($p < 1\%$).

Result 3: On average, participants agreed that all anti-cheating techniques were fair, but a significant percentage of participants were neutral or disagreed.

On a scale of -2 to $+2$, for perceived fairness, the sample means were 0.64 for plagiarism detection software, 0.33 for lockdown browsers, and 0.86 for test centers, i.e., they were all positive. Simple t-tests confirm that these averages are all statistically significantly greater than zero ($p < 1\%$). However, in each case, a considerable proportion of participants responded neutrally or negatively: 39% for plagiarism detection software, 51% for lockdown browsers, and 33% for test centers.

These data indicate that while UK students generally perceive these methods to be fair, this view is far from unanimous, and a considerable proportion are either ambivalent or hold negative attitudes.

Result 4: Participants perceived a centralized test center to be the most fair anti-cheating technique, followed by plagiarism detection software and lockdown browsers.

In terms of ranking, a test center is clearly deemed to be the most fair option, followed by anti-plagiarism software, and then lockdown browsers. All the bilateral comparisons differ in a statistically significant way according to a simple t-test (p < 1%).

Result 5: The extent to which students perceive an anti-cheating technique to be fair is significantly lower than the extent to which they perceive it to be effective.

In all three methods, the average agreement level with the technique's effectiveness is higher than with its fairness (difference of 0.19 for plagiarism detection software and lockdown browsers, 0.54 for test centers), and all are statistically significant according to a simple t-test (p < 1%).

As explained above in the section on survey questions, the data we gathered do not allow us to infer the explanations for this pattern of responses. In the synthesis section below, we provide some speculative hypotheses, and we also hope that some follow-up focus-group sessions will allow us to draw more definitive conclusions as part of our future research.

We also investigated how the aforementioned results were affected by breaking down the participants by demographic group. In the interests of parsimony, a full factorial analysis is beyond the scope of this chapter. Accordingly, we here focus on three dimensions: gender, age, and degree level (bachelor's versus master's).

Result 6: Results 1 to 5 are unaffected by gender; however, male participants expressed lower levels of agreement with the perceived effectiveness and fairness of both plagiarism detection software and lockdown browsers.

As an illustration, when asked about the effectiveness of lockdown browsers, the sample mean for women was 0.64, compared to 0.28 for men. In contrast, in the case of test centers, the difference between men and women was statistically insignificant.

Result 7: Results 1 to 5 are unaffected by age group; however, younger students expressed lower levels of agreement with the perceived effectiveness and fairness of lockdown browsers.

Defining young students as those aged 18–24, the sample mean for lockdown browser effectiveness was 0.47 for young and 0.64 for old (p = 3%), and for lockdown browser fairness it was 0.23 for young and 0.52 for old (p < 1%). For all other responses, the difference was statistically insignificant.

Result 8: Results 1 to 5 are unaffected by the degree level.

For all six questions, a simple t-test reveals that the responses of those undertaking a bachelor's degree are statistically insignificantly different from those of students enrolled in a master's degree. This result is important because it suggests that Results 1 to 5 are not significantly affected by student ability: master's students tend to be drawn from the highest performing bachelor's degree students, owing to a combination of superior academic skills and a more serious attitude toward their studies.

4 Discussion

The first observation that merits discussion is the differences in the perceived effectiveness of the three anti-cheating methods under consideration. Despite considerable advances in remote learning technology, including the prevalence of lockdown browsers, students regarded test centers to be considerably more effective than the alternatives.

That test centers should perform better is virtually inevitable since the ways in which one can cheat in a test center are almost a strict subset of the ways one can cheat online; however the magnitude of the difference was somewhat surprising, and likely reflects a need for further refinement of online anti-cheating measures.

In the case of lockdown browsers, the reality is that they only constrain what the student can do on their computer, and they remain largely able to cheat via peripheral devices, such as mobile phones, or via the assistance of others who cannot be seen by an online proctor. In the case of anti-plagiarism software, the ease of procuring ghostwriting services massively limits the software's effectiveness as a way of ensuring attribution.

The second observation concerns why perceived effectiveness exceeds perceived fairness in all cases. Here, we suspect that student inequality plays an important role, especially when many dimensions of the inequality have been accentuated by the pandemic. Thus, while a test center is an effective way of preventing plagiarism, students differ in their flexibility, in their ability to leave children and/or the infirm unattended at home, in their working hours if they are employed, and so on, creating a preference for asynchronous assessment such as take-home exams.

Similarly, in the cases of lockdown browsers and anti-plagiarism software, income is a strong determinant of a student's ability to circumvent the measure, though they may not choose to deploy that ability. Students with greater financial means can afford better quality ghostwriters and tutors to illicitly assist them during assessments. And, as mentioned above, some algorithms are inadvertently discriminatory, meaning that minority students waste more time and endure more stress dealing with the artificial intelligence system.

The third observation is the demographic differences in responses. That male respondents had a significantly lower evaluation for the effectiveness and fairness of lockdown browsers and anti-plagiarism software than did female ones may reflect gender-based differences in attitudes and behavior in the cyber realm. For example, there is a large literature indicating that men have a larger propensity to commit cyber crimes than do women, and this phenomenon is present even for the subsample that is college students [9].

In the case of the finding that the young had lower evaluations of the effectiveness and fairness of lockdown browsers, this may reflect differential exposure, since younger students are much more likely to have had to endure lockdown browsers; or it could reflect the reliance that the young tend to have on computers, which makes them regard lockdown browsers as more invasive than do their older counterparts.

What do our findings imply for the assessment-related investment decisions that educators and academic administrators need to make? Based on our results, we have the following remarks and accompanying recommendations.

First, there is significant heterogeneity in student perceptions toward the different assessment methods. In aggregate, though students on average regarded the various methods considered as effective, a large percentage saw them as ineffective. Moreover, there were significant differences by demographic group, including an age effect that was present despite small age differences within the sample.

The implication of this heterogeneity in perceptions is that educators need to be flexible and adaptive, and should avoid being overly reliant on one method of assessment. While a professor's current class may enjoy take-home exams and regard them as effective and fair assessment methods, there is a good chance that next year's class has a different attitude. It is an educator's responsibility to respond to their students' needs, and in the case of assessment, that means removing technical and infrastructural barriers to change.

Second, perceived effectiveness and perceived fairness are distinct concepts, and this distinction is evident in students' survey responses. Certain assessment methods may be highly effective at preventing cheating, but at the cost of marginalizing an element of the student corps. Therefore, educators need to pay attention to both phenomena when they make their investment decisions.

Notably, this is not merely an issue of ethical concern for the marginalized. The integrity of the education system also requires that assessments be fair and be perceived as such; otherwise, those who are disadvantaged may drop out of the system and seek alternatives.

Third, despite considerable advancements in the technical sophistication of remote learning assessment methods, such as anti-plagiarism software and lockdown browsers, the area remains a work in progress. A significant proportion of students still feel that these methods are ineffective, especially compared to traditional test centers. Accordingly, there is a need for more innovation before remote learning assessment systems can be regarded as a complete substitute for traditional ones.

Moreover, innovators need to pay attention to developing methods that address inequality-related concerns. It is tempting to focus on purely technical refinements, but doing so risks exacerbating the extent to which certain groups are marginalized in the education process, such as those with lower financial means. Until that happens, educators and academic administrators need to maintain a high level of flexibility and adaptability.

5   Conclusion

The Covid-19 pandemic has caused unprecedented disruption to education, most notably via a forced transition to remote learning. At the time of writing this chapter (summer 2021), it is unclear how long this disruption will continue. However, even in the event that a favorable conclusion to the pandemic arrives swiftly, there are various indications that remote learning will play a persistently larger role in the future of education.

Under these circumstances, educators must make careful judgments about the investments they make in assessment methods. While many aspects of education remain in flux, one consistent element has been the need for students to be able to demonstrate what they have learned in a transparent and reliable manner that is highly resistant to cheating. Remote learning opens many new avenues for prospective cheaters, underscoring the importance of a measured approach to assessment.

This chapter presents a series of recommendations to administrators based on a survey of almost 1,000 UK students. Participants were asked to evaluate the effectiveness and fairness of three anti-cheating methods (anti-plagiarism software, lockdown browsers, centralized test centers). We chose to gather these novel data because it is incumbent upon administrators to take student opinions into account when making assessment decisions [15]. As central stakeholders, students have a right to be heard; and they also possess important inside knowledge regarding the operational effectiveness of various assessment methods.

Our survey uncovered significant variation in student perceptions regarding the effectiveness of the chosen anti-cheating methods. Test centers were considered to be significantly more effective than anti-plagiarism software, which in turn was considered to be significantly more effective than lockdown browsers. This latter option was also perceived to be ineffective by a large percentage of participants.

Our survey also indicated that students drew a clear distinction between the effectiveness of an anti-cheating measure and its fairness. In general, fairness perceptions were lower than those regarding effectiveness, and this likely reflected the tendency for certain methods, especially lockdown browsers, to be inadvertently discriminatory toward marginalized groups.

Based on these findings, and building on the literature, we made three recommendations to educators. First, the need to make investments that allow for flexibility and adaptability in light of the large degree of heterogeneity in student attitudes. Second, the importance of distinguishing between effectiveness and fairness when evaluating competing assessment methods. Third, the need to avoid exclusive reliance on remote assessment methods, as the technologies remain a work in progress.

While our survey provided us with rich data, follow-up research should focus on gaining a better understanding of student opinions. Due to resource constraints, our questions were restricted to expressions of attitudes toward certain choices, and they did not include any questions about why a certain belief was held. For example, the data did not provide a clear-cut answer to questions like: why do students dislike lockdown browsers?, or why do some students perceive test centers to be unfair?

In principle, both quantitative and qualitative methods can be used as part of this follow-up research. However, in light of the unique and unprecedented nature of the pandemic's impact on education, we believe that qualitative methods should be the primary choice, as they allow respondents to provide answers that might not be anticipated by the research team. A combination of semi-structured interviews and focus group sessions would provide scholars with valuable insights regarding the bases of student opinions, which would in turn furnish academic administrators with more precise data as they make important assessment decisions going forward.

References

1. Alschuler, A. S., & Blimling, G. S. (1995). Curbing epidemic cheating through systemic change. *College Teaching, 43*(4), 123–125.
2. Bailey, S., Hendricks, S., & Applewhite, S. (2015). Student perspectives of assessment strategies in online courses. *Journal of Interactive Online Learning, 13*(3).
3. Beech, N., & Anseel, F. (2020). COVID-19 and its impact on management research and education: Threats, opportunities and a manifesto. *British Journal of Management, 31*(3), 447.
4. Bloxham, S., & Boyd, P. (2007). *Developing effective assessment in higher education: A Practical guide: A practical guide.* McGraw-Hill Education (UK).
5. Card, D. (1999). The causal effect of education on earnings. *Handbook of Labor Economics, 3*, 1801–1863.
6. Daniel, J. (2020). Education and the COVID-19 pandemic. *Prospects, 49*(1), 91–96.
7. Daniels, L. M., Goegan, L. D., & Parker, P. C. (2021). The impact of COVID-19 triggered changes to instruction and assessment on university students' self-reported motivation, engagement and perceptions. *Social Psychology of Education, 24*, 299–318.
8. Dill, D. (2007). Quality assurance in higher education: Practices and issues. in-Chief Barry McGaw, Eva Baker and Penelope P. Peterson. Elsevier Publications.
9. Donner, C. M. (2016). The gender gap and cybercrime: An examination of college students' online offending. *Victims & Offenders, 11*(4), 556–577.
10. Faucher, D., & Caves, S. (2009). Academic dishonesty: Innovative cheating techniques and the detection and prevention of them. *Teaching and Learning in Nursing, 4*(2), 37–41.
11. Frasineanu, E. S., & Ilie, V. (2017). Student-centered education and paradigmatic changes. *Revista de Stiinte Politice, 54*, 104.
12. George, M. L. (2020). Effective teaching and examination strategies for undergraduate learning during COVID-19 school restrictions. *Journal of Educational Technology Systems, 49*, 23–48.
13. Guangul, F. M., Suhail, A. H., Khalit, M. I., & Khidhir, B. A. (2020). Challenges of remote assessment in higher education in the context of COVID-19: A case study of Middle East College. *Educational Assessment, Evaluation and Accountability, 32*(4), 519–535.
14. Hussein, E., Daoud, S., Alrabaiah, H., & Badawi, R. (2020). Exploring undergraduate students' attitudes towards emergency online learning during COVID-19: A case from the UAE. *Children and Youth Services Review, 119*, 105699.
15. Jankowski, N. A. (2020). Guideposts for assessment during COVID-19. *Assessment Update, 32*(4), 10.
16. Küppers, B., Kerber, F., Meyer, U., & Schroeder, U. (2017). Beyond lockdown: towards reliable e-assessment. Bildungsräume.
17. Ledwith, A., & Rísquez, A. (2008). Using anti-plagiarism software to promote academic honesty in the context of peer reviewed assignments. *Studies in Higher Education, 33*(4), 371–384.
18. Levine, J., & Pazdernik, V. (2018). Evaluation of a four-prong anti-plagiarism program and the incidence of plagiarism: A five-year retrospective study. *Assessment & Evaluation in Higher Education, 43*(7), 1094–1105.
19. Lines, L. (2016). Ghostwriters guaranteeing grades? The quality of online ghostwriting services available to tertiary students in Australia. *Teaching in Higher Education, 21*(8), 889–914.
20. McCabe, D. L., & Trevino, L. K. (1996). What we know about cheating in college longitudinal trends and recent developments. *Change: The Magazine of Higher Learning, 28*(1), 28–33.
21. O'Brien, H. L., & Toms, E. G. (2010). The development and evaluation of a survey to measure user engagement. *Journal of the American Society for Information Science and Technology, 61*(1), 50–69.

22. Prince, D. J., Fulton, R. A., & Garsombke, T. W. (2009). Comparisons of proctored versus non-proctored testing strategies in graduate distance education curriculum. *Journal of College Teaching & Learning (TLC), 6*(7).

23. Roig, M., & Marks, A. (2006). Attitudes toward cheating before and after the implementation of a modified honor code: A case study. *Ethics & Behavior, 16*(2), 163–171.

24. Rust, C. (2002). The impact of assessment on student learning: How can the research literature practically help to inform the development of departmental assessment strategies and learner-centred assessment practices? *Active Learning in Higher Education, 3*(2), 145–158. https://doi.org/10.1177/1469787402003002004

25. Seyfried, M., & Pohlenz, P. (2018). Assessing quality assurance in higher education: Quality managers' perceptions of effectiveness. *European Journal of Higher Education, 8*(3), 258–271.

26. Swauger, S. (2020). Our bodies encoded: Algorithmic test proctoring in higher education. *Critical Digital Pedagogy.*

27. Teixeira, A. A., & Rocha, M. F. (2010). Cheating by economics and business undergraduate students: An exploratory international assessment. *Higher Education, 59*(6), 663–701.

28. Wang, Y. M. (2008). University student online plagiarism. *International Journal on E-learning, 7*(4), 743–757.

29. Williams, D. (2021). *College Administrators Strategies to Implement Online Education Services for Increasing Enrollment Revenues* (Doctoral dissertation, Walden University).

The Future of Education After Covid-19, What is the Role of Technology

Ritik Verma and Neetesh Soni ⓘ

1 Introduction

1.1 Background of Future Education

Nowadays the technological advancement of human civilization is astonishing as compared to the ancient and medieval periods. The inventions of more advanced technologies are happening with each passing day. New milestones are achieved in all fields of technologies from farming to space research, from the calculator to robots, but when it is coming to the education system there is no significant improvement has been observed from the past century. This Covid-19 provided the opportunities and demonstrate the necessities of improving our education system by highlighting the disadvantages of our existing education system. Covid-19 points out the areas where the improvement in the education system is required to take it to the next level and build the education system where it can available to everyone. Through this futuristic approach, it can make the education system more efficient and provide a more effective way of learning. By implementing this futuristic approach, the pathway for possible approaches towards the complex problems with better understanding and concept clarity for primary, secondary, and higher education is achieved [1].

The conventional education system consists of books, notebooks, and the concept of classrooms. Here is a step towards the future by removing the stress of education and making it more joyful and interacting. Hesitation in students is removed by filling the gap between teacher and students, social media is being used by teachers for personal interaction with their students. Removing the extra burden of school

R. Verma
G.L. Bajaj Institute of Technology and Management, Greater Noida, Uttar Pradesh, India

N. Soni (✉)
School of Materials Science and Engineering, SWPU, Chengdu, China

© The Author(s), under exclusive license to Springer Nature Switzerland AG 2022
A. Hamdan et al. (eds.), *Technologies, Artificial Intelligence and the Future of Learning Post-COVID-19*, Studies in Computational Intelligence 1019,
https://doi.org/10.1007/978-3-030-93921-2_2

bags from the shoulders of the upcoming generation, by providing the best in class educational material on their palms in the form of smart tablets. Which include the lectures by teachers from top universities to best books published in respective subjects. Through this education system, the level of education provided to the student of a particular age group should be the same nationwide. This improved system is eco-friendly in many aspects, about 15 billion trees are chopped down in the making of books each year and it affects our ecosystem very badly by raising global issues like global warming and pollution [2]. By adopting this improved education system, the cutting of trees for making paper can be reduced by a significant amount [3].

However, the future education system needs to be developed by using present technology and future-based demands of the education curriculum. This chapter mainly focuses on AI-based technology development for future education.

1.2   Present Online Education

In response to covid-19, the schools and colleges are closed and over 1.725 billion students are suffered worldwide [4]. This enormous loss of education is a very serious concern thus the education system is adapting towards the current scenario and struggling to cope up with the covid-19. To provide the education amenities and try to recover the education loss from this covid-19 pandemic. There are very few options available for the educational institutes to continue providing education in a conventional way. The conventional ways like social distancing in classes, masks, hygiene, and sanitization. However, managing them is a very challenging task thus taking physical classes seems impossible. As a result, the option of distance/remote learning with online classes, video lectures, assessments, live lectures, app-based education platforms are implemented as the only solution. This e-learning system is new for students and teachers, this sudden shift from conventional classroom to the online digital platform is challenging for most people, many people face different problems in an early phase of e-learning [5]. It is clearly presented in Fig. 1.

Surprisingly, e-learning opens new doors to the education system [6]. Through e-learning students can get better educational content from around the globe, the students can now able to connect with the world's best professors through online. The top institutes are announcing various online courses regularly so that interested students from anywhere around the world get enrolled, get certified and build their careers. The remote degree and remote diploma are also announced by various universities for graduation, this is very helpful for those who want to do graduation through online means [5]. E-learning also promotes self-learning, tracking ability, and self-assessment of students. This e-learning increases the studying hours of students by removing traveling time, and studying at home gives students a comfortable environment for study [7].

Fig. 1 Post pandemic effect on education

1.3 Future of AI-Based Education

Through Artificial Intelligence the approach to fulfill future goals in the education system can be accomplished [8]. The use of AI in educating students and assisting teachers in their teaching methods. AI is very beneficial in assisting teachers in various ways. The AI-based robots teach by Collaborative Learning Model (CLM) comprising teacher, student and the robot, all of them together collaborate in the classroom to deliver a lesson. Teachers can use smart gadgets based on AI during lectures, building smart classrooms with the use of digital boards, smart projectors and smart glasses can enhance the teaching experience and also improve the quality of education thus the efficiency of knowledge transfer from teacher to the student is improved by a great extent. The AI is used in the analysis of answer sheets and making a grading evaluation more precise. The deep analysis of answer sheets of paper can reveal the hidden talents in the students and it also helps to guide the students for improvements in the respective subjects. The AI is very helpful in the analysis of the answer sheets by figuring the student's approach towards the answer for a particular problem. Moreover, AI can be used for the analysis of student behavior in the class by analyzing facial expressions. Moreover, AI helps maintain the record of student interaction with other classmates. AI glasses (wearable smart glasses for teachers in the classroom) can analyze the student's facial expression and detect the behavior during the lecture for example- the stress meter, confusion in a related topic, and attention to the lecture. Through AI smart projectors teachers can display the 3D models and interactive videos so the students can easily grasp the knowledge,

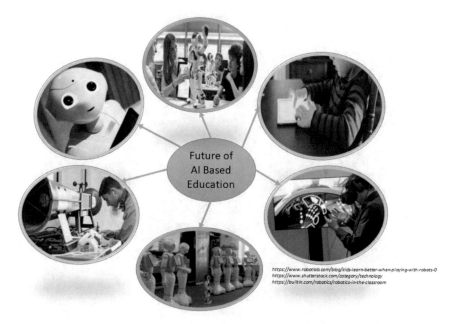

Fig. 2 Several types of AI-based future technology

even the complex topics can be taught very easily. It shows in research that the retention memory and concept clarity is enhanced by a significant rate if it is taught by interactive videos and through the use of 3D content. With the use of proper data management systems with integration of network across the education facility. The data and the security of the facility have been ensured with proper security features. All the classrooms and security cameras are connected to a single database thus ensuring a single AI system manages the whole facility as shown in Fig. 2.

1.4 Future Technology Involvement

Future technology like integration of Artificial Intelligence in various fields of education such as teaching, management, and robotics [9]. The use of AI can highly increase the education quality and significantly reduce human efforts. All management will be done and performed by automated machines. The educational robots integrated with Artificial Intelligence can be used to teach students at the primary level and provide them basic education, for example, the Saya robot from Japan which can perform human expressions and communicate with students [10]. The AI robots are used as assistants in various labs because they can solve complex mathematic problems and calculations in a blink of an eye. The Virtual reality classroom helps the students learn difficult concepts by using the 3-D virtul world and thus students get a better understanding of the subject. The use of AI is very helpful in creating virtual Reality machines, these machines can create virtual labs as well virtual worlds. The students can perform various complex experiments in virtual laboratories like working on harmful chemicals, mechanism of complex machines, build advanced infrastructure and design, experiments on space anomalies even spacewalk on International Space Station. Robots can perform experiments in harsh environments by applying neural network thinking to perform tasks like deep ocean imaginary or space exploration and can report the user if they found an anomaly. Moreover, AI is extensively used on various simulations to provide the best training experience and mimic real-life conditions like airplane simulation, driving simulation, Space shuttle simulation, and many more.

2 AI-Based Implementation

To improve the education system, it is necessary to develop more advanced technologies and integrating current technology with the use of Artificial intelligence [11]. The major fields to focus on the implementation of Artificial intelligence through the use of Machine Learning and Deep learning are Robotics, Speech Recognition, e-learning technologies, Natural language processing, and Computer vision.

2.1 Robotics Based Implementation

Robots can provide assistance for improving education in various ways, the term used here is educational robots (ER) [11, 12]. These robots are automated with the ability to make decisions and able to assist humans with performing various tasks for example robot assistant Yuki in Germany [13]. Smart Robots are useful in three important opinions such as 1. Managing digital libraries, 2. Provide assistance during examinations, 3. Checking answer sheets and analyzing the results. Automated robots can scan tons of books in the library and make digital books from the collected data and these digital books are further transformed into audiobooks by e-readers. The AI in e-reading software is used to make e-reading more efficient and more like a human speech. This can be achieved by applying Deep learning concepts in analyzing different human recordings, to create nuances of human speech such as breath, causes, emotions, and excitement for making it more natural. AI robots can also assist teachers and help them during the exams by preventing cheating in exams by continuous monitoring of students by using motion capture sensors. Through the face-detection technique and biometric signature, the validity of the examinee is also determined. Now, To check answer sheets and evaluate the answers, the answer sheets are scanned and analyzed by using Optical Character Recognition (OCR) furthermore AI concepts arc used, like Deep Learning and AI-based evaluation methods through image processing techniques and neural networks algorithms (Fig. 3).

Fig. 3 Technology and AI relation development for education

Robots serve as toys in the early phase of student life [14], it can help the students learn basic education interactively like children talking robots or drawing bots. The robot seeds curiosity in student's minds to learn new things and hence generate creativity and critical thinking as well as problem-solving skills in the students. By use of robots, students lean towards knowledge and start experimenting with new things with robots which is the root for new inventions. The robots create a personal database of each student by remembering past communications with the respected student i.e. personalized robots which is possible by the use of machine learning methods. Robots can teach students personally in absence of a teacher and provide solutions to the problems by the use of supervised learning and Natural language processing technologies. Personalized robots can keep a record of student's daily routines and analyze their daily lifestyle which helps the student to find their passion. This Individualized approach for each student is very helpful in determining the interested fields of the student. Robots are very helpful in the development of autistic children and make them more sociable and interactive in human society. Several other emotions are also performed by robots such as disgust, anger, surprise, fear, happiness, and sadness which is key for better communication towards the student. The personalized use of robots helps students to develop interaction with machines and essential qualitative skills. The early robotic interaction encourages students to learn the technical knowledge.

2.2 E-learning Technologies

As the usage of smartphones and electronic devices is increased, many people are getting connected by the internet as a result, the growth rate of e-learning is also improved worldwide [15]. The Internet plays a significant role in the boom of e-learning. As people get the internet e-learning begins, every search they do give them knowledge and information. The passing of knowledge from the digital world to the human mind is known as e-learning. E-learning includes the use of websites, webpages, mobile applications, and digital books. Many elements together contribute to the e-learning platform for example videos, simulations, recorded and live lectures, quizzes, games, webinars, and many more interactive activities. The highlights of the e-learning environment are as follows, it is full of qualitative content, it can be learned from any place with no time restrictions, self-evaluation, economically affordable, provides global connectivity, and the main necessity is an internet connection with a digital device. As e-learning getting more popular day-by-day, because all of the students want to be updated with the latest technological advancements, hence they prefer an online education platform so, their daily routine is not affected and they can study in their free time. Just imagine the importance of e-learning that if it was not available then it would be very difficult to find out alternatives for the books, schools, and teachers for any education.

The pandemic has assisted to support this marketplace situation: Google's E-learning technology platform, Google Classroom, saw its user base double from March 2020 into April from the resource of Google as shown in the graph clearly in Fig. 4.

The tech giant donated 4,000 Chromebook laptops and pledged free internet availability to US 100,000 households' citizens. Microsoft has also participated in the space, provided that $1 M to Kano, a company intensive on educating kids in what way to code can be done [27].

2.3 Digital Education System

Digital education is differentiated into several categories with respect to the availability of different digital learning components [16]. The face to face learning is

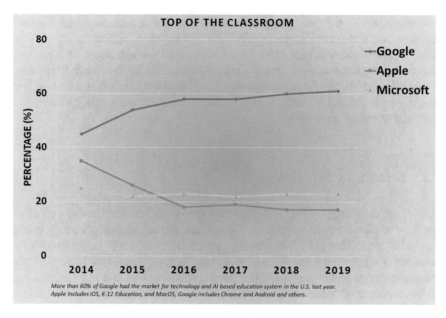

Fig. 4 Role of AI and Technology based education system in U.S. schools

the most basic form of education in this the student attends the class with the presence of a teacher, just like conventional schooling attending regular lectures with the assistance of teachers and a proper study environment. In self-learning, the student learns by books without the assistance of a teacher. Self-learning promotes self-assessment and self-questioning of problems by building their understanding of the subjects and develop necessary problem-solving skills. In the asynchronous type of the learning is done remotely through the use of various communication devices, this is an improved version of self-learning, but in this, there is very limited guidance that can be provided by the mentor. Synchronous learning is one of the best types of remote learning, in this the student and teachers both are virtually present, and live video interaction through the use of smart devices, doubt clearance is much easier than asynchronous type of learning. In hybrid-asynchronous learning, the live video communication is disturbed due to some issues like network connectivity, because of this irregular presence the knowledge transfer is affected. In hybrid-synchronous learning, the integration of smart devices in the physical presence of the classroom to boost learning efficiency. As shown in Fig. 5.

The digital education classification is a pioneering way of using digital technology in teaching and providing education, it is also known as digital learning as shown in Fig. 2. Digital learning is integrated by the use of the latest technologies that is replacing traditional educational practices. Furthermore, digital technologies offer modern infrastructure to educators so that they are able to make interactive learning options in the courses they teach. Teachers can upload content and distribute it through online courses and digital programs. The digital education system includes

Fig. 5 Detailed layout of digital education arrangement

the use of projectors, digital boards, cameras, tablets, laptops, and smartphones in teaching. With the use of speech recognition and natural language processing the smart gadgets like voice-based AI-powered digital assistants can become handy in daily life for example -amazon's Alexa, Cortana and google assistant [17]. Virtual reality also promotes digital education by simulating the virtual environment and help in improving learning and conceptual knowledge.

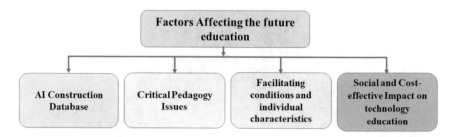

Fig. 6 Variables of future education

3 Factors Affecting the Future Education

3.1 Education Variable

The present cataclysmic wave shows the importance of the requirement of a new education system. This system is able to complete the future demands and cannot be affected by the future pandemics and build a secure education system for upcoming generations. The experiences from Covid-19 and future technology both having strong relations for laying the foundation of the future education system. However, several factors affect future education like the availability of the internet and essential facilities in the educational institutes as shown in Fig. 6. There are many other social, cultural, technological, economic, and political issues that may arise in the future. People's acceptance of the use of robots and AI in the future is an additional concern [18].

3.2 AI Database Construction

AI database construction is aiming to increase the computational speed of AI-based operations. The efficiency and organization of the database are very crucial for reducing the processing load on AI hardware and increasing output efficacy. AI databases include data warehousing, advanced analytics, and algorithms for in-memory databases. AI database performs various functions simultaneously for example analyze, explore, process, visualization, and perform complex data handling within milliseconds [19]. For the database construction, the immersion of appropriate datasets with the use of system-oriented algorithms can significantly improve the learning processes and improve the decision-making abilities of the machines. The main points to focus on while construction of database are as follows:

- Build Database according to purpose
- Enhanced data ingestion
- Parallel processing

- Data refining
- Several layer data securities
- User data encryption.

3.3 Critical Pedagogy Issues

Critical pedagogy is a teaching technique through which highly educated teachers can guide and encourage students to critique different structures of society. This theory involves the awareness of the surroundings and asking questions from the responsible authorities. In the future, critical pedagogy plays a significant role in the evaluation of advanced AI software, educational robots and also highlight the dark side of using AI-based technologies. The reviews are very helpful in showing the harmful effects of AI on the human future [20].

3.4 Facilitating Conditions and Individual Characteristics

As the technology becomes more complex and resource-expensive, the transferring of AI from experimentation to adoption is a very challenging task. It is essential to fulfilling the high demand for computing resources and infrastructure development to support the requirements of AI. In the Large-scale infrastructure modernization and ensuring better network coverage power outlets. AI requires large and robust databases with hybrid cloud technologies to support a high amount of data transfer. Additionally, strict assurance of data security because AI handles sensitive data for example personal information, banking, and financial data. AI infrastructure must be secured from end-to-end encryption with the latest technology and multilayer security solutions [21]. To ensure better energy solutions for powering AI systems corresponding to the use of solar power and more eco-friendly solutions. However, AI-based robots have to be environment friendly and better suited to operate around the world. These AI robots must have several build qualities to withstand harsh environments for example water resistance, dust resistant, anti-rusting, withstand at different temperatures, and mobility of robots. The need for appropriate infrastructure for construction, recycling, and dismantling of robots is set up and must ensure no adverse effect on the atmosphere [22].

Furthermore, for using AI-based solutions in the education system, professors have to learn how to interact with AI and make themselves familiar with the AI inter-face by getting specialized training. Through specialized training, professors learn to use various AI interfaces and input several data into AI databases such as videos, audio, notes, books, digital animations, and student information, and essential data of the course. The training for professors to use AI in their work to improve the efficiency and quality of the education provided [23]. AI-based solutions are also helpful to educators in their schedule and time management to organize a personalized timetable for the teachers. These AI solutions supportive to the professors in their professional career development and also in their research work by assisting in laboratories.

4 Factors Social and Cost-Effective Impact on Technology Education

Social and cultural practices of different communities whether the communities accept the AI interface in their daily lifestyle [24]. Communities' acceptance or rejection towards AI determines the behavior of humans in the direction of AI. The adaptation of AI has a great effect on the human social lifestyle. Humans have been more interacting with AI robots thus spending less time in normal human-to-human contact. AI benefits should be available to all, AI systems should not be biased thus provide similarity regardless of age, race, gender, and other human diversities. Some principles should be set up for AI like altruism, social values, transparency, accountability.

To make AI in reach of most people and ensure the availability of AI applications for both rich and poor people. It is required to make sure the manufacturing of AI is cost-efficient and things that escalate the manufacturing cost of AI-based systems cost of raw materials and manufacturing processes. The early research and development also upsurge the cost of AI systems. The AI requires hi-tech modern infrastructure, development, as well as deployment of these infrastructures require a lot of coinage. The setting up of big data centers, cloud storage, network connectivity, and power requirements, etc. is shown in Fig. 7.

If AI-based technologies are expensive, only rich people can afford them and enjoy the benefits. To adopt AI in education and revolutionize the whole education system to the next level, it is required to find cost-effective solutions and discover more advanced methods to make AI more accessible and affordable for all sectors of society [25].

4.1 Corporate, Expertise and Education Relation for Future

The accomplishment of AI-based technologies and their implementation are based on three main opinions:

How the corporate commercializes these robots and do marketing of these AI solutions? [26] Corporate brands and business giants use various methods to take advantage of innovative technology. These corporates heavily invest in AI technologies to provide best-in-class solutions for the Innovation, development, and manufacturing of AI systems. The competition between the companies provides better products

Fig. 7 AI-based systematic arrangement

that are qualitative as well as cost-efficient. Moreover, commercialization includes attractive advertisements and tie-ups with various schools and colleges worldwide. Finally, the sole aim is to make AI attractive, popular and make it a sort of necessity.

How the profession based on AI work? How do the Experts use AI to serve the education needs of the future? The professionals who want to teach students as a profession through remote learning by AI services and digital education. Implementation of AI-based options such as creating video content, provide online lectures, provide 3D animation, and even use of virtual reality. The training for handling personalized robots and techniques to use them at their optimum capacity is more focused. The careers in managing these AI services are in great demand in near future like- data scientists, data analysts, big data engineers, and robot scientists [27–29].

How the higher education is flourish under the umbrella of AI Services? Higher education is getting many benefits from the adaptation of AI. There are many scientific experimentation and laboratories which demonstrate a higher level of studies while using virtual reality to accomplish our goals and get more accurate results. However, using personalized robots in the research and development, these robots can perform complex calculations and provide assistive analysis for the observation. Therefore, workload management is organized and thus, overall efficiency is improved; For example—A senior surgeon can teach various junior doctors how to perform a certain surgery by using virtual reality and AI resources [30].

5   Conclusion

- The future of education by the use of Artificial Intelligence in modernizing the education infrastructure.
- Different technological approaches to make AI capable to conquer this challenge.
- The requirement of modern education is inevitable for the demands of future generations.
- The AI is implemented in various education platforms by different means such as robotics, digital devices, smartphone, laptops, etc.
- The sole aim is to provide education to everyone through the use of smart technology.
- The benefits of using Artificial intelligence in modernizing education are limitless.

Author Views

Impact of Covid-19 Discussion

Humans never imagined that a virus can bring so much destruction to the World, there are millions of deaths recorded worldwide due to Covid-19, even the powerful Nations are badly affected by the virus. This shows humans are not ready for a pandemic like this. COVID-19 put a challenge to humankind and it shows us the power of Nature and teaches us how to respect nature and prepare ourselves for future biohazards. It is unimaginable how covid-19 drastically affects the lives of people in this hi-tech modern world and puts a scary situation in front of Humans. As the primitive measures, the lockdown is obligatory to prevent the spread of the virus.

In India, COVID-19 is spreading like a wildfire and the students are like the birds on the tree, that cannot fly due to fire in the Jungle. As a student myself I observe there is a very harsh impact of Covid-19 on the education and future of students. The government proposed an online way of teaching so the students can learn from home and get the education through online lectures and live classes on various platforms. But this system of education is new for all of us and we are not habitual for this type of education system. The main problem with this system is, not everyone is associated with the internet and not everyone has smartphones and Personal computers to access online education. In India, the internet coverage is about 50% of our total population and around 55% of the total population has smartphones, moreover, the internet quality is not good in most of the rural areas of the nation. How the government ensures education is reaching everyone?

In the rural areas, the condition of primary and secondary education is on the brink of collapse already. How the government provides Online education in these areas. The Covid-19 put the future of the students in danger and innocent students are not getting proper education and most of them belong from a family below the poverty line so they cannot afford the online way of teaching too.

Online education seems beautiful and effective but, it is not that what it seems to be. Students in the urban areas have faced many problems in education due to Covid-19. The main thing is the connection issue, in most of the areas the internet is not good and efficient, the main source of internet is through the mobile sim and the signal is weak in most of the places. Another thing is limited internet data like 1 GB/day is not enough for attending regular classes for 6–8 h, the 1 Gb data only cover 2–3 h of online lecture thus rest of the lectures missed. The internet speed is not high enough to support online lectures so the students are likely to miss online lectures so they learn from the recorded lectures and are unable to clear their doubts on the spot. The lack of a study environment becomes the main concern among students the classroom, the desks and benches, the blackboard, and colleagues make a pleasant study environment. The lack of teacher-student connection, the person-to-person connection between teacher and students is most important for the flow of information as the teacher is the only one who knows his student weaknesses and strength. In online classes the clarifying of doubts becomes a difficult task because of limited class time and asking doubts in front of the whole class is like a shameful act (everyone thinks you are dumb). I observe most of the students are not taking classes seriously they join the class and put the teacher on mute and busy themselves in their business-like watching TV or playing games. Due to Covid-19 most of the students are not getting books and education material as the markets are closed, So they must study from online books and PDF the sharp alphabets and prolonged usage of smartphone and PC is hazardous for the eyes of the students. As you know there are so many disturbances already present on the smartphone like social media platforms Facebook, YouTube, and different apps and game to attract students and distract them from education. One of the main effects, I observe in online education is the lack of laboratory experiments, live experiments, and projects. The group projects and group experiments and lab studies and extracurricular group activities.

I really appreciate the efforts of the teachers they are trying hard to reach every student through online classes and provide them the best they can. The loss of focus on the subjects is due to disturbance as most of the students do not have their room. The Covid-19 panic in the family most of the time the news channel so exaggerates the Covid-19 that it creates a dense environment in the family due to which it becomes hard to focus on study. Even one sneeze can cause a havoc reaction from the family members and creates panic. As a student our career and future put into danger due to the Covid-19. Our traditional way of education was better than the present AI based online education system.

References

1. Teräs, M., Suoranta, J., & Teräs, H. (2020). Post-Covid-19 education and education technology 'Solutionism': A seller's market. *Postdigital Science Education, 2*, 863–878. https://doi.org/10.1007/s42438-020-00164-x
2. University of Michigan. (2021). Retrieved from http://sustainability.umich.edu/environ211/reduce-textbook-waste/promote-environmental-benefits.
3. Ran. (2021). Retrieved from https://www.ran.org/the-understory/how_many_trees_are_cut_down_every_year/.
4. Reddy, V., Soudien, C., Winnar Desiree, L. (2020) Impact of school closures on education outcomes in South Africa. Retrieved May 25, 2020, from www.theconversation.com.
5. Online studies Home page. (2020). Retrieved from https://www.onlinestudies.com/article/the-future-of-online-learning/.
6. Michael, M. L., Madden, G. (2020). A pedagogy of data and Artificial intelligence for student subjectification. *Teaching Higher Education, 25*(4), 456–475. https://doi.org/10.1080/13562517.2020.1748593
7. Td home page. Retrieved from https://www.td.org/talent-development-glossary-terms/what-is-e-learning
8. Zawacki-Richter, O., Marín, V. I., & Bond, M. (2019). Systematic review of research on artificial intelligence applications in higher education—Where are the educators? *International Journal of Education Technology High Education, 16*, 39. https://doi.org/10.1186/s41239-019-0171-0.
9. Dr. Aldosari, S. A. M., & Dr. Aldosari, S. A. M. The future of higher education in the light of artificial intelligence transformations. https://doi.org/10.5430/ijhe.v9n3p145
10. Hashimoto, T., Kato, N., & Kobayashi, H. Development of educational system with the android robot SAYA and evaluation. https://doi.org/10.5772/10667.
11. Hrastinski, S., Olofsson, A. D., & Arkenback, C. (2019). Critical imaginaries and reflections on artificial intelligence and robots in postdigital K-12 education. *Postdigital Science Education, 1*, 427–445. https://doi.org/10.1007/s42438-019-00046-x
12. Sánchez, H., Martínez, L. S., & González, J. D. Educational robotics as a teaching tool in higher education institutions: A bibliographical analysis. https://doi.org/10.1088/1742-6596/1391/1/012128.
13. https://www.dw.com/en/meet-germanys-first-robot-lecturer/av-47653794.
14. Çetin, M., & Özlen Demircan, H. (2020). Empowering technology and engineering for STEM education through programming robots: A systematic literature review. *Early Child Development and Care, 190*(9):1323–1335. https://doi.org/10.1080/03004430.2018.1534844.
15. Craig, A., Coldwell-Neilson, J., Goold, A., & Beekhuyzen, J. A review of e-learning technologies – opportunities for teaching and learning. In *CSEDU 2012—4th International Conference on Computer Supported Education, [INSTICC]*, Porto, Portugal (pp. 29–41).
16. Negash, S., & Wilcox, M. (2008). E-learning classifications: differences and similarities. https://doi.org/10.4018/978-1-59904-964-9.ch001.
17. Mishra, A., Makula, P., Kumar, A., Karan, K., & Mittal, V. K. (2015). A voice-controlled personal assistant robot. In *2015 International Conference on Industrial Instrumentation and Control (ICIC)*, (pp. 523–528). https://doi.org/10.1109/IIC.2015.7150798.
18. Maheshwari, G. (2021). Factors affecting students' intentions to undertake online learning: An empirical study in Vietnam. Education and Information Technologies. https://doi.org/10.1007/s10639-021-10465-8.

19. Xiao, C., & Liu, Y., & Akhnoukh, A. (2018). Bibliometric review of Artificial Intelligence (AI) in Construction Engineering and Management. 32–41. https://doi.org/10.1061/978078448172 1.004.
20. Snir, I., & Eylon, Y. (2016). Pedagogy of non-domination: Neo-republican political theory and critical education. *Policy Futures in Education, 14*(6), 759–774. https://doi.org/10.1177/147 8210316650603
21. Li, S., Tryfonas, T., & Li, H. (2016). The Internet of Things: a security point of view. *Internet Research, 26*, 337–359. https://doi.org/10.1108/IntR-07-2014-0173.
22. Hassija, V., Chamola, V., Saxena, V., Jain, D., Goyal, P., & Sikdar, B. (2019). A survey on IoT Security: Application areas, security threats, and solution architectures. *IEEE Access*, 1–1. https://doi.org/10.1109/ACCESS.2019.2924045.
23. Lane, J. M. (2012). Developing the vision: Preparing teachers to deliver a digital world-class education system. *Australian Journal of Teacher Education, 37*(4), Article5. Retrieved from http://ro.ecu.edu.au/ajte/vol37/iss4/5.
24. Tai, M. C. (2020). The impact of artificial intelligence on human society and bioethics. *Tzu Chi Medical Journal, 14*;32(4), 339–343. https://doi.org/10.4103/tcmj.tcmj_71_20. PMID: 33163378; PMCID: PMC7605294.
25. Feijóo, C., Kwon, Y., Bauer, J. M., Bohlin, E., Howell, B., Jain, R., Potgieter, P., Vu, K., & Whalley, J. (2020). Xia. *Journal of Telecommunication Policy, 44*(6), 101988. https://doi.org/10.1016/j.telpol.2020.101988
26. Singh, S. (2017). Digital marketing in online education services. *International Journal of Online Marketing,7*. https://doi.org/10.4018/IJOM.2017070102.
27. Research Berief Home page. Retrieved September 02, 2020, from https://www.cbinsights.com/research/back-to-school-tech-transforming-education-learning-post-covid-19/.
28. Kohan, L., Moeschler, S., Spektor, B., Przkora, R., Sobey, C., Brancolini, S., Wahezi, S., Anitescu, M. (2020). Maintaining high-quality multidisciplinary pain medicine fellowship programs: Part I: Innovations in pain fellows' education, research, applicant selection process, wellness, and ACGME implementation during the COVID-19 pandemic. *Pain Med 1, 21*(8), 1708–1717. https://doi.org/10.1093/pm/pnaa168.
29. Shah, S., Diwan, S., Kohan, L., Rosenblum, D., Gharibo, C., Soin, A., Sulindro, A., Nguyen, Q., & Provenzano, D. A. (2020). The technological impact of COVID-19 on the future of education and health care delivery. *Pain Physician, 23*(4S), S367–S380.
30. Verma, A., Verma, S., Garg, P., Yadav, S., & Banoth. (2021). Webinar as future of continued medical education: a survey. *British Indian Journal of Surgery,* May 17, 1–2. https://doi.org/10.1007/s12262-021-029295.

Technology Role in Education to Overcome COVID-19 Challenges

Manal Alfayez, Thaira Al Shirawi, Amira Karam, and Allam Hamdan

1 Introduction

The COVID-19 pandemic has conveyed the pivotal role of technology in combating the spread of the virus without the sacrifice of education and knowledge. Technology has resulted in any impartial reason for outside gatherings to desist as there is seemingly an application for a wide range of necessities such as: groceries, classrooms, entertainment, as well as accessing financial services and any governmental documents. This is a clear connotation that technology is no longer a simple luxury, but a necessity when considering crisis management.

The Coronavirus pandemic has created the largest disruption in education systems around the world in history, affecting nearly 1.6 billion students in 190 countries and regions around the world, according to a report issued by the United Nations on the impact of the pandemic on education [2].

The Covid-19 virus has been the most substantial inhibition in worldwide education systems in human history. In accordance with a UN report on the effects of the pandemic in regards to education, it has been stated that there have been approximately 1.6 billion affectees spanning 190 individual countries globally.

The temporary physical shut down of educational establishments has been the tactic employed by most governments as an attempt to damper the spread of the Covid-19 virus. This tactic has also consequently affected a large statistic of all current students worldwide. Some countries were more vulnerable to the effects of the pandemic and may have more disadvantaged groups than other countries, which

M. Alfayez (✉)
Ministry of Education, Manama, Bahrain

T. Al Shirawi · A. Hamdan
Ahlia University, Manama, Bahrain

A. Karam
Bahrain Institute for Banking and Finance BIBF, Manama, Bahrain

© The Author(s), under exclusive license to Springer Nature Switzerland AG 2022   37
A. Hamdan et al. (eds.), *Technologies, Artificial Intelligence and the Future of Learning Post-COVID-19*, Studies in Computational Intelligence 1019,
https://doi.org/10.1007/978-3-030-93921-2_3

is the reason UNESCO is striving to accomplish extra support to the groups that require it the most as a method to maintain the education continuity.

On the other hand, the effects of this pandemic have not all been negative, for it has been an opportunity to resolve classroom density issues in schools as well as universities that are typically found in less economically developed countries. This issue has been observed as a roadblock to a successful learning environment in which the students may learn to maximum fruition. The improvised system of social learning has carried out a huge burden for solving this issue for classrooms worldwide. This issue would have otherwise been neglected, or not been resolved if the classrooms were to operate as normal.

Technology is fundamental in many aspects of the modern-day human. It is, after all, the nifty solution that was able to be conjured swiftly as the world descended into the chaos of the pandemic. The hindrance in all of our lives whether it is work-related, personal lives, or even scientific research was easily provided for with the alternative of technology. Bahrain, in testimony of this fact, is a fluid technology user as it is common to find in all governmental and private institutions alike, as well as in all other sectors including education, tourism, and so on [3].

The Covid-19 outbreak brought a barrier alongside it, the barrier of space. Although seemingly harmless initially, the mental factors will catch up to the student and begin affecting the quality of the learning. "The overrun of my place made the absence of fixed spatial barriers an excuse to rise to different worlds through the vast Internet, and the overrun of my time possessed the tools to get rid of the routine of coming and going and crowding out others in search of speedy access to a space that was perhaps narrower than the spaciousness of minds could bear". Total reliance on EdTech educational technology to confront the educational crisis left by the Corona epidemic, cannot represent a solution, especially in the short term. Likewise, the current trend to invest in creating new educational curricula and purchasing electronic devices on a large scale is inappropriate.

Here, the summary indicates possible opportunities to invest in, alongside areas of support for which teachers and students with the hope of improving the education process in the short term (within weeks of school closures), and in the long term (within one year, either during school closures or when countries need to address educational gaps that have emerged).

Therefore, the current research problem can be crystallized by answering the following question: To what extent can modern technology and information technology contribute to meeting the challenges posed by the Corona pandemic (Covid 19) in the education sector in the Kingdom of Bahrain?

From this question are divided into the following questions:

- What is the reality of employing technology to confront the Corona pandemic in the education sector in the Kingdom of Bahrain?
- What are the challenges facing the education sector in light of employing technology to confront the Corona pandemic?

– What are the proposals and recommendations that can contribute to enhancing and improving the employment of technology in the education sector to confront the Corona pandemic?

2 Research Scientific

The importance of research is explored in two aspects:

Scientific theoretical importance

This research is considered as an important input, linking multiple scientific fields namely: technology, education, and crisis management. This appears through the theoretical framework prepared by the research for the benefit of researchers and those interested in the field of education and its relationship to technology.

Applied scientific importance

The research sheds light on the role of the Ministry of Education in managing the crisis related to the Corona pandemic (Covid 19) through the optimal employment of information technology. In its later stages, this research also provides several recommendations that can benefit officials in the field of education in terms of optimizing technology to face the challenges of the Corona pandemic (Covid 19).

3 Research Aims

The research aims to achieve the following objectives:

– Uncovering the reality of technology employment in facing the challenges posed by the Corona pandemic in the education sector in the Kingdom of Bahrain.
– Knowing the challenges facing the employment of technology in the field of education in light of the Corona pandemic in the Kingdom of Bahrain.
– Formulating some recommendations for the benefit of officials in the field of education in the Kingdom of Bahrain to face the repercussions of the Corona pandemic through the ideal employment of technology.

4 Theory and Literature Review

Under the heading "Education Disruption due to the New Coronavirus and Responding to It", a UNESCO report stated that "the spread of the virus has set a record for children and youth who have stopped going to school or university. As of March 12, 61 countries have been announced in Africa, Asia, Europe and the East." Central, North America, and South America have closed schools and universities, or implemented the closure, as 39 countries closed schools throughout, affecting more than 421.4 million children and youth, and an additional 14 countries closed schools in some regions to prevent the spread of the virus or to contain it. If these countries decide to halt schools and universities nationwide, the education of more than 500 million other children and youth will be disrupted (according to the organization). With all the audio-visual resources, illustrations, and animations, distance education has transformed from an "indoctrination" method to an "interactive" method accompanied by visual and audio effects that make the "static" educational process more attractive.

Previous Studies

This study maintains a primary focus on the Kingdom of Bahrain along with the effects of the pandemic, however, there has been multiple studies conducted on alternative countries and the results prove to be different:

As aforementioned in this study, there has been a considerable shift from physical classes to online classes. A study was conducted in order to track the newfound online environment. This study was conducted by the esteemed Zach Moore, Jake Stallard, Ashley Tittemore as well as Jessica Y. Lee in accordance with the Commission on Dental Accreditation (CODA). The study was conducted at the University of North Carolina (UNC) and explored the psyche of both the faculty as well as the students to see how the educators, as well as the educated, are striving in the very recently discovered learning environment.

The online learning environment that was observed, has been surveyed and the results prove that the clear majority of students, as well as teachers, are coping with the online learning environment [4].

On the other hand, if the situation is assessed from across the globe, in Ethiopia the transition into online proved to be a difficult one according to a study done by Wondwosen Tamrat. Tamrat has documented the difficulty in the alteration from standard physical classes to online education. Tamarat has also documented the use of SMS as opposed to more advanced systems such as zoom as the means of communication during online education.

If the issue of online education were to be assessed as a whole, it would prove to be a failing model or a barely contained catastrophe depending on the country. This boils down to the digital divide and how well countries are technologically versed as opposed to other countries.

Case of Bahrain

As a response to the spread of the Corona pandemic, and as an implementation of the instructions and decisions of the Coordinating Committee, a meeting took place in March 2020 for the approval of the Ministry of Education's plan to activate e-learning, as it became clear to the concerned authorities, the development of distance learning structures in light of the emergency situation has become a mandatory requirement In order to advance the educational process and its success. The spokeswoman noted that the ministry had not stopped issuing directives, but has actually contributed to its implementation by supporting all government institutions of higher education and special. The General Secretariat has also supported these efforts by providing the necessary administrative and health instructions and holding many of them Meetings with sector institutions. Higher education institutions have responded by employing all infrastructures and strategies for using technology to avoid disrupting the educational process. In this context, the spokeswoman praised the cooperation. The model demonstrated by all types of higher education institutions and their success in securing a fast, smooth, and effective transition to a system Distance learning and finding appropriate alternatives to cover curricula in a way that ensures the continuation of the educational process and the safety of students and staff Administrative and academic and expanding the range of facilities directed to students beyond academic support to financial aspects.

These institutions have trained lecturers and students and provided services for platforms on smartphones, to ensure that they reach the widest Possible segment of learners. The sector institutions also organized many electronic events to exchange experiences in the management of This emergency crisis. In her evaluation of the experience of the Kingdom of Bahrain, the spokeswoman concluded that what happened was the development of existing and applied systems Originally, institutions benefited from the availability of electronic infrastructure available to them, and their accumulated experience in the application of education Electronic. One of the indicators of the success of the experiment is the high student attendance rates. And the number of lectures that were implemented electronically (approx7500 lectures in the sector institutions only), and copies of the materials of these lectures are available for benefit from them at any later time. Based on these results, the electronic programs represented a lifeline to continue providing education services And avoid the educational disaster that UNESCO warned about. Looking ahead to the future of education in the Kingdom, she noted the speaker said that traditional education remains the most appropriate option especially for the early stages due to its collaborative nature and its role in refining Learner personality, A distance learning option must be provided. However, duty bearers are encouraged to be flexible in visualizing The future of education systems according to changing circumstances, it is expected that the future will witness a combined formula that mixes the two types of education. This feeling was reflected in the ministry's generalization by transferring the study from its prescriptive classroom form to e-learning from Through the various available e-learning platforms, and issuing instructions related to determining the mechanism for evaluating educational achievement in a system Distance learning, and the General Secretariat directing higher education institutions to work on implementing distance learning at a level that meets the Standards for institutional academic accreditation and its call to present plans that respond to various possible scenarios [5].

Challenges facing technology employment in education in light of the Corona pandemic

We call for the need for educational policy-makers to be realistic, because the pre-drawn education goals will not be achieved to an ideal level during the application of distance education, as the educational process will be different during the crisis, and

EdTech will not be able to cope with the disruption of the educational system alone; Therefore there is a need to strengthen the educational process, and prepare for the return phase of the educational classes in schools following the end of the Corona crisis, while working to address the educational gaps that emerged at the start of the crisis.

Having mentioned that it is essential to note that the huge investment into the field of distance education without sufficient studies will possibly yield obstacles, and among the most prominent of these obstacles are:

The weakness of the data collection and education management system is exacerbated by the expansion of the application of distance learning

Education systems around the world are decentralized, which provides more spaces for increased innovation in educational methods, and creates challenges related to how to delegate decision-making powers. The closure of schools causes a lack of data and information related to the educational process locally, so the decision-making process is faltering, and it is difficult to improve it because following up the distance learning process cannot accurately measure its results.

The slowdown in the progress of students in the formal educational curriculum

The urgent closure of schools did not leave sufficient time to prepare a suitable transition strategy to implement distance learning. The school curricula that used to adopt the testing system upon completion of their teaching are now invalid and need urgent development in light of the distance learning system. There is a state of anxiety and anticipation among students regarding how they will move to the next educational year under the new system, as well as how to adapt to the distance education system [6].

Certain groups of students are more affected, and their academic progress is delayed.

They are primary school students and students with weak educational attainment. Lack of internet access among students is a very noteworthy point. The digital divide in low-income countries will be a hindrance in the way of completing the school year for the distance learning system for many students. There exists a disparity in access to the Internet and mobile networks among people, for example, more than 80% Of the population of some Southeast Asian countries has access to the Internet, while the percentage does not exceed 39% in Vietnam and some African countries. Here a state of inequality materialises. Those who have access to the Internet will be able to follow the progress of education remotely, while those who do not have electricity, smartphones, and computers will be deprived of continuing education.

The difference in the interaction of teachers with the distance education system.

The capabilities of the teachers themselves vary during the application of the distance education system, some of them will be effective in the production of electronic files and educational videos and the establishment of video conferences while maintaining interaction with their students through social media applications. Others will find it difficult to deal with educational technology, and they will feel overwhelmed when they are asked to comply with technological standards that they are not familiar with. Therefore, heads of educational systems must be familiar with teacher levels when setting goals for distance education systems, because the educator's capabilities are one of the main determinants of the success of the educational process.

5 Research Methods

This section will be reflecting upon the research by presenting: the methodology, design, research population, selection process, definitions of the research sample, the research tools used in data collection and analysis, as well as the research variables and determinants.

Methodology and design

The research followed the descriptive, analytical, quantitative approach, through the method of collecting data from the population. Examinations using questionnaires that the student prepared and designed by examining the studies and previous and theoretical literature, and this approach is considered the most appropriate for the research objectives and its questions, as it is the approach capable of measuring the relationship between variables He described this relationship an accurate description.

Population and sample research

All workers in the education sector are considered a target population for current research, whether the workers are in the administrative or academic staff in pre-university education schools in the Kingdom of Bahrain. As for the research sample, a random sample of 100 male and female employees working in private schools in the Kingdom of Bahrain was chosen.

Measurement

The research was conducted in the form of a questionnaire that inquired about separate variables. The questionnaire consisted of 10 questions that measured two dimensions. Namely: Employment of technology in education, and teacher job performance.

The validity of the questionnaire

The research relied on measuring the accuracy and reliability of the questionnaire by the Alpha Cronbach coefficient, by applying it to an experimental sample from the same population.

Table 1 Cronbach's Alpha coefficient for variables

Dimensions	Software	Infrastructure and equipment	Teachers' job performance	Overall
Alpha coefficient	0.846	0.798	0.884	0.897

6 Data Analysis

This chapter of the research is concerned with presenting statistical procedures and analyzing the data and information gathered from the research sample with scientific measurement tools, then analyzing them through some different statistical methods, verifying the validity and reliability of those tools, and discussing the results.

6.1 Validity and Reliability Test

The next step of the testing process is the validity test to measure whether the item truly measures what it is supposed to measure. Validity tests were conducted using both content validity and construct validity tests.

The Alpha Cronbach coefficient was used to verify the validity of the questionnaire by applying the questionnaire to a primary sample representative of the research community, which numbered 30 teachers in the elementary stage, and the results of the analysis indicated a high level of validity and reliability, as the value of the Alpha Cronbach coefficient reached (0.897), which means the possibility of reliance on The questionnaire, its validity, and its reliability, in order to verify the research hypotheses (Table 1).

The following table shows the stability values for the questionnaire dimensions: **Descriptive Statistics** *table.*

Descriptive Statistics *for* Software, *(n = 100).*

No	Item	Descriptive statistics		Ranking according to mean
		Mean	Std deviation	
1	The school uses advanced software in the educational process	3.98	1.344	4
2	The school relies in its communication with students and parents on safe and appropriate social media programs	4.11	1.214	3

(continued)

(continued)

No	Item	Descriptive statistics		Ranking according to mean
		Mean	Std deviation	
3	The school owns the ownership rights to the electronic programs it uses in the distance education process	4.20	0.984	2
4	The school trains the educational staff on the skills of using technological programs to activate the distance learning process	4.32	0.764	1
Overall		**4.15**	**1.07**	

The descriptive statistics of the electronic software axis used by schools to implement distance learning processes during the Corona pandemic (Covid-19) illustrate, as the general average value increased, which amounted to (4.15), which means that schools provide suitable electronic software for distance learning (Table 2).

The descriptive statistics of the axis of electronic and technological equipment, equipment and tools provided by schools to implement distance learning processes

Table 2 Descriptive Statistics for Infrastructure and equipment, (n = 100)

No	Item	Descriptive statistics		Ranking according to mean
		Mean	Std deviation	
1	The school has a suitable infrastructure to implement the distance education process	4.18	1.054	2
2	The school has tools, means and equipment that help teachers to implement the goals of distance education	4.22	0.953	1
3	The internet used in the school is suitable for distance education in terms of speed and efficiency	3.98	1.213	3
4	The school has fully equipped classes with electronic devices and various learning methods	3.44	1.543	5
5	Teachers are trained to use technological devices on an ongoing basis	3.75	1.354	4
Overall		**3.914**	**1.2234**	

Table 3 Descriptive statistics for teachers' job performance, (n = 100)

No	Item	Descriptive statistics		Ranking according to mean
		Mean	Std deviation	
1	Teachers in the school possess technical capabilities and skills that help them implement the distance education process	4.10	0.993	2
2	Teachers run the classroom effectively through distance learning programs	3.54	1.543	4
3	Teachers perform their educational tasks according to educational quality standards	3.79	1.054	3
4	The performance of teachers did not differ during and before the Corona pandemic	3.01	1.321	5
5	Teachers' performance is assessed on an ongoing basis and provided with feedback on their performance	4.32	0.872	1
Overall		**3.752**	**1.1566**	

during the Corona pandemic (Covid-19) show, as the general average value increased, which amounted to (3.914), which means that schools provide technological and electronic equipment suitable for distance learning (Table 3).

The descriptive statistics of the column of teachers' job performance in relation to distance education during the Corona pandemic (Covid-19) illustrate, as the general average value increased, which amounted to (3.752), which means that schools train teachers, raise their efficiency and develop their performance to complete tasks related to distance learning.

Test of hypotheses

H1: There is a positive relationship of statistical significance at (0.05) between the employment of technological equipment and infrastructure and the job performance of teachers.

To verify this hypothesis, Pearson Labs was used to ensure that there is a correlation between teachers' job performance and the school's electronic infrastructure and equipment, and it is explained by the following countries:

	Technological equipment and the job performance of teachers
Person coefficient	0.765**
p-value	0.000

**Sig. at (0.01)

The value of the correlation coefficient, which amounted to (0.765), showed the existence of a positive, high, and statistically significant correlation at the level of (0.01) between the electronic and technological equipment provided by the school and the job performance of teachers in light of the Corona pandemic (Covid-19), which means that teachers' job performance In schools, it increases as the appropriate technological equipment is available for the process of distance education.

H2: There is a positive relationship with statistical significance at (0.05) between the employment of technological software and the job performance of teachers.

To verify this hypothesis, Pearson Labs was used to ensure that there is a correlation between teachers' job performance and employment of technological software, and it is explained by the following countries:

	Employment of technological software and the job performance of teachers
Person coefficient	0.885**
p-value	0.000

** Sig. at (0.01)

The value of the correlation coefficient, which amounted to (0.885), showed the existence of a positive, high, and statistically significant correlation at the level of (0.01) between the electronic software provided by the school for communication between teachers and students during the Corona pandemic, and between the job performance of teachers, which means that the job performance of teachers in schools It increases with the availability of cartoon software and technology suitable for the distance education process.

Recommendations

First—in the short term

1. Creating electronic platforms for teachers. In light of the Covid-19 crisis, an electronic platform can be created through which teachers can share their ideas and suggestions on developing educational technology. For example, the Indian platform DIKSHA, which has hosted educational plans and materials proposed by teachers over the years, with the support of the central authority.
2. Contribution of platforms to improving the abilities of educators to use educational technology.
3. Improving the quality of curricula through teachers' efforts to create advanced educational content in local languages, as in the case of the Tusome platform in Kenya, which contributed to bridging learning gaps in resource-limited settings.

4. The use of low-tech technological solutions in low-income countries, such as the use of SMS text messages.
5. Broadcast educational lessons on television and radio platforms to reach students without internet use:
6. Existing infrastructure and cheap technology can be leveraged to set up initiatives quickly, on a large scale, and with little investment, as educational technology applications are designed to work through regular phones via SMS (textTETEA in Tanzania, or Eneza Education in Kenya, Ghana and Côte d'Ivoire).
7. Create comprehensive educational resources on the Internet sites. It includes various educational courses and is free of charge, and these sites can be accessed by students themselves or with the help of their parents. An example is the Khan Academy.
8. The need for policy makers to realize that Internet courses across global platforms differ in quality, culture and language, and for this reason, each country should invest in developing e-course content in line with its culture and local language to maximize educational benefit, with the necessity of making educational content free for students, as well as making partnerships with official, non-formal and international institutions to improve the digital educational content provided, whether in the short or long term.

Second—in the long run

1. Development of the education system through play. If schools are closed for a long time, students will lose their enthusiasm for the educational process, and parents will not be able to encourage them on their own to continue learning. Therefore, there is a need to develop educational applications in the form of games, containing various educational lessons, and broadcast over the Internet or Radio or television, with awarding points for advanced students over their peers.
2. The launch of an educational database. This database will contribute to providing basic data on students, their needs, and their social conditions, while defining "the technology available within each family",
3. Mobile applications can be relied on to obtain the teachers' evaluations of their students, and to collect the necessary data on the educational process in general geographically.

References

1. United Nations. (2020). Policy Brief: Education during COVID-19 and beyond, August.
2. United Nations Development programme (UNDP). (2020). COVID-19 and human development: Assessing the crisis, envisioning the recovery.
3. Aljamal, S. (2018). The practice of transformational management and its role in achieving institutional excellence from the point of view of workers in the directorates of education in

Hebron. *International Journal of Business Ethics and Governance, 1*(1), 61–83. https://doi.org/10.51325/ijbeg.v1i1.12

4. Alves, G. (2018). The impact of culture and relational quality in the cooperation between export companies and local distributors. *International Journal of Business Ethics and Governance, 1*(2), 1–19. https://doi.org/10.51325/ijbeg.v1i2.13

5. UNESCO. (2020). Global Education Monitoring (GEM) Report, 2020: Inclusion and education: all means all.

6. Adeosun, O. T., & Owolabi, K. E. (2021). Gender inequality: Determinants and outcomes in Nigeria. *Journal of Business and Socio-economic Development, 1*(1), 125–138. https://doi.org/10.1108/JBSED-01-2021-0007

Stress-Coping Strategy in Handling Online Classes by Educators During COVID-19 Lockdown

D. Binu Sahayam⬤, G. Bhuvaneswari⬤, S. Bhuvaneswari, and A. Thirumagal Rajam⬤

1 Introduction

Teaching is frequently considered one of the most stressful jobs, and it comes with its own set of obstacles. "One-third of teachers find their current teaching environment extremely stressful" [1]. Teaching is more prone to stress than other occupations since dealing with children of all ages throughout the day is demanding in and of itself. Teachers experience higher levels of psychological distress and poorer levels of job satisfaction than the general population [2]. Teachers, according to research, are subjected to a variety of stressors. Teaching uninspired kids, maintaining classroom discipline, a rigorous schedule, frequently changing plans, and assessment based on interpersonal relationships unfavourable working conditions are a few [3]. Every educator experiences unique stress factors, according to [3], and are dependent on the individual interaction between the teacher's personality, values, talents, and circumstances. Furthermore, various elements such as coping mechanisms and tactics, personality traits, and environmental features might interact to influence a teacher's judgement of how stressful a scenario is. Teachers utilize coping methods such as cognitive, emotional, and behavioural soothing and adaptation to stressful situations to deal with unpleasant events and decrease feelings of pain [3, 4].

D. Binu Sahayam (✉) · G. Bhuvaneswari
School of Social Sciences and Languages (SSL), Vellore Institute of Technology, Chennai, India
e-mail: binusahayam.d@vit.ac.in

S. Bhuvaneswari
Vellore Institute of Technology, Chennai, India
e-mail: bhuvaneswari.s@vit.ac.in

A. Thirumagal Rajam
Department of Social Work, Madras School of Social Work, Chennai, India
e-mail: thirumagal@mssw.in

© The Author(s), under exclusive license to Springer Nature Switzerland AG 2022
A. Hamdan et al. (eds.), *Technologies, Artificial Intelligence and the Future of Learning Post-COVID-19*, Studies in Computational Intelligence 1019,
https://doi.org/10.1007/978-3-030-93921-2_4

Teachers deal with a wide variety of new pressures due to the COVID-19 outbreak, including adapting to online teaching. Neither the parents nor the teachers were well equipped to deal with these developments' numerous issues. Teachers struggled to maintain relationships with their pupils and missed out on school advice and support. With the emergence of COVID-19, the reasons for occupational stress have grown. A quick shift to online delivery has exacerbated workloads that were earlier considered excessive. Many teachers were unprepared, but whose consequences are expected to continue for years. Many teachers struggle to balance personal and professional obligations and work from home requirements that blur the physical, temporal, and psychological balance between school and home. Space to be shared by the along with significant others who might have health concerns, social and physical distance, travel restrictions, closed borders, shortages of essentials and uncertainty about when life would return to "normalcy". Faith is the assurance that what 'normal' implies today or in the future has been undermined; the long-term implications for teachers and teaching are unknown. Some children require their attention. The nature of the global pandemic, in particular, offers a unique set of COVID-19 challenges. There's no doubting that living with COVID-19 has become more tough and complicated for everyone, especially the teachers, all of a sudden and unavoidably. Teaching, unlike other professions, emphasizes interpersonal interaction, body language, and a kinesthetic setting. The 'new normal' online educational environment has entirely shattered this paradigm. This paper tries to find out how the teachers deal with the crisis when much of the world is shut down, and education systems are disrupted.

2   The Rationale of the Study

Online education has brought many educational challenges between educators and learners. There is no doubt that the COVID-19 pandemic has a severe effect on the education sectors worldwide. It has caused a lockdown of educational activities and a sudden transition to online learning. The inadequate knowledge on the usage of digital techniques and tools, lack of technical support, security issues, fluctuating power supply, low internet bandwidth, poor network connectivity, unable to predict learners comprehending course content, student's participation and motivational level, unequal access to students are the hitches faced by the educators and learners in the virtual classroom. Online classes have been challenging for 84% of teachers, as stated in the India Today web desk [5]. Thus, this study focuses on the educators' stress level, coping strategies, and educational challenges during this COVID-19 pandemic and ways for possible interventions in generating an undisrupted learning process.

3 Objectives

- To study the demographic details of the respondents
- To examine the level of stress faced by the educators during the COVID-19 pandemic in conducting online classes
- To analyze the various forms of coping strategies implemented by the educators during the COVID-19 pandemic in conducting online classes.

4 Research Questions

- What is the level of stress faced by the educators?
- What coping strategies do the educators use in managing stress?
- What are the challenges faced by educators during online teaching?

5 Literature Review

Stress is your body's reaction to changes that have occurred by a person, situation or event. Stress can be caused due to physical or psychological effects. When an individual is overstressed, it will affect his/her physical and psychological well-being. Stress is a normal reaction that occurs in everyone life. There are several various definitions of stress [6, 7] used by psychologists, doctors, and management consultants. Stress is made up of various experiences, pathways, responses, and outcomes triggered by a variety of events or circumstances [6, 7]. Stress is a well-studied psychological concept that refers to a psychological reaction to environmental situations (stressors) that can result in bodily arousal and dangers to one's well-being [8, 9]. Stress is a circumstance in which a human reacts to or faces something different to a new opportunity, the constraints, and the effort required to meet the demand [10]. In behavioural sciences, stress is defined as the "perception of threat, with resulting anxiety discomfort, emotional tension, and difficulty in adjustment". The majority of stressful events fall into one of three categories: harm-and-loss circumstances, threat circumstances, or challenge situations" [11]. Stress is defined as "an imbalance between an individual's perceived expectations and his or her perceived ability to meet those expectations [12]. According to Lazarus, a unique interaction between the person and the environment, putting one's well-being at risk, is also called stress.

The way we evaluate an incident, according to Lazarus, impacts how we react emotionally. Thus, the subject's perception of the expectations in the workplace and his or her ability to meet those influences work-related stress [13]. To put it the other way, workers must judge their circumstances and the demands of their surroundings as stressors and perceive themselves as lacking the resources to deal with them, resulting in negative reactions to their well-being [14]. Since the stress phenomenon is complex and characterized by several psychological, social, and biological processes involving human-environmental interaction, recent studies inspired by the Positive Psychology movement have sought to understand how these processes can affect people's health and well-being. When demands and pressures do not match the employee's knowledge and pose a threat to survival in a workplace, it results in stress [15]. Stress can visibly effect emotionally, behaviourally and physically, and it differs from person to person [16].

Coping refers to a person's approach to dealing with a stressful situation. Stress coping styles [17] are "a coherent set of individual behavioural and physiological variances in the reaction to a stressor that stay consistent across time and context." "Ways of Coping" was a metric designed [18] to differentiate between two basic kinds of coping, problem-focused and emotion-focused coping. Problem-focused coping focuses on resolving issues or altering the source of stress [19]. On the other hand, emotion-focused coping seeks to lessen or regulate the emotional suffering connected with the event. Even while most stressors elicit both coping styles [20], problem-focused coping tends to predominate when people believe they can do something positive [19]. When people believe that the stressor must be endured, emotional-focused coping takes precedence. Efforts to manage stressful demands, regardless of the outcome, must be included in definitions of coping. This means that no single technique is thought to be superior to the others. A strategy's quality (efficacy, appropriateness) is assessed solely by its results in a specific encounter and its long-term implications.

6 Theoretical Framework

The study's theoretical framework is based on Lazarus and Folkman's [21] Transactional theory of stress and coping [21]. This theory helps to understand the prevailing situation in a deeper sense and suggest insightful study aspects. There are many reasons for giving this theoretical construction. The first reason is to understand the stress encountered by the educators in imparting online education. The second reason is to understand the stressors encountered by the educators in conducting online classes. The third reason is to understand the coping strategies adopted by the educator they practice to manage stress. Stress is a common factor encountered by every individual. We feel stressed when we do continuous and monotonous physical activity, which leads to emotional and psychological disturbances. When an individual is involved in continuous and monotonous physical activity, it leads to emotional and psychological turbulences. The transactional model of stress and coping reveals that transaction occurs when stress results from an imbalance of demands and resources.

The steps adopted in overcoming the stress.

The flowchart (Fig. 1) depicts the method used by educators to deal with pressures. The flowchart depicts the numerous types of pressures that educators endure. To deal with their stress, the first and most important step is to make a list of all the

Fig. 1 Steps adopted in overcoming the stress

stressors, then understand the sources of stress, and last come up with possible coping mechanisms and abilities to deal with the stressors.

This paper focuses on the stress caused by various stressors experienced by the educators in handling online classes during the COVID-19 pandemic (Fig. 2). A few of the stressors affecting the physical and psychological well-being of the educators are technical difficulties, a lack of motivation in the learners', work pressure and keeping up to deadlines and monitoring the students' attention. Eye pain, neck

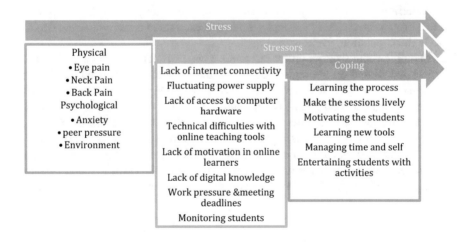

Fig. 2 Theoretical framework stress-coping strategy by educators in handling online classes

pain, back pain, body pain, numbness, and cramps are physical effects. In contrast, factors such as anxiety, irritation, anger, isolation, peer pressure, environmental surroundings arise due to psychological effects. Learning the process, motivating sessions for students, managing self and time, spending time with family and friends, eating and engaging in extracurricular activities, organizing debates and discussion sessions, role plays, application of animations in theory classes, game-based learning techniques, doing yoga/mediation, exercises, digital understanding knowledge and exploring interactive technology to make classes livelier and an adventurous one are some of the coping strategies adopted by the educators in handling online sessions during COVID-19 pandemic. Accessing online classes may be a new platform for educators and maybe a contented one or may not. Still, the COVID-19 pandemic has made "Online classes" a popular one among the academic community. Therefore, this model appears to be a good fit for this research. Educators teaching online classes experience stress in numerous forms, which has associated hurdles, and so gives coping strategies for overcoming the stress during the COVID-19 pandemic.

7 Materials and Methods

This is a descriptive study conducted during the COVID-19 pandemic lockdown among educators working in colleges and universities. The sample size of the study is 100 (50 female and 50 male) respondents. The "Stratified Disproportionate Random Sampling" method was applied in selecting the respondents.

Stratified sampling is one of the probability sampling methods. It is a method of sampling that divides the population into smaller sub-groups known as strata [22]. In a Stratified sampling method, stratification can either be proportionate or disproportionate. In this study stratified disproportionate sampling method is applied. In a disproportional stratified sample, the size of each strata is not proportional to its size in the population, as shown in Fig. 3.

An online Faculty development programme was conducted in which around 167 participated. Based on the registration form, the researcher ghettoized the female and male respondents. With the application of the lottery method, the researcher finalized 50 male and 50 female for the study. Data was collected using the mailed-questionnaire method with demographic details and two standardized scales (1) Depression, Anxiety and Stress Scale—21 Items (DASS-21) and (2) Coping self-efficacy scale was prepared in collecting information from the educators and data was analyzed using the Statistical Packages for Social Sciences. Cronbach's Alpha test assessed the reliability of variable used in this research.

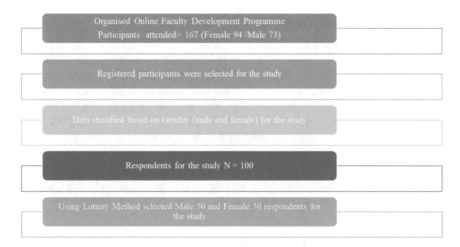

Fig. 3 Sampling process

8 Hypothesis

1. There is a significant difference between respondents' male and female respondents regarding their stress and coping with self-efficacy.
2. There is a significant relationship between age and hours spent in an online class of the respondents about their stress.
3. There is a significant relationship between the respondent's age about their coping with self-efficacy.
4. There is an association between the marital status of the respondents about stress and coping with self-efficacy.

9 Reliability Test

The reliability test of stress and coping with self–efficacy was done from the data collected. From Table 1, it is obvious that the Cronbach's Alpha value is 0.874 for stress and 0.967 for coping with self-efficacy, which is more than 0.6. This infers that the given data is reliable for further study.

Table 1 Reliability test for stress and coping with self-efficacy

Stress	Cronbach's Alpha	N of Items
	0.874	7
Coping with self-efficacy	Cronbach's Alpha	N of Items
	0.967	26

10 Results and Discussion

Globally, educational institutions tuned towards online learning platforms to continue with the process of educating students [23]. The COVID-19 pandemic has initiated digital learning or online learning in wider perspectives. Educators and learners adapting to the new normal in educational sectors have increased online learning platforms. Based on the objectives, the findings of the study are explained in details. A total of 100 responses were received for this investigation.

Table 2 shows the demographic characteristics of the respondents. Out of 100 respondents, 50% were females, and 50% were males with ages ranging from 21 to 55. The majority (84%) of the respondents were from Tamil Nadu state, and the rest were from the following states Kerala, Karnataka, Telangana, Pondicherry, Uttar

Table 2 Demographic details of the respondents

Variables	No. of respondents (n = 100)	Percentage (%)
Gender		
Male	50	50
Female	50	50
State		
Tamil Nadu	84	84
Kerala	6	6
Karnataka	4	4
Telangana	1	1
Pondicherry	3	3
Uttar Pradesh	1	1
West Bengal	1	1
Hours spent in online class		
2 h	37	37
3 h	23	23
4 h	23	23
5 h	17	17
Marital status		
Single	23	23
Married	77	77
Educational qualification		
Graduate	2	2
Post-Graduation	15	15
Professional	1	1
Doctorate	82	82

(continued)

Table 2 (continued)

Variables	No. of respondents (n = 100)	Percentage (%)
Age		
21–30 years	6	6
31–40 years	56	56
41–50 years	36	36
51–60 years	2	2

Pradesh and West Bengal. The majority of them, 82%, had doctorates, while the rest were primarily postgraduates and graduates. Faculty members were drawn from both the public and private educational institutions. The majority of them (85%) chose to remain at home with their families. The findings also reveal that the majority, 77%, were married and the educators handled a minimum of two hours of online classes in a day, i.e. 37% of them expressed that they handle two hours a day, 56 percent (23% said three hours and 23% said four hours) and 17% said five hours a day. During the lockdown, stress was created by a lack of technical and social support, extreme workload using computers and social media. Teachers utilized both functional and dysfunctional coping techniques, with functional approaches being more prevalent [24]. Furthermore, we looked into the impact of socio-demographic characteristics on stress levels, coping techniques and encountered impediments [25].

Hypothesis—1

H0: There is no significant difference between male and female respondents about their stress level and coping with self-efficacy.
H1: There is a significant difference between respondents' male and female respondents about their stress and coping with self-efficacy.

The Table 3 explains the stress faced by both the gender and also explains their coping with self-efficacy. The table about gender and stress highlights that the P-value is greater than 0.05, the null hypothesis is accepted. Thus the findings reveal that irrespective of gender, both male and female undergo stress. The table also explains

Table 3 Genders of the respondents about their stress and coping with self-efficacy

Variable Gender	Mean	Standard deviation	Statistical inference
Stress			
Male	12.84	9.698	$Z = 1.386$ $P > 0.05$ Not significant
Female	10.32	8.440	
Coping with self efficacy			
Male	194.42	32.442	$Z = 1.180$ $P > 0.05$ Not significant
Female	184.82	47.512	

a significant relationship between gender and coping with self-efficacy; the P-value is greater than 0.05. Hence the null hypothesis is accepted. Therefore both the gender adopt coping techniques to overcome their stress which is significant from the study. Females used functional coping strategies more frequently than males, according to a study by [26], although gender did not influence coping strategy implementation.

Hypothesis—2

H0: There is no significant relationship between age and hours spent in an online class of the respondents about their stress.
H1: There is a significant relationship between age and hours spent in online class of the respondents regarding their stress.

From the above Table 4, it is observed that regardless of age, every person undergo stress, and it is inferred from the table that the P-value is greater than 0.05; hence the null hypothesis is accepted. Furthermore, the data collection reveals that the respondents' age ranges from 21 years of age till 55 years of age, therefore whether young or elder, all persons undergo stress in various forms.

The number of hours spent in online class may increase the stress and decreases and depends on the hours. The table highlights that P values are less than 0.01, it is highly significant and shows a positive relationship between two variables (number of hours spend and stress); hence the null hypothesis is rejected. When the number of hours increases, the stress increases, and when the number of hours decreases and the stress decreases.

Poor internet connectivity, power fluctuations, inadequate knowledge on online tools, lack of IT support, poor environment status, lack of electronic gadgets, poor response from the learners in online classes or no response from the learners, work pressures and meeting deadlines, monitoring learners are some of the factors that cause stress for the educators in handling online classes [27].

Hypothesis—3

H0: There is no significant relationship between the respondents' age about respondents coping with self-efficacy.
H1: There is a significant relationship between the respondents' age about respondents coping with self-efficacy.

Age is not a factor in this study. Whatever the age is, the respondents are trying to cope up with the situation. It is observed from Table 5 that the P-value is greater than 0.05. Hence the null hypothesis is accepted. The teacher spends more time online for

Table 4 Age and hours spent in an online class of the respondents about their stress

Variables	Correlation value	Statistical inference
Age with stress	−0.057	P > 0.05 Not significant
Number of hours spent in online class with stress	0.579**	P < 0.01 Highly significant

Table 5 Respondents age about their coping with self-efficacy

Variables	Correlation value	Statistical inference
Age with coping with self efficacy	0.012	P > 0.05 Not significant

teaching, which significantly impacts the disruption of an individual's self-efficacy, regardless of age [28]. Various forms of coping tactics adopted by the educators are talking positively, involving in extracurricular activities, praying/meditating, and self-distracting hobbies like watching TV and listening to music, spending time with family, friends etc.

Hypothesis—4

H0: There is no association between the marital status of the respondents about stress and coping with self-efficacy.
H1: There is an association between the marital status of the respondents about stress and coping with self-efficacy.

In this paper, the marital status of the respondents was subdivided into the following categories—single and married. Table 6 portrays that whether married or single, any person faces the stress and can cope with the stress in handling the online classes. On the other hand, stress has no impact on understanding whether a person is married or single. Thus, the marital status of the respondents highlights that any individual irrespective of their status, undergo stress and try to adopt coping strategies to overcome their stressors. The P-value is greater than 0.05; hence the null hypothesis is accepted.

Table 6 Marital status of the respondents about stress and coping with self-efficacy

Level of stress	Marital status		Statistical inference
	Single	Married	
Normal	17	57	$X^2 = 6.255$
Mild	3	10	df = 4
Moderate	3	2	P > 0.05
Severe	0	1	Not significant
Extremely severe	0	7	
Level of coping with self efficacy	Marital status		Statistical inference
	Single	Married	
Low	14	43	$X^2 = 0.182$
High	9	34	df = 1 P > 0.05 Not significant

Table 7 Gender of the respondents about stress and coping with self-efficacy

Level of stress	Gender	
	Male	Female
Normal	34	40
Mild	9	4
Moderate	2	3
Severe	0	1
Extremely severe	5	2
Level of coping with self efficacy	Gender	
	Male	Female
Low	32	25
High	18	25

Gender and Level of Stress and Level of Coping with Self Efficacy

The role of gender in academic sectors plays an immense role [29]. Both male and female undergo stress and tires to adopt coping techniques to overcome their stresses. From the table, it is clear that females (40) undergo normal levels of stress and male (7) respondents undergo extremely severe stress levels (Table 7). In coping with self-efficacy, both male and female respondents try to balance low and high level in their performances. About female respondents, 25 try to use their coping techniques at a higher level, whereas 25 respondents apply their coping strategies at a low level. Female respondents try to apply coping techniques as much as possible to overcome their stress. Few female respondents shared that they do multitasking work, and this is also one of the major reasons for them to balance any situations with their application of coping techniques. Thirty-two male educators use the coping strategies at a very low level, and only eighteen respondents use their coping strategies at a high level to overcome their stress.

According to a study by Schwarzer [30], human self-efficacy is the result of a broad and stable sense of personal competencies to deal and act effectively with a variety of stressful situations, so decided to link it to the level of self-efficacy of teachers and they're proactive coping as an effective coping strategy. It has been discovered that the better a teacher's self-efficacy, the better their proactive coping. A proactive instructor is also a source of encouragement and achievement for his or her students [31].

11   Limitations of the Study

In this study, only those educators who attended the online faculty development programme were considered. Due to time constraints, only 100 respondents were selected for the study. Transgender were excluded from the study.

12 Implications and Conclusions

The noteworthy findings regarding the effect of gender on the amount of stress generated by stress causes and the perceived efficiency of coping mechanisms are presented in Table 3.

The study findings show a link between educators stress levels [induced by a variety of stressors], their perceptions of the effectiveness of various coping mechanisms, and their personal and professional qualities. The findings also show that personal and professional characteristics have varying impact on educators' perceptions of various stressors and coping mechanisms. Gender, age, and marital status do not make a significant difference in experiencing higher stress.

Teachers at various levels of education have diverse perspectives on the efficacy of modifying behaviour and seeking help as coping techniques. For example, the effectiveness of behavioural change as a coping mechanism is rated higher by female educators than by male educators.

They also believe that asking for assistance is a more effective method than males. In addition, when compared to their counterparts, female educators believe that getting help as a coping approach is more beneficial. Gender, age, and marital status have little bearing on whether or not a person experiences increased stress levels and used coping techniques. Behavioural change, seeking social support, and segregating work at school were all scored higher by women than by males. Gender inequalities are caused by a variety of gender roles and societal customs addressing various gender-related tasks. Women are more capable than men when it comes to coping abilities.

The findings also reveal a relationship between teachers' stress levels [caused by a range of stressors], their perceptions of the effectiveness of various coping techniques, and their personal and professional traits. The data also demonstrate that teachers' views of various stressors and coping techniques vary depending on their personal and professional traits.

The findings show that people who spend more time online are more stressed due to health issues such as straining eyes, spine-related difficulties, sleep loss, melancholy, and anxiety. In addition, communicating in a classroom where one does not meet individuals face to face and make eye contact to form a connection, where one is unable to employ socializing skills, and where one has little control over pupils causes educators a great deal of stress [32].

The findings also show that instructors who spend less time online value the effectiveness of coping through behavioural change and emotional control more than those who spend more time online, which could be taken as a lack of confidence in their ability to change. As a result, teachers who spend less time online believe that coping through self-work is more effective.

Conflict of Interest No conflict of interest exists.

References

1. Borg, M. (1990). Occupational stress in british educational settings: A review. *Educational Psychology, 10*, 103–126.
2. Travers, C.J., & Cooper, C.L. (1993). *Mental Health, Job Satisfaction and Occupational Stress among the UK.*
3. Kyriacou, C. (2001). Teacher stress: Directions for future research. *Educational Review, 53*(1), 27–35.
4. Admiraal, K., & Wubbles, A. (2000). Effects of student teachers' coping behavior. *British Journal of Educational Psychology, 7*(1), 33–52.
5. 84% of teachers facing challenges during online classes: Survey (online) in India Today Web desk. New Delhi. Retrieved from https://www.Indiatoday.in/educationtoday/latest-stu dies/story/84-of-teachersfacing-challenges-during-online-classes-survey-1780816-2021-03-1 8. (2021).
6. Taylor, S. E. (1999). *Health psychology* (4th ed.). McGraw-Hill.
7. Steptoe, A. (1997). *Stress and disease, The Cambridge handbook of psychology, health and medicine.* Cambridge University Press.
8. Lazarus, R. S. (1996). *Psychological stress and the coping process.* McGraw-Hill.
9. Lazarus, R. S. (2006). *Stress and emotion: A new synthesis.* Springer.
10. Naturale, A. (2007). *Secondary traumatic stress in social workers responding to disasters.*
11. Anspaugh, D. J., Hamrick, M. H., & Rosato, F. D. (2003). *Wellness: Concepts and applications* (5th ed.). McGraw-Hill.
12. Cox, T. (1978). *Stress.* Baltimore, MD: University Park Press.
13. Häusser, J. A., Mojzisch, A., Niesel, M., Schulz-Hardt, S. (2010). Ten years on: A review of recent research on the Job Demand-Control (-Support) model and psychological well-being. *Work Stress, 24*(1):1–35.
14. Malik, S., & Noreen, S. (2015). Perceived organizational support as a moderator of affective well-being and occupational stress among teachers. *Pakistan Journal Commerce Social Science, 9*(3), 865–874.
15. Steve, W. (2011). Managing workplace stress: a best practice blueprint, Volume 1 of Fast track series, Publisher *Wiley*, Original from Cornell University, 78–84.
16. Friedman, H. S., & Booth-Kewley, S. (1987). The" disease-prone personality": A meta-analytic view of the construct.
17. Tudorache, C., Braake, A., Tromp, M., Slabbekoorn, H., & Schaaf, M. J. (2015). Behavioral and physiological indicators of stress coping styles in larval zebrafish. *Stress, 18*, 121–128.
18. Folkman, S., & Lazarus, R. S. (1985). If it changes, it must be a process: A study of emotion and coping during three stages of a college examination. *Journal of Personality and Social Psychology, 48*, 150–170.
19. Carver, C. S., Scheier, M. F., & Weintraub, M. F. (1989). Assessing coping strategies: A theoretically based approach. *Journal of Personality and Social Psychology, 56*(2), 267–283.
20. Folkman, S., & Lazarus, R. S. (1980). An analysis of coping in a middle-aged community sample. *Journal of Health and Social Behaviour, 21*, 219–239.
21. Lazarus, R. S., & Folkman, S. (1984). *Stress, appraisal, and coping.* New York, NY, USA: Springer Publishing Company.
22. Adams, J., Khan, H. T., Raeside, R., & White, D. (2007). *Research methods for graduate business and social science students.* SAGE Publications India Pvt Ltd.
23. Klapproth, F., Federkeil, L., Heinschke, F., & Jungman, T. (2020). Teachers' experiences of stress and their coping strategies during COVID- 19 induced distance teaching. *Journal of Pedagogical Research, 4*(4).
24. Carbonaro et al. (2008). Integration of e-learning technologies in an interprofessional health science course. *Medical Teaching, 30*(1):25–33 (2008), Feb. https://doi.org/10.1080/014215 90701753450

25. Dagada, R., & Chigona, A. (2015). Integration of E-Learning into curriculum delivery at university level in South Africa. *International Journal of Online Pedagogy and Course Design, 3*(1), 53–65.

26. Bhuvaneswari, G., Swami, M., & Jayakumar, P. (2020). Online classroom pedagogy: Perspectives of undergraduate students towards digital learning. *International Journal of Advanced Science and Technology, 29(04)*, 6680–6687. Retrieved from http://sersc.org/journals/index.php/IJAST/article/view/28069.

27. Huang, Q. (2018). Examining teachers' roles in online learning. *The EUROCALL Review, 26* (2). https://doi.org/10.4995/eurocall.2018.9139.

28. Kituyi, G. (2013). A framework for the integration of e-learning in higher education institutions in developing countries. *International Journal of Education and Development using Information and Communication Technology (IJEDICT), 9*(2), 19–36.

29. Bhuvaneswari, G. (2016).'Teacher cognition.' *International Journal of Economic Research, 13* (3), 693–695.

30. Mathew. Stress and Coping Strategies among College Students. *IOSR Journal of Humanities and Social Science (IOSR-JHSS)* 22, 8, 40–44 (2017). <dbe2e1ad06aab3910b24b30b4c29b6f29373.pdf>

31. Schwarzer, R. (1992). Self-efficacy in the adoption and maintenance of health behaviors: Theoretical approaches and a new model. In R. Schwarzer (Ed.), *Self-efficacy: Thought control of action* (pp. 217–243). Hemisphere Publishing Corp.

32. G. B., & Vijayakumar, S. (2021). Emotional intelligence and values in digital world through emoticons among indian students and faculty. *International Journal of Asian Education, 2*(2), 267–276. https://doi.org/10.46966/ijae.v2i2.142

E-learning and Its Application in Universities During Coronavirus Pandemic

Abdulsadek Hassan

1 Introduction

The era in which we live is characterized by rapid science progress today and it has a clear development in the various scientific and technical fields. Modern technology has imposed itself in various areas of life, and among these areas is the field of education, as technological progress has led to the emergence of new methods of education that depend on employing technological innovations to achieve the required education. What distinguishes this period of history are the new means and methods with which information can be changed and processed, the increasing speed with which it is dealt with and used, and the latest computer capabilities are fundamental changes in all areas of life, especially communications, information and learning. In view of those changes that the world is targeting with the entry of the information age and the communications revolution, there is an urgent need at this particular time to develop programs for educational institutions that keep pace with these changes.

E-learning is one of the most important technological applications in the fields of education and its methods, so it can be said that it represents the new model that works on changing the entire form of traditional education in an educational institution contributes to global cooperative learning, continuing education, continuous training, and training of professionals in all educational and scientific fields [1].

In light of the rapid technological changes, transformations and developments in all aspects of life, the urgent need to keep pace with this development, especially in the field of education, which has taken in its various forms and institutions and due to its characteristic mutual influence with society, has become a spacious field to benefit from the data of humanity in today's world, so many forms are adopted technology is either according to the capacity of educational institutions and their

A. Hassan (✉)
Ahlia University, Manama, Bahrain
e-mail: aelshaker@ahlia.edu.bh

© The Author(s), under exclusive license to Springer Nature Switzerland AG 2022
A. Hamdan et al. (eds.), *Technologies, Artificial Intelligence and the Future of Learning Post-COVID-19*, Studies in Computational Intelligence 1019,
https://doi.org/10.1007/978-3-030-93921-2_5

available capabilities or in line with technological developments. Hence, talking about e-learning is not an idea and an option as much as it is a prerequisite for entering the age of knowledge.

Our current era is known as the era of the technological revolution and the explosion of knowledge, the last decade of the twentieth century and the beginning of the twenty-first century witnessed tremendous progress in the field of information and communication technology, and modern technological means have transformed the world into a small global village, and this development has been reflected in many areas, education is the most prominent benefiting from it, as modern technologies have become one of the most important elements that help learners for acquire learning experiences efficiently, effectively, quickly and enjoyably in what is currently called e-learning [2]. As a result of this revolution in educational methods and technologies, there is an urgent need for faculty members at the university level to possess the competencies of e-learning, in order to fulfill the academic roles assigned to them, in line with the requirements of the age of technology and informatics [3].

In spite of the importance of the material and technical elements in any educational organization, the human factor remains the central element, who can employ and use all these elements to achieve the greatest amount of productivity, efficiency and effectiveness, whatever tools, machines, devices and programs have been developed, whatever theories and philosophies that appeared in the field of education; The quality of education itself can only be achieved by providing he competent lecturer, and as technology entered the educational field, this led to the emergence of many concepts and terms that have not been in circulation for years, such as: Borderless Higher Education, Virtual University, and Internet University (Online University) and Virtual Classroom, which required faculty members to possess technological competencies due to their increasing importance in education [4].

The progress in the field of information and communication technology has led to changes and challenges that have imposed themselves on the educational system with all its components, from lecturer, curriculum and learner, teaching strategies and evaluation methods, which made educational institutions strive to accommodate these changes, awareness of their importance, and work to find a new quality of the lecturers, possessing the skills to deal with it, and being able to produce and employ knowledge, which helps to create an information society instead of being a knowledge consumer [5].

E-teaching is one of the trends that have begun to gain interest among educators in the modern era, and it is a teaching system for processes and activities designed according to information and communication technology, the characteristics and models of e-learning, the principles of both educational and communication technology and education systems based on competence [6].

That e-teaching can be understood through two meanings. First: It is a system designed to improve lecturer performance, self-regulation and motivation, the second: e-teaching services are designed with the aim of supporting lecturer performance effectively in the e-learning environment. E-learning is learner-based, while e-teaching is based on the needs that focus on the lecturer [7].

In this age that is described as digital and informatics, and the role of the techno-logical component in education is increasing, the need to find new formulas capable of meeting many of the technological requirements in the educational environment has increased, as the lecturer has transformed from a source of knowledge to a guide, and he becomes a practitioner [8]. Therefore, new responsibilities have been placed on the lecturer, not only in his field of specialization and teaching style, but also in the extent of his understanding, development of awareness and comprehension of the requirements for employing this technology, as the role of the learner has increased, and his dependence on technology has increased [9]. Rather, their importance lies in how they are employed in educational situations, and this can only be achieved through a lecturer who is fluent in dealing with these innovations [10]. As the shift from traditional classrooms, classroom activities to virtual classes, and electronic activities, lecturers and students have positive trends towards the use of computers and its technologies in the teaching and learning processes [5].

On the importance of electronic teaching, indicating that it represents a great challenge, as lecturers face a new generation of students known as the Millennial Generation or the Digital Generation, who rely mainly on developing their knowl-edge on the Internet and mobile phones, and on multiple methods of electronic communication [11].

E-teaching represents a basic structure for the lecturer in order to teach success-fully, and it is a prerequisite for the success of e-learning, especially with the emer-gence of the second-generation technologies for Web2, which shifted interest from e-learning to e-teaching [12].

E-learning and its role in higher education: Higher education comes to embody a qualitative leap in the knowledge of the learner in its personal aspects and to meet his modern needs, and because higher education embodies the top of the pyramid of education for all societies, it seeks to provide it with all the experiences and gains necessary for a better life in the present and professional formation in the future, and in order to reach this message, this happen by throwing and indoctrination and providing some experiences to learners, or introducing technology as a technique and working to employ it to develop the learning process and provide the most appropriate education for each student, especially since the progress criterion for nations is measured by the level of their human resources[13]. In general, the relationship between educational technology and higher education institutions can be drawn as follows.

Renewing its educational goals in line with the age of knowledge: on the basis that the university embodies a cognitive space for scientific ideas in various directions, and because the challenges before societies today are primarily cognitive challenges, they are required to reconsider their composition and philosophy in order to be able to effectively contribute to the production, management and access to the comprehensive development of society [14].

Modernizing the intellectual and cognitive environment at the university: through the inclusion of new disciplines that are in line with international scientific developments that occur in the field of science and technology in various aspects of life, and in line with their capabilities and requirements of those disciplines [15].

Innovative education: Through the fact that e-learning is a method of education using modern communication mechanisms, it in turn provides an opportunity for learners to creatively and creatively deal with educational situations in terms of providing innovative solutions and suggestions [16].

Making technology part of its educational and learning system: this is through the interest in employing all technological innovations in quantitative terms, and because educational technology provides an effective and appropriate tool for the university to enter the world of information and benefit from its advanced systems such as electronic universities and virtual universities [17].

Introduction to educational quality: the introduction of modern technology in the field of higher education is one of the main pillars that quality in education calls for, which describes its structure "a set of standards and characteristics that should be available in all elements of the educational process [18], whether they are related to inputs, processes or outputs that meet society's needs, requirements, learners' desires and needs, and achieving those standards through the effective use of all material elements [19].

2   Modern Types of Education

Internet learning: It is the umbrella or broad term that is used to describe educational and study activities that take place over the Internet in general, that is, those that are accessed remotely through a device, such as a tablet or phone, or on a website or through an application [20].

E-learning: a general term used to describe the use of technology primarily in the implementation and teaching of an online program or course, using resources available on the Internet [21].

Virtual learning: It uses online tools in the form of sound, image, video, or electronic books and computer programs to provide an educational program. Virtual learning may also include a virtual learning environment (VLE). It often combines personal and virtual elements [22].

Blended learning: one of the teaching or learning formats in which e-learning merges with traditional classroom learning in one framework, using a mix of online tools and resources with educational methods 75%–25% in favor of lecturer-based education [23].

Hybrid education: Hybrid learning is a model for course design in which a part of the time is devoted to regular face-to-face learning inside the classroom, and part of the time is devoted to e-learning outside the classroom, it is the integration of direct and electronic learning into an organized learning experience, that is, with a balanced rate of 50%. 50% [24].

Distance learning: This is a course or program that is taken exclusively online without using any of the teaching methods on campus [25].

3 Types of E-learning

In light of the difference between e-learning and distance education, it is stated that there is more than one educational environment in which the concept of e-learning can be achieved, which includes the following:

Direct online education: Here, the traditional and recognized form of university is abandoned, as the student gets here all the knowledge, he wants through various Internet sites or through educational channels [26].

Blended learning: It is an ideal online learning method. Since it combines some forms of traditional education and some forms of e-learning in a more advanced and positive way, it includes a good degree of interaction and participation between the student and the lecturer in the course of the educational process [27].

Supportive education: through this educational environment; The student searches for the information by himself, and then discusses it with the lecturer in order to achieve a greater degree of knowledge and understanding [22].

Advantages of E-learning Technology

E-learning came to embody a creative way to present an interactive environment centered around learners, well-designed and accessible to any individual, anywhere and anytime [28], using the characteristics and resources of the Internet and digital technologies in conformity with the principles of educational design appropriate for an open, flexible and distributed learning environment [29], and the most important advantages of e-learning can be summarized in the following:

Ease of mobility: as it is possible to navigate with portable computers anywhere, on which educational materials are saved, and the student can retrieve them at any time he requests [30].

Convenience and ease of use: As the online lessons became easier to use than attending the lecturers in the real lessons [28].

Strategic employment: There is a diversity of educational materials available electronically for students, which makes it easier for them to choose the most suitable one for them and employ them in practical life strategically, which are provided by traditional methods of education [31].

Flexibility: Electronic lessons can be used during the real semester, as lecturers follow the practical application of scientific theories within the classroom, correct students' mistakes immediately and follow them directly [32].

Conveniences: The videos provide students with the complete conviction of seeing and applying scientific theories in reality [33].

Simplicity: Simplifying educational materials through electronic lessons allows students and learners to focus on learning more [34].

Low cost: electronic lessons mean less expensive than traditional educational methods [35].

E-learning Strategies

E-learning environments vary to suit the diversity of learners, as well as the diversity of the courses and objectives, but it should not be dealt with e-learning without specifying the strategies used in teaching, and it is intended by which the method by which the educational content is presented to the learners [36], as the e-learning system includes the design of different learning strategies including the second generation of web services and electronic tools for transferring content and making the learning process happen [37].

Learning strategies include a number of procedures to present educational content in a way that helps learners achieve educational goals and these strategies vary with the diversity of goals [38], as there are many electronic educational strategies that vary and prepare, and as a result there is a diversity in the activities that the lecturer and the student undertake [12]. It cannot be said that certain strategies are better than other strategies, and there may be a better strategy than others depending on the learning environment, educational conditions and within the limits of certain material or human capabilities [39].

E-learning strategies can be presented as follows:

Electronic Lecturer Strategy (E-lecturer)

The lecturer is a way to present facts and information that can be presented through audio files, video files, text files, or through one of the multimedia presentation authoring systems such as Flash or Power Point and make them available to the learner during the course so that they can be downloaded, heard and viewed at any time, and it can also contain some links related to the topic of the lesson [40]. The lecturer strategy is implemented in electronic learning environments through some files that display the academic topic in different types and ways, and these files of all kinds are uploaded to the Internet in order to be re-run by the user on his computer [30].

E-discussion Strategy

The strategy of electronic discussions includes electronic conversations based on mutual interactions between the participants and cooperation in presenting information, expressing opinions in the process of education, and helping to overcome temporal and spatial problems in the timing of discussion or psychological problems that impede the implementation of facing studying situations and participating actively and seriously [41].

The discussion strategy is one of the most important tools for communication and interaction in the e-learning environment, through which many educational goals are achieved. It can be defined as a strategy that allows users to communicate by sending topics to members for each other to read and comment on either in a linear manner, or in threaded manner [42].

E-programmed Education Strategy

In which the content is divided into small educational units linked with each other in a way that identifies multiple paths for the learner to interact with the parts of the course which depends on his answer to the various questions through self-correcting tests [43].

E-problem Solving Strategy

The problem-solving method aims to help the learner to be able to grasp the basic cognitive concepts in solving the educational problems that he may encounter, and it also helps the learner to direct his behavior and abilities, and the problem-solving strategy can be applied in e-learning by posing a research problem to students through the course page, so that they are asked to employ what they have learned to solve the problem, but individually, and each student can discuss the lecturer by e-mail or direct dialogue [25].

It is also possible to present a research problem that the teacher chooses and discusses with the learners about it, leaving each learner alone to present his point of view to solve it, then collecting solutions and putting them on the discussion board discussion boards so that extensive controversial discussions revolve around it by all the learners to take opinions on the determination of the most appropriate solutions and develop adequate justifications to adopt the most appropriate solution, then reach a final decision on this solution and circulate it to students [44].

E-discovery Learning Strategy

This strategy is one of the best ways to obtain learning based on understanding, as the student in the discovery position is an active learner, and acquires effective and fruitful learning [45], as well as the research skills and skills of observation, classification, prediction, measurement, interpretation, estimation, design, recording notes, interpreting information and other skills [46].

Discovery is the process of organizing information in a way that enables the learner to go beyond the information presented to him. It can also be said that it is the way in which the verbal formulation of the concept or generalization to be learned is postponed until the end of the educational situation through which the concept or generalization is taught [47].

Instructional Games Strategy

It aims to teach study topics through entertaining games with the aim of generating excitement and suspense that endear learners in learning these topics, and also develop their ability to solve problems, decision-making, flexibility, initiative, perseverance and patience [45], and each game contains a number of components, including the content of the game, and educational objectives of the game, the rules of the game and the role of players, instructions of playing and how to calculate the gain and loss, so that they are known to the learner before playing [5].

E-simulation Strategy

Simulation is the representation of the situation or a group of real situations that are difficult for the learner to study in reality, so that it is easy to view and delve into them to explore their secrets, and to closely identify the potential results when representing a particular situation in reality, because of its cost or danger such as nuclear experiments, dangerous chemical reactions, etc. [48].

Project Based E-learning Strategy

The project-based e-learning strategy is one of the strategies that can be used in training and preparing students, as it is characterized by the possibility of employing and using electronic interaction tools via the web to achieve cooperation and partic- ipation in the implementation of these projects[49], and to make use of all elec- tronic resources available through the web in obtaining and exchanging electronic information between Students, without resorting to the supervisor of projects [50].

The project-based e-learning strategy is one of the strategies that occupy the learner's mind in acquiring knowledge and skills through inquiry processes around questions related to the curriculum for final product that is evaluated in light of its achievement of the learning objectives through a set of tasks that are followed by the learner and carefully designed by the lecturer [45].

4   Employment of E-learning Within Teaching

E-learning can be fully utilized within teaching by employing it in several ways, namely:

1. **The Complementary model**
   It uses types of technologies that support the traditional educational process, and this takes place inside or outside a study room, such as directing lecturers to students to peruse and search for a specific lesson via the Internet [51].
2. **The blended model**
   This model includes combining the traditional and electronic method of teaching, inside the classroom or inside places prepared with techniques for the electronic method of learning [52].

   It has the characteristics of ancient and modern education, and the role of lecturers in it is limited to guidance, as well as managing educational situations, and students have a positive role [53].
3. **The pure model**
   With it, e-learning is used instead of traditional, because it is possible to learn at any time or in any place by students [27].

The role of the network is to be the main mediator to fully present the educational process.

Examples include that students study the electronic curriculum alone. Students learn with a group of colleagues, by completing a project by relying on electronic educational tools such as chat rooms [54].

5 E-learning Tools

Simultaneous E-learning Tools

Simultaneous e-learning tools are tools that provide the opportunity for the user to communicate directly with other users on the network [49], and these tools include:

Conversation: It is the ability to speak with other users over the Internet at the same time by using a program that constitutes a virtual station that brings all users from all over the world on the Internet to speak in a form of writing and voice [5].

Voice conferencing: It is an electronic technology based on the Internet, where it uses an ordinary telephone, and a conversation mechanism in the form of telephone lines that connects the speaker to a number of receivers in different places of the world [55].

Video conferences: allow communication between a group of individuals separated by a distance through a high-power television network using the Internet, where everyone can ask questions, and hold any discussion with the speaker, and he can also see the speaker [56].

Whiteboard: It is a blackboard similar to the traditional blackboard, as it enables the user to make drawings and annotations that are transmitted to others [27].

Satellite Programs: They are programs related to computer systems, and linked directly to the communication network, which facilitated the possibility of making use of visual and audio channels in the teaching and learning processes [57].

6 Asynchronous E-learning Tools

Asynchronous e-learning tools are tools that allow the user to communicate with other users indirectly, as they do not require the presence of the user and other users on the network at the same time, and its tools include:

E-mail: It allows the possibility of exchanging letters and over the Internet [58].

The web: it is an information system that displays various information on linked pages. It also allows the user to access various Internet services [59].

Mailing Lists: They are a list of postal addresses added by the institution or person, where messages are forwarded from a single postal address [60].

Transfer files: The function of this tool is to transfer files from one computer to another connected to it via the Internet [61].

Interactive video: It is the technology that provides interaction between the learner and the presented material, with the aim of making the teaching more interactive [62].

Compact Disks (CDs): They are the disks through which the curricula are prepared, uploaded to students' computers, and then refer to them when needed [63].

7 E-learning Difficulties

E-learning, like other methods of education, has obstacles that hinder its implementation, and these obstacles include:

1. **Standards Development**:
   E-learning faces difficulties that may impede its rapid spread. The most important of these obstacles is the issue of approved standards. When we see some educational curricula and curricula in universities or universities, we will find that they need to make many adjustments and updates as a result of different developments every year [33], sometimes even every month. If the university has invested in purchasing educational materials in the form of books or CDs, you will find that it is unable to modify anything in it unless these books and discs are rewritable, which is a complicated matter even if it is possible [64].

2. Compensatory regulations and incentives from the requirements that motivate and encourage students to e-learning. As e-learning still suffers from a lack of clarity in the systems, methods in which education is carried out in a clear manner, and the failure to decide the issue of incentives for the educational environment which is one of the obstacles that hinder the effectiveness of e-learning [65].

3. **The safe and effective of the educational environment, includes**:

   – Lack of support and cooperation provided for the effective nature of education.
   – Lack of standards for developing and operating an effective and independent program.
   – Lack of incentives to develop content [66].

4. **Methodology**:
   Technical decisions are often taken by technicians depending on their uses and personal experiences, and the user's interest is often not taken into consideration, but when it comes to education, we must develop a program standard plan because this directly affects the lecturer (how he knows) and the student (how to learn) [1]. This means that most of those involved in e-learning are specialists in the field of technology, or at least most of them, while the specialists in the field of curricula and education do not have a view in e-learning, or at least

they are not the decision-makers in the educational process [38]. Therefore, it is of utmost importance to include educators, lecturers, and trainers in the decision-making process [40].

5. **Privacy and confidentiality**:
   The occurrence of attacks on the main sites on the Internet, affecting lecturers and educators, and put in their minds many questions about the impact of this on e-learning in the future, so the penetration of content and examinations is one of the most important obstacles to e-learning [12].

6. **Digital filtering**:
   It is the ability of persons or institutions to determine the perimeter of communication and time for people and whether there is a need to receive their communications, then are these communications restricted or not, and whether they cause harm and damage? and that is by setting filters to prevent communication or close it for unwanted communications, as well as the matter with advertisements [67].

7. The students' response to the new pattern and their interaction with it [10].

8. Monitoring the ways of interaction in the classroom with E- learning and making sure that the curriculum is going according to the plan set for it [42].

9. Increase the focus on the lecturer and make him feel his personality and importance to the educational institution and make sure that he does not feel an unimportant and that it has become something traditional heritage [14].

10. Community members' awareness of this type of education and not standing passively from it [26].

11. The constant need to train and support learners and administrators at all levels, as this type of education requires continuous training in accordance with technical renewal.

12. The need to train learners on how to teach using the Internet [7].

13. The need to publish high-quality content, as competition is global [35].

14. Amending all the old rules that hinder innovation and developing new methods that promote innovation in every place and time to advance education and demonstrate competence and ingenuity [15].

8 Models for E-learning Applications

There are many experiences at the local and international levels for many universities that have applied e-learning techniques, whether this application is in bachelor's degree or even in the postgraduate stage [68]. There are some universities that have completely relied on e-learning applications, and some have relied on a combination of the traditional method and e-learning applications [69].

9   Teams Application for Distance Learning

The experience of Teams Application is one of the pioneering experiences in the United States of America in applying e-learning. E-learning techniques were applied from the primary stage to the secondary stage [51].

The activities of the Teams Application in universities covered twenty-one American states, and more than one hundred and forty-five thousand learners are applying e-learning techniques in these universities and more than seven thousand five hundred lecturers [29].

The activities of Teams Application began in the year 1990 AD, through financial support provided by the Education Department in the United States, and it relied on providing its educational services in its infancy on the use of satellites [7].

Teams Application established its first website on the Internet in the year 1994 AD, and thus it was one of the first educational institutions to establish a website for it on the Internet, providing all students, lecturers and parents with the opportunity to enter and benefit from the site's services. The following figure shows the Teams Application in the classroom system [70]

Zoom Application

The Zoom program is one of the programs specialized in the field of making audio and video calls, or as it is called video calls, as one of the parties hosts the call, and all the powers are in his hands, and it is possible that many callers participate in the call, and it is considered one of the programs suitable for use in group work meetings, which includes a host and participants, so that each of them can share his screen image at any time, which would facilitate the process of communication between them, and make it more quick and easy [52].

Zoom program has many features that distinguish it from other similar programs, as it is distinguished as a universal program that targets many different platforms, so that participants in the call can share their screen [68]; Either a phone screen or a computer; With others, in addition to that it provides them with the ability to write comments, and clarify in written or audio form on them, knowing that it is compatible with Android, Mac, Blackberry, and Windows devices, and it is also compatible with the iOS operating system, and many others, so it can say that it is an ideal option for formal get-togethers, cloud conferences, interviews, and group correspondence in universities [30].

Virtual Learning

Virtual learning is among the most important developments that have affected the fields of contemporary education and moved it a qualitative leap towards more brilliance and human service, and the achievement of the educational and educational goals of all institutions interested in education [71]. Where it plays an important role in facilitating the educational learning process and achieving its goals and upgrading society in general and the student community in particular, and the Virtual learning is considered as the most important tools that can be employed in the normal situation in all societies [41].

Virtual learning in general use the virtual and computer media in the process of transferring and communicating information to the learner. This usage may be in the simple form, such as the use of electronic means to assist in the process of displaying information or to give lessons in traditional classrooms, as well as the optimal investment of virtual and computer media in building virtual classroom through Internet and Interactive TV Technologies [37].

Learning can be ideally defined as: expanding the concept of the teaching and learning process beyond the boundaries of traditional classroom walls and launching into a rich multi-resource environment in which remote interactive education technologies have a fundamental role in which the role of both the lecturer and the learner is reformulated, and this is evident through the use of computer technology [46]. Automated support, selection and management of the teaching and learning process, at the same time, Virtual learning is not a substitute for the lecturer, but rather enhances the course as a supervisor, directed and organizer to manage the educational process in line with the developments of the modern era [41].

In order for Virtual learning to be effective, this requires increasing and strengthening the existing skills than developing new capabilities [19].

Virtual Learning Requirements

1. Comprehensive infrastructure consisting of fast communication means, modern computer equipment and laboratories [72].
2. Qualifying and training lecturers on the uses of technology and getting acquainted with the developments of the times in the field of education [19].
3. Investing in building e-learning curricula and materials [73].
4. Set up systems and legislation that contribute to supporting the educational process in its contemporary form [58].
5. Set up information systems capable of managing the education process in its new form [5].

What are the Virtual Learning Environments?

The Virtual Learning Environment (VLE) is a set of teaching and learning tools designed to enhance a student's learning experience using computers and the Internet in the learning process [74]. The main components of virtual learning environments include courses and curricula which are broken down and presented in different ways. These systems track the student and provide online support to both the lecturer and the student [71], via electronic communications such as e-mail, discussions, chat rooms, online posting and external links to external curriculum resources. Typically, users of virtual learning environments are assigned a login ID, either as a lecturer or a student [5].

It is also known as systems that aim to support teaching and learning by establishing a learning environment that mimics the real educational environment, and the virtual learning environment works naturally via the Internet, taking advantage of the tremendous development in communication technologies and the spread of rapid access to the Internet and the competition of Internet service providers to offer competitive offers for subscribers and the emergence of the Internet connection service via mobile phones, Wi-Fi and 3G technologies, the spread of the iPhone and iPad, and other devices and applications that have made the world closely related to everything technical [17].

Virtual Classroom

Virtual classes that allow lecturers and students to communicate with each other over the Internet. Information about the class, teaching materials, and assignments are usually provided over the web. Students can log into the class website to view this information and can also download required assignments and reading materials to their computers [75]. Some of these environments allow you to do online homework and tests. In virtual classrooms, the lecturer can communicate with students in real time using video conferencing or web conferencing technology [76], and this type of communication is usually used for giving lecturers and question and answer sessions. If the lecturer only needs to post some assignments or assignments, he can simply post it on the class website. Students may also receive an email notification to see what's new. If students have questions about homework, they can participate in online forums or ask the lecturer individual questions [73].

The virtual learning environment is the online education platform; They are used as an extension of regular classroom lessons and contain many tools to help students learn the topics of study [37], for example: The lecturer can put the lesson material in the virtual classroom in the form of electronic files, and then students open the files and complete the task required of them [77].

Students can upload their homework into the virtual learning environment for lecturers to see and correct.

Virtual learning environments can contain tests for student use, some tests can be validated electronically with the lecturer and the results can be seen instantly.

Students can share work, enabling them to work together on a project [19].

Emails can be sent by lecturers and students to each other [37].

Social media can be built like forums, wikis, and blogs [48].

Students can access the virtual environments from home by logging into the system. This enables them to do their homework or complete their projects from home [37].

10 Effective Use of Virtual Learning Environments

The use of rich virtual learning environments raises several important issues for universities. Without addressing the issues of effective learning, the use of these technologies may exacerbate past mistakes and produce negative experiences for learners, leading to a flattening of the learning process [67]. The use of appropriate tools and the context in which learning takes place are the main factors affecting overall success in institutions of higher education and lifelong learning. Effective use of tools enables learners and lecturers to organize learning and form learning capable of solving practical situations or problems in or outside university life, expressing the creativity of the learner [37], and which can be placed in a balanced context with other resources within the system, and re-use and share with peers or lecturers. The following table illustrates the best context for using the tools provided by virtual learning environments [5].

Methods and methods of teaching and learning	Virtual environment tools for real education
Brainstorming	Blogs, Chat, Survey, Feedback, Forums, Word Glossary, Mind Map,
Collaborative groups	Tasks and Assignments, Calendar, Blogs, Feedback, Instant Messaging, Workshop,
Small discussion groups	Chat, Messages, Blogs, Quizzes, Workshops
Examples analysis	Assignments, blogs, exams, workshops, links
Working groups	Calendar, Chat, Lessons, and Workshops
Individual work	Assignments, feedback, lessons, mind maps
Problem Solving	Tasks, chat, feedback, workshops
A reference study	Tasks, blogging, database
Self-test	Questionnaire, mind maps, tests

Successful Experiences of Distance Education

Distance education is one of the most important concepts and modern technologies for education at all levels, and this type of education has become an important pillar of the knowledge economy, and it is worth noting that distance education, or what is sometimes called computerized e-learning or online learning, does not mean teaching curricula and storing them on CDs, but the essence of distance education is the interactive style [12], as it means the existence of mutual discussions between students and each other, and interaction with the lecturer, there is always a lecturer communicating with students, and determining their tasks and tests [78].

There are several mechanisms for distance education, either through the "video-conference" technology, or "Live" lecturers, or professors and specialists recording a number of lecturers and placing them on a specific website. In this context [79], some successful experiences in this field can be reviewed as follows:

First: The United States: The United States is the undisputed leader in the field of online education. Hundreds of online colleges and thousands of online training courses are available for students. A study conducted in 2011 by the Sloan Consortium, an American institution, indicated there is, 6 million students in distance education in the United States take at least one online course [25].

As a result of the increasing demand for courses offered through the distance education system, this encouraged prestigious American universities such as Stanford, Berkeley, Princeton, the University of California, and a number of other American educational institutions to offer online educational courses for those who prefer this method and cannot participate in the classroom in its traditional form [80].

Second: South Korea: The strong technological infrastructure in South Korea allowed the spread of distance education, as it has the most powerful infrastructure in the world, as it provides one of the highest Internet speeds in the world, and Internet services are available even in rural areas, which made the situation favorable [73]. Due to the flourishing of this type of education, and in this context, South Korea witnesses every year an increase in the number of students enrolled in distance education courses by more than students enrolled in traditional educational institutions [58].

Third: Australia: Distance education has become an increasingly popular option for Australians who want to return to university without leaving their jobs. Over the past five years, the online education market in Australia has grown by nearly 20%, and greater growth is expected in distance education programs offered by Australian universities, especially with more Asian students, making Australia one of the leading providers of distance education services [81].

Fourth: India: Online learning in India has grown faster than traditional educational institutions, as the country suffers from a major educational crisis as a result that more than half of the population has received a limited education, and in many cases Indian citizens do not have the means to complete their education, either Due to cost factors, or geographical factors, which are the long distances between universities and remote villages in India, and thus e-learning has allowed a wide range of Indian students to complete their education in its various stages [51].

Fifth: Malaysia: Malaysia is progressing at full speed in terms of opening new opportunities for learning via the Internet. The Asia e- University, based in Kuala Lumpur, is one of the most important technological universities in Malaysia. This university has worked to support citizens in the regions that the university suffers from the lack of availability [70], but it has access to the Internet, which facilitated the spread of the distance education system among Malaysian citizens and even Asians as well, as this university offers online educational courses for students of 31 different Asian countries, and the university has entered into partnerships with other universities to offer programs Online educational degrees are available, as, for example, an MBA program has been developed in cooperation with the International Business university in Denmark [27].

Sixth: China: China is one of the most important countries in which the distance education sector thrives. China has more than 70 online institutions and colleges, and as a result of intense competition for jobs there, students seek to obtain more degrees and training courses in several fields, so that they can get better jobs. So the economic necessity has generated a great opportunity for the growth of the distance education sector in China [35].

Seventh: The Arab World: As for the Arab world, a report prepared by the research company "Frost & Sullivan" entitled "Augmented Reality: The Middle East as a Launchpad" indicated that virtual reality technologies will spread widely in the Gulf countries by 2025, which enhances the opportunity for students [70]. Arabs are keeping pace with these changes, knowing that Effat University in the Kingdom of Saudi Arabia, King Abdullah University of Science and Technology, and Qatar University have already begun to experiment with this technology [2].

11   The Difference Between Virtual Learning and E-learning

Distance education is somewhat different from e-learning, as through distance education the learner relies on searching through the web for various information and delving into understanding on his own without presence or commitment to the place [82], and the student can also obtain various information that he needs Also, through some manuals or various sources and references by himself, and distance education is one of the types of self-learning [6].

What is unknown to many is that some universities have applied the concept of education for a long time, but this name is not explicitly used for the method of education [23], examples of which are students who learn through the university affiliation system; Where the student is the one concerned with the research alone, and an example of this is also the open education system that some universities allow through the student obtains the educational material and the educational process takes place remotely without attendance [35].

If it appears that there are some similarities between e-learning and distance education; However, there are some fundamental differences between the two systems, as educational experts have indicated that the difference between e-learning and distance education comes as follows:

Degree of Flexibility

Both e-learning and distance education carry a measure of flexibility, but nevertheless; Distance learning is related to submitting assignments and following up attendance and absence with the university and the lecturer, while e-learning is not related to place or time, but the learner can learn what he wants when he loves [83].

Adherence to the Appointment

Some people are afraid to be lazy as long as there is no one to follow the educational process with them, and here it is better for them to learn through the method of distance learning that adheres to a specific time of attendance, while e-learning has a tremendous amount of freedom; Whereas the learner is the one who determines the final time to study, the time to take the test, and others [65].

Accreditation

E-learning is accredited because it does not abandon the basic system of education, but only some methods of receiving science and scientific knowledge are changed, while the distance education offered by some universities may not be accredited. Therefore, it is necessary to make sure that the university that provides services or courses via distance education [51].

Support from the Lecturer

In the case of e-learning; There will be a supportive lecturer for the student to help him understand and provide him with the answer to the necessary inquiry, while distance education lacks continuous support from the lecturer and the learner may be exposed to a delay in obtaining answers about his inquiries [78].

In conclusion, after we presented the difference between e-learning and distance education, and despite the presence of some differences that favor one of them from the other [84]; However, every person must be well aware of his nature and the desirable methods of education and what he will be able to acquire, learn and know through, and accordingly he determines whether he wants to learn through e-learning or distance education [80].

Virtual University

It is an academic institution that aims to secure the highest levels of higher education for students in their places of residence via the Internet, by creating an integrated electronic learning environment based on an advanced network [37].

The difference between a traditional university and a virtual university is that a virtual university does not need classrooms within walls, or direct instruction from the professor to the student, or students gather in exam halls, or the student's coming to the university for registration and other procedures. Rather, students are grouped into virtual classes that are done. Communication between them and the professors through their own website on the Internet, and taking examinations remotely by evaluating the level of research submitted by university affiliates during the period of their studies [19].

If the traditional educational institution allocates a physical place for the student (seat - class - library ….), then the student's seat in the virtual institution is in front of the computer screen, and his class is on the Internet, and his library is not limited to a limited number in a room. Rather, he can see Millions of books with the fastest and easiest way to search and follow. And in the virtual university, students do not exchange ideas with a computer, Rather, they are interviewing a group of people from all over the world via the computer to interact with a global group of professors and students, from different cultural backgrounds and national affiliations [23].

The process of selecting majors offered by virtual universities is a dynamic and continuous process that is directly related to the needs of the labor market in general, and includes many disciplines such as information and communication technology [74], business administration, computer science and industrial intelligence, tourism facilities management, agricultural genetic engineering, educational technology, educational administration, and these All majors are offered at several levels: Diploma - Bachelor - Master – PhD [41].

Virtual Universities are academic institutions that aim to secure the highest levels of international university education for students from their place of residence via the Internet, by creating an integrated electronic learning environment that relies on a highly developed network and offers a range of university degrees from the world's most prestigious internationally recognized universities [84]. It also provides all kinds of support and assistance to students under the supervision of a virtual network that includes the best experts and university professors in the world, and the difference between a traditional university and a virtual university is that a virtual university does not need classrooms within walls [84], or direct instruction from the professor to the student or the student gathering in examination halls or the student's coming to the university for registration and other procedures, but students are grouped into virtual classes that communicate between them and the professors through their own website on the Internet, and tests are conducted remotely by evaluating the level of research presented by the university's affiliates during their studies [77].

Just as the term virtual university and virtual classroom are commonly used, the term virtual learner has also become common, and if we have accepted the inadequacy of using the term virtual education, then it is better to acknowledge the inappropriateness of using the term "The virtual learner" [9], and therefore many scholars see this term being mistaken and correcting it with the term "the electronic learner" because the student (human) will not change his type with the technology or tool he uses to learn, but rather what changes how he learns [57].

It may be necessary to point out that the term electronic learner or the virtual learner is an unstable term, as this term may be called and meant as the real learner, and it may be called and intended as the (virtual learner) or the (virtual student) in this case [74], What is meant here is what is known as the (Virtual Agent) or (Cyber Agent) that replaces the student in educational sessions when he is unable to attend them, or the virtual companion, the student can choose a virtual student to problems solutions, and exchange roles with him, and just as there is a virtual student [41], there is also a virtual tutor and a lecturer assistant [85].

Basic Tools in Virtual Classrooms

There are a number of basic tools used in the virtual classroom, this tool allows direct and instant communication between two or more people through the participating computer networks and through group discussions, brainstorming exercises, and problem-solving activities that they exchange using this tool [17].

Live Audio with Visuals

It is represented in the ability to speak with lecturer through the Internet, in which direct voice can be used in real time with visuals, and in it the importance of changing the sound, tone and speed is highlighted [41].

Application Sharing

Shared applications are intended to enable the students to participate with others in working on an application such as electronic spreadsheet or one of the presentations designed with a program (PowerPoint) or using the electronic whiteboard on the network [19].

The Electronic Whiteboard

It is the main tool in common applications, and it is exactly like the well-known whiteboards that give students the ability to write, make notes, draw and paste on, in addition to the ability to save their contents, transfer them or send them by e-mail to the lecturer [73].

Quizzes and Opinion Poll

This tool gives the lecturer in the virtual class the ability to conduct a short test or opinion poll that measures the success of the session and the extent of achieving its goals at the end of the session and can obtain results directly and easily [55].

Internet Browsing

This tool gives the ability to browse the Internet through the virtual classroom by typing the desired address (URL) in the space provided for it [7].

Breakout Rooms

This tool gives the lecturer the ability to divide those in the classroom into breakout groups (cooperative learning groups), to exchange views and interact with them [35].

Desktop and File Sharing

This tool enables the lecturer to share the desktop with the participants after they give you permission to do so and then exchange files with them, save or print them, and there are tools for human interaction with students in the classroom by expressing their feelings such as approval, rejection, raising the hand, permission, or applause, laughter, resentment … etc. [6].

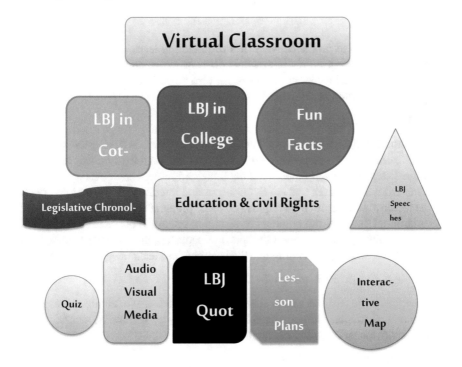

12   Methods of Distance Education

There are many methods of distance education, and each of these methods expresses a specific stage of educational interaction, and due to the increasing development in information and communication technology [86], this has reflected on the expansion of its educational uses and the emergence of new, more effective methods of distance learning, of all kinds [57], and among the most important methods that have been proven Its advantage in distance education are:

Correspondence Learning Style

The lecturer sends the printed material to the learner, and then the learner comments on it, poses questions and inquiries about it, and then returns it to the lecturer, and e-mail is now the main method in the work of the Internet and this method is one of the traditional methods of distance learning, as separating the lecturer from the learner is a spatial space, This is in order to fill the educational void, and this method can give learners the opportunity to university education, as well as providing them with a database at their workplace [37].

Multimedia Style

This method is based on the delivery of information to learners, through audio and video recordings using floppy or compact disks or by telephone, and radio or television broadcasts. Educational materials can also be sent to students in the form of printed materials such as references, study guides, etc. [27].

Video Conferencing Style

It is a method similar to the style of learning that takes place in the classroom, except that the learners are far (separated) from their lecturers and their colleagues as they are linked to high-capacity electronic communication networks, everyone can see and hear from the lecturer, and to ask questions and interact with the topic proposed by the lecturer. But this method requires advance preparation and a longer time than the traditional class, as the numbers of scientific material and media are required, as well as training the lecturer on the speed of capturing the learner's attention and interest, while training the lecturer and the learner to use technology effectively [57].

The Style of the Printed Material

This method is the basis on which all systems or methods have relied for presenting educational curricula. Printed materials vary, such as textbooks, course plans, exercises, abstracts, tests, and others [81].

The Default Learning Method

In this method, the scientific material and the communication between the lecturer and the learner are transmitted through the web and e-mail, and although the age of this educational method is old, it is steadily increasing to the point that distance education is not intended in most cases except for this technology, and the communication may be between lecturer and learner synchronously or asynchronously [87].

CD Style

It is one of the good and important means of transferring information, and it is characterized by its ability to store the largest possible amount of information and data and to re-run them in a high-quality manner, so it has been widely used in distance education, but the study materials remain restricted within the limits set by the program designer. The learner cannot correct the method, and it helps to self-learning, but its production and preparation requires more time and cost [81].

Remote Interactive Learning Method

This method is based on the overall interaction between the lecturer and the distance learner through audio-visual communications and educational channels that are broadcast through or by satellites [88].

The Future of Distance Education After the Corona Pandemic

The global Internet and general and specialized search engines allowed students and faculty alike to complete research and access research resources within a very short period [89], a task that used to take weeks or months in the past, and training course management systems provided lecturers with the ability to set up videos and graphics [79]. During lessons, there are creating spaces for dialogue, having interactive video chats, storing lessons and assignments, and retrieving them easily at any time.

However, with the increasing trend towards e-learning, and its increasing adoption by educational institutions around the world, the need to develop these technologies and rely on the tremendous capabilities that the fourth industrial revolution can provide, mainly represented by artificial intelligence and virtual reality technologies.

Educational games powered by artificial intelligence are one of the main trends surrounding education [90], due to their ability to engage students in an engaging way during their education, and improve their interaction, and these technologies are expected to be able to teach a variety of subjects, such as mathematics, languages, physics, law and medicine [91].

Artificial intelligence technologies can improve their ability to recognize and understand human voices, to improve discussion skills, answer questions, and help students in performing their homework, as voice chats have become the latest technology trend in many institutions [92].

For example, IBM's "Watson" artificial intelligence system answers questions that students ask in the language of nature. Watson analyzes inputs on a deep level to provide accurate answers to questions that humans generate [93].

The use of artificial intelligence helps to automate routine tasks that increase the burden on lecturers, and to speed up administrative and organizational tasks, such as checking homework, classifying papers, looking at absence records, preparing reports, and issuing results, which are tasks in which lecturers spend most of their time and artificial intelligence can do it without errors in just a few minutes [94].

Artificial intelligence can help digitize textbooks and references, create smart content for students that is customizable to age groups, and help them memorize and learn [7].

Virtual reality technologies can also be taken advantage of in the direct lessons, by creating virtual meeting rooms that make discussions more attractive and flexible and simulate face-to-face meetings, and augmented reality can clearly contribute to e-learning, by interacting with images and models so that they appear as real, Especially in the field of space sciences, biology and chemical experiments [80].

Finally, if the Corona pandemic, with the temporary rules it imposed, has directly affected the world economy, and temporarily changed a lot of societies' lifestyles, especially in the field of work and education, then educational institutions must now think more seriously about investing in e-learning and virtual training [76], when It achieves great benefits for students and lecturers, overcoming obstacles that hinder its spread, and making use of technology to provide flexible and low-cost education while preserving the quality of its outputs [12].

The spread of distance education during the epidemic crisis opened a global dialogue and difficult questions about a number of issues, starting with the philosophy of university and university campus in shaping the personality of students, as many academics believe that online education will create more un- healthy graduates and more frustration with the idea of social communication [87].

Another issue that arises in the context of the ongoing debate is about the cultural and psychological aspect of faculty members who are the slowest among all other professional segments to accept change [67]. Moving into the digital world makes them feel threatened and asks them to abandon the skills they practiced for decades, and the issue is not limited to uploading educational materials to the World Wide Web, but a change in the way of thinking and learning new technical skills and software, and the new reality needs students to conduct assessments one by one, instead of doing it as a group activity [29]. The talk is about a difficult transitional phase, because the issue is not individual, but a large human group that has been practicing this type of education for hundreds of years, and the rehabilitation and training of faculty members requires effort, time and financial investment [95]. Despite this, it seems that humanity has no choice but to expand the use of what the digital revolution has provided in terms of opening the horizons of education to wide ranges, and testing under the pressure of the crisis provided the opportunity for development, as technical problems have arisen in the applications of the proposed educational platforms with the usage of the most popular application Zoom which faced serious problems for users [57], especially in countries that lack fast internet, and also with Chinese online platforms, including those of the huge local technology company, Tencent, faced a downtime after receiving tenfold leaps In browsing and usage [15].

However, the digital transformation in education is still limited in its extent. Even with the successes that some universities have achieved in terms of online teaching, good virtual alternatives to field of academic exchanges do not appear on the horizon, not to mention the social and cultural attractions and personal experience of campus life that cannot be transformed into a virtual world [96].

Many academics point to the need to resort to the co-educational system more, in a way that balances the educational needs that suit each institution and its programs separately [97]. However, this does not cancel the fact that well-established educational institutes and universities have announced that they will fully implement the idea of e-learning, through integrated solutions to transfer the idea of education to digital platforms, on which major international universities depend on it such as Harvard, London University, Sydney, Johns Hopkins and others [17]. The entry of universities and institutes with a global academic reputation to the field of full e-learning enhances expectations of revitalizing and developing the digital education industry faster than experts estimated, when they anticipated this qualitative shift in the post-2030 phase [83].

There are two fundamental observations regarding the issue of the future of education. The first is that the major technological revolution that the world is experiencing today, including the applications of artificial intelligence which has become today an essential component in the future of many industries, economic and financial management, and the mechanisms of political decision-making [12]. The effects and applications of this revolution will be reflected in all levels of education, whether some institutions move to fully electronic or mixed education, which maintains the university campus but benefits greatly from the dimensions opened by the digital age [98]. The second issue: It is that digital or Virtual learning is the basic block of what is known in political literature as the democracy of education and knowledge. We have begun to witness the redefinition of technical competence, knowledge and expertise not related to a university degree, where major companies such as Microsoft and Google are now employing competencies on the basis of knowledge and achievement. E-learning has become accessible to everyone of all classes and age groups [53].

New Roles for Both Lecturer and Student Through E-learning in Virtual Classrooms After Coronavirus Pandemic

New Roles for the Lecturer After Coronavirus Pandemic

The lecturer turns from the wise and lecturer who provides students with answers to the expert in stirring up arguments to guide and supply educational resources [99].

Educators become designers of educational experiences while providing students with the first batch of work, increasing their encouragement for self-direction, and looking at topics with multiple perspectives with an emphasis on salient points as well as competition between lecturers providing content for quality access [100].

The lecturer is the center of strength in the structure of changes, as he transforms from the insider member in his total monitoring of the learning environment to a member of the learning team participating in the learning environment as a companion for the learners [33].

New Roles for the Student After Coronavirus Pandemic

The student is transformed from receptacles that memorize facts and deal with the lowest level of knowledge, to develop solutions to complex problems that build knowledge [49].

Students revise their questions and search for answers on their own, seeing the topics in multiple perspectives according to their work in groups, and performing collaborative assignments noting that group interaction leads to an increase in learning experiences [40].

Emphasizing the urging of students to self-independence while urging them to manage their time and their learning processes and benefit from learning resources [34].

The Saudi Electronic University as a Model During Coronavirus Pandemic

The Saudi Electronic University needs to fulfill its mission and reach its vision for an integrated administrative entity and infrastructure that meets the purposes and nature of teaching and learning in it, and although the majority of the educational process is carried out in a large proportion in virtual environments on the global network [101], but the university administration follows-up the design and development of its curricula that requires the presence of a large headquarters with smaller sub-centers to monitor learners' access to the same educational services [55], in addition to meeting the requirements for evaluating academic achievement that is implemented in computer labs that require the presence of learners themselves to accomplish [20]. And being electronic means the use of an e-learning model based on e-learning techniques and distance education, and it does not mean that the university has no headquarters or centers to manage the educational process, but its need is not comparable to the traditional university that provides headquarters and halls that accommodate all students and its lecturers, in addition to the lack of full-time faculty members in it, and the lack of need for student housing and some other support services [98]. This does not mean the small number of employees for the university, as the virtual environment has its requirements such as an increase in technical and academic support services around the day and the needs of content development [69].

13 Conclusion

The pattern of e-learning represents a comprehensive mechanism for all sectors that does not pertain to a specific field or elite, and for the sake of updating and developing our behaviors and ideas in line with the age of knowledge, we cannot remain at their doors as it is imperative for our universities to adopt them so that they do not miss all opportunities and this can only happen through Intensifying our endeavors and efforts and striving to prepare for it by preparing well for our teachers, encouraging our students, and accelerating the fight against all aspects of information illiteracy in our educational institutions from local and national networks linking the various parties in order to reach the knowledge society [25, 102].

E-learning is closely related to virtual education. Without the use of electronic media, there can be no virtual education, as it is a modern scientific revolution in the methods and technologies of education that harness the latest technology, including devices, programs and the Internet in the service of stimulating and developing the educational process [45, 103]. Virtual learning helps cover a large number of students in different geographic regions, as it reduces the financial costs that may be an obstacle to providing university buildings, providing lecturers, and other educational facilities required by traditional classes, so that the teaching and learning process is no longer confined to a place or time or controlled by strict schedules [88, 104].

E-learning tools have contributed to the emergence of modern methods and technologies for teaching and learning, including virtual education, where the so-called virtual classrooms, virtual reality, the virtual lecturer and the virtual library appeared. This development in the field of education came as a result of the information revolution and the development of communication means, and this paper will deal with e-learning and how its modern tools have been used in stimulating virtual education, as well as some of the advantages of this education and some of its disadvantages, and the new roles for both lecturer and student in the virtual classroom.

The e-learning used in the virtual classroom is similar to the education inside the traditional classroom in terms of the presence of the lecturer and the students, but the difference is that it is on the global network of information where it is not restricted by time or place, as the virtual learner is a real learner but in an electronic environment, e-learning plays a large role in promote virtual learning in particular, and this is what is known about now and is applied by virtual universities that provide distance education through modern electronic media such as the Internet, e-mail, channels and satellites that are used in the transmission of lecturers, programs, courses and student evaluation [58, 105, 106].

The new Corona virus crisis forces universities to face long-standing challenges in the higher education sector, among them - for example - the very high educational expenditures, and the perceptions of limiting education to the elite. Some of the changes that result from meeting these challenges may be permanent. In the long term, universities may move to teaching many classes online (a trend already in progress), receive fewer students from other countries, and even re-design their programs to suit local and national communities more, with the aim of solving problems. It is urgent, and to demonstrate its importance at a time when experts and public institutions are facing increasing criticism.

Recommendations

1. Investing in positive directions for students and faculty members towards e-learning, developing plans and programs to benefit from these directives, and giving training courses in the field of e-learning for both students and faculty members.
2. Training and encouraging teachers to communicate with students through electronic pages and e-mail, given that many students have internet service at home.
3. Emphasizing the need for the university to pay attention to introducing e-learning in university education, and to spread electronic culture among students to achieve the greatest amount of interaction with this type of education.
4. Providing an appropriate educational structure for the implementation of e-learning at the university and removing all human, material and technical obstacles that prevent its spread in the educational system in various stages and fields.
5. The university must conduct more studies and research to know the effectiveness of e-learning in the presence of harsh conditions and hold conferences and seminars for the development and advancement of e-learning.
6. The necessity for the university to offer materials that acquire the student's e-learning skills and technologies in order to facilitate the process of interaction and benefit by students with the educational materials presented electronically.

References

1. Affouneh, S., Salha, S., N., & Khlaif, Z. (2020). Designing quality e-learning environments for emergency remote teaching in coronavirus crisis. *Interdisciplinary Journal of Virtual Learning in Medical Sciences, 11*(2), 1–3.
2. Aljaber. (2018). E-learning policy in Saudi Arabia: Challenges and successes. *Research Comparative Intearnational Education, 13*(1), 176–194.
3. Shahzad, A., Hassan, R., Abdullah, N. I., Hussain, A., & Fareed, M. (2020). COVID-19 impact on e-commerce usage: An empirical evidence from Malaysian healthcare industry. *Humanities & Social Sciences Reviews, 8*(3), 599–609.

4. Purarjomandlangrudi, A., & Chen, D. (2020). Exploring the influence of learners' personal traits and perceived course characteristics on online interaction and engagement. *Educational Technology Research and Development, 68*, 2635–3265.
5. Torres Martín, C., Acal, C.;,El Honrani, M. Mingorance Estrada, Á. C. (2021). Impact on the virtual learning environment due to COVID-19. *Sustainability, 13*, 582.
6. Allam, S. N. S., Hassan, M. S., Mohideen, R. S., Ramlan, A. F., & Kamal, R. M. (2020). Online distance learning readiness during Covid-19 outbreak among undergraduate students. *International Journal of Academy Research Business Society Science, 10*, 642–657.
7. Alqahtani, A. Y., & Rajkhan, A. A. (2020). E-learning critical success factors during the covid-19 pandemic: A comprehensive analysis of e-learning managerial perspectives. *Education in Science, 10*, 216.
8. Göğüş, A., & Ertek, G. (2020). A scoring approach for the assessment of study skills and learning styles. *International Journal of Information Education Technology, 10*, 715–722.
9. Ahmad, S., Sultana, N., & Jamil, S. (2020). Behaviorism vs constructivism: A paradigm shift from traditional to alternative assessment techniques. *Journal of Applications Linguistics Language Research, 7*, 19–33.
10. Konak, A., Kulturel-Konak, S., & Cheung, G. W. (2019). Teamwork attitudes, interest and self-efficacy between online and face-to-face information technology students. *Team Performance Management, 25*, 253–278.
11. Yang, X.; Zhang, M.; Kong, L.; Wang, Q.; Hong, J.-C. (2020). The effects of scientific self-efficacy and cognitive anxiety on science engagement with the "question-observation-doing-explanation" model during school disruption in COVID-19 pandemic. *Journal of Science Education Technology*, 1–14.
12. Drijvers, P. (2019). Head in the clouds, feet on the ground–A realistic view on using digital tools in mathematics education. In Vielfältige Zugänge zum Mathematikunterricht; Büchter, A., Glade, M., Herold-Blasius, R., Klinger, M., Schacht, F., & Scherer, P., (Eds.), Springer Spektrum: Wiesbaden, Germany, p. 163.
13. Fulton, C. (2020). Collaborating in online teaching: Inviting e-guests to facilitate learning in the digital environment. *Information Learning Science, 121*, 579–585.
14. Raaper, R., & Brown, C. (2020). The Covid-19 pandemic and the dissolution of the university campus: Implications for student support practice. *Journal of Professional Capital Community, 5*, 343–349.
15. Block, P., Hoffman, M., Raabe, I. J., Dowd, J. B., Rahal, C., Kashyap, R., & Mills, M.C. (2020). Social network-based distancing strategies to flatten the COVID-19 curve in a post-lockdown world. *Nature Human Behavior*, 588–596.
16. Toader, D.-C., Boca, G., Toader, R., Măcelaru, M., Toader, C., Ighian, D., & Rădulescu, A. T. (2020). The effect of social presence and chatbot errors on trust. *Sustainability, 12*, 256.
17. Hsiu-Mei, H., & Liaw, S. S. (2018). An analysis of learners' intentions toward virtual reality learning based on constructivist and technology acceptance approaches. *International Review Research Open Distributed Learning, 19*, 90–115.
18. Ahmed, G., Arshad, M., & Tayyab, M. (2019). Study of effects of ICT on profesional development of teachers at university level. *European Online Journal of National Society Science, 8*, 162–170.
19. Kliziene, I., Taujanskiene, G., Augustiniene, A., Simonaitiene, B., Cibulskas, G. (2021) The impact of the virtual learning platform EDUKA on the academic performance of primary school children. Sustainability, 1–13.
20. Naveed, Q. N., Muhammad, A., Sanober, S., Qureshi, M. R. N., & Shah, A. (2017). A mixed method study for investigating critical success factors (CSFs) of e-learning in Saudi Arabian universities. *Methods*, 1–8.
21. Watermeyer, R., Crick, T., Knight, C., & Goodall, J. (2020). COVID-19 and digital disruption in UK universities: Afflictions and affordances of emergency online migration. *Higher Education, 81*, 1–19.
22. Le, D. A., MacIntyre, B., & Outlaw, J. (2020). Enhancing the experience of virtual conferences in social virtual environments. In *Proceedings of the 2020 IEEE Conference on Virtual Reality*

and 3D User Interfaces Abstracts and Workshops (VRW), Atlanta, GA, USA, 22–26 March 2020, pp. 485–494.

23. Bozkurt, A., & Sharma, R. C. (2020). Emergency remote teaching in a time of global crisis due to CoronaVirus pandemic. *Asian Journal of Distance Education, 15*, i–vi.
24. Huang, R., Tlili, A., Yang, J., Chang, T.-W., Wang, H., Zhuang, R., & Liu, D. (2020). *Handbook on facilitating flexible learning during educational disruption: The Chinese experience in maintaining undisrupted learning in COVID-19 outbreak.* Smart Learning Institute of Beijing Normal University.
25. Ribeiro, C., Maia, M., & Duarte, S. S. (2020). Distance education or emergency remote educational activity: In search of the missing link of school education in times of COVID-19. *Research Society Development, 9*, 1–29.
26. Suryasa, W., Zambrano, R., Mendoza, J., Moya, M., & Rodríguez, M. (2020). Mobile devices on teaching-learning process for high school level. *International Journal of Psychosocial Rehabilitation, 20*, 330–340.
27. Fidalgo1Patricia, T., Joan, O., & Lencastre, J. A. (2020). Students' perceptions on distance education: A multinational study. *International Journal of Educational Technology in Higher Education, 17*, 18.
28. Liesa, M., Latorre, C., Vázquez, S., & Sierra, V. (2020). The technological challenge facing higher education professors: Perceptions of ICT tools for developing 21st century skills. *Sustainability, 12*, 5339.
29. Portillo, J., Garay, U., Tejada, E., & Bilbao, N. (2020). Self-perception of the digital competence of educators during the COVID-19 pandemic: A cross-analysis of different educational stages. *Sustainability, 12*, 1–13.
30. Wiederhold, B. K. (2020). Connecting through technology during the coronavirus disease 2019 pandemic: Avoiding "Zoom Fatigue." *Cyberpsychology Behavior. Society Network, 23*, 437–438.
31. Thoring, A., Rudolph, D., & Vogl, R. (2018). The digital transformation of teaching in higher education from an academic's point of view: An explorative study. In *Proceedings of the Applications of Evolutionary Computation*, Las Vegas, NV, USA, 15–20 July 2018; pp. 294–309.
32. Zhou, L., Wu, S., Zhou, M., & Li, F. (2020). "School's out, but class's on", The largest online education in the world today: taking China's practical exploration during the COVID-19 epidemic prevention and control as an example. *Best Evidence China Education, 4*, 501–519.
33. Almaiah, M. A., Al-Khasawneh, A., & Althunibat, A. (2020). Exploring the critical challenges and factors influencing the E-learning system usage during COVID-19 pandemic. *Education and Information Technologies, 25*, 5261–5280.
34. Hodges, C., Moore, S., Lockee, B., Trust, T., & Bond, A. (2020). The difference between emergency remote teaching and online learning. *Educational Review, 27*, 1–12.
35. Duarte, S. S., Ribeiro, C., & Maia, M. (2020). Distance education in the digital age: Typologies, variations, uses and possibilities of e-learning. *Research Social Development, 9*, 1–17.
36. Al Shehab, N. (2020). How to increase knowledge retention in elearning during Covid-19 pandemic? In *Proceedings of the ECEL 2020 19th European Conference on e-Learning* (pp. 10–15). Berlin, Germany, 29–30 October 2020.
37. Kondratavi˘ciene, R. (2018). Individualization and differentiation of the content of primary eEducation by using virtual learning ˙ environment "EDUKA class." *Pedagogika, 130*, 131–147.
38. Pal, D., Vanijja, V., & Patra, S. (2020). Online learning during COVID-19: students' perception of multimedia quality. In *Proceedings of the 11th International Conference on Advances in Information Technology*, Bangkok, Thailand, 1–3 July 2020; pp. 1–6.
39. Ouadoud, M., Chkouri, M. Y., & Nejjari, A. (2018). Learning management system and the underlying learning theories: Towards a new modeling of an LMS. *International Journal of Information Science, 2*, 25–33.
40. Al-Kumaim, N. H., Alhazmi, A. K.; Mohammed, F., Gazem, N. A., Shabbir, M.S., Fazea, Y. (2021). Exploring the Impact of the COVID-19 Pandemic on university students' learning

life: An integrated conceptual motivational model for sustainable and healthy online learning. *Sustainability, 2*–21.

41. Bogusevschi, D., & Muntean, G. M. (2019). Virtual reality and virtual lab-based technology-enhanced learning in primary school physics. In *Communications in Computer and Information Science, Proceedings of the International Conference on Computer Supported Education, CSEDU 2019*, Crete, Greece, 2–4 May 2019; Lane, H.C., Zvacek, S., Uhomoibhi, J., Eds.; Springer: Cham, Switzerland, 2019; p. 467.

42. Dwivedi, A., Dwivedi, P., Bobek, S., & Zabukovšek, S. S. (2019). Factors affecting students' engagement with online content in blended learning. *Kybernetes, 48*, 1500–1515.

43. Lin, M. H., & Chen, H. G. (2017). A study of the effects of digital learning on learning motivation and learning outcome Eurasia. *Journal Mathematical Science Technology Education, 13*, 3553–3564.

44. Vershitskaya, E. R., Mikhaylova, A. V., Gilmanshina, S. I., Dorozhkin, E. M., & Epaneshnikov, V. V. (2020). Present-day management of universities in Russia: Prospects and challenges of e-learning. *Education and Information Technologies., 25*(1), 611–621.

45. Hui, D. S., Azhar, E. I., Madani, T. A., Ntoumi, F., Kock, R., Dar, O., Ippolito, G., Mchugh, T. D., Memish, Z. A., Drosten, C., et al. (2020). The continuing 2019-nCoV epidemic threat of novel coronaviruses to global health—The latest 2019 novel coronavirus outbreak in Wuhan China. *International Journal of Infectious Disease, 91*, 264–266.

46. Monroy, F. A., Llamas, F., Fernández, M. R., & Carrión, J. L. (2020). Digital technologies at the pre-university and university levels. *Sustainability, 12*, 10426.

47. Muhajirah, M. (2020). Basic of learning theory:(Behaviorism, Cognitivism, Constructivism, and Humanism). *International of Journal of Asian Education, 1*, 37–42.

48. Robinson, H., Al-Freih, M., & Kilgore, W. (2020). Designing with care: Towards a care-centered model for online learning design. *International Journal of Information and Learning Technology, 37*, 99–108.

49. Allo, M. D. G. (2020). Is the online learning good in the midst of Covid-19 Pandemic? The case of EFL learners. *Journal of Sinestesia, 10*, 1–10.

50. Van de Velde, S., Buffel, V., Bracke, P., Van Hal, G., Somogyi, N. M., Willems, B., & Wouters, E. (2021). The COVID-19 international student well-being study. *Scandinavian Journal of Public Health, 49*, 114–122.

51. Kruszewska, A., Nazaruk, S., & Szewczyk, K. (2020). Polish teachers of early education in the face of distance learning during the COVID-19 pandemic—The difficulties experienced and suggestions for the future. *Education, 3–13*, 1–12.

52. Serhan, D. (2020). Transitioning from face-to-face to remote learning: students' attitudes and perceptions of using Zoom during COVID-19 pandemic. *International Journal of Technology Education Science, 4*, 335–342.

53. Popa, D., Repanovici, A., Lupu, D., Norel, M., & Coman, C. (2020). Using mixed methods to understand teaching and learning in COVID 19 times. *Sustainability, 12*, 8726.

54. Vagg, T., Balta, J. Y., Bolger, A., & Lone, M. (2020). Multimedia in education: What do the students think? *Health Professional Education, 6*, 325–333.

55. Abouelnaga, H. M., Metwally, A. B., Mazouz, L. A., Abouelmagd, H., Alsmadi, S., Aljamaeen, R., & Eljawad, L. (2019). 'A survey on educational technology in Saudi Arabia.' *International Journal of Applied Engineering Research, 14*(22), 4149–4160.

56. Li, K. W. (2021). Switching to a synchronous mode of Chinese calligraphy teaching during the period of COVID-19 pandemic: An experience report. *The Electronic Journal of E-Learning, 19*(1), 18–20.

57. Cicha, K., Rizun, M., Rutecka, P., & Strzelecki, A. (2021). COVID-19 and higher education: First-Year students' expectations toward distance learning. *Sustainability*, 1–19.

58. Marek, M. W., Chew, C. S., & Wu, W. V. (2021). Teacher experiences in converting classes to distance learning in the COVID-19 pandemic. *International Journal of Distance Education Technology, 19*, 89–109.

59. Asanov, I., Flores, F., McKenzie, D., Mensmann, M., & Schulte, M. (2021). Remote-learning, time-use, and mental health of Ecuadorian high-school students during the COVID-19 quarantine. *World Development, 138*, 1–21.

60. Basilaia, G., Dgebuadze, M., Kantaria, M., & Chokhonelidze, G. (2020). Replacing the classic learning form at universities as an immediate response to the COVID-19 virus infection in Georgia. *International Journal for Research in Applied Science and Engineering Technology (IJRASET), 8*, 101–8. Retrieved June 16, 2020, from https://doi.org/10.22214/ijraset.2020.3021.
61. Favale, T., Soro, F., Trevisan, M., Drago, I., & Mellia, M. (2020). Campus traffic and eLearning during COVID-19 pandemic. *Computer Networks, 176*, 107290.
62. Gabaree, L., Rodeghiero, C., Presicce, C., Rusk, N., & Jain, R. (2020). Designing creative and connected online learning experiences. *Information Learning Science, 121*, 655–663.
63. Murphy, M. P. A. (2020). COVID-19 and emergency eLearning: Consequences of the securitization of higher education for post-pandemic pedagogy. *Contemporary Security Policy, 41*(3), 492–505. https://doi.org/10.1080/13523260.2020.1761749
64. Saroughi, M., & Kitsantas, A. (2021). Examining relationships among contextual, motivational and wellbeing variables of immigrant language-minority college students. *Alternative Higher Education, 46*, 1–19.
65. Dhawan, S., & Learning, O. (2020). A Panacea in the time of COVID-19 crisis. *Journal of Educational Technology Systems, 49*(1), 5–22.
66. Wieser, D., & Seeler, J. T. (2018). Online, not distance education. In A. Altmann, B. Ebersberger, C. Mössenlechner, & D. Wieser (Eds.), *The disruptive power of online education* (pp. 125–146). Emerald Publishing Limited.
67. Bao, W. (2020). COVID-19 and online teaching in higher education: A case study of Peking University. *Human Behaviour Emerging Technology, 2*, 113–115.
68. Alqahtani, A. Y., & Rajkhan, A. A. (2020). E-learning critical success factors during the COVID-19 pandemic: A comprehensive analysis of e-learning managerial perspectives. *Education in Science, 10*, 2–16.
69. Hargitai, D. M., Pinzaru, F., & Veres, Z. (2021). Integrating business students' E-Learning preferences into knowledge management of universities after the COVID-19 pandemic. *Sustainability, 13*(2478), 1–21.
70. Basilaia, G., Dgebuadze, M., Kantaria, M., & Chokhonelidze, G. (2020). Replacing the classic learning form at universities as an immediate response to the COVID-19 virus infection in Georgia. *International Journal for Research in Applied Science & Engineering Technology, 8*(III).
71. Alsalhi, N. R., Eltahir, M., Daw, E., Abdelkader, A., & Zyoud, S. (2021). The effect of blended learning on the achievement in a physics course of students of a dentistry college: A case study at Ajman University. *The Electronic Journal of E-Learning, 19*(1), 1–17.
72. Melo, E., Llopis, J., Gascó, J., & González, R. (2020). Integration of ICT into the higher education process: The case of Colombia. *Journal of Small Business Strategy, 30*, 58–67.
73. Yoon, P., & Leem, J. (2021). The influence of social presence in online classes using virtual conferencing: Relationships between group cohesion, group efficacy, and academic performance. *Sustainability, 13*(1), 2–19.
74. Christopoulos, A., Kajasilta, H., Salakoski, T., & Laakso, M. J. (2020). Limits and virtues of educational technology in elementary school mathematics. *Journal of Educational Technology Systems, 49*, 59–81.
75. Arghode, V., Brieger, E., & Wang, J. (2018). Engaging instructional design and instructor role in online learning environment. *European Journal of Training Development, 42*, 366–380.
76. Gupta, S. B., & Gupta, M. (2020). Technology and E-learning in higher education. *Technology, 29*(4), 1320–1325.
77. Affinito, S. (2018). *Literacy coaching: Transforming teaching and learning with digital tools and technology* (pp. 1–168). Portsmouth, UK: Heinemann.
78. Correia, A. P. (2020). Healing the digital divide during the COVID-19 pandemic. *Quarterly Review Distance Education, 21*, 13–21.
79. Pinto, M., & Leite, C. (2020). Digital technologies in support of students learning in higher education: Literature review. *Digital Education, 37*, 1–18.

80. MES. (2020). Ministry of Education, Science, Culture and Sport of Georgia. 'Ministry of Education, Science, Culture and Sport of Georgiato strengthen distance learning methods. Retrieved February 16, 2021, from https://www.mes.gov.ge/content.php?id=10271&lang=eng.
81. Littlefield, J. (2018). The difference between synchronous and asynchronous distance learning. Retrieved from https://www.thoughtco.com/synchronous-distance-learning-asynchronousdistance-learning-1097959.
82. Bracci, E., Tallaki, M., & Castellini, M. (2019). Learning preferences in accounting education: A focus on the role of visualization. *Meditari Accountancy Research, 28*, 391–412.
83. Khan, M. A., Nabi, M. K., Khojah, M., & Tahir, M. (2021). Students' perception towards E-Learning during COVID-19 pandemic in India: An empirical study. *Sustainability, 13*, 57.
84. Chiu, T. K., Jong, M. S. Y., & Mok, I. A. (2020). Does learner expertise matter when designing emotional multimedia for learners of primary school mathematics? *Educational Technology Research and Development, 68*, 2305–2320.
85. Chung, S., & Cheon, J. (2020). Emotional design of multimedia learning using background images with motivational cues. *Journal of Computer Assisted Learning, 36*, 922–932.
86. Crawford, J., Butler, K., Rudolph, J., Glowatz, M., Burton, R., Magni, P., & Lam, S. (2020). COVID-19: 20 countries' higher education intra-period digital pedagogy responses. *Journal of Application Learning Teaching, 3*, 1–21.
87. Nash, M., & Churchill, B. (2020). Caring during COVID-19: A gendered analysis of Australian university responses to managing remote working and caring responsibilities. *Gender Work Organization, 27*, 1–14.
88. König, J., Jäger-Biela, D. J., & Glutsch, N. (2020). Adapting to online teaching during COVID-19 school closure: Teacher education and teacher competence effects among early career teachers in Germany. *European Journal of Teacher Education, 43*, 608–622.
89. Crawford, J., Butler-Henderson, K., Rudolph, J., Malkawi, B., Glowatz, M., Burton, R., Magni, P. A., & Lam, S. (2020). COVID-19: 20 countries' higher education intra-period digital pedagogy responses. *Journal of Application Learning Teaching, 3*, 9–28.
90. Gama, J. A. P. (2018). Intelligent educational dual architecture for University digital transformation. In *Proceedings of the 2018 IEEE Frontiers in Education Conference (FIE)*, San Jose, CA, USA, 3–6 October 2018; pp. 1–9.
91. Presicce, C., Jain, R., Rodeghiero, C., Gabaree, L. E., & Rusk, N. (2020). WeScratch: An inclusive, playful and collaborative approach to creative learning online. *Information Learning Science, 121*, 695–704.
92. Theoret, C., & Ming, X. (2020). Our education, our concerns: The impact on medical student education of COVID-19. *Medical Education, 54*, 591–592.
93. Toquero, C. M. (2020). Emergency remote education experiment amid COVID-19 pandemic. *International Journal of Education Research Innovation, 15*, 162–172.
94. Xu, F., & Du, J. T. (2018). Factors infuencing users' satisfaction and loyalty to digital libraries in Chinese universities. *Computers in Human Behavior, 83*, 64–72.
95. Young, S., Young, H., & Cartwright, A. (2020). Does lecture format matter? Exploring student preferences in higher education. *Journal of Perspectives Application Academy Practice, 8*, 30–40.
96. Yusuf, B. N. (2020). Are we prepared enough? A case study of challenges in online learning in a private higher learning institution during the Covid-19 outbreaks. *Advance Social Science Research Journal, 7*, 205–212.
97. Hash, P. M. (2021). Remote learning in school bands during the COVID-19 shutdown. *Journal of Research in Music Education, 68*, 381–397.
98. Haythornthwaite, C. (2019). Learning, connectivity and networks. *Information Learning Science, 120*, 19–38.
99. Joia, L. A., & Lorenzo, M. (2021). Zoom In, Zoom Out: The impact of the COVID-19 pandemic in the classroom. *Sustainability, 13*(2531), 1–19.
100. Karagiannopoulou, E., Milienos, F. S., Kamtsios, S., & Rentzios, C. (2020). Do defence styles and approaches to learning 'fit together' in students' profiles? Differences between years of study. *Educational Psychology, 40*, 570–591.

101. Muhammad, A., Shaikh, A., Naveed, Q. N., & Qureshi, M. R. N. (2020). Factors affecting academic integrity in e-learning of saudi arabian universities. An investigation using delphi and AHP. *IEEE Access, 8*, 16259–16268.

102. Al Kurdi, O. F. (2021). A critical comparative review of emergency and disaster management in the Arab world. *Journal of Business and Socio-economic Development, 1*(1), 24–46. https://doi.org/10.1108/JBSED-02-2021-0021

103. Abdulla, H., Ebrahim, M., Hassan, A., Hashimi, K. A., Hamdan, A., Razzaque, A., & Musleh, A. (2020). Measuring the perception of knowledge gained during the virtual learning: Business research method course case study. Paper presented at the Proceedings of the European Conference on e-Learning, ECEL, 2020-October 1–9. https://doi.org/10.34190/EEL.20.143.

104. Razzaque, A., & Hamdan, A. (2020). Internet of Things for learning styles and learning outcomes improve e-learning: A review of literature. Advances in Intelligent Systems and Computing, 2020, 1153 AISC, pp. 783–791

105. Elali, W. (2021). The importance of strategic agility to business survival during corona crisis and beyond. *International Journal of Business Ethics and Governance, 4*(2), 1–8. https://doi.org/10.51325/ijbeg.v4i2.64

106. Mokadem, W., & Muwafak, B. M. (2021). The difference of the theoretical approach of corporate social responsibility between the European Union and United States of America. *International Journal of Business Ethics and Governance, 4*(1), 124–136. https://doi.org/10.51325/ijbeg.v4i1.22

Digital Learning Crisis Management Competencies Among Leaders During the Covid-19 Pandemic

Juliza Adira Mohd (a) Ariffin and Ahmad Aizuddin Md Rami

1 Introduction

The Covid-19 virus was discovered in December 2019 in Wuhan, China, highlighting the impact of globalization on health. In a matter of months, the virus spread throughout the world, forcing all countries to face and deal with the same issues. Covid-19 was declared a public health emergency of international concern by the World Health Organization (WHO) on January 30, 2020, and a pandemic on March 11, 2020. The Covid-19 pandemic has been referred to as a health crisis and has prompted many organizations to adopt new norms in their operations.

Living in a VUCA (Volatile, Uncertainty, Complexity, and Ambiguity) world teaches us to be prepared for uncertainty. The disruption or crisis is not limited to the Covid-19 health crisis, earthquakes, floods, natural crises, civil conflicts, cultural constraints, and social restrictions. Still, there are many more that we may face in the future [8]. It is a new normal that is real and maybe around for a long time [35], and organizations need to make sure that they are well prepared to face any possible crisis in the future. Between the 1980s and the 2000s, Mitroff and Alpaslan [27] conducted studies involving 500 companies, revealing that only 5% to 25% of companies were prepared for a crisis. Only a very small number of companies had a crisis management plan. According to Khalil et al. [19], news and social media have exposed the failure of some leaders to save their organizations and jobs during times of crisis. What is lacking is leadership preparation to deal with the complexities of a crisis. Even though an individual is an experienced leader, the experience is negligible if the leader cannot handle panic or critical situations during a crisis. How a leader reacts in a

J. A. Mohd (a) Ariffin (✉) · A. A. Md Rami
Faculty of Educational Studies, Universiti Putra Malaysia, 43400 Seri Kembangan, Selangor, Malaysia

A. A. Md Rami
e-mail: ahmadaizuddin@upm.edu.my

© The Author(s), under exclusive license to Springer Nature Switzerland AG 2022
A. Hamdan et al. (eds.), *Technologies, Artificial Intelligence and the Future of Learning Post-COVID-19*, Studies in Computational Intelligence 1019,
https://doi.org/10.1007/978-3-030-93921-2_6

crisis and their decision-making capabilities in a crisis will affect the organization. Thus, as Black [6] stated, any organization must have competent leaders to maintain their good reputation and uphold performance standards. Wooten and James [46] also agree that effective crisis management is dependent on their leadership behaviours.

According to Schwantes [36], despite extensive research and literature, little research has been conducted over the years to identify leadership competencies in crisis management systematically. In today's VUCA world, since a crisis can unexpectedly occur, an organization must ensure that it is equipped and ready to face a crisis. In disseminating knowledge regarding leadership competencies during a crisis, it is important that the research and literature regarding a crisis, especially in leadership, expand.

With this in mind, this study seeks to discover the most needed leadership competencies during a crisis. Before studying the competencies required of leaders in dealing with a crisis, a description of what constitutes a crisis and the importance of crisis management competence among leaders is provided. Examples of the proposed skills and opportunities, and challenges leaders face in implementing those skills are duly provided. The skills proposed in this paper are based on proven models, theories, and previous research conducted by well-known scholars. This article is concluded with a discussion of the lessons that can be learnt from the past crises, and the contribution of this paper to literature for relevant fields of study is also highlighted.

2   Understanding the Nature of Crisis

The term crisis refers to the high value, rational, sound judgments and decisions that a leader must make immediately, using all of his power and knowledge [5]. The concept of a crisis has existed since the dawn of time. For example, in the nineteenth century, synthetic chemicals were developed, and the twentieth century experienced the development of nuclear power. These incidences raised the prospect of toxic chemical disasters and radioactive dust crises. The crises such as floods and earthquakes that existed since time immemorial have continued into the present time. However, new crises and disasters have become an addition to the old form of crises [33]. A crisis is a one-of-a-kind, unforeseeable, and disruptive event. Some of the obvious organizational consequences of a failure to deal with a crisis are loss of reputation, financial loss, or even death [21]. A crisis is an event caused by external factors such as terrorism or natural disasters or internal factors such as product defects or corruption [7, 37].

Furthermore, Van Wart and Kapucu [44] define a crisis as an event or problem that threatens the organization's high-priority values, presents a limited time for a response, and is unexpected or unanticipated [45]. Natural causes of crises include unusually large events or extremely rare occurrences. Major floods, large storms, and other widespread natural disasters are anticipated, planned for, and mitigated to some extent. Still, their scope can also cause crises such as infrastructure damage, transportation chain disruption, communication disruption, and so on. An organizational crisis can occur when a crisis occurs within an organization. Furthermore, organizational crises are difficult to predict, have serious negative consequences, and require quick and dynamic responses to mitigate their effects. Many types of major crises that a company may face [28] as mentioned in Table 1.

3 The Importance of Leaders with Competencies During Crises

There are numerous challenges that leaders may face during a crisis. These include failing to communicate a problem, not having enough information to make an informed decision, effectively leading, adopting an offensive versus a defensive posture, and failing to take time to care for oneself [25].

A crisis can have a negative impact on an organization if it is not handled properly. Some of the effects of the crisis on the organization and its employees, according to Pelin [32], include panic within the organization, high labour turnover, loss of key personnel and knowledge, a lack of morale and motivation resulting in poor performance, rising healthcare costs, a negative public image, misinformation, leadership and reassurance lost, difficulty in understanding psychological problems in the responding and recovery phases, cancellation of recruitment and also cancellation of scheduled training programs.

One example of the psychological impact on employees during a crisis is emotional exhaustion, which can affect work motivation, performance, and mental health, lower employee engagement with their work and lead to absenteeism and the intent to quit [31], between the 1980s to the 2000s, Mitroff and Alpaslan [27] conducted studies engaging 500 companies. The studies showed that between 5 and 25% of these companies were prepared for a crisis and a very small number of the companies have a crisis management plan.

A crisis requires effective crisis management conducted by an effective leader and organization. Even though a leader has ample leadership experience, it is sometimes ineffective if the leader cannot handle panic or critical situations during a crisis. Crisis management necessitates making quick decisions in critical situations and putting decision-makers in an urgent decision-making situation with the requirement of minimizing potential consequences [24]. How leaders respond to the crisis will have an impact on their subordinates' economic and social well-being as well as their health. In this situation, some leaders will be able to adapt to the challenge while others will fail.

A competent leader is skilled at his or her job. Studies suggest that there is enough evidence to argue that the performance of an organization is derived from competent leaders who become an important asset to the future of the organization [6, 29, 41]. After conducting a self-reported survey of executive public leaders in Turkey, Kapucu and Ustun [17] discovered that core leadership competencies positively correlate with

Table 1 Type of major crises a company may face [28]

Crisis cluster	Types of crises involved
Economic	1. Strikes 2. Labor revolt 3. Labour shortage 4. A significant drop in stock price and fluctuation 5. Market drop 6. A drop in major earnings 7. Takeovers in hostile environments
Informational	1. Misappropriation of proprietary and confidential information 2. Untrue information 3. Attempting to tamper with computer records 4. Loss of critical computer information about customers, suppliers, and so on
Physical, infrastructure	1. Loss of significant equipment, plants, and raw materials 2. Failures of significant equipment and plants 3. Loss of significant facilities 4. Significant plant disruptions 5. Explosions 6. Defective or poor product design 7. Product flaws 8. Lack of quality control
Human resource	1. Loss of key executive 2. Loss of key personnel 3. An increase in absenteeism 4. An increase in vandalism and accidents 5. Violence in the workplace 6. Inadequate succession planning 7. Corruption
Reputational	1. Slander 2. Gossip 3. Stupid jokes 4. Rumours 5. Negative impact on corporate reputation 6. Attempting to tamper with corporate logos 7. False rumours
Psychopathic acts	1. Tampering with the product 2. Kidnapping 3. Hostage-taking 4. Terrorism 5. Violence in the workplace 6. Criminal/terrorist 7. Psychopathic behaviour
Natural calamities	1. Earthquakes 2. Fires 3. Floods 4. Typhoons 5. Hurricanes 6. Mudslides

crisis management effectiveness. Their study established the importance of core leadership competencies in crisis leadership effectiveness, and their study's hypothesis confirmed the positive impact of core leadership competencies on crisis management effectiveness. Limsila and Ogunlana [23] defined competencies as knowledge, attitudes, skills, and various personal characteristics that affect the job, have a relationship with job performance, be measured against well-accepted standards, and can also be measured improved by training and staff development programs.

4 Different Skills That a Leader Should Have During Crises

Table 2 shows three different skills leaders should possess according to the level of their positions as described by Katz [18]. During a crisis such as the Covid-19 pandemic, leaders need to enhance these basic skills and utilize whichever is best suited. This study will discuss the most important skills necessary during a crisis that leaders need to possess, as introduced by Katz [18].

4.1 Technical Skills

Knowledge and proficiency in a specific type of work or activity are referred to as technical skills. During this pandemic, one of the most important technical skills leaders must have is using and implementing work through information and communication technologies. It necessitates specialized knowledge, analytical abilities, and the ability to employ appropriate tools and techniques.

4.1.1 The Importance of Information and Communication Technologies (ICT) Skills During Crises

Many countries' efforts to stop the pandemic chain during the Covid-19 pandemic are through quarantine, social distancing, isolation of infected populations, travel restrictions, and many more. For organizations, implementing the work from the home policy has become the best method to avoid interrupting business activities.

Table 2 Different skills leaders should have based on their job levels [18]

Top-level of management	Technical skills (Low)	Human skills (**High**)	Conceptual skills (High)
The middle level of management	Technical skills (High)	Human skills (High)	Conceptual skills (High)
Supervisory level of management	Technical skills (High)	Human skills (High)	Conceptual skills (Low)

Working from home necessitates a high level of ICT. In other words, information and communication technologies (ICTs) are devices that can digitally store, retrieve, manipulate, transmit, or receive data. ICT includes personal computers, televisions, telephones, email systems, robotic and smart devices, and other internet-enabled systems such as traditional media and social media [22].

In their study, Baruch and Nicholson [4] stated four elements need to be fulfilled to make working from home more feasible and effective. The four elements are as follows: (1) the home or work interface, which includes a variety of factors ranging from the quality of family relationships to the type of physical space and facilities available, (2) the nature of the job and the availability of technical support, (3) personal qualities of employees in conducting homeworking, and (4) how supportive the organizational culture is in supporting homeworker employees. This study highlighted the importance of technology in implementing virtual work and agreed that homeworking would become inefficient without proper technology suited to the job requirements.

According to Strielkowski et al. [39], effective management will depend not only on an effective demand-side response but also on information and communication technology availability and efficiency. Therefore, to be competent in tasks, especially during virtual work, a leader needs to have and be familiar with using ICT to perform his or her job. According to Arnfalk et al. [1], being a leader with virtual work organization responsibilities is generally more difficult than leading traditional teams. Susannah [40] emphasized the importance of leaders in constantly seeking relevant information about the impact of the crisis from credible sources such as health professionals, researchers, managers, etc. While intuition is important, leaders must also act according to credible expertise and advice. With the help of ICT, this entire task becomes possible and easy, especially during a pandemic.

4.1.2 Opportunities in Using ICT During the Covid-19 Pandemic

Telework was widely implemented by many organizations around the world when the movement control order was implemented during the Covid-19 pandemic. This is done to keep the Covid-19 virus from spreading. Telework is the use of ICT for work done outside the office [16]. For example, UNESCO has estimated that as many as 192 countries have implemented full school closures [43]. Many countries have decided to implement online learning. With the availability of ICT facilities, all activities can be carried out without face-to-face interaction. Businesses can run as usual with meetings held in the new norms, and students do not lag in lessons.

Furthermore, many organizations have enhanced their capabilities in online service. For example, banks have introduced tools for customers in the banking sector where users need not visit banks. They can access banking services from the comfort of their own homes. Banks have also introduced loan moratoriums available through an online application on the banks' websites [20]. In addition, countless made-for-profit organizations have begun using online platforms to sell products and services. This provides customers with more purchasing choices and saves time to compare prices and products by conducting a simple online survey.

According to Rachmawati et al. [34], telework can be done with more benefits such as cost efficiency, reduced allowance, efficiency in office expenditures (electricity, paper, ink, etc.), reduced pollution, reduced garbage production, and reduced risk of traffic accidents and disease spread. Furthermore, telework has helped to improve family harmony. Employee performance has also become measurable because all work done virtually can be recorded and logged. It is further expected that the reduction in time spent communicating with colleagues will increase employees' productivity.

4.1.3 Challenges in Using ICT During the Covid-19 Pandemic

Although ICT plays a vital role in the survival of organizations during a pandemic, it also has its own set of challenges. The wide use of ICT in various sectors has had different effects and impacts. Even though ICT is not something new and most organizations were already utilizing it before the pandemic, most of them were not ready to fully implement working online, either in terms of competencies or facilities. For example, a study conducted by Soleiman et al. [38] discovered that issues exist in terms of cultural and social contexts when implementing virtual teaching throughout the pandemic, such as a lack of teacher readiness in virtual learning, accessibility issues to virtual learning facilities, issues with student attendance, difficulty in applying distance training for all ages, and a lack of smartphone access. According to survey results, 26 percent of children under 12 and 10 percent of those with children aged 12–17 work from home as a physical distancing measure [12]. These employees juggle work and caregiving responsibilities, and they are confronted with new and challenging dynamics in managing their work-life balance. To avoid this obstacle negatively impacting the organization and their employees, leaders play a huge role when providing a solution. Proper guidelines and skills in handling challenges are needed, especially when telework has been implemented to avoid tension among employees and leaders. This is when and how human skills become crucial for a leader as this set of skills can promote harmony and promote both the employees and leader's well-being.

4.2 Human Skills

The second skill is human skills, which are concerned with interpersonal abilities. These include abilities that enable leaders to effectively collaborate with subordinates, peers, and superiors to achieve organizational goals. A leader's interpersonal skills allow him or her to assist group members in cooperating to achieve common goals. Leaders with strong interpersonal skills can create a trusting environment in which employees feel at ease, secure, and encouraged to contribute to organizational

Table 3 Situational leadership model [15]

Characteristics	Description
Intuition	A situational leader recognizes their employees' needs and then modifies their leadership style to meet those needs
Flexibility	To meet current demands, a Situational Leader seamlessly transitions from one leadership style to another
Belief	A Situational Leader earns their employees' trust and confidence
Problem solving	A Situational Leader solves problems by using the best leadership style for the situation
Coaching	A situational leader can assess their employees' maturity and competence and implement the best strategy to improve their employees' skill sets and goals

planning. Human skills are still important in teleworking, and leaders must adopt them.

4.2.1 Applying Situational Leadership Models in Human Skills During Crises

In managing individuals, one of the models that can be applied in a crisis is the Situational Leadership Model introduced by Hersey [15]. This model's underlying concepts revolve around adaptive leadership styles tailored to the maturity or developmental level. Situational leadership is adaptable and flexible, and it considers variables and provides leaders with the tools they need to suit current circumstances best and achieve their objectives. A leader who is capable of taking this approach will show five characteristics as listed in Table 3.

Previous research has shown that situational leadership can assist leaders in being adaptive while remaining committed, being an effective decision-maker, and being a good planner [2, 3, 14, 30].

Christopher and Alvin [11] also agreed that, in understanding problems and devising effective solutions in the face of a prevailing crisis like the Covid-19 pandemic, the Situational Approach is one of the best options for organizations. These organizations will use a situational approach to make decisions for their organizations' survival. They are open to changes and ready to adapt to ensure the success of their business during this difficult time.

4.2.2 Challenges Faced by Leaders in Managing Subordinates During Crises

When managing subordinates during a crisis, leaders may face numerous challenges, particularly in virtual working conditions. According to Charalampous et al. [10], telework can professionally and socially isolate workers, reducing the support employees receive from their organizations in terms of professional and personal development, creating job insecurity, and making it difficult to focus due to distractions at home. Furthermore, they find it difficult to take a vacation or annual leave from work.

Furthermore, because people can connect via ICT at almost any time of day or night, telework can sometimes have no limits in working hours. Emotional exhaustion is one type of psychological impact on employees during a crisis that can affect work motivation, performance, and mental health, lowering employee engagement and leading to absenteeism and the intent to quit [31]. To avoid employee tension and stress, leaders and organizations must address all of these issues by amending organizational policies to mitigate their impact while working from home.

4.3 Conceptual Skills

The third skill is conceptual ability. This refers to the mental labour of defining the significance of organizational or policy issues and comprehending what a company stands for and where it is or should be heading. A leader with strong conceptual abilities is at ease discussing the ideas that shape an organization. According to [18], conceptual skills can help leaders reason and make correct decisions. They are good at planning and achieving the organization's goals and can interpret the economic principles that will affect the organization.

4.3.1 Crisis Management Models as Guidelines for Leaders in Handling Crisis

A crisis management model is a conceptual framework that includes all aspects of crisis preparation, prevention, response, and recovery. Leaders and organizations can gain context and become better at applying the best practices by viewing events through a model. They can prepare a plan for confrontation, resilience and recovery from crises based on the model proposed by scholars.

4.3.2 Crisis Management Model Introduced

Burnet and John [9] proposed a crisis management model that consists of steps in dealing with crises. This model suggests goal formulation, environmental analysis, strategy formulation, strategy evaluation, strategy implementation, and strategic control. Leaders with strong conceptual skills can implement these steps appropriate for the nature of the crisis and align with the organizations' policies. A leader with strong conceptual skills can also create a crisis strategy and help the organization respond to the crisis through strategy implementation and strategic control.

Table 4 Relational model of crisis management [42]

Elements in crisis	Activities
Crisis preparedness	Planning
	Preparing a system or manuals as guidelines
	Simulations and training
Crisis prevention	Scanning and early warning
	Management of issues and risks
	Response to an emergency
Crisis incident management	Recognizing a crisis
	Response to system activation
	Crisis management
Post-crisis management	Recuperation and resumption of business
	The post-crisis effects
	Impact assessment and modification

4.3.3 Relational Model of Crisis Management Introduced

Critical processes and activities such as crisis prevention and planning, according to Tony Jacques [42], frequently overlap or occur concurrently, and crisis management is not a linear process with sequential phases. He presented a model that included four major components: crisis planning, crisis prevention, crisis incident management, and post-crisis management. As shown in Table 2, each of these elements is associated with clusters of activities and processes. He concluded that understanding the relationship between these elements and how it relates to the organization's context would reduce crisis-related losses, as shown in Table 4.

4.3.4 Fink's Crisis Model

Figure 1 depicts Steven Fink's four-stage crisis model, including the prodromal, acute, chronic, and resolution stages. The prodromal stage, according to Fink, encompasses the period between the first signs of a crisis and its eruption. During this stage, leaders should proactively monitor, attempt to identify signs of a crisis and find the best solution to prevent or limit the scope of the crisis. The acute stage is the second stage in this model. This stage starts when a trigger sets off the events of a crisis. Crisis managers and their plans are activated during this phase. The chronic stage is the third stage and includes the long-term effects of the crisis, such as after the Covid-19 pandemic, when an organization must analyze how the nature of the crisis-affected their organization as a whole in every way. Finally, the resolution stage denotes the

Fig. 1 Fink's crisis model
[13]

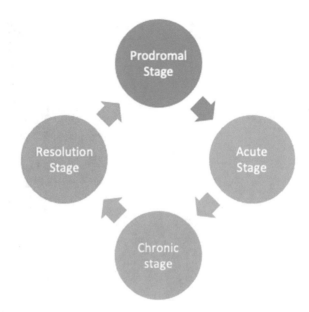

end of the crisis and the need to internalize what was discovered through a root cause analysis and implement changes to ensure that it does not happen again.

4.3.5 Mitroff's Five-Stage Crisis Management Model

Mitroff [26] described five crisis stages. The first two concern signal detection, identifying signs of potential crises within an organization and determining ways to prevent them. They include the proactive steps that leaders and organizations can take before a crisis event. The damage containment stage follows, focusing on the actions taken in the aftermath of the crisis event. The final two stages, known as the recovery and learning stages, aim to get an organization back to its normal operations as soon as possible while also allowing it to incorporate what can be learned from the crisis to develop new policies or guidelines. The flow of this five-stage crisis management model can be simplified as Fig. 2.

5 Conclusion

Even though most organizations are aware of the negative consequences of an organizational crisis, they have failed to provide formal training to their leaders. Because of their leaders' lack of on-the-job learning experience, they tend to fail in crisis management. In today's VUCA world, the organization must ensure that besides having a complete set of guidelines or standard operating procedures in managing

Fig. 2 Five-stage crisis
management model [26]

crisis, they also need to assist their leaders in developing crisis management competencies that will help them effectively manage crises. Learning and development
programs in crisis management also need to be implemented to create a culture of
crisis awareness among employees to ensure all employees, especially leaders are
always ready to face any uncertainty in the future. In this study, the authors agree
with Katz [18] that three main skills, namely the technical, human and conceptual
skills, are very important for leaders in organization to be effective in both a normal
environment and a crisis environment. For technical skills, ICT skills are crucial,
especially in a crisis such as the Covid-19 pandemic. It has been seen that many
organizations and schools implemented remote working and online learning during
this pandemic. For human skills, a situational approach seems to be best suited for
complex situations like a crisis [15]. According to Hersey, this approach is flexible depending on the situation and environment. In regards to conceptual skills, the
authors highlighted three models introduced by well-known scholars. Since conceptual skills involve skills in organizational policy and goal development, it is assumed
that all three models can be guidelines and provide knowledge regarding the multi-
stages of crisis. In addition, the second major issue highlighted was the challenges
and opportunities related to these skills, which can become guidelines for preparation for future crises in an organization. With all of the information, it is hoped that
this study enhances the current literature regarding crisis management, especially
in leadership competencies. It helps policymakers or organizations redesign their

current policies regarding crisis management and assist practitioners in enhancing their competencies, especially in managing crises.

References

1. Arnfalk, P., Pilerot, U., Schillander, P., & Grönvall, P. (2016). Green IT in practice: Virtual meetings in swedish public agencies. *Journal of Cleaner Production, 123*, 101–112.
2. Asio, J. M. R., Riego de Dios, E. E., & Lapuz, A. M. E. (2019). Professional skills and work ethics of selected faculty in a local college. *PAFTE Research Journal, 9* (1), 164–180.
3. Asio, J. M. R. (2020). Effect of performance review and faculty development to organizational climate. *International Journal of Management, Technology and Social Sciences, 5*(2), 1–10.
4. Baruch, Y., & Nicholson, N. (1997). Home, sweet work: Requirements for effective home working. *Journal of General Management, 23*(2), 15–30.
5. Benaden, F., et al. (2016). A metamodel for knowledge management in crisis management: 2016 49th Hawaii International Conference on System Sciences (HICSS), paper read at the 2016 49th *Hawaii International Conference on System Sciences* (HICSS) (pp. 126–135).
6. Black, S. A. (2015). Qualities of effective leadership in higher education. *Open Journal of Leadership, 4*(02), 54.
7. Boin, A., & Van Eeten, M. J. G. (2013). The resilient organization: A critical appraisal. *Public Management Review, 15*, 429–445.
8. Bozkurt, A., et al. (2020). M. A global outlook to the interruption of education due to COVID-19 pandemic: Navigating in a time of uncertainty and crisis. *Asian Journal of Distance Education, 15*(1), 1–126.
9. Burnet, T., & John, J. (1998). A strategic approach to managing crises. *Public Relations Review, 24*(1998), 475–488.
10. Charalampous, M., Grant, C. A., Tramontano, C., & Michailidis, E. (2019). Systematically reviewing remote E-workers wellbeing at work: A multidimensional approach. *European Journal of Work and Organizational Psychology, 28*(1), 51–73.
11. Christopher, & Alvin. (2020). Emergence of a situational leadership during COVID-19 pandemic called new normal leadership. *International Journal of Academic Multidisciplinary Research (IJAMR), 4*(10), 15–19.
12. Dublin. (2020). https://www.eurofound.europa.ea. Last accessed 27 June 2021. Euro found: Living, working and COVID-19: First findings – April 2020.
13. Fink, S. (1986). *Crisis management: Planning for the inevitable*. AMACOM.
14. Francisco, C. D. C., & Celon, L. C. (2020). Teachers' instructional practices and its effects on students' academic performance. *International Journal of Scientific Research in Multidisciplinary Studies, 6*(7), 64–71.
15. Hersey, P. (1985). *The situational leader*. Warner Books.
16. Digitalization and Decent Work: *Implications for Pacific Island Countries*. Suva: ILO Office for Pacific Island Countries (2019).
17. Kapucu, N., & Ustun, Y. (2018). Collaborative crisis management and leadership in the public sector. *International Journal of Public Administration, 41*(7), 548–561.
18. Katz, R. L. (1955). Skills of an effective administrator. *Harvard Business Review, 33*(1), 33–42.
19. Dirani, K. M., Abadi, M., Alizadeh, A., Barhate, B., Capuchino Garza, R., Gunasekara, N., Ibrahim, G., & Majzun, Z. (2020). Leadership competencies and the essential role of human resource development in times of crisis: A response to Covid-19 pandemic, Human Resource International, *Human Resource Development International, 23*(4), 380–394.
20. Khanna, S. (2020). ICT enabled learning—A tool in crisis management. *Aptisi Transactions on Technopreneurship (ATT), 2* (2), 82–93.
21. Knowles, D., Ruth, D. W., & Hindley, C. (2019). Crisis as a plague on organization: Defoe and a journal of the plague year. *Journal of Organizational Change Management, 32*(6), 640–649.

22. Lee, Y. C., Malcein, L. A., & Kim, S.C. (2021). Information and communications technology (ICT) usage during COVID-19: Motivating factors and implications. *International Journal of Environment Research Public Health 2021, 18*, 3571.
23. Limsila, K., & Ogunlana, S. (2008). Performance and leadership outcome correlates of leadership styles and subordinate commitment. *Engineering Construction Architect Management, 15*, 164–184.
24. Al Shobaki, M. J., Abu Amuna, Y. M., Abu Naser, S. S. (2016). Strategic and operational planning as approach for crises management field study on UNRWA. Faculty of Engineering & Information Technology, *Al-Azhar University, Gaza, Gaza-Strip, Palestine.*
25. Micheal and Rachel. (2020). Leading in the COVID-19 crisis: Challenges and solutions for state health leaders. *Journal of Public Health Management Practice, 26*(4):380–383, July/August.
26. Mitroff, I. I. (1994). Crisis management and environmentalism: *A Natural Fit. California Management Review, 36*(2), 101–113.
27. Mitroff, I. I., & Alpaslan, M. C. (2003). *Preparing for Evil.* Harvard Business School Publication.
28. Mitroff, I. I. (2004). Think like a sociopath, act like a saint. *Journal of Business Strategy, 25*(5), 42–53.
29. O' Connell, P. K. (2014). A simplified framework for 21st century leader development. *The Leadership Quarterly, 25*(2), 183–203.
30. Hersey, P., & Blanchard, K. (1982). *Management of organizational behavior: Utilizing human resources.* Prentice-Hall.
31. Charoensukmongkol, P., & Phungsoonthorn, T. (2020). The effectiveness of supervisor support in lessening perceived uncertainties and emotional exhaustion of university employees during the COVID-19 crisis: The constraining role of organizational intransigence. *The Journal of General Psychology.*
32. Vardarlier, P. (2016). Strategic approach to human resources management during crisis. In *12th International Strategic Management Conference, ISMC 2016,* 28–30 October 2016, Antalya, Turkey.
33. Quarantelli, E. L., Lagadec, P., & Boin, A.: A heuristic approach to future disasters and crises: New, old, and in-between types. *Handbook of Disaster Research,* 16–41.
34. Rachmawati, R.; Choirunnisa, U.; Pambagyo, Z. A.; Syarafina, Y. A.; Ghiffari, R. A. (2021). Work from home and the use of ICT COVID-19 pandemic in Indonesia and its impact on cities in the future. *MDPI Sustainability, 13*, 6760.
35. Ramakrishnan, R. (2021). Leading in a VUCA world. *Ushus Journal Business Management, 20*(1), 89–11.
36. Schwantes, M. (2021). 4 Signs to instantly identify a great leader during crisis. Inc., March 24. Retrieved June 27, 2021 from https://www.inc.com/marcel-schwantes/great-leader-time-of-crisis.html.
37. Sherman, W. S., & Roberto, K. J. (2020). Are you talkin' to me? The role of culture in crisis management sense-making. *Management Decision, 58*(10), 2195–2211.
38. Soleiman et al. (2020). Transition to virtual learning during the coronavirus disease–2019 crisis in Iran: Opportunity or challenge? *Society for Disaster Medicine and Public Health,* Inc..
39. Strielkowski, W., Firsova, I., Lukashenko, I., Raudeliu¯niene˙, J., Tvaronavic˘iene˙, M. (2021). Effective management of energy consumption during the COVID-19 pandemic: *The Role of ICT Solutions. Energies* 2021, 14, 893.
40. Erwin, S. (2020). Leadership during the COVID-19 pandemic: Building and sustaining trust in times of uncertainty. *Sustainability, 13*(5).
41. Sutton, A., & Watson, S. (2013). Can competencies at selection predict performance and development needs? *Journal of Management Development, 32*(9), 1023–1035.
42. Jacques, T. (2007). Issue management and crisis management: An integrated, non-linear, relational construct. *Public Relations Review.* June 2007, 147–157.
43. United Nations Education Scientific and Cultural Organization. Retrieved June 27, 2021 from https://www.un.org: COVID-19 *Educational Disruption and Response.*

44. Van Wart, M., & Kapucu, N. (2011). Crisis management competencies: The case of emergency managers in the USA. *Public Management Review, 13*(4), 489–511.
45. Wisittigars, B., & Siengthai, S. (2019). Crisis leadership competencies: The facility management sector in Thailand. *Facilities, 37*(13–14), 881–896.
46. Wooten, L. P., & James, E. H. (2004). When firms fail to learn: The perpetuation of discrimination in the workplace. *Journal of Management Inquiry, 13*(1), 23–33.

Utilization of Digital Network Learning and Healthcare for Verbal Assessment and Counselling During Post COVID-19 Period

Ravi Kumar Gupta

1 Introduction

Throughout the continuing Coronavirus pandemic, hospitals have been on the verge of being overburdened by the number of patients who have been seriously ill. People have been encouraged to self-care for COVID-19 symptoms at home as much as feasible. As a result, more practical techniques to assisting medical workers in the diagnosis and triage of patients are urgently needed. Various methods, including natural language processing techniques, have lately been developed, for example [1–3]. Furthermore, in China, social media search queries linked to COVID-19 symptoms have been demonstrated to be a good predictor of infection numbers [4].

Utilizing therapist healthcare online posts, we propose a later part Natural Language Processing process for independently misdiagnosing and classifying COVID-19 instances in this paper. The triage may help decision-makers understand the severity of COVID 19, and the diagnosis may aid in determining the incidence of infections in the general community. The creation of a high-quality human-labelled dataset on which to base our workflow is a major challenge. We'll go over our process and how we built our dataset in the following sections.

R. K. Gupta (✉)
Humanities and Management Science Department, Madan Mohan Malaviya University of Technology, Gorakhpur, India

© The Author(s), under exclusive license to Springer Nature Switzerland AG 2022　　　117
A. Hamdan et al. (eds.), *Technologies, Artificial Intelligence and the Future of Learning Post-COVID-19*, Studies in Computational Intelligence 1019,
https://doi.org/10.1007/978-3-030-93921-2_7

The initial phase in the pipeline is to create an annotation application that detects and emphasizes COVID-19-related symptoms in a social media post, together with their intensity and length, which will be referred to as concepts. Relationships between symptoms and other relevant ideas are automatically found and annotated in the second stage. Breathing hurts, for example, is a symptom associated with the upper chest area of the body.

As illustrated in Fig. 1, one author manually tagged our data with ideas and relations, allowing us to provide postings with highlighted concepts and relations to three experts, along with many questions. The first question asked the experts to categorize a patient into one of three groups: stay at home, see a doctor, or go to the hospital. The second question used a Likert Scale of 1 to 5 to diagnose the likelihood of COVID-19 [5].

The three specialists are UK-based foundation doctors who were reassigned to COVID-19 wards during the first wave of the pandemic, which occurred between March and July 2020. Their responsibilities included diagnosing and managing COVID-19 patients, including those who were extremely ill and required non-invasive or invasive ventilation. For doctors working on COVID-19 wards, some

---

Hi im currently the same day 27 since my symtoms started , deep breathing hurts [upper chest area][throat] which is upper chest area into throat , breathing [slightly][laboured] is slightly laboured time to time , dry cough on and off , also have major fatigue weakness took a course of Amoxcillian given by GP which made no change to me , have asthma so take my inhalers which aint making no change , never been so unwell in my life ! ! !

Question 1: Please specify recommendation from one of the options below:

○ Stay at home

○ Send to a GP

○ Send to hospital

Question 2: How would you rate the chance of this person having COVID-19 on a range of 1 to 5?

○ 1 (Very unlikely)

○ 2 (Unlikely)

○ 3 (Uncertain)

○ 4 (Likely)

○ 5 (Very likely)

Question 3: Was the highlighted text sufficient in reaching your decision?

○ Yes

○ No

Fig. 1 Symptoms (light green), afflicted body areas (pale blue), duration (light yellow), and severity levels are marked on a patient-authored social media post (pink). When the distance between a symptom and a body part/duration/severity was more than 1, the sentences in square brackets suggest relationships between the two. Three doctors were given this annotated post to triage and diagnose the post's author by answering Questions 1 and 2

training courses were organized. However, because there was little information about the virus and how to treat it, these were only delivered near the conclusion of the first wave. A lot of learning took place "on the fly," with doctors learning from previous experience and pattern identification of a collection of symptoms, which is similar to machine learning. The doctors used a similar method to analyze the medical forum posts they were given. We also asked the doctors if the highlighted text supplied was helpful to determine its use when we included it in the annotation interface. According to the doctors' responses to Question 3 in Fig. 1, the annotations were judged to be useful in as much as 85 percent of the posts.

The doctors' labels were then utilized to create two types of predictive machine learning models using Support Vector Machines (SVM) [6, 7], as described in Step 4 of the Methods section. The triage models use hierarchical binary classifiers that consider the doctors' risk aversion or tolerance [8]. The diagnostic algorithms begin by calculating the likelihood that a patient has COVID-19 based on doctor ratings. The probabilities are then used to create three alternative decision functions for classifying COVID and NO COVID classes described in the Methods section of the Problem setup chapter.

We used two separate methods to train the SVM models: ground truth annotations and predictions from the concept and relation extraction process described above. Conditional Random Fields (CRF) [9] are used to make predictions from the concept extraction process; see Step 1 of the Methodology Sub Section in Methods for implementation details. An unsupervised Rule-Based (RB) classifier [10] is used to extract relations from these anticipated concepts; see Step 2 in the Methods section.

We also compare the most common symptoms from [3] and our dataset to the feature importance acquired from the developed COVID-19 diagnostic models. Anosmia/ageusia (loss of taste and smell) were determined to be among the top five most important features but not among the top five most common symptoms; see Discussion. Overall, we make the following contributions:

1. We show that using natural language narratives from patients, machine learning models may be built to triage and diagnose COVID-19. No previous research has attempted to triage or diagnose COVID-19 from social media posts, to our knowledge.

2. We also use automated idea and relation extraction to create an end-to-end NLP pipeline. Our investigations show that models created using concept and relation extraction predictions give results comparable to those created utilizing ground-truth human concept annotation. Evident in the mobile and pervasive computing, including (Fig. 1).

2 Background

Classification and Regression Trees have previously been used to estimate the likelihood of symptom severity of respiratory infections such as influenza in clinical settings [11]. Decision trees and other machine learning methods have shown promise in detecting COVID-19 from blood test findings [12].

Experiment with COVID-19 diagnosis utilizing Linear Kernel SVR regression to identify the important predictive variables in the training set. We employed the Symptom-Only Vector Representation, which was built from the ground truth. We averaged feature weights from seven models and three decision functions for each si in s0, s1,..., sn; see Methods. After that, the traits are mapped to the categories present in [3]'s Twitter COVID-19 lexicon. Cough, Anosmia/Agusia, Dyspnoea, Pyrexia, and Fatigue are the top five critical features in our dataset. [22] identified four of these symptoms as the most common Coronavirus symptoms, which is consistent with our findings.

We calculated the matching frequencies of our 5 most essential symptoms to compare our importance ranking with that of Sarker et al. [3]'s common categories. In Fig. 2, the normalized weights and frequencies are displayed. The stacked bar chart on the upper left compares our five most critical qualities to Sarker's frequencies. Cough is the most common symptom in our sample, and it is also the most essential. Anosmia/Aguesia is ranked second in importance and seventh in frequency on our list. Pyrexia is ranked first in both the frequency and importance categories.

Figure 2 displays a comparison between Sarker's frequent ranking and our importance rating in the top-right chart. We've taken the top five most common symptoms from Sarker's list and normalized them. Pyrexia, Cough, Body Ache, Fatigue, and Headache are the symptoms. We calculated the significance weights of those symptoms and compared them. The stacked bar chart was created. Headache is ranked 22nd in our importance ranking and 5th in our frequency ranking. There is a significant disparity between the two rankings, showing that the most common symptoms are not the same. Those aren't always the most significant.

Then, using the methods outlined above, we compare our most essential feature weights to the frequency ranking of our dataset. Figure 2's bottom-left stacked bar chart shows that Anosmia/Aguesia is ranked 11th in terms of frequency. Cough ranks second in our dataset's frequency ranking, much like Sarker's.

Finally, the bottom-right chart in Fig. 2 compares the frequency and relevance rankings of the related symptoms in our dataset. Anxiety is ranked 4th in the most common list. However, it is ranked quite low in the most important list, i.e. 23rd.

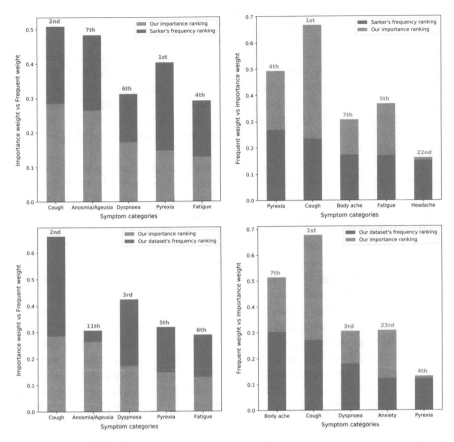

Fig. 2 Comparing our most significant aspects to Sarker's most common symptoms (top row) and our most important traits to our most frequent symptoms (bottom row) (bottom row). The feature significance rankings are calculated using the symptom-only vector representation and an SVM linear kernel.

3　Related Work

In existing, the findings are based on the experiment with COVID-19 diagnosis utilizing Linear Kernel SVR regression to identify the important predictive variables in the training set. Employed the Symptom-Only Vector Representation, which was built from the ground truth. The ground truth labels are always used to test trained models. In the proposed study, the findings show that social media posts can be used to triage and diagnose COVID-19 patients. Using machine learning models to forecast a patient's health status by utilizing ground truth labels and using predictions from the NLP pipeline. Automated NLP pipeline used to triage and diagnose patients. Health care providers and researchers could use triage models to assess the severity of COVID-19 cases in the population and diagnostic models to estimate the pandemic's prevalence.

This research aims to extract as much usable information as possible from the loud world of social networking. Posts on medical social media, in particular, may give an additional source of information about the severity of symptoms and the incidence of COVID-19 in the general community. To triage and diagnose COVID-19, we focus on features collected from a textual source. Features from online symptom tracker applications, telehealth visits, and structured and unstructured patient/doctors notes from Electronic Health Records are used in the research connected to our work (EHR). COVID-19 clinical prediction models can be divided into risk, diagnosis, and prognosis [13].

Various mobile app-based self-reported symptom tools have emerged since the onset of the COVID-19 epidemic to follow emerging symptoms [14]. The mobile app in [15] used Logistic Regression (LR) to forecast the percentage of likely infected patients among all app users in the United States and the United Kingdom combined.

In [16], patients were divided into four risk groups using a portal-based COVID-19 self-triage and self-scheduling tool: emergent, urgent, no-urgent, and self-care. By combining demographic data, clinical symptoms, blood tests, and computed tomography (CT) scan findings, the online telemedicine system in [17] was able to identify low, moderate, and high-risk patients.

Schwab et al. [2] created several machine learning models to predict patient outcomes based on clinical, laboratory, and demographic data from EHR [18]. Gradient Boosting (XGB), Random Forest, and SVM were found to be the most effective models for predicting COVID-19 test results, as well as hospital and ICU admissions for positive patients. [19], An ensemble of XGB models to construct predictive models for inpatient mortality in Wuhan provides a complete list of clinical and laboratory variables. Similarly, XGB classifiers were used to predict death and critical events for patients in [20, 21]. Finally, [13] provides a critical analysis of several COVID-19 diagnostic and predictive models used in clinical settings.

COVID-19 symptoms were retrieved from unstructured severe observations in the EHR of patients who had COVID-19 PCR testing. The authors of [22] used primary care EHR data to conduct a statistical analysis to determine the longitudinal dynamics of symptoms before and during infection. COVID-19 SignSym [23] was also created to extract symptoms and related attributes from free text automatically. Furthermore, to diagnose COVID-19, the study in [24] uses radiological text reports from lung CT scans. Lopez et al. [24] used a common medical ontology [25] to extract concepts and then used word embeddings [26] and concept vectors [24] to generate a document representation. However, our methodology differs from theirs in terms of extracting relationships between concepts. Our dataset, which includes postings from medical social media, is more difficult to work with because social media posts have more language variety than radiological text reports.

Finally, we compare our work to a COVID-19 symptom lexicon collected from Twitter [3].

4 Methods

4.1 Data

From a community called patient [27], we gathered social media messages about COVID-19 medical issues. This is a public forum established at the start of the Coronavirus outbreak in the United Kingdom. We gained permission from the site administrator and gathered posts from April to June of 2020. After the posts were anonymized and duplicates were deleted, we chose 500 unique posts at random. The initial author annotated these postings with the classes displayed in Fig. 3. The class labels reflect symptoms and related notions such as I length, (ii) intensifier, which enhances symptom severity, (iii) severity, (iv) negation, which indicates whether the symptom or severity is present or absent, and (v) afflicted body parts. At the phrase level, we additionally identified relationships between symptoms and other concepts. The relationship between a symptom and a severity idea, for example, is denoted as (SYM, SEVERITY). As illustrated in Fig. 1, the posts were then labelled with concepts in various colours, and the relations were inserted right after the symptom in square brackets. A web program was used to deliver each marked post to the doctors, and they were each asked three questions separately; see Fig. 1. The responses to questions 1 and 2 are referred to as the COVID-19 symptom triage and diagnosis, respectively. As a result, we have three distinct responses for each post.

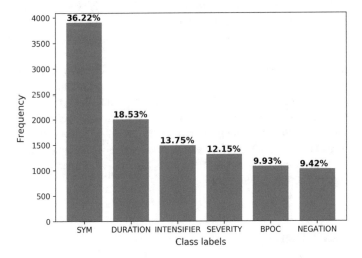

Fig. 3 The frequency distribution of the text's annotated classes/concepts are shown. Symptoms, duration, intensifiers, severity, body parts, and negations are represented by SYM, DURATION, INTENSIFIER, SEVERITY, BPOC, and NEGATION. After discounting the OTHER labels, we additionally present the percentage of each class. The average number of tokens per post is 130.17, with a standard deviation of 97.83. From three doctors, which we denote as A, B, and C, respectively, these correspond to the last three authors of the paper and have been assigned randomly

Table 1 For Questions 1 and 2, a pair-wise agreement between pairs of doctors' answers is shown in Fig. 1 as an example. Linear adjusted, Quadratic, Quadratic adjusted weights, and Correlation are acronyms for Linear adjusted, Quadratic, Quadratic adjusted weights, and correlation, respectively

Question1					Question 2					
Pair	p_o	Linear	Quadratic	Correlation	p_o	Linear	Linear Adj	Quad	Quad. Adj	Corr
		K	K	Pearson		κ	κ	κ	K	Pearson
AB	0.77	0.35	0.46	0.48	0.72	0.21	0.38	0.33	0.39	0.42
AC	0.74	0.30	0.34	0.39	0.71	0.12	0.22	0.22	0.24	0.41
BC	0.80	0.17	0.21	0.29	0.74	0.27	0.45	0.4	0.37	0.45

4.2 Measurement of Agreement

We used Cohen's [28] and Pearson's correlation coefficient [29] to measure the agreement and association between the three doctors' answers (recommendations and ratings) to Q1 and Q2 of Fig. 1; the strength of agreement for different ranges of and the interpretation of the correlation coefficient are respectively given. For the three pairs: AB, AC, and BC, the correspondence of answers is measured. In the case of linear weighted, we notice a poor agreement between BC for Q1 and the same for AC for Q2; see Table 1. We calculated the quadratic weighted average for both questions, which resulted in fair agreements for all pairs, as shown in Table 1. We changed the linear and quadratic weights for Q2 to distinguish between neighbouring diagnoses, resulting in a good linear adjusted weighted agreement for AC of Q2. We see relationships ranging from weak to moderate in the correlations between the pairs [29].

Using the conventional measurements of Cohen's and Pearson's correlation, we found no significant agreement between doctors' triage and diagnosis. This is most likely because COVID-19 is a novel virus for which doctors had no prior experience or training before the pandemic's initial wave. Furthermore, there are likely to be variances in risk tolerances among the clinicians, resulting in potentially divergent conclusions and diagnoses. We notice a high result when we calculate the proportion of observed agreement, po, as proposed by [30, 31], who specify a measure of dependability rather than agreement.

5   Problem Setting

5.1   Triage Classification for Question 1

The alternatives Stay at home, Send to a GP, and Send to hospital are translated to the values 1, 2, and 3, respectively, by mapping the doctors' recommendations from Q1 to ordinal values. We take the average of two or more doctors' recommendations before combining them. This number can be rounded to an integer in two ways: taking the floor or taking the ceiling. We classify the average ceiling as risk-averse, denoted by AB(R-a), and the average floor as risk-tolerant, designated by AB(R-b) (R-t). As a result, for Q1, we have eleven recommendations from three doctors for each patient's post. For each of these recommendations, we create a hierarchical classification model to classify a post into one of three categories.

5.2 Diagnosis Classification for Question 2

To diagnose whether a patient has COVID-19 from his or her post, we first estimate the probability of having the disease by normalizing the rating, i.e. given a rating, r, the probability of COVID-19, Pr(COVID r), which we term as the ground truth probability (abbreviated GTP), is:

$$Pr(COVID|r) = \frac{r-1}{4}$$

Given our ground truth probability estimates are discrete we investigate three decision boundaries based on a threshold value of 0.5 to classify a post as follows:

LE: If $Pr(COVID|r) <= 0.5$, then *NO_COVID*, else *COVID*. *LT*: If $Pr(COVID|r)$ < 0.5, then *NO_COVID*, else *COVID*. *NEQ*: If $Pr(COVID|r) < 0.5$ then *NO_COVID*, else if $Pr(COVID|r) > 0.5$ then *COVID*.

Note NEQ differs from the other decisions in that we ignore those cases on the 0.5 boundaries.

6   Methodology

Figure 4 depicts a diagram of our strategy for triaging and diagnosing patients based on their social media posts. The circles in this diagram represent the pipeline steps. We'll go through each of these processes in more depth now.

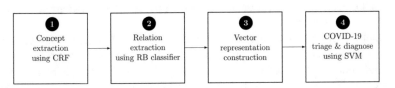

Fig. 4 A block schematic of the COVID-19 text processing pipeline for triage and diagnosis. Conditional Random Fields, Rule-Based Classifier, and Support Vector Machine are the acronyms for Conditional Random Fields, Rule-Based Classifier, and Support Vector Machine

Step 1: Excavation of concepts

In the first phase, we use the GATE software's built-in Natural Language Processing (NLP) pipeline to separate each patient's post into sentences and tokens. We create discrete features for each token in a sentence that indicate if the token belongs to one of the following dictionaries: I Symptom, (ii) Severity, (iii) Duration, (iv) Intensifier, and (v) Negation. The dictionaries were created by analyzing and annotating the posts. We also employ the MetaMap method [25] to map tokens to three helpful semantic categories: sign or symptom, disease or syndrome, and body part, organ, or organ component, providing that it contains all of the essential technical terminologies. The system does not expect any new additional terms due to the assumption regarding medical terms. Thus, we are justified in extracting ideas and relations in pre-processing phases. The pre-processed text is then utilized to create an idea extraction module that uses a CRF [9] to recognize the classes, as Fig. 3. Manning and Schutze [32] contains a full discussion of our CRF training approach. The retrieved concepts are then employed in the next stage, which is to recognize relationships between them.

Step 2: Excavation of Connection

An unsupervised RB classification technique is used to clarify the semantic relationship between a symptom and additional concepts, which we formally refer to as modifiers. To establish a relation, we first filter all symptom and modifier pairings from a phrase within a preset distance, then select the closest modifier to a symptom. We found five different types of relationships: (SYM, SEVERITY), (SYM, DURATION), (SYM, BPOC), (SYM, NEGATION), and (SYM, ?).

On a scale of 1 to 5, the severity modifiers are mapped. The scale's semantic meaning is as follows: very mild, mild, moderate, severe, and very severe. The duration modifiers are also linked to real-world values in weeks' worth of time. For example, the value 1.42 is mapped to 10 days.

Step 3: Interpretation in dimensions

Fixed length vector representations suitable as input for SVM classifiers are built as follows.

Symptom-only vector representation: Let $(s0, s1..., s_n)$ be a The number of unique symptom words/s in our dataset is n = 871, and the vector of symptoms is created from the symptom vocabulary. We extract the concept, SYM, and the relation (SYM, NEGATION) and set si to 1, 0, or −1, depending on whether the symptom is present, not present, or negated.

Symptom-modifier relation vector representation: The symptom-modifier relation vector is much larger than the symptom-only relation vector, and it contains three appended vectors that contain I the absence or presence of 110 unique body parts, (ii) the absence of value of symptom duration, and (iii) the absence, negation, or value of symptom severity.

Step 4: Triage and diagnosis

We use SVM classification and regression models to triage and diagnose patients' posts from the vector representations given above. For Q1, the class name of the post is the advice from a doctor or a group of doctors; for an explanation, see the section Problem setup in Methods. To create a binary classifier, we first aggregate the Send to a GP and Send to Hospital recommendations into a single class called Send. We call this SVM Classifier 1 because it has been trained to discriminate between the Stay at home and Send options. The posts labelled as Stay at home are then removed, and SVM Classifier 2 is developed to categorize the Send to GP and Send to hospital suggestions using the remaining posts. For COVID-19 triage, this results in a hierarchical classifier.

We use SVM variation called Support Vector Regression (SVR) [6] to predict COVID-19 instances' probability of diagnosing COVID-19 cases. As explained above, SVR takes a high-dimensional feature vector as input, such as a symptom-only or symptom-modifier relation vector representation. As the dependent variable, we utilize the GTP calculated from the responses to Q2. The three decision functions discussed previously, LE, LT, and NEQ, are used to classify the data.

7 Results

7.1 Evaluation

We use the conventional measurements of precision (P), recall (R), and Macro- and Micro-averaged F1 scores to evaluate the performance of the CRF and SVM classification algorithms [33]. Micro-averaged scores are computed by examining all of the classes together, whereas macro-averaged scores are derived by considering each class separately and then taking the average. We give Micro-averaged scores for SVR classification because our dataset is generally balanced with COVID and NO COVID classes. On the other hand, the other class has the upper hand when it comes to concept extraction. As a result, for the CRF classification results, we report the Macro-averaged scores.

7.2 Experimental Setup

We present threefold cross-validated Macro-averaged results for the CRF. We used a Python wrapper [34] for CRFsuite to train each fold, as shown in [35]. We used our unsupervised rule-based system to extract relationships from 500 posts, and we calculated F1 scores by altering distances in the two situations with and without stop words.

We built SVM binary classifiers, SVM Classifier 1 and SVM Classifier 2, with both Linear and Gaussian Radial Basis Function (RBF) kernels [7], using the Python wrapper for LIBSVM [36] provided in Sklearn [37]. Similarly, Linear and RBF kernels are used in the SVR [38], which is implemented using LIBSVM. Grid search was used to find the hyperparameters (C = 10 for the penalty, = 0.01 for the RBF kernel, and s = 0.5 for the threshold).

For COVID-19 triage and diagnosis, we created two scenarios. The predictive performance of SVM and SVR models trained using the ground truth is first examined when they are deployed as stand-alone applications. Second, they resemble an end-to-end NLP application when trained using CRF and RB classifier predictions. The models were always evaluated against the ground truth to ensure a comparable result. We report Macro and Micro-averaged F1 scores for SVM classifiers and SVR, respectively, to measure performance.

7.3 Evaluation Outcomes

Table 2 shows the results of the idea and relation extraction stages, which give excellent and very good prediction performance. Table 3 shows the triage classification findings from Q1. The results of SVM Classifier 1 and 2, when trained using the Symptom-modifier vector representations from the ground truth, are in the range of 72–93 percent and 83–96 percent, respectively. The results of the symptom-only vector representations vary from 71 to 94 percent and 79 to 95 percent, respectively. These findings imply that for classifying Stay at home and Send and Send to a GP and Send to hospital, we can get extremely strong predictive performance.

Table 2 Three-fold cross-validation was performed on the idea extraction using CRF and the relation extraction using the RB classifier

Concept extraction using CRF					Relation extraction using RB classifier						
						With stop words			Without stop words		
Label	P	R	F_1	Support	Distance	P	R	F_1	P	R	F_1
SYM	0.94	0.97	0.95	1300	2	0.74	0.63	0.68	0.74	0.64	0.69
SEVERITY	0.80	0.79	0.79	437	3	0.75	0.67	0.71	0.75	0.67	0.71
BPOC	0.92	0.83	0.87	356	4	0.75	0.69	0.72	0.75	0.69	0.72
DURATION	0.87	0.91	0.89	667	5	0.75	0.71	0.73	0.74	0.71	0.73
INTENSIFIER	0.88	0.97	0.92	494	6	0.74	0.72	0.73	0.74	0.72	0.73
NEGATION	0.83	0.89	0.86	338	7	0.73	0.73	0.73	0.73	0.73	0.73
OTHER	0.99	0.98	0.98	16,892							
Macro-average	0.89	0.89	0.89								

Table 3 Problem 1: Using the symptom-modifier relation vector, hierarchical classification results for the RBF kernel

Symptom-modifier relation vector

	Trained on the ground truth						Trained on the CRF predictions					
Model	SVM Classifier 1			SVM Classifier 2			SVM Classifier 1			SVM Classifier 2		
	P	E	F_1	P	R	F_1	P	R	F_1	P	R	F_1
A	0.82	0.91	0.86	0.73	0.95	0.85	0.81	0.89	0.85	0.72	0.91	0.80
B	0.73	0.77	0.75	0.81	0.99	0.89	0.74	0.74	0.74	0.81	0.99	0.89
C	0.85	0.98	0.91	–	–	–	0.85	0.96	0.90	–	–	–
AB(R-a)	0.70	0.75	0.72	0.80	0.96	0.88	0.73	0.71	0.71	0.81	0.96	0.88
AB(R-t)	0.84	0.96	0.89	0.85	1.00	0.92	0.84	0.94	0.88	0.84	1.00	0.92
BC(R-a)	0.72	0.75	0.73	0.92	1.00	0.96	0.74	0.71	0.72	0.92	1.00	0.96
BC(R-t)	0.86	0.99	0.92	–	–	–	0.88	0.98	0.93	–	–	–
AC(R-a)	0.79	0.87	0.83	0.89	1.00	0.94	0.81	0.85	0.83	0.89	1.00	0.94
AC(R-t)	0.88	0.98	0.93	–	–	–	0.88	0.98	0.93	–	–	–
ABC(R-a)	0.70	0.76	0.73	0.89	0.99	0.93	0.72	0.72	0.72	0.89	1.00	0.94
ABC(R-t)	0.88	0.99	0.93	–	–	–	0.89	0.98	0.93	–	–	–

	Trained on the ground truth						Trained on the CRF predictions					
Model	SVM Classifier 1			SVM Classifier 2			SVM Classifier 1			SVM Classifier 2		
	P	R	F_1	P	R	F_1	P	R	F_1	P	R	F_1
A	0.83	0.91	0.87	0.74	0.85	0.79	0.84	0.89	0.87	0.74	0.82	0.78
B	0.71	0.81	0.76	0.81	0.98	0.89	0.74	0.79	0.77	0.82	0.98	0.89
C	0.87	0.97	0.92	–	–	–	0.86	0.95	0.90	–	–	–
AB(R-a)	0.69	0.75	0.72	0.83	0.96	0.89	0.72	0.76	0.73	0.83	0.92	0.87
AB(R-t)	0.85	0.94	0.89	0.85	1.00	0.92	0.87	0.93	0.90	0.84	0.98	0.90
BC(R-a)	0.71	0.79	0.75	0.92	0.99	0.95	0.72	0.78	0.75	0.92	0.99	0.95
BC(R-t)	0.88	0.98	0.93	–	–	–	0.87	0.97	0.92	–	–	–
AC(R-a)	0.80	0.86	0.83	0.89	1.00	0.94	0.80	0.86	0.83	0.89	1.00	0.94
AC(R-f)	0.90	0.98	0.94	–	–	–	0.89	0.95	0.92	–	–	–
AEC(R-a)	0.68	0.74	0.71	0.90	1.00	0.95	0.71	0.76	0.73	0.89	0.99	0.93
ABC(R-t)	0.90	0.98	0.94	–	–	–	0.90	0.95	0.92	–	–	–

Risk-tolerant models, on average, outperform risk-averse models in terms of performance. However, we cannot report some models since posts with the label Send to hospital are absent in the test set (as shown in Fig. 4). Because Q1 is framed as a choice issue with a priori equal weights for the classes, we report Macro-averaged F1 results. For both representations and classifiers, the results obtained after training with CRF predictions are in similar ranges. This is significant since it implies that an end-to-end NLP application will give similar predicted results.

Regarding Q2, the outcomes of COVID-19 diagnosis are in the range of 72–87 percent, 61–76 percent, and 74–87 percent for the LE, LT, and NEQ decision functions, respectively, when we trained the models with the Symptom-modifier vector representation from ground truth; see Table 4. The symptom-only vector representation yields 70–88 percent, 59–79 percent, and 74–87 percent outcomes, respectively. The exclusion of questionable cases where the GTPs are exactly 0.5 improves the performance of NEQ models in general. Figure 5 shows the support ratios for each model for various decision functions. When we used the Symptom-modifier vector form from the CRF predictions to train the models, the results for the LE, LT, and NEQ decision functions were 68–86 percent, 64–76 percent, and 73–87 percent, respectively (Fig. 6). This means that an end-to-end NLP application will likely perform similarly to stand-alone programs in diagnosis and triage. We present Micro-averaged F1 scores because NO COVID cases predominate in our dataset; this closely mimics the natural distribution in the population, where people who test positive for Coronavirus make up a small percentage of the total population even when the virus's incidence is high.

Finally, we used a Linear kernel to train our models, but we discovered that RBF prevails in most circumstances; however, Linear kernels effectively determine feature importance [39].

8   Conclusion

This study shows that social media posts can be used to triage and diagnose COVID-19 patients. In the existing method, the findings are based on the experiment with COVID-19 diagnosis utilizing Linear Kernel SVR regression to identify the important predictive variables in the training set. Employed the Symptom-Only Vector Representation, which was built from the ground truth. The ground truth labels are always used to test trained models. In the proposal, we demonstrated a proof-of-concept system that uses machine learning models to forecast a patient's health status. Narrative. The models are trained in two ways: I utilizing ground truth labels and (ii) using predictions from the NLP pipeline. In both situations, we got good results, indicating that an automated NLP pipeline may be used to triage and diagnose patients based on their narrative; see the Evaluation results in the Results section for more information. In general, health care providers and researchers could use triage models to assess the severity of COVID-19 cases in the population and diagnostic models to estimate the pandemic's prevalence. There are two major drawbacks to our manually annotated dataset. Because we only have three specialists, the quality of our labeling is limited, even though we consider this study to be a proof of concept. In a follow-up study, a larger number of specialists, particularly more senior doctors, might be useful. The robustness of our findings could be further enhanced by expanding our dataset and including posts from a variety of different sources. It's worth repeating that social media posts, notorious for being noisy, are not comparable to a patient's appointment with a doctor. Because the posts are from social media, it's unclear

Table 4 Problem 2: Micro-averaged F1 outcomes for various models and decision functions (Question 2). A, B, and C are three medical professionals (abbreviated as Dr) participating in the study

Model	Trained on the ground truth						Trained on the CRF predictions					
	Symptom-modifier			Symptom-only			Symptom-modifier			Symptom-only		
	LE	LT	NEQ	LE	LT	NEQ	LE	LT	NEQ	LE	LT	NEQ
A	0.72	0.61	0.78	0.70	0.59	0.74	0.68	0.64	0.76	0.50	0.79	0.74
B	0.78	0.61	0.76	0.73	0.62	0.77	0.76	0.64	0.77	0.78	0.57	0.74
C	0.87	0.75	0.87	0.88	0.75	0.87	0.86	0.75	0.87	0.87	0.74	0.86
AB	0.72	0.66	0.74	0.74	0.65	0.75	0.70	0.65	0.73	0.71	0.66	0.74
BC	0.84	0.76	0.84	0.85	0.79	0.86	0.83	0.76	0.83	0.85	0.78	0.86
AC	0.81	0.73	0.81	0.83	0.74	0.83	0.80	0.74	0.82	0.80	0.73	0.81
ABC	0.74	0.67	0.76	0.75	0.67	0.77	0.72	0.69	0.76	0.74	0.69	0.77

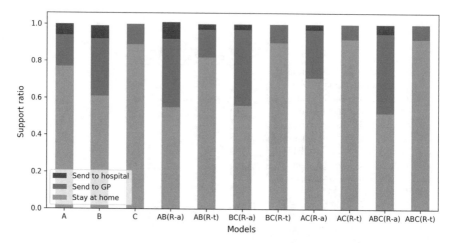

Fig. 5 For Problem 1 classification tasks, the support ratio of triage classes varies between models. In test sets, the absolute values for the Send to hospital class are as follows: A = 10, B = 12, AB(R-a) = 14, AB(R-t) = 5, BC(R-a) = 6, CA(R-a) = 5, ABC(R-a) = 9; the remaining models have a value of zero

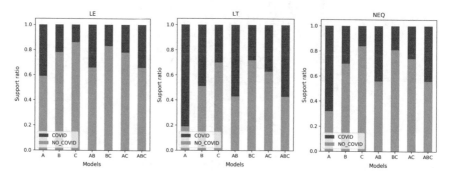

Fig. 6 For Relevant queries classification tasks, the support ratio of diagnosis classes across models and three decision functions were used

whether the results may be used in a diagnostic system without being combined with actual consultations. It's worth mentioning, though, that medical social media, such as the posts we used here, could reveal new information about COVID-19. The pandemic of the Coronavirus has brought attention to the need for automated methods to offer extra data to academics, health professionals, and decision-makers. Medical social media offers a wealth of timely information that could be used for this purpose, such as indicating the severity of a condition and estimating its prevalence in the community using outputs from our triage and diagnosis system. Despite the varied character of ordinary social media messages, we have proved that an automated triage and diagnosis system may be created.

References

1. Obeid, J. S., Davis, M., Turner, M., et al. (2020). An artificial intelligence approach to COVID-19 infection risk assessment in virtual visits: A case report. *Journal of the American Medical Informatics Association, 27*(8), 1321–1325.
2. Schwab, P., DuMont Schütte, A., Dietz, B., & Bauer, S. (2020). Clinical predictive models for COVID-19: Systematic study. *Journal of Medical Internet Research, 22*(10), e21439.
3. Sarker, A., Lakamana, S., Hogg-Bremer, W., et al. (2020). Self-reported COVID-19 symptoms on Twitter: An analysis and a research resource. *Journal of the American Medical Informatics Association, 27*(8), 1310–1315.
4. Qin, L., Sun, Q., Wang, Y., et al. (2020). Prediction of the number of 2019 novel coronavirus (COVID-19) using social media search index. *International Journal of Environmental Research and Public Health, 17*(7), 2365.
5. Norman, G. (2010). Likert scales, levels of measurement and the "laws" of statistics. *Advances in Health Sciences Education, 15*(5), 625–632.
6. Drucker, H., Burges, C. J., Kaufman, L., et al. (1996). Support vector regression machines. *Advances in Neural Information Processing Systems, 9*, 155–161.
7. Marsland, S. (2014). *Machine learning: An algorithmic perspective* (second ed.). CRC.
8. Arrieta, A., García-Prado, A., Gonzalez, P., & Pinto-Prades, J. L. (2017). Risk attitudes in medical decisions for others: An experimental approach. *Health Economics, 26*, 97–113.
9. Sutton, C., & McCallum, A. (2012). An introduction to conditional random fields. *Found Trends Mach Learn, 4*(4), 267–373.
10. Bach, N., & Badaskar, S. (2007). A review of relation extraction. *Literature Review Language Statistics, II*(2), 1–15.
11. Zimmerman, R. K., Balasubramani, G., Nowalk, M. P., et al. (2016). Classification and Regression Tree (CART) analysis to predict influenza in primary care patients. *BMC Infectious Diseases, 16*(1), 1–11.
12. Brinati, D., Campagner, A., Ferrari, D., et al. (2020). Detection of COVID-19 infection from routine blood exams with machine learning: A feasibility study. *Journal of Medical Systems, 44*(8), 1–12.
13. Wynants, L., Van Calster, B., Collins, G. S., et al. (2020). Prediction models for diagnosis and prognosis of COVID-19: systematic review and critical appraisal. *British Medical Journal, 369*.
14. Zens, M., Brammertz, A., Herpich, J. et al. (2020). App-based tracking of self-reported COVID-19 symptoms: Analysis of questionnaire data. *Journal of Medical Internet Research, 22*(9):e21956.
15. Menni, C., Valdes, A. M., Freidin, M. B., et al. (2020). Real-time tracking of self-reported symptoms to predict potential COVID-19. *Nature Medicine, 26*(7), 1037–1040.
16. Judson, T. J., Odisho, A. Y., Neinstein, A. B., et al. (2020). Rapid design and implementation of an integrated patient self-triage and self-scheduling tool for COVID-19. *Journal of the American Medical Informatics Association, 27*(6), 860–866.
17. Liu, Y., Wang, Z., Tian, Y. et al. (2020). A COVID-19 risk assessment decision support system for general practitioners: Design and development study. *Journal of Medical Internet Research, 22*(6), e19786.
18. Einstein Data4u.Diagnosis of COVID-19 and its clinical spectrum AI and Data Science supporting clinical decision ((from 28th Mar to 3rd Apr)). Retrieved February 02, 2021 from https://www.kaggle.com/einsteindata4u/covid19.
19. Wang, K., Zuo, P., Liu, Y., et al. (2020). Clinical and laboratory predictors of in-hospital mortality in patients with coronavirus disease-2019: A cohort study in Wuhan China. *Clinical infectious diseases, 71*(16), 2079–2088.
20. Vaid, A., Somani, S., Russak, A. J. et al. (2020). Machine learning to predict mortality and critical events in a Cohort of patients with COVID-19 in New York City: Model development and validation. *Journal of Medical Internet Research, 22*(11), e24018.

21. Wagner, T., Shweta, F., Murugadoss, K. et al. (2020). Augmented curation of clinical notes from a massive EHR system reveals symptoms of impending COVID-19 diagnosis. *Elife,* 9:e58227.
22. Mizrahi, B., Shilo, S., Rossman, H., et al. (2020). Longitudinal symptom dynamics of COVID-19 infection. *Nature Communications, 11*(1), 1–10.
23. Wang, J., Abu-el Rub, N., Gray, J. et al. (2021). COVID-19 SignSym—A fast adaptation of general clinical NLP tools to identify and normalize COVID-19 signs and symptoms to OMOP common data model. *Journal of American Medical Information Association.*
24. Aakanksha Singhal, D.K. Sharma. (2020). Generalized 'Useful' Rényi & Tsallis information measures, some discussions with application to rainfall data. *International Journal of Grid and Distributed Computing, 13*(2), 681–688.
25. Bodenreider, O. (2004). The Unified Medical Language System (UMLS): Integrating biomedical terminology. *Nucleic Acids Research, 32*(1), 267–270.
26. Mikolov, T., Sutskever, I., Chen, K. et al. (2013). Distributed representations of words and phrases and their compositionality. *In Advances in Neural Information Processing Systems,* pp. 3111–3119, May.
27. Patient. Retrieved January 18, 2021 from https://patient.info/forums/discuss/browse/corona virus-covid-19--4541.
28. Landis, J. R., & Koch, G. G. (1977). The Measurement of observer agreement for categorical data. *Biometrics,* 159–174.
29. Schober P, Boer C and Schwarte L. A. Correlation coefficients: Appropriate use and interpretation. Anesthesia & Analgesia, 126(5):1763–1768, 2018. de Vet H. C, Mokkink L. B, Terwee C. B et al. Clinicians are right not to like Cohen's κ. *British Medical Journal,* 346:f2125, (2013).
30. Hooda, D. S., & Sharma, D. K. (2010). Exponential survival entropies and their properties. *Advances in Mathematical Sciences and Applications, 20,* 265–279.
31. Hooda, D. S., Upadhyay, K., & Sharma, D. K. (2015). On parametric generalization of 'Useful' R- norm information Measure. *British Journal of Mathematics & Computer Science, 8*(1), 1–15.
32. Manning, C., & Schutze, H. (1999). *Foundations of Statistical Natural Language Processing.* MIT Press.
33. Python-crfsuite. Retrieved March 14, 2018, from https://python-crfsuite.readthedocs.io/en/lat est/.
34. Kumari, R., & Sharma, D. K. (2019). Generalized 'useful non-symmetric divergence measures and Inequalities. *Journal of Mathematical Inequalities, 13*(2), 451–466.
35. Chang, C.-C., & Lin, C.-J. (2011). LIBSVM: A library for support vector machines. *ACM Transaction Intelligent System Technology, 2*(3), 1–27.
36. Pedregosa, F., Varoquaux, G., Gramfort, A., et al. (2011). Scikit-learn: Machine learning in Python. *Journal of Machine Learning Research, 12,* 2825–2830.
37. Support Vector Machines. Retrieved January 19, 2021 from https://scikit-learn.org/stable/mod ules/svm.html.
38. Weston, J., Mukherjee, S., Chapelle, O. et al. (2000). Feature selection for SVMs. *In Advances in Neural Information Processing Systems.*
39. Zhou, N. R., Liang, X. R., Zhou, Z. H., & Farouk, A. (2016). Relay selection scheme for amplify-and-forward cooperative communication system with artificial noise. *Security and Communication Networks, 9*(11), 1398–1404.

Distance Learning Solutions and the Role of Artificial Intelligence: A Review

Ismail Noori Mseer⊙

1 Introduction

Is there any ambiguity in the educational process? Is there a drop in academic achievement? How did students react to the concept of distance learning? Given the direct contrast with electronic universities' affairs [4], which are considered the least important in university life, concerns about survival and resilience, failure and achievement, and anxiety associated with this experience have increased significantly since the beginning of the experiment in March 2020. Many academics consider it a ruse, an untrustworthy experience.

This isn't a case based on a single person's opinion or the acceptance or rejection of one party by another. The entire process focuses on deliberating the essence of the material, perceiving facts, and diagnosing what is valuable, fruitful, and harmful. The key aim in all of this is focused on the way of looking at the educational process, with all of its information related to the educational system, in terms of academic, student, and administrative apparatus, [5] and the official institutions in charge of formulating educational policy, as well as the relationships between international organizations that decide the global approach's direction. The historical, cultural, and social obligation to the educational process is a prerequisite for coping with humanity's future. This emphasizes the significance of describing the scientific vocabulary associated with e-learning, which can be used as a catalyst in the educational process or as a replacement for traditional education [6]. Is it possible to believe that e-learning has become a reality and that we must now announce the death of the well-known traditional educational model? Is the movement toward programming anything and interacting through a digital framework that organizes anything and all over, and has the campus function, its annexes, an administrative apparatus, and an infrastructure ended?

I. N. Mseer (✉)
Ahlia University, Manama, Bahrain

Let's be honest: these ideas are starting to gain traction in academia. Instead, a growing number of students emphasize the importance of e-learning because it allows them to be more flexible in attending and engaging with their universities. How convenient it is to participate in the lecture from the comfort of your own house, on your sleeping sofa, or in one of the capital's sidewalk cafes. Except for pressing the login button and communicating with the lecturer after you have provided a system with good specifications, nothing costs you anything [7] can access the Internet quickly and efficiently.

Does the story come to a close with this condensed image? Before reading this scene, we should take a moment to consider the key terms associated with the subject, such as on-campus education, distance learning, virtual learning, e-learning, and distance learning. This is in terms of attempting to follow the strategy of using the direct timeframe imposed by the Covid 19 pandemic's conditions [8]. On the other hand, significant expectations are more closely linked to broader and deeper definitions like formal education, private education, public education, special education, lifelong learning, and open education.

It is impossible to deal with the education sector using current descriptions, simplify terminology, and encourage principles imposed by extraordinary circumstances such as a pandemic, catastrophe, or emergency incident. Education is a very fertile human environment focused on assumptions and hypotheses that must take into consideration. They are well-established cognitive relationships resulting from behavioral beliefs, social construction [9], and subtle knowledge. The issue is focused on universally agreed-upon primary principles and a long history of experiences, understandings, consensus, and international protocols. The most critical educational standards have been developed. As a result, a distinction was established between high-level universities, how students are received in particular disciplines, and the foundations and rules for that educational institution's success and its uniqueness and recognition from others.

2 Communication Foundations

It isn't just a matter of interacting with students that are at issue. However, there is a pattern in terms of sobriety, differentiation, the desire to plan, investing in information capital, follow-up, and selecting a skilled academic staff capable of fulfilling the role of teaching [10] as a supreme objective, and not just a job through which to engage with a group of people who want to obtain Just a testimonial? The distinction between the two models remains apparent, even though the former is attempting to fish in muddy waters. It focuses on university-based education, where the roots and transparent and consistent principles define sober universities' work, and the other model is based on distance education as an undoubted aid [5]. However, if distance education is a viable alternative to the traditional model, further discontinuation is needed. How many phony colleges have attempted to propel them ahead in their academic careers by encouraging distance learning and approving any application that comes in after the fee is paid? Recognized universities should invest in distance education to further advance the blended learning mechanism [11]. Universities are investing in seminars, lectures, and courses to address this problem. It was marketed to encourage education and learn without putting the academic community under the strain of scientific validity arguments.

The Scientific Union, the primary imperative for consolidating the ties of work in a field of science, according to Thomas Kuhn [12], is the paradigm. Without this, manipulating and weakening the original material on which the academic activity is centered would be all too simple. There is no question that digital education has made significant progress in engaging with those who want to learn in different parts of the world. We can draw on EDX experience [13], whose slogan promotes information importance. Enroll today and learn something new; this experience, as well as many others, is evolving, and they are performing a valuable scientific role in providing mature information services [14] at competitive rates, mainly since the science courses are supervised by major universities such as Harvard, MIT, Princeton, Pennsylvania, and others. The educational experience at edX is based on cutting-edge cognitive science. An approach that includes more than two dozen distinct learning features to assist in achieving objectives is based on three main principles: Immersive video lectures and dynamic graphics to data visualizations and interactive elements, learn new knowledge and skills in several ways. Demonstrating understanding is an essential aspect of learning. Quizzes, open response tests, simulated worlds, and other tools are available in edX courses and programs. Learning on edX changes the way think and what I can do, and it translates into the real world right away—immediately apply learned skills in the sense of work [15].

Due to the massiveness and openness of MOOCs, there is the potential for great diversity among learners. As a result, MOOCs are a viable option for offering high-quality online learning to people with disabilities. "MOOCs" provide great opportunities to improve the quality of life of older people by facilitating lifelong learning and inclusion in learning communities; This promotes cognitive enhancement, social interaction, and a sense of belonging. However, that "MOOCs" may present certain accessibility obstacles that can prevent elderly students from fully participating. Furthermore, non-native speakers face challenges while taking MOOCs [16] due to their lack of proficiency in the course language, which causes them to become learners with cognitive issues. This diversity necessitates meeting the accessibility requirements of students with various disabilities.

It is, however, possible to boost the capacity of existing MOOC platforms to facilitate the production and publishing of open content. Furthermore, it is even simpler to build new MOOC platforms because they are designed to be available for everybody. It simply means (recognition) attempting to determine the proper distinction between academic and intellectual. There's no harm in doing so. Let's introduce the word "learner" to the mix. Do you believe it's possible to herald the end of the professional academic era and argue that the digital age has put a lot to monopolize knowledge? [17] Such a statement is not without congested and prejudiced reports issued by losers in riding known universities and attempting to compensate by enrolling in educational courses offered by major universities as if they were the best option.

Nobody has a monopoly on knowledge; even in the age of paper culture, anyone can read and learn from books. Nobody would be able to stop you from practicing your right. More overlap, chaos, and poor design are needed to make this knowledge equal to the right to ascend the steps at the academic community's expense. Simultaneously, the existing reality is also focused on the use of products from the digital revolution [18], which benefit from bridging distances, promoting collaboration, and uncovering new approaches that contribute to the overall advancement of the educational process. As a result, the cultural, social, and economic influence of digital education has become apparent.

However, this effect is limited to determining the rights and responsibilities to which the academic community is committed. Digital channels have been dealt with in the climate, the field, and the official educational field that the approved institution has accepted. They are now attending in full force until they have become an important means for broadcasting lectures and engaging with students with complete clarity versatility [10]. Instead, the relationship has progressed to the point that the entire academic community (professors, students, and administrators) is searching for the best, most accurate, and stable investment. The mobile and the method of accessing Internet access are no longer only for social purposes. As much as it has become a primary tool at the heart of the natural learning process, one party stands out from the other. The focus here is on tracking the desired and desired outcomes rather than the social effect. Rather, the cultural, intellectual, and educational effects have become the reference for these results' investment approaches. This is reflected in how students approach this service, which can be abbreviated (e-learning) [9].

3   Both Substance and Appearance are Essential

The e-learning experience opened my eyes to the vast geographic horizon it covers. As a result, more prestigious universities with a strong reputation for investing in this feature and relentlessly trying to deal with this field through collaboration with institutions specializing in this field began to emerge. Instead, correspondence did not end at the monopoly's boundaries or when interacting with a single institution. Instead, when you read a course advertisement for a major university, and you have topped the promotional scene on this or that platform. Still, the invasion of personal e-mail has become Present with extraordinary intensity [19]. It encourages people to enter and register while providing the registrant with more benefits, not unnecessary simplification. Under the pressure of capital, even some of these ancient universities are risking their illustrious reputations?! For just one hundred dollars, anybody can get a certificate signed by the largest and most well-known academic name, complete with the official logo's presence. To demonstrate seriousness, the institution needs the associate to achieve an average score of 65 percent on short exams administered over six weeks, reflecting the study course's length. It reveals that the only way to get the answers is to copy and paste them, whether from the blog or Google.

Humanity is at a fork in the road when coping with fundamental social values and perceptions. Artificial intelligence (AI) and the Internet of Things (IoT) products add to the mystery and perplexity of the situation [20]. What was once called science fiction has become a present fact that is not based on skepticism but rather on self-evident certainty. And what you learn from software skills and advances in computers and smartphones vanishes in the blink of an eye. Marathon goods for the digital world were gasped and accelerated. Who would obliterate all of humanity's history and personal memories and transform them into a source of scarcity and weakness?!

Instead, high-ranking academics are becoming embroiled in the deterioration of the accelerating conditions in the development of difficult-to-catch-up programs, as well as the work required to execute them, before they can enlist the assistance of their younger students. The latter is better suited to this fast-paced setting. The direct educational truth has devolved into a scandalous revelation of the key and crucial issue regarding education's future. As a result, it is severe and fair to address the issue of teaching and learning theory, priorities, dreams, and perceptions now, rather than in the future? The academic situation is nothing more than a bitter and cruel (ordeal) of what the future holds in surprises [21]. And if it was to adapt to the smartboard and interact with students using the Moodle and Adreg programs, as well as the educational package's overall vocabulary. The next move takes place in the heart of the educational scene, where the electronic forum, which pits academics against each other in unusual ways, proves its worth.

The reality of education, teaching strategies, method of interacting with students, desired habits, anticipated goals, and desired outcomes are all challenges that are constantly being planned, accompanied by other, more direct challenges. And his approach to coping with his upbringing's conventional values about the university campus's definition, the sacredness of the classroom, the form of direct contact with students, and how to connect with them. And there's a way to stop making unintentional errors. And, for the time being, the vast majority of academics still regard e-learning as a supplement to traditional schooling. It is nothing more than a way of resolving the crisis brought on by the Covid 19 pandemic's conditions.

On the other hand, a sizable number of academics continue to connect through e-learning platforms to deliver scientific lectures in various universities worldwide, both East and West, and to participate in high-level scientific conferences via media that has grown and become widely recognized. Instead, colleges in the Middle East with limited financial resources have realized their dream of gaining international recognition in the academic classroom. Chomsky is the most well-known linguistics philosopher globally, and Fukuyama is the creator of the end-of-history theory. By submitting free e-mail and giving a lecture, the university could obtain a decent speed Internet connection.

4 What are the Benefits of E-learning?

Let us start with the most direct question: why are nations and communities eager to develop specific educational strategies? Why is education so important in the minds of people? And control the lion's share of the federal budget? Typically, the response has something to do with influencing the future. Here we are confronted with the key issue we're discussing, which is related to the potential implications that have become reliant on the digital revolution [22]. And we won't be able to hide our heads in the sand and forget what happened to the fields and industries that were once thriving but are now nothing more than ruins. We can read about the paper press's conditions, which fell into the digital media's hands, and the end of the giant media institutions in this. The one-person replacement emerged, requiring only a (smartphone) to access his YouTube channel, which he broadcast to millions of viewers and followers. This is how the research centers' awareness reports show the abolition of thousands of conventional jobs and the creation of new employment opportunities imposed by the digital revolution and its goods [23]. The education sector is no exception, and it cannot conceal or avoid the consequences of this tidal wave.

It is no longer possible to rely solely on conventional educational inputs and outputs, and it is past time for universities to rethink how they interact with facts. It is no longer helpful to discuss the establishment of additional masters and doctoral programs. This is a hypothetical discussion that could take place in museums, not reality?! The problem is solely about the labor market's requirements. It is no longer helpful to speak about the ultimate purpose, or that education is a holy message aimed at human development. Whoever advocates for this slogan should encourage his family members to enroll in bachelor's degrees that are not in demand in the labor market?! Instead, these proverbs' proponent is quick to lament his bad luck, as he has worked in a field that isn't considered a priority in the labor market.

As a result, job prospects will be restricted. Allow us to be confronted with an undeniable suffocating existential crisis. Otherwise, how can he live and communicate with truth unless he has an explicit promise of job security and is willing to get interested in the minutiae of life?

Here, education is based on a pragmatic perspective rather than romantic quotations. Education is a critical component of the state's ability to create and retain momentum in carrying out its primary functions. Still, it is not a high-profile or well-received sector. As the number of unemployed graduates has risen, we can see the severe systemic challenges that most countries across the world face. And the ramifications on a social, political, economic, and cultural level. Instead, dissatisfaction reaches a dangerous peak when this graduate becomes a ticking time bomb. Will have the effect of jeopardizing the stability and social harmony. Universities reproduce without a specific aim or vision, and specializations are recorded without economics, management, or planning. A growing number of people with advanced degrees have little direct influence on fact. Obtaining certification and therefore expecting prestige and social dominance is a dream come true [24, 26]. From its profound and original substance, it is a system of empty faces. Simultaneously, the economic reality necessitates more skilled trainees to fill current practical and genuine labor market positions. Whoever develops educational strategies must consider the overall change of details related to resources, capital, and consumer movement, how to cope with the capabilities and sovereignty exerted by transnational corporations, the consequences of globalization, environmental impact, and climate change. The truth is that I will not write a master's or doctoral thesis to answer this problem. Their findings and recommendations would only profit from the production of more paper files on university library shelves. The issue here is one of coping realistically without precedents or ready-made solutions. All of this is contingent on how the principles are expressed in the spirit of future reconciliation.

5 Conclusion

Realistic logic and the rapid pace of change in the world have prompted us to reconsider the idea of education in terms of decision-making, management, organization, vision, purpose, and objectives. A logic focused on identifying, tracking, and diagnosing the learning system's strengths and weaknesses and evaluating the components with an open and flexible vision. All of this is happening simultaneously as people's interaction with social and cultural perceptions shifts away from conventional slogans and banners. It is a formulation focused on reformulating meaning for the entire educational process by the primary needs dictated by economic, social, political, and cultural realities.

The intensity of the higher diploma's symbolic capital is the critical issue on which education is dependent. The case has little to do with the countries of the East or the least developed. The advanced West, which says top academics are modest and asks students to address them by their abstract names, places a great deal of reliance on the symbolic capital of diploma and degree. It is a means of establishing social and class distinctions and rank, wealth, and authority. The issue is based on a clear contrast between a Harvard University professor's status and a Harvard University employee?!

The disparity between a professor at a prestigious university and a professor at a small college is apparent. In light of the rise of e-learning, which would inevitably overturn the thrones and inequalities that the present world resides in, all of these concepts and ramifications will be a distant memory. Institutions and businesses are surpassing traditional job expectations. It has also skewed in favor of the applicant who completed applied educational training courses rather than the recipient of an academic degree. Whereas the distinction is based on the fact that these businesses need a workforce who can keep up with business demands and needs, [25] the individual who obtains an academic degree is a significant financial burden on the operating institution, in addition to the psychological and cultural symptoms that differentiate both models. The employee with hands-on experience easily integrates into the work environment. Employees with theoretical expertise, on the other hand, suffer from a lack of prestige and rank. Nonetheless, theorizing can often obstruct work, while fact necessitates more concrete, realistic responses.

References

1. Hamdan, A., Musleh, A. M. A., Al-Sartawi, R. K., Anaswah, M., & Hassan, A. (2018). Board interlocking and IT governance: Proposed conceptual model. *EMCIS, 2018*, 457–463.
2. Habes, M., Salloum, S.A., Elareshi, M., Ganji, S. F. G., Ziani, A. -K., & Elbasir, M. (2020). The influence of youtube videos on ELA during the COVID-19 outbreaks in Jordan. In *2020 Sixth International Conference on e-Learning* (second), Sakheer, Bahrain, 2020, pp. 133–138. https://doi.org/10.1109/econf51404.2020.9385501.
3. Razzaque, A., Hamdan, A. Internet of things for learning styles and learning outcomes improve e-learning: A review of literature. ACV 2020: 783–791
4. Alhabeeb, A., & Uygur, S., & Hamdan, A. (2020). Management of organization and religion interactions through business excellence model: The case of Islam and large Saudi private firms (February 10, 2020). In *Proceedings of the Industrial Revolution & Business Management: 11th Annual PwR Doctoral Symposium (PWRDS) 2020*. Available at SSRN. https://ssrn.com/abstract=3659101 or https://doi.org/10.2139/ssrn.3659101.
5. Badawi, S., Reyad, S., Khamis, R., Hamdan, A., & Alsartawi, A. M. (2019). Business education and entrepreneurial skills: Evidence from Arab universities. *Journal of Education for Business, 94*(5), 314–323. https://doi.org/10.1080/08832323.2018.1534799
6. Abdalmuttaleb, M. A. Musleh Al-Sartawi, Hamdan, A. (2019). Social media reporting and firm value. I3E 2019: 356–366.
7. Hamdan, A., Sareaa, A., Khamisb, R., Anaswehc, M. (2020). A causality analysis of the link between higher education and economic development: empirical evidence Author links open overlay panel. *Heliyon, 6*(6), June 2020, e04046. https://doi.org/10.1016/j.heliyon.2020. e04046.
8. AL-Hashimi, M., & Hamdan, A. (2021). Artificial intelligence and coronavirus COVID-19: Applications, impact, and future implications. The Importance of New Technologies and Entrepreneurship in Business Development: In The Context of Economic Diversity in Developing Countries: The Impact of New Technologies and Entrepreneurship on Business Development, 194, 830–843. https://doi.org/10.1007/978-3-030-69221-6_64.
9. Musleh Al-Sartawi, A. M., Razzaque, A., Alhashimi, M., & Hamdan, A. (2021). Knowledge management and big data analytics for strategic decision making. IGI Global. https://doi.org/10.4018/978-1-7998-5552-1.
10. Al-Sartawi, A. M., Hussainey, K., Hannoon, A., & Hamdan, A. (2020). Global approaches to sustainability through learning and education. IGI Global. https://doi.org/10.4018/978-1-7998-0062-0.
11. Aho, J. M. et al. (2015). Mentor-Guided self-directed learning affects resident practice. *Journal of Surgical Education, 72*(4), 674–679. Alshaikh, I. Y., Razzaque, A. & Allawi, M., 2017. Positive emotion and social capital affect knowledge sharing: case of the public sector of the kingdom of Bahrain. Coimbra, Portugal, Springer Lecture Notes in Business Information Processing volume.
12. Hollinger, D. (1973). T. S. Kuhn's theory of science and its implications for history. *The American Historical Review, 78*(2), 370–393. https://doi.org/10.2307/1861173.
13. https://www.edx.org/.
14. Alamri, A., Muhammad, G., Al Elaiwi, A., Al-Mutib, K., & Hossain, M. (2014). Media content adaptation framework for technology enhanced mobile e-Learning. *Journal of Universal Computer Science, 20*(15), 2016–2023.
15. Alexander, B. (2008). Connectivism course draws night, or behold the MOOC. http://infocult. typepad.com/infocult/2008/07/connectivism-course-draws-night-or-behold-themooc.HTML.

16. Bohnsack, M., & Puhl, S. (2014). Accessibility of MOOCs. In Proceedings of international conference on computers helping people with special needs (pp. 141–144). Springer International Publishing.
17. Said, N. M., Uthamaputhran, S., Zulkiffli, W., Hong, L. M., & Hong, C. W. (2021). The mediating effects of entrepreneurial education towards antecedents of entrepreneurial intention among undergraduate students. The importance of new technologies and entrepreneurship in business development. *In The Context of Economic Diversity in Developing Countries: The Impact of New Technologies and Entrepreneurship on Business Development, 194*, 2029–2040. https://doi.org/10.1007/978-3-030-69221-6_146
18. Mseer, I.N. (2021), Ethics of artificial intelligence and the spirit of humanity. In Hamdan, A., Hassanien, A. E., Razzaque, A., Alareeni, B. (Eds.), *The fourth industrial revolution: Implementation of artificial intelligence for growing business success*. Studies in Computational Intelligence, vol 935. Springer, Cham. https://doi.org/10.1007/978-3-030-62796-6_19.
19. Elareshi, M., Ziani, A.-K., & Al Shami, A. (2021). Deep learning analysis of social media content used by Bahraini women: WhatsApp in focus. *Convergence, 27*(2), 472–490. https://doi.org/10.1177/1354856520966914
20. Saad ALShaer, D., Hamdan, A., Razzaque, A. (2020). Social media enhances consumer behaviour during e-Transactions: An empirical evidence from Bahrain. *Journal of Information Knowledge Management, 19*(1):2040012:1–2040012:23
21. Fuad, A., Al-Hashimi, M., & Hamdan, A. (2020). Innovative technology: The aviation industry and customers preference. *ACV, 2020*, 696–707.
22. Razzaque, A., & Hamdan, A. (2020). Internet of things for learning styles and learning outcomes improve e-learning: A review of literature. *ACV, 2020*, 783–791.
23. Al-Sartawi, A. M., Badawi, S., & Reyad, S. (2019). The role of knowledge-oriented higher education institutions in sustainable education practices: Knowledge and sustainable education. In Albastaki, Y. A., Al-Alawi, A. I., & Abdulrahman Al-Bassam, S. (Ed.), *Handbook of Research on Implementing Knowledge Management Strategy in the Public Sector* (pp. 275–288). IGI Global. https://doi.org/10.4018/978-1-5225-9639-4.ch015
24. Badawi, S., Reyad, S., & Al-Sartawi, A. M. (2019). Cross-functional teams in support of building knowledge capacity in public sector. In Albastaki, Y. A., Al-Alawi, A. I., & Abdulrahman Al-Bassam, S. (Ed.), *Handbook of Research on Implementing Knowledge Management Strategy in the Public Sector* (pp. 221–230). IGI Global. https://doi.org/10.4018/978-1-5225-9639-4.ch011
25. Al-Sartawi, A. M., & Razzaque, A. (2020). Cyber Security, IT governance, and performance: A review of the current literature. In Albastaki, Y. A., & Awad, W. (Ed.), *Implementing Computational Intelligence Techniques for Security Systems Design* (pp. 275–288). IGI Global. https://doi.org/10.4018/978-1-7998-2418-3.ch014
26. Elali, W. (2021). The importance of strategic agility to business survival during corona crisis and beyond. *International Journal of Business Ethics and Governance, 4*(2), 1–8. https://doi.org/10.51325/ijbeg.v4i2.64

Online Education During Covid-19 Pandemic in Kingdom of Saudi Arabia

Haifa Frad and Ahmed Jedidi

Abbreviations

ICT	Information and Communication Technology
LMS	Learning Management System
KSA	The Kingdom of Saudi Arabia
CMS	Course Management Systems
MOE	Ministry of Education
MOHE	Ministry of Higher Education
NCEL	National Centre for e-learning
SWOC	Strengths, Weaknesses, Opportunities, and Challenges
CIC	Computer and Information Center
IT	Information Technology
SEU	Saudi Electronic University

H. Frad (✉)
Imam Abdulrahman Bin Faisal University, Dammam, Saudi Arabia
e-mail: hmhfrad@iau.edu.sa

A. Jedidi
College of Engineering, Ahlia University, Manama, Bahrain
e-mail: ajedidi@ahlia.edu.bh

National School of Engineering, Sfax University, Sfax, Tunisia

© The Author(s), under exclusive license to Springer Nature Switzerland AG 2022   145
A. Hamdan et al. (eds.), *Technologies, Artificial Intelligence and the Future of Learning Post-COVID-19*, Studies in Computational Intelligence 1019,
https://doi.org/10.1007/978-3-030-93921-2_9

1 Introduction

In 2020, suddenly the world faced the Covid-19 pandemic and various precautions were taken by governments to face this pandemic. Especially, Social distance was imposed, and the education system was the first affected. all schools and colleges impose online work. Two major trends have characterized the education landscape: the advancement of e-learning and the rapid global expansion of technology [1]. In the first trend, e-learning has led to entirely new opportunities for students to access new courses and improve their knowledge of using technology include the use of a computer, the use of software, and communication interaction [2, 3]. E-learning has become an indispensable part of the competitive services domain [4]. Several empirical studies have examined the e-learning critical success factors [5, 6, 7, 8] (Lee et al., 2009). The critical success factors for online learning have been explored and grouped into 4 categories as presented in Fig. 1 [9].

In the second trend, the global expansion of technology has led to the explosive growth of the information technology system and learning environment that are

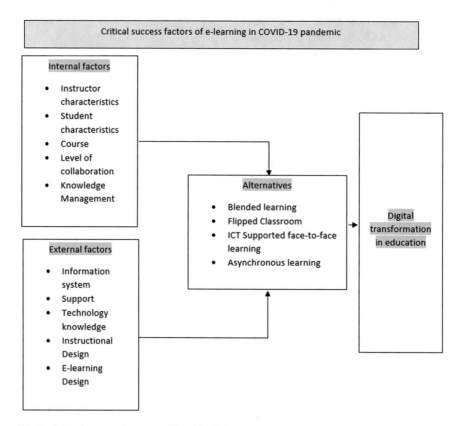

Fig. 1 Critical success factors problem hierarchy

provided for both learners and teachers [10]. Certainly, IT environments and Web technologies in education domain offer new solutions for implementing new social culture in learning contexts. Thus, the internet provides learner's serval learning opportunities adapted to their needs [11]. Platforms like the blackboard can stimulate student interest in learning and successfully improve the educational effect [12]. A new language is created by intelligent computers and artificial intelligence, technology could be one of the essential tools soon. For that, the link between learning and technology is important to understand [13]. However, before the pandemic, investments in education technology reach $ 18.66 billion in 2019 and due to the Coronavirus, the global online education market is expected to reach up to 350 billion dollars by 2025 [14].

So, it is important to analyze the experience offered by the quarantine period and understand the concepts and things that have happened, good and bad, and understand how to implement the content remotely without mistakes. In Saudi Arabia, Internet use has increased intensely, in 2020 (Figs. 2 and 3), the number of Internet users was 31,856,652, almost 91.6% of the population (Table 1), (IWS 2021). According to the Internet World Stats (IWS), the number of Facebook subscribers in KSA touched 26,350,000 in June 2016, with a 75% penetration rate.

The chapter will be organized as follow. Section 2 will present the Distance Learning in the Saudi Educational system. In Sect. 3, we will describe the E-learning in Saudi Arabia during the COVID-19 pandemic. Section 4 will develop an SWOC analysis of e-learning during Covid-19 in Saudi Arabia. Section 5 the conclusion.

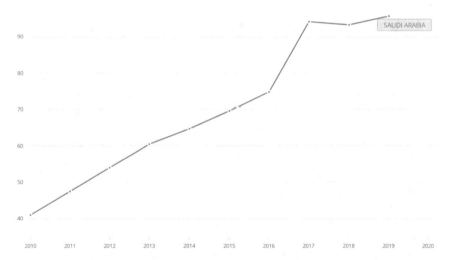

Fig. 2 Individuals using the Internet (% of the population) (https://data.worldbank.org/)

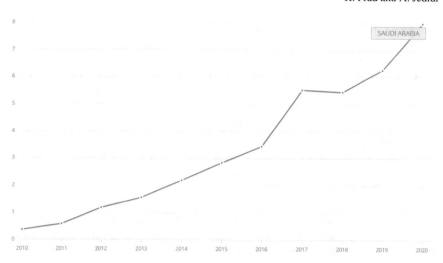

Fig. 3 Secure Internet servers (per 1 million people) (https://data.worldbank.org/)

Table 1 Saudi Arabia usage and population statistics (IWS 2021)

Country	Population (2021)	Users (2000)	Internet usage (2020)	% population (Penetration)	Facebook (2020)	% penetration
KSA	34,783,757	200,000	31,856,652	91.6%	26,350,000	75.8%

2 Distance Learning in the Saudi Educational System

In Saudi Arabia's education system, students and staff are completely segregated by gender, requiring separate buildings and staff for their students [15]. E-learning technologies are opening a new way of teaching for many Saudi educational institutions, allowing students to use a learning option that best suits their teaching style. The government allowed the public to access the Internet in 1997 after linking research and medical institutions for the first time. In 1996, the Ministry of Higher Education (MOHE) established the Computer and Information Center (CIC) which provides a range of ICT (Information and Communication Technology) services to schools and educational centers to design new curricula and develop the capabilities of both teachers and students.

The Ministry of Education then launched an ambitious computer project in 2000 It aims to cover all schools in Saudi Arabia. This was followed by the National Schools Network project launched in 2001, to connect schools and educational directorates through a wide area network (WAN) covering the entire country. To facilitate e-learning, the Deanship of Distance Learning was established at King Abdulaziz University in 2007 [16]. The strategic plan of the e-learning project began in 2009 at Imam Muhammad bin Saud Islamic University (IMAMU), and the university is translating and publishing materials in Arabic to spread Islamic knowledge. The virtual institution The Arab Open University (AOU), a branch of the UK Open University, started as a traditional institution for distance learning and developed alongside technology in education and is a project of Saudi Prince Talal bin Abdulaziz Al Saud, originally using technologies such as broadcasting Television and radio, CD and DVD lectures, postal correspondence, and telephone tutoring. The institution also uses online resources with a mix of blended online courses. The Saudi Arabia branch of AOU reported 15,455 current students and 14,590 postgraduate students, studying in 130 courses in 2019 (the Arab Open University). The branch of the Arab Open University in Saudi Arabia offers courses in Business Studies, Computer Studies, Language Studies, and Education Studies.

Founded in 2007 with programs in English and Arabic, the University offers 4-year bachelor's degrees in Islamic Studies (English) and Bachelor's degrees in Sharia and Qur'anic Studies (KIU 2017). The course includes watching online videos, completing reading materials, and then passing online multiple-choice exams. [17, 18]. Also, The Saudi Electronic University was established in 2011 in cooperation with one of the oldest virtual universities in the United States, the University of Phoenix, along with universities in Minnesota and Ohio. Operating on a combined model, SEU offers bachelor's degrees (and one MBA) in computing and informatics, management and financial sciences, health sciences, science, and theoretical studies. The SEU is seen as an example of a new learning model that uses online education and online learning and provides certification equivalent to that of other Saudi universities. Its graduates have the same opportunities as regular Saudi university graduates in terms of future careers in the employment market. Currently, SEU is in four cities: Riyadh, Jeddah, Dammam, and Al-Medina.

In addition, and to develop the education system, the MOHE has established the National Centre for e-learning and Distance Learning (NCEL). This Centre has as objectives to expand the use of online education in the education system, and support online education and help both teacher and learner. In addition, the NCEL aims to provide online teaching and consulting and arrange seminars, workshops, and conferences. This Centre in his turn has established many projects to assist Saudi universities. The Saudi Centre for Support and Counselling (SANEED), JUSSUR, Saudi Digital Library, and Maknaz and Training Programs (NCEL, 2012) present services provided by the NCEL to improve the ICT infrastructure and facilitate online classes. A summary of the services is shown in Table 2.

The integration of technology into university courses, and the increasing use of the Internet have enabled Saudi educational institutions to offer a convenient and appropriate method for the advancement of courses [19]. The MOHE provides the necessary funds to develop the education system and encourages universities to develop and improve their ICT infrastructure to ensure that distance learning functions well when needed.

Today, the Ministry of Education manages and provides all universities with access to a Learning Management System (LMS) such as Blackboard, Moodle, WebCT and Lotus Notes [20]. LMS is a very useful technology for both students and instructors in online learning environments. With the development of e-learning systems, LMS has become a reliable means for education and training. LMS is used in most universities to manage courses and educational content. Teaching through the LMS system guarantees success in both teaching practice and student learning [21].

Table 2 Descriptions of services provided by the NCEL

Services	Descriptions
SANEED	A Centre for support that: – Provide education and academic support – Assist all beneficiaries of e-learning – Solve technical problems via email, live voice
JUSUR	A learning Management System (LMS) that: – Register students – Organize and arrange courses – Schedule courses – Control and evaluate the advancement of students – Provide tests and exams
MAKNAZ	An electronic national portal that: – Improve digital courses – Provide educational resources
Saudi digital library	A digital library that: – Encourage the scientific research – Assist educational program with digital content – Provide educational resources to teacher and learner – Provide an electronic system to review all available databases with the ability to search
Training programs	A national project that: – Provide training to both faculty members and administrators – Aim to provide a basic program to improve the trainers – Collaborate with a professional expert from inside or outside to give an advanced program

3 E-learning in Saudi Arabia During the COVID-19

Regarding the World Health Organization (WHO) data, COVID-19 has been stated in more than 216 countries and there are areas with millions of confirmed cases. This has led to the development of e-learning platforms in many countries [22]. Saudi Arabia's large geographical area, large population, and geographical disparities between modern urban cities such as Riyadh and Jeddah and rural desert areas, some of which are still close to indigenous Bedouin lifestyles, pose a major challenge to the Saudi government's achieve justice [23]. On February 1st, 2020, the KSA has updating emergency, and after two days, on 3rd February a Royal Decree # 35,700 has been signed concerning measures to combat the COVID-19 epidemic and on March 8th, 2020, the Minister of Education has issued its decision to close schools (Starting with the governance of Qatif) and to shifting to Online education within 24 h (Fig. 4). However, online education had to be implemented as an alternative to face-to-face education. The student had adequate access to educational programs. And the MEO supervises all the directorates of education which in turn supervises the educational operation (252 Educational Offices, 32,553 Schools, 13,584 Boy's School and 18,967 Girl's School) [24].

Fig. 4 Shifting steps from traditional education to online education due to the COVID-19 crisis

3.1 Online Education for Secondary, Intermediate, and Elementary School During COVID-19 Pandemic

The Saudi Minister of Education said: "*Many norms have changed with Coronavirus, there is a great trend for online education.*" In the following, we summarize most of the platforms that have been made available to students to complete their educational process:

- Among the most important solutions were AIN's YouTube satellite lessons, which allowed more than six million students full access to the lessons. Al Ain National Gateway Platform for Education Enrichment has provided reliable and efficient online education services. All students in schools in the KSA have benefited from this portal. In addition, Technical learning environments, such as Zoom and Microsoft Teams, which are among the most popular CMC tools for learners and teachers are used. On this portal the learners obtained their course, homework, quizzes and semester and final exams (Fig. 5).
- (vschool.sa) The unified education system is one of the proposals offered by the Ministry of Education as another option for online education. Moreover, to inform students and teachers of its educational resources, the ministry announced the availability of this educational means by SMS in the form of advice messages.
- The "Future Gate" portal assisted middle and secondary schools spread over more than 33 educational directorates across the kingdom using a CMC tools for both teachers and students (Fig. 6).
- Virtual Kindergarten: The Ministry of Education has offered virtual classes for online education kindergartens to children from three to seven years old under the direction of their parents, providing a variety of educational elements, instructions, and educational content (Fig. 7).

All the efforts of the MOE and of the families are aimed at offering more options and more solutions to save the school year 2019/2020 and to control six million

Fig. 5 Home page of the Ain platform

Fig. 6 Home page of the future gate

Fig. 7 Online education tools during Covid-19 epidemic

Table 3 Online education system statistics

Domestic views for AiN lessons	60 million
International views for AiN lessons	600 thousand
Discussion sessions	1,3 million
Electronic content items	2 million
Subscribe on YouTube	600 thousand
Lessons through TV	4493
Online tests	715 thousand
Online assessments	1.8 million
Broadcasting hours	2835
Reviewed cases	1107
Technicians	100
Virtual classrooms	413
Ain TV Channels for secondary school	11
Ain TV Channels for intermediate school	3
Ain TV Channels for elementary school	6
Statistics for the technical support for online education System	
Live electronic chat	36 thousand
Beneficiaries	217 thousand
Service requests	48 thousand
Phone calls	133 thousand

Source www.moe.gov.sa

students simultaneously. The ministry has sent messages of advice to students, teachers, and parents to join the educational resources. The launch of the online education system has shown instant effectiveness. During the coronavirus crisis (COVID 19), 60 million views for the Ain satellite TV course, 413,000 virtual classes, 715,000 published tests, 1.8 million published assignments, …. (Table 3) statistics that reflect the efficiency of the online education system in the kingdom which adapts effective learning technologies to advance online education.

Some challenges and difficulties were rise during this educational journey some of them are listed in Table 4.

3.2 Online Education in High Education During COVID-19 Pandemic

The Minister signed a decision on March 9, 2020 which announced the implementation of online education throughout the school closure. So, all universities begin to adopt the necessary procedures and make the essential choices to fight the pandemic.

Table 4 Challenges and difficulties during Covid-19 crisis

Difficulties	Descriptions
Lack of computers	With cooperation with the Takaful foundation, the ministry has gave more than 20,000 tablets, laptops, and Desktop computers to students who need them
Lack of internet service	Equal access to education is a major objective for the Ministry of Education. For this, The MOE works in collaboration with the Ministry of Communications and Information Technologies to assist students against technical problems and obstacles by providing free access to educational sites and platforms of Distance Learning
Out-of-the-box online education	The culture of online education is certainly new and strange to Saudi society initially, and learners were not ready to embrace it. This challenge was met by offering user guides to help educate distance learning platforms and other available options
The infrastructure capacity	Infrastructure capacity: Suddenly, the infrastructure received many users simultaneously. This challenge was met by using icloud services

The Saudi Electronic University (SEU) and the National Center for E-Learning have cooperated to make a switch from traditional education to e-learning in all Saudi universities. Some technical challenges started to emerge after the launch of online learning tools and were effectively eliminated after improving technical support teams in universities.

On April 4, 2020, the MOE met virtually with the presidents of the universities to discuss the mechanism of the final exams during the crisis period, so several recommendations were taken:

- Distance learning will be continued, and students continue their education online.
- Motivate faculty members and prepare them to use available resources and LMS technologies to support online education.
- Directing faculty members to use activities and tasks appropriate to the current situation and to help students and develop their skills and solve their difficulties during this crisis.
- All exams for all types of education will be done on time, April 28th.
- Change the scale used before exceptionally and become 80 points for midterm, assignments, projects, and other tasks and 20 points for the final exams.

The MOE with the Saudi Electronic University (SEU) and the National Center for E-Learning put together plans to support the total transformation to distance learning around Saudi universities despite some difficulties cited in the following Table 5.

The MOE and Saudi Electronic University have joined hands to reassure the switching. More than four hundred servers were added to the technical center of the education's system. This increased the total number of servers to over 750 servers for efficient usage of the system and training over 500 people for the operation of

Table 5 Challenges and difficulties during Covid-19 crisis in high education

Difficulties	Descriptions
Devices' lack:	Some students do not have online access via devices or tablets. These students were quickly identified, and their problems were successfully resolved
Internet complications	Some students were unable to use the Internet, so the Ministry of Education and the Ministry of Communications cooperated to guarantee continued education for all students in the KSA
Training issues	Some instructors were not qualified to handle virtual classroom and platforms, so guides and training workshops were made available to these users
LMS problems	Blackboard was the main LMS technology used during the crisis. Users encountered technical issues that required technical solutions from the vendor. Other CMC solutions have been certified to overcome any failure, such as Zoom and Microsoft Teams to offer simultaneous communications between teachers and learners Students can reach live classes and participate in those classes by using the recording and playback abilities, real-time quizzes. Through the Blackboard the instructors can record synchronous classes and upload them that students can download anytime
Worried about the change of system	The culture of online education is new and comes suddenly, many students are worried about the change. They were also afraid of the lack of communication with their instructors, for that a virtual consultancy initiative and working hours have been launched at Saudi universities to address these subjects

this system. These are considered the most important challenges the ministry has confronted in starting online teaching.

Thus, the Kingdom succeeded in making the sudden change in teaching go smoothly, since more than 16,586,048 digital contents passed in the first week of the transition (9th Educational week), to arrive in the fourth week of online education (12th Educational week) 21,186,279 digital content, as shown in the following Fig. 8.

MOE has provided an e-learning support to reduce the effect of COVID-19 on education domain and achieve the best possible outcomes. The statistics are good and reflect the great effort made by the government which:

– Conduct over 2.6 million virtual classrooms targeting over 1.4 million students.
– Organize 2.8 million instructional hours with more than 75,000 Faculty members via the LMS platform with more than 4.5 million E-tests.
– Design and implement over 420,000 educational Subjects.
– Create guides and training manuals for the e-learning tools.

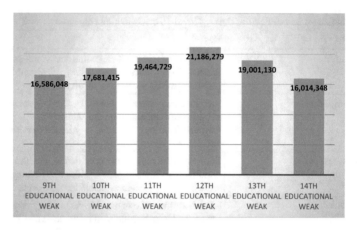

Fig. 8 Statistics during school closure

- More than 1,400 master and doctoral theses are supported via LMS Solutions.
  (www.moe.gov.sa).

4 The SWOC Analysis to KSA E-learning Experience

In general, an SWOC analysis, carried out at the launch of a company or a new
product, should allow the manager to set up a roadmap, by identifying the strengths
and opportunities on which he can rely and the weaknesses with which he will
have to compose. SWOC analysis allows a general development of the company by
crossing two types of data: internal and external. The internal information taken into
account will be the strengths and weaknesses of the company. As for external data,
it will relate to nearby opportunities and challenge. We develop an analysis of online
learning during the Corona Virus pandemic in KSA (Fig. 9).

The most important advantage of e-learning is to save the education process
in such difficult times as the case of the Covid-19 pandemic. E-learning provides
the equilibrium between time and location, the student found herself centered with
flexibility in time and location. E-learning also makes it possible to modify the
contents and regulate the learning processes while respecting the needs of the learner.
Furthermore, and to create a collaborative and interactive learning environment, e-
learning uses important online tools to ensure effective learning, such as the use of
a combination of audio, video, and text to reach students [6].

luckily, nowadays the technology is well developed and offers innovative and
flexible solutions in times of crisis like the Anywhere–Anytime e-learning feature
which helps people to work with almost no need for face-to-face interaction [25].

The human contact between the learner and the teacher was the most important
characteristic of traditional teaching. However, online learning has some weaknesses
as it can hinder direct communication between learner and teacher [26].

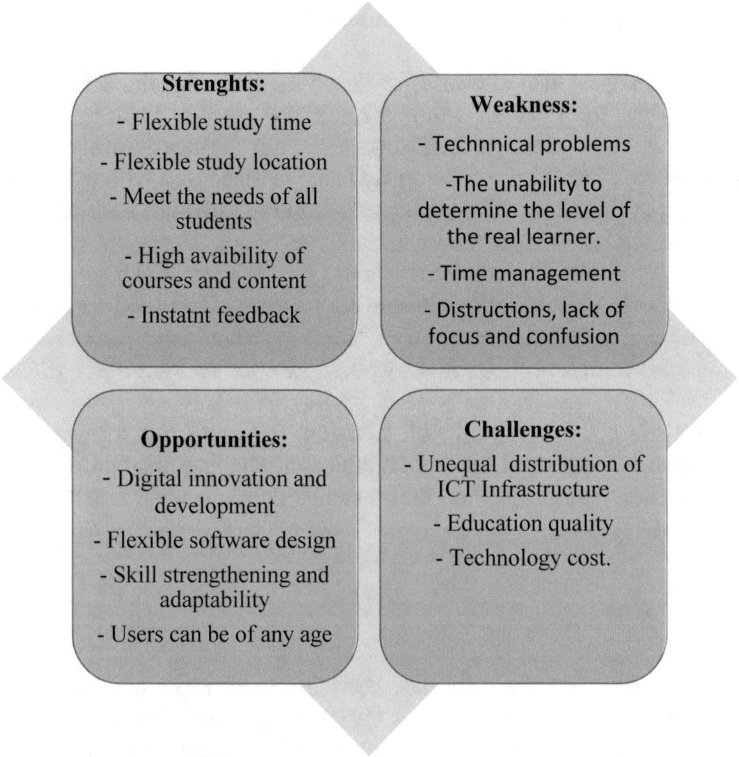

Strenghts:
- Flexible study time
- Flexible study location
- Meet the needs of all students
- High avaibility of courses and content
- Instatnt feedback

Weakness:
- Technnical problems
- The unability to determine the level of the real learner.
- Time management
- Distructions, lack of focus and confusion

Opportunities:
- Digital innovation and development
- Flexible software design
- Skill strengthening and adaptability
- Users can be of any age

Challenges:
- Unequal distribution of ICT Infrastructure
- Education quality
- Technology cost.

Fig. 9 An SWOC analysis during COVID-19

Some students encounter many technical problems that may disable both the teaching and learning process. The elasticity of time and space, although it is considered a strong point of E-learning, can also be fragile and creates problems due to the unserious behavior of the pupils because the pupils are not the best ones. Same, they differ in their abilities and skill levels [27]. Some of them don't feel comfortable learning online, which increases frustration and confusion. Also, a lack of necessary digital tools, lack of internet connections, or insufficient Wi-Fi connections can cause many problems because many students miss out on learning opportunities [28].

However, the opportunities offered by online education are numerous, among which: the development of the educational process in terms of touching modern learning methods and breaking away from traditional methods [26]. Innovations and digital developments allow teachers to practice techniques and design various flexible programs to better provide information to students [29]. In addition, users of all ages can use online platforms and enjoy the benefits of E-learning's flexibility in terms of The Anywhere–Anytime e-learning feature. In addition, innovative technology-based educational programs can be developed by teachers. For this, EdTech startups have many opportunities to revolutionize education and go very far with technology [12].

A SWOC analysis also allows us to detect existing challenges in the practice of learning. Thus, the transition from the traditional mode of teaching to the online mode is seen as the most difficult challenge, especially the idea of involving student's education process [30]. Despite the many benefits associated with e-learning, it is important to pay attention to the quality of e-learning programs seen as a real challenge, as there is a lack of quality norms, quality regulator, improvement of electronic resources, and transfer of electronic content [31]. Another challenge belonging to institutions should make efforts to ensure easy access to the necessary educational resources either by the teacher or the student. They should also make sure that all e-learning apps work on cellphones as well, in case the students don't have laptops [32].

We can learn a lot in this difficult situation. e-learning has many strengths to develop, weaknesses to avoid, and opportunities and challenges to consider. Teachers are required to select and implement the best tool to deliver education to their students.

5 Conclusion

E-learning interest into the KSA has been increased more and more after the successful experience in times of the COVID-19 epidemic. Saudi Arabia has met the unexpected challenge and has certainly fully embraced e-learning as an international best practice in education. So, following this study, we recommend some advices for a successful distance learning future:

- A balance must be found between the quantity and the quality of the e-learning resources to be distributed to the learners.
- The online learning resources are considered insufficient to meet all the educational needs of the student.
- Consideration should be given to students living in remote areas.
- There is a need to develop regulatory frameworks, education administration, and teacher training programs.
- Solve the plagiarism problem which was visible in the online discussions of the students and the exams and tests. Certainly, computer environments and Web technologies in the field of education offer new solutions to implement new didactic intentions in learning contexts.
- Be certain that the internet is smoothly running with good speed, which is responsible for offering learners learning opportunities adapted to their needs.

During the COVID-19 crisis, new standards are set in education. The experience of KSA has been fruitful and successful and e-learning will take an important place in the future with a more accessible and more qualified system. Among the most important solutions were AIN's YouTube satellite courses, which have given more than six million students full access to courses. Since 2002 Saudi Arabia has been involved in e-learning platforms such as Backboard or Moodle. The kingdom has already adopted communication strategies and challenges, including virtual teamwork and conflict management. The Covid-19 was "a good opportunity" for these countries to test their capacity to manage the crisis. Teacher and student experienced this new online environment during the Coronavirus crisis. So, e-learning is on the right road to success. Except that Teachers should focus on online teaching methods and realize the future vision of education before implementing e-learning programs and platforms, they should also focus on their development and their use of the software. Teachers need to learn how to create, facilitate and evaluate online courses. They have to learn how to interact with students in exclusively online systems without forgetting to work on the emotional and psychological side of their students. In addition, teachers should use more interactive methods to attract students with different learning styles. Also, without forgetting the motivation of the student who is the subject of the learning operation. The student must develop his skills and knowledge to ensure his understanding and evolve rapidly. Like, for example, implementing flexible learning schedules, using reward systems, and creating a pleasant virtual environment.

References

1. Lee, Y. C. (2008). The role of perceived resources in online learning adoption. *Comput. Educ., 50*, 1423–1438.
2. Namisiko, P., Munialo, C., & Nyongesa, S. (2014). Towards an optimization framework for e-learning in developing countries: A case of private universities in Kenya. *J. Comput. Sci. Inf. Technol., 2*, 131–148.
3. Pudjianto, B., Zo, H., Ciganek, A. P., & Rho, J. J. (2011). Determinants of e-government assimilation in Indonesia: An empirical investigation using a TOE framework. *Asia Pacific J. Inf. Syst., 21*, 49–80.
4. Kelly, T. M., & Bauer, D. K. (2003). Managing intellectual capital -via e-learning at Cisco. *Handbook on Knowledge Management,* 511–532
5. Alqahtani, A. Y., & Rajkhan, A. A, (2020). E-learning critical success factors during the COVID-19 pandemic: a comprehensive analysis of e-learning managerial perspectives. *Education sciences, 10*, 216. https://doi.org/10.3390/educsci10090216.
6. Crawford, J., Butler-Henderson, K., Rudolph, J., Malkawi, B., Glowatz, M., Burton, R., Magni, P. A., & Lam, S. (2020). COVID-19: 20 countries' higher education intra-period digital pedagogy responses. *Int. Perspect. Interact. Educ., 3*, 1–20.
7. Naveed, Q.N., Muhammed, A., Sanober, S., Qureshi, M.R.N., & Shah, A. (2017). Barriers effecting successful implementation of e-learning in Saudi Arabian universities. *International Journal of Emerging Technologies in Learning (iJET).* https://online-journals.org/index.php/i-jet/article/view/7003.
8. Basak, S. K., Wotto, M., & Bélanger, P. (2016). A framework on the critical success factors of e-learning implementation in higher education: A review of the literature. *World Academy of Science, Engineering and Technology International Journal of Social, Behavioral, Educational, Economic, Business and Industrial Engineering , 10*(7).
9. Selim, H.M. (2007). E-learning critical success factors: An exploratory investigation of student perceptions. *International Journal of Technology Marketing, 2*(2). https://doi.org/10.1504/IJT MKT.2007.014791.
10. Smith, J. (2001). Content delivery: A comparison and contrast of electronic and traditional MBA marketing planning courses. *Journal of Marketing Education, 23*(1), 35-46.

11. Tuomi, I. (2006) Open educational resources: what they are and do they matter. OECD Report (2006) Reviewed from http://www.meaning#processing.com/personalPages/tuomi/art icles/OpenEducationalResources-OECDreport.pdf.

12. Zhao, Z., & Yang, J. (2019). Design and implementation of computer aided physical education platform based on browser/server architecture. https://doi.org/10.3991/ijet.v14i15.11146.

13. Viner, R.M., Russell, S.J., Croker, H., Packer, J., Ward, J., Stansfield, C., Mytton, O., Bonell, C., & Booy, R. (2020). School closure and management practices during coronavirus outbreaks including COVID-19: A rapid systematic review. *Lancet Child Adolesc. Health., 4*, 397–404.

14. Campbell, C. (2020) IGI global newsroom "Experts Weigh in on the Future of Education Post-COVID-19".

15. Krieger, Z. (2007). Saudi Arabia puts its billion behind Western-style higher education. *The Chronic of Higher Education, 54*(3), A1.

16. King Abdulaziz University: E-learning and online education (2011). Reviewed from http://ele arning.ksa.edu.sa/Content.aspx?

36 Al Zoubi, M.I. (2013). The Role of Technology, Organization, and Environment Factors in Enterprise Resource Planning Implementation Success in Jordan.

37 Basheer, S.G., Ahmad, S.M., Tang, Y,C. (2013). A Conceptual Multi-agent Framework Using Ant Colony Optimization and Fuzzy Algorithms for Learning Style Detection. ACIIDS 2013: *Intelligent Information and Database Systems*, 549–558.

19. Moisio, S. (2020). State power and the COVID-19 pandemic: The case of Finland. *Eurasian Geogr. Econ, 61*, 598–605.

20. Sallum, S.A. (2008). Learning management system implementation: Building strategic change. www.iiis.org/cds2008/cd2008sci/EISTA2008/PapersPdf/E2955V.pdf.

21. Habib, F., & Islam, S. (2011). *Is Saudi ready for E-learning? A case Study*. Najran University Press.

22. Mishra, L., Gupta, T., Shree, A. (2020). Online teaching-learning in higher education during lockdown period of COVID-19 pandemic. *Int. J. Educ. Res. Open, 1*, 100012.

23. Mayer, J. D., & Lewis, N. D. (2020). An inevitable pandemic: Geographic insights into the COVID-19 global health emergency. *Eurasian Geogr. Econ., 61*, 404–422.

35 MOE. (2021). https://moe.gov.sa/en/knowledgecenter/dataandstats/Pages/educationindica tors.aspx..

25. Tanveer, M., Bhaumik, A., Hassan, S., & Haq, I. U. (2020). Covid-19 pandemic, outbreak educational sector and students online learning in Saudi Arabia. *J. Entrep. Educ., 23*, 1–14.

26. Favale, T., Soro, F., Trevisan, M., Drago, I., Mellia, M. (2020). Campus traffic and e-Learning during COVID-19 pandemic. *Computer Networks, 176*, 107290.

27. Gonzalez, T., De La Rubia, M.A., Hincz, K.P., Comas-Lopez, M., Subirats, L., Fort, S., Sacha, G.M. (2020). Influence of COVID-19 confinement on students' performance in higher education. *PLoS ONE, 15*, e0239490.

28. Abdulrahim, H., & Mabrouk, F. (2020). COVID-19 and the digital transformation of Saudi higher education. *Asian J. Distance Educ., 15*, 291–306.

29. Krzysztofik, R., Kantor-Pietraga, I., & Spórna, T. (2020). Spatial and functional dimensions of the COVID-19 epidemic in Poland. *Eurasian Geogr. Econ., 61*, 573–586.

30. Kebritchi, M., Lipschuetz, A., Santiague, L. (2017). Issues and challenges for teaching successful online courses in higher education. *Journal of Educational Technology Systems, 46*(1), 4–29. Google Scholar | SAGE Journals.

31. Affouneh, S., Salha, S., & Khlaif, Z. N. (2020). Designing quality e-learning environments for emergency remote teaching in coronavirus crisis. *Interdisciplinary Journal of Virtual Learning in Medical Sciences, 11*(2), 1–3.

32. Liu, J., & Fu, R. (2018). Development of an accounting skills simulation practice system based on the B/S architecture. *International Journal of Emerging Technologies in Learning, 13*(10), 134–145. https://doi.org/10.3991/ijet.v13i10.9459.

33. Bendahmane, M., El Falaki, B., & Benattou, M. (2019). Toward a personalized learning path through a services-oriented approach. *International Journal of Emerging Technologies in Learning*. https://doi.org/10.3991/ijet.v14i15.10951.

34. Lee, B. C., Yoon, J. O., & Lee, I. (2009). Learners' acceptance of e-learning in South Korea: Theories and results. *Computers & Education, 53*(4):1320–1329.
35. Flores, M. A., & Gago, M. (2020). Teacher education in times of COVID-19 pandemic in Portugal: National, institutional and pedagogical responses. *J. Educ. Teach., 46*, 507–516.
36. Wang, L. (2017). Personalized teaching platform based on web data mining. *International Journal of Emerging Technologies in Learning, 11*(11): 15-20. https://doi.org/10.3991/ijet.v11 i11.6253.
37. Azzi-Huck, K., & Shmis, T. (2020). Managing the impact of COVID-19 on education systems around the world: How countries are preparing, coping, and planning for recovery". *Education for Global Development.*

Online Examination Systems, Preventing Cheating and the Ethical, Social and Professional View

Online Examination Systems, Presenting Cheating and the Ethical, Social and Professional View of Them

Abeer AlAjmi⬲

1 Introduction

One of the consequences of the COVID-19 pandemic, is the effect it had on the traditional education systems. Teaching methods took on a decidedly different form and a new pathway of teaching has evolved. Educational institutions had to adapt to new methods in order to deliver course materials and conduct online exams.

New, effective and smart teaching platforms were implemented. This was orchestrated in a very short and time in an attempt to allow the educational wheel to keep spinning, especially during the early stage of the COVID-19 pandemic.

Early in 2020 a new global pandemic appeared and quickly spread all over the world. The effects were devastating and affected every aspect of life, from humanity from simple daily life activities, including going to school. Technology turned out to be a blessing, facilitating life rhythm during that time. Although many educational institutions all over the world used online teaching systems before the pandemic, it now became a staple. Face to face interaction has always been preferred in a normal teaching system. Unfortunately, education in this manner was completely disrupted.

Necessity is the mother of invention, so the pandemic was responsible for a new revolution in the field of education. New smart online platforms were designed to fit the needs and the purposes of different educational needs. Teachers can see students and can discuss course materials, assessments can be uploaded and graded. However, online teaching has proven to have both advantages and disadvantages.

One of the primary advantages of the online teaching system, from my own perspective as an instructor, is that it meets the basic needs of students with special needs and educational limitations. The online teaching system presents handy data and many creative tools where students, regardless their situation, can access the

A. AlAjmi (✉)
Box Hill College Kuwait, 29192, 13152 Safat, Kuwait
e-mail: Abeer.AlAjmi@bhck.edu.kw

© The Author(s), under exclusive license to Springer Nature Switzerland AG 2022
A. Hamdan et al. (eds.), *Technologies, Artificial Intelligence and the Future of Learning Post-COVID-19*, Studies in Computational Intelligence 1019,
https://doi.org/10.1007/978-3-030-93921-2_10

information and submit their work on a safe database. There has been so many implementations of smart technologies and artificial intelligence which has made interaction between students and their instructors easy and functional.

Hunting information and self-instructing is another advantage of online learning. Digital learning shifted some load from teachers to students; online courses tend to be more generally theoretically presented which reduces research importance.

Online examination methods took a dramatically different path during the pandemic, new solutions have been devoted and designed to maintain most ethical academic learning environment. There are many types on online examinations and many improved digital online proctoring and grading systems. Authentication methods are the most important technical tools that help sustaining strong ethical online examination domains.

2   The Phenomena of Online Learning Post Covid-19

Due to the COVID-19 pandemic of 2020, the world witnessed a new revolution of technology and the implementation of many online services. For many educational institutions it has been crucial to use functional online platforms to pursue studying and deliver course materials. For many students, in different educational stages, this has been a new experience which required preparation and the acquisition of new skills. For students, the main challenge at this time was to learn how to access information online. However, for faculty members the biggest challenge has been the security and integrity of submitting assessments and exams.

The online examination environment required an extra awareness of exam format and grading methods. Taking tests at home does not emulate in class tests in many ways. Many test formats had to be modified for many reasons. One reason is the time zone and the location of the test taker. This requires the faculty member to enhance the exam electronically so students from different locations and time zones can access the exam and have the same fair time to answer questions and deliver their response. Another reason is the quantity of the course materials that needs to be covered during the pandemic. Many institutions had altered their teaching strategies, including presentations, assessment briefs, grading rubrics and exams. However, for many students this turned out advantageous, as they were not stressed. Students taking exams do not have to be in class and feel the tension of exam day. For many students, taking online courses mean depending on themselves to find the information, although teaching platforms allows many communication tools with faculty members. Emails, direct messages, voice notes, pictures coma videos and other available avenues can be sent and received both ways between instructors and students at any time, which allows a faster response to questions and information.

Post the pandemic, educational institutions that continued to offer online courses developed their online platforms, and that encouraged students to pursue their studies. Students of great distance or international students have registered in many universities and colleges. The advantage here is the savings of part of their educational expenses. Many educational institutions took into considerations the situation of what the world is going through and facilitated the payments along with other student requirements. Moreover, special needs students were able to use online platforms as many artificial intelligence tools were applied to these platforms. For example, there have been many audio recording tools that helps students to record their answers during the exam. Another example is that artificial intelligence tools can identify facial expressions by using the webcam to identify students. This allows students, especially speech case students, to submit their assessments and take exams securely. This limit cheating and plagiarism by being able to identify student ID or code using their faces or other identification methods thus ensuring that it matches the student on the system.

Higher educational institutions offering different majors had to develop different exam methods and had to be flexible with the way these exams are administered. "In March, Imperial College London successfully assessed their 280 sixth-year medicine undergraduates via two online exams." According to Dr. Amir Sam, Imperial's head of undergraduate medicine; "To the best of our knowledge, this is the first digital 'open book' exam delivered remotely for final-year students." [1].

Conducting a remote exam requires special security tools and software systems. To achieve that, most universities use their official websites, dashboards or blackboards, uploading many guidelines for students to follow before taking their exams. There are many integrity tools that have been customized, such as test monitoring software or browsers. Technically, the aim of these browsers or software is to control and assure a maximum and flawless test environment.

In order to understand how these tools are implemented and function, while students are taking the exam, it is imperative to first understand the type of online tests and the submission methods.

Through the web, there are many types of online exams that students can participate in. Each requires a certain delivery, proctoring and grading system. In many cases students are taking a mock test or a practice exam prior to taking the real exam to check their strengths and weaknesses. Along with this, students have an opportunity to experience any web limitations. This gives them a chance to plan their time, organize their ideas and get familiar with the platform. Also, it gives the students a chance to find errors and computer constraints which can help void any last minute problems before the exam.

Before the exam, faculty members need to highlight some important instructions and details for students prior to the exam day. It is important to state the date and the time of the exam according to the country where the educational institute is located. This is important especially for students taking exams from other countries with different time zones. Also, it is important to specify the format of the exam and required materials as well as proctoring and grading techniques.

To solve the time zone and location problems, many universities allow for a twenty-four-hour standard time window for students to complete their exams. This does not necessarily mean that the students take the entire 24 h on one exam, instead this gives students extra time to finalize their exams. This is a great advantage especially for students living in places that lack good internet connection. Also, another factor is the many religious or ritual practices that exist in different cultures. For the students who are done with their exams before the end of the 24 h time window closes, they can submit their exams and thus possibly reducing stress. This is especially true if they are submitting multiple exams at the same time.

As students from different locations and different time zones might take the same exam, many can finish and submit the answers before their colleagues. Thus, universities are advising and forbidding students from discussing exam questions with their colleagues. Students may be accused of cheating attempts if the university discovers and verifies this. As a solution to this dilemma, many universities are preparing different exam models to minimize cheating and plagiarizing attempts.

Essentially, online examination instructions vary from one institution to another, and from one department to another. However, there is some common or general instructions that most universities have implemented for their online users. Students are reminded about the value of required browsers such as Google Chrome, Firefox or Internet Explorer.

3 Types of Online Exams

3.1 Live Online Exam

COVID-19 pandemic has altered the world significantly. Both faculty members and students, especially in universities, had to learn online platforms such as Zoom and Teams in a short amount of time. Such platforms are customized to simulate the classroom environment. During live exams, instructors can proctor students and communicate with them instantly. Students can raise their hands, ask questions and privately chat with instructors without disturbing their colleagues. Also, instructors can supervise the exam using webcams to make sure the students are not cheating.

These platforms however are only a communication method or the portal the student can access to view the exams. However, to make sure that the students are not cheating there are other integrity software that protects against the likeness of plagiarism while students are taking the online exam.

Through artificial intelligence and smart tools these security or integrity software assist on live online exams to monitor the exam takers.

First the webcam can identify the students and compare their facial expressions to their registered ID's on the system. The webcam can also detect suspicious facial expressions, even if the student is talking to someone else in the room even when the mic is off. Also, in many cases before students start the exam the instructor requires the student to take a 360-degree view of the room to make sure there are no one else but the student in the room, as well as no open notes or books. Such software is superior in a way that can also detect suspicious movement in the room where the students is taking the exam in, such as other people or moving objects like books.

In real time exams students are given a limited amount of time to answer all test questions, regardless of the style of the exam. Security software can assist the student with many reminders like saving the answers or showing altering them if the time is almost up. It also facilitates the saving format just in case the students were not able to auto save the documents on a certain format. The security software can provide support in that area or at least save a draft copy.

Security software disables many tools on the computer while the student is taking the exam online such as the printing, screen capture, copy, paste and browsing the net through any search engine. The live exam can be an open book exam, multiple choice exam or any other test format instructors wish to customize through previously mentioned software. Moreover, if the student agrees in many cases to take the live exam, the software might allow the instructor to share the students test screen.

3.2 Recorded Exams

These exams require the student with security software to save either the video or audio test. The procedure of such test format depends on the artificial intelligent tool embedded within the test software that recognizes and monitors visual and acoustic indicators.

The integrated features of AI allow the test taker, using the webcam, to explore surrounding environment. However, the recorded exams may be difficult to supervise by the instructor as the student is recording the video ahead and might use blind spots or attempt to find a loophole through the computer.

Moreover, recorded exams are prepared ahead, which means that the student might find a way to copy, paste, cut or even get help in preparing for the exam. Many instructors tend to use this exam method if the exam is practical, so no one but the student has the knowledge about the exam topic. An example might be Interior Design students and Architecture students when they present a final project. This exam method grading rubrics can be analytical, it is preferred to be graded after it gets critiqued by instructors.

4 Online Examination Proctoring Systems

Before the Covid 19 pandemic, many universities and colleges were the pioneers in adapting remotely administrated online examination systems. Specifically, educational institutes depended, to a certain extent, on technology to accommodate international students. As well, many institutions have satellite campuses all over the world. As the demand on online education increases, there has been an embarked use of effective exam proctoring and grading systems. Thanks to technology and AI, education has become a handy tool, in spite of some challenges.

In their article "Towards Greater Integrity in Online Exams" that was published back in 2016, both Sinjini Mitra and Mikhail Gofman form California State University are highlighting proctoring methods of online exams using the computer's webcam and biometrics-based proctoring that monitors a student's mouse movement. As well, there are other vital signs such as head or eye movements, which can be useful in detecting attempts at cheating [2].

Remotely proctored exams start with verifying the identity of the exam taker. How do we make sure that the registered student is the one taking the online exam?

As mentioned earlier in this chapter, there are many integrity software, such as widely used LockDown Browser, to secure online examination environment.

"The traditional approach to secure assessment involves physical control over the examination environment and identity verification. Human invigilators verify identity documents against physical appearance of the document holders prior to granting access to the examination materials. In case of remote proctoring, user identification may be performed using an array of technologies including video cameras and physical or behavioral biometrics." [3].

Audio and video authentication methods has been used in online examination using AI and advanced software technologies to assure maximum performance quality and integrity during e-examinations (Fig. 1).

According to Mary Clotilda, Creatrix Campus online examination software is an example of online exam preparing platforms that secures online exams for higher institutions, remote digital proctoring services and auto proctoring.

In her article, published in June 2020, Clotilda explains eight online exam security features that will help assure secure online examination and proctoring environment during and after Covid-19. This starts first with an exam cheating prevention system, this system is installed to detect errors and suspicious actions during the exam. Then, with a major dependency on the webcam, instructors or proctors can share the screen with students and administrate the exam process. A proctor can view exam monitor, screen capture and see the test takers surrounding environment while maintaining a high level of student privacy.

One major key responsible for the success in this loop is the LockDown Browser. This tool will freeze the exam taker's device wither it's a computer, I pad, smart phone or any other device. This will not enable the device user to bypass security features and will not be able to shuffle between web windows. The tool will detect any attempt to copy, paste or print from other programs on the same device as well. This will help in detecting plagiarism giving the instructor or proctor the permission to end the exam or shut down the exam window on the student device.

Moreover, the administrator of the online exam can shuffle questions and answers if for any reason they are alerted that the exam taker might be copying answers during the exam. For example, if the exam is multiple choice question format, the administrator can change the order or questions and make the exam look different so students can't copy answers from each other under any circumstances.

The installed software not only protect exams from being plagiarized and prohibit students from cheating, it also safely secures student data and submitted projects.

Online examination system

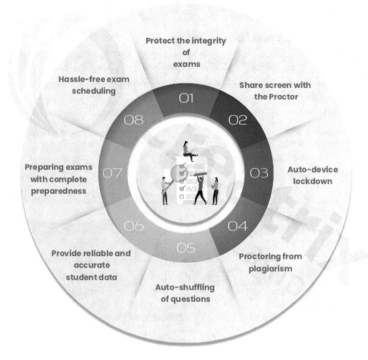

Fig. 1 Example of secured online examination system to minimize academic dishonesty

Instructors can instantly grade student performance and confidentially assess their work. Student can have future feedback and reflect to it.

Figure 2 demonstrates the format style of LockDown Browser. It looks like a standard screen window of a test with minimum options. Options in the red box are the only available tools for the test takes. Administrators or instructors can customize the exam to be multiple choice, essay, or any other format. LockDown works with both Windows and MAC operating systems. It is also available for iPads and other ISO systems. However, devices with reduced systems will not work appropriately with the browser. The browser needs to be downloaded before the beginning of the semester and has integrated webcam and microphone tools to help students, especially special cases, to pursue their exams.

To use such a browser, especially for the first time, student need to be trained on how to use the browser. They need to know how to log in, log out, save the exam and refresh in case the system lags. Holding remotely digital exams is constrained with technology efficiency. If the Internet is disconnected, then the exam needs to be automatically saved so students do not fail.

According to NUKSks, National Union of Kuwaiti Students, on their official twitter page, some students in Kuwait Universities mentioned that in many cases the

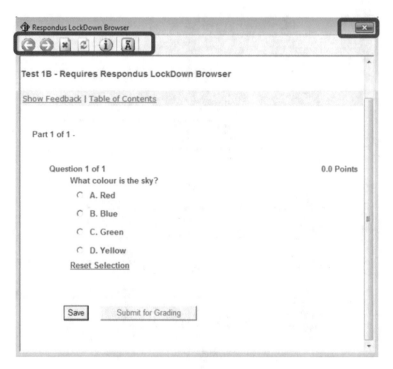

Fig. 2 Example of LockDown browser online exam page

exam window lagged as they tried to save and move on to log out. In fact, the system returned them back to the previous step. The reason behind this might be the load on the exam browser by the huge number of students and users. The browser allows the student to have an instant chat with proctors and address any unexpected incident. In this case proctors can extend exam time or check the system errors [4].

In her article, Clotilda is also touching on one important perspective when it comes to online examinations which is "Exams Question Bank". She is illustrating that Creatrix Campus is recruiting AI to help automatically pull exam questions in different formats after student's log in to the system and verify their identities. Next, the software can automatically save the answers and analyze them. The exam can be graded, and the student might get feedback reports instantly. She also emphasizes the ease this software brings with time exams, so students have the flexibility to access scheduled exams and answer them within a controlled environment. Exam validations extensions are more capable to deal with over any other online or device interruptions.

5 Validating Exam Taker Identity

The usual username and password authentication tool can be paired with a biometric authentication tool. In normal class exams, instructors would have already developed a relationship with students during the semester which is considered a moral connection. Instructors can see the students' facial expressions, discus class issues and develop an impression on each student. Where in online courses, instructors face difficulties to reach out to students' impressions unless the students reacted and responded to the teacher.

Thus, the biometric authentication tool has helped recognize behavioral characteristics of individuals in class. Physiological and physical signs such as an individuals' voice recognition, fingerprints or facial expressions would be detected and constantly asked for any time during the exam. This tool helped to measure reactions by students and also was useful in analyzing face and body movement.

Another exam taker authentication tool is the keystroke. This is a knowledge-based tool where students can be asked random personal questions about themselves. This is similar to questions asked by a bank. Over the phone, the bank may ask for the account number, or personal and private questions. This is done in order to maintain a high level of quality service and to also assure the identity of the customer.

In many academic institutions, several previously mention authentication tools are used together in many cases. This is necessary as plagiarism is the most pressing problem during online examinations, according to scholars.

First let's define plagiarism. "The practice of taking someone else's work or ideas and passing them off as one's own." https://www.lexico.com/definition/plagiarism. It is a challenge by itself for faculty to proctor online exam and monitor students through webcams. Therefore, plagiarism detecting tools are required. These tools detect similar answers between students and at the same time detect copied answers from the web. Planarization detecting tools work by utilizing algorithms and calculations to compare similarities in submitted exams or assessments. This limits cheating attempts in research-based exams yet might not be helpful in different exam formats.

Plagiarism detecting tools has limitations in securing many e-exams or e-assessments format such as multiple-choice questions format. Moreover, these tools work more efficient with essay or project-based e-examinations. These tools do not eliminate the role of the supervisor or the instructor, yet it highlights existing errors in the answers.

Through analyzing answers and comparing it to database, plagiarism detecting tools enable instructors to detect repeated patterns of cheating attempts, which shows an index and importance of the validity of these tools in today's online educational framework.

Most educational institutes are appreciating the role of technology and started adapting these technologies to maintain the continuity of education quality.

Why do students cheat in online exams? This chapter has presented tools to prevent online examination cheating attempts and also presented validation systems that support reducing these attempts. Yet, there must also be discussions on the reasons explaining why students cheat during online examinations.

Most likely, the obvious reason majority will agree on is that the student did not have enough time to study or is after better grades. However, this cannot be a generalize that as there are many students with integrity, creativity and smart enough to complete the course without cheating.

Instructors have the role of reinforcing the personal development of student's mentality and innovation. If the instructor made the entire semester experience all about grades, students will automatically feel free of creativity and motivation to compete with themselves. There is an important role teachers and instructors play in the student's journey. This can minimize the students desire to go easy with cheating rather than learning and improve their potentials. Jesse Stommel, senior lecturer in digital studies at the University of Mary Washington and co-founder of the Digital Pedagogy Lab, said "We have to start by trusting students and using approaches that rely on intrinsic motivation, not policies, surveillance and suspicion." [5].

So what could motivate students to cheat? First, instructors need to make students believe that a grading system is required only to measure their understanding level of course materials and not their qualified potentials. That is not the goal of learning, yet it will be accomplished eventually if the student has the passion to learn and make a difference. In many cases, students feel that it is a must to take each course without knowing why it was listed as a requirement in the curriculum. The students need to saturate knowledge surrounding their majors as it is the key to a successful profession. Not only the instructors, but the entire education system needs to make the student uncaged by the classroom walls and free to explore, discuss, negotiate, and have an opinion about what they are studying.

"Cheating in some professions can hurt or injure people. In architecture, it might result in an ill-constructed building. Or it could cause permanent damage in surgery. So I wanted to find a way to help students learn something for the betterment of themselves, for the betterment of their profession and the betterment of society as a whole."—Megan Krou, Analytics and Research Technology Specialist at Teachers College's Digital Futures Institute.

Moreover, by building their confidence, student will not feel bad about making mistakes in a safe learning environment. These mistakes will only make them better and stronger when facing challenges.

Students, on the other hand, need to fight learning boredom. Interesting teaching methods and self-reliance will result in less cheating attempts as student will always feel independent and inspired to learn.

Finally, "loosen the rope of learning." Students need to feel their worth, trust and reliance by providing them choices and putting them in charge of their own learning responsibilities rather than telling them what to do as machines. A blooming mind needs a nourishing environment.

What about the academic staff? Do they get to have a role in students cheating attempts? In many cases, unfortunately, the answer is yes!

In many academic departments, especially in online examinations, many faculties though it is useless to change the exam questions bank or method as online students are far from contacting previous students and can't interact with each other. Unfortunately, leniency in this matter exacerbates cheating in exams, this will not encourage the students to try and study, contrary, it will make them less interested in understanding the course material.

In their article, Achievement Motivation and Academic Dishonesty: A Meta-Analytic Investigation, Megan R. Krou, Carlton J. Fong and Meagan A. Hoff conducted 79 studies where "meta-analytic results indicated that academic dishonesty was negatively associated with classroom mastery goal structure, individual mastery approach goals, intrinsic motivation, self-efficacy, utility value, and internal locus of control".

"According to the International Center for Academic Integrity, academic integrity is defined "as a commitment, even in the face of adversity, to six fundamental values: honesty, trust, fairness, respect, responsibility, and courage" [6].

There is no flawless system at all, even in education. All systems might face errors and corruptions yet there will always be modification attempts. The new phenomenal of studying online and e-learning is still facing a challenge to pursue efficiently and genuinely. In normal situation of in class exam, the faculty might be exposed to different cheating methods such as student sneaking unallowed materials like notes, pieces of papers or calculators. In many exams, faculties strip students from all smart devices like smart phones, iPads or even smart watches as they all can be linked to unwired earphones. As high- tech devices are getting widely available in market with cheap prices, lazy students might take advantage to the easy accessibility of these cheap handy tools to cheat. For example, smart phones can be connected to smart watches and students can share pictures during the exam, text and audio messages.

The use of high-tech device for the purpose of cheating are not only limited on paper-based exams or in class exams. Today, with various presented online classes, lazy students can recruit technology to cheat. One of the worst fears most educational institute been fighting against is the online paid tutoring.

As e-learning still with all the connection tools between instructors and students lacks sense of in class presence, this might create a gap in which student feel lazy and careless to complete assessment and exams. Covid 19 pandemic and other circumstances that forces student to pursue their study online created kind of sensational disconnection form normal learning environment. Many online tutoring websites take the advantage of that separation and invested well through helping cheating students in their assessment and exams.

This is another reason of why universities and higher educational institutes are forcing and magnifying the role of authentication and identity tools on their online platforms.

6　Online Education Post Covid-19

Due to the pandemic situation of Covid-19 early in 2020, virtual education faced a lot of challenges. The biggest challenge was the need for the immediate shift of all courses, especially in higher level education, to online. As well, there was a need to change the methodology of assessments and exams submissions within a short time. As distance learning is not a recent trend, all participants need to be educated with the new designed platforms.

As e-learning started to spread there has been some concerns regarding many websites that offers help to students in their assessment and exams. Online tutoring, substances attendee to online exams, plagiarism in research and other concerns need to be taken under consideration as it is hard to control remote education.

According to Colleen Flaherty, in his article published on May 11, 2020, Boston University and Georgia Institute of Technology are two examples of Universities who took a step forward investigating websites offer help to students such as Chegg to determine attempts of "illicit help" [5].

There has been an opposing contrary perspective with the role of plagiarism software in online learning systems. Turnitin is widely use "plagiarism detection software" as some call. It is an auto detecting tool that might find text matches in students answers with the Internet Web. Many faculties have concerns that there is a risk as this might establish an adversarial relationship between students and teachers. This might evolve a negative impact on both parties as instructors will always predict mistakes, act like a surveillance team, and see the student as possible cheaters. Students will invariably always try to prove their honesty.

So, what about large number of students in some courses? How can faculties proctor online exams with a class that has a three-digit number of students?

Andrew Robinson, instructor of physics at Carleton University in Canada, applied the principle of "deep question bank" with "randomized variables in calculations". Robinson aligned with the idea of creating different format of the same exam on the same day and just try to make it difficult for student to cheat that way. He said "You can either implement surveillance tech to prevent that, which will inevitably be bypassed eventually, or come up with different assessments" [5, 7].

Covid-19 pandemic is a demonstrative situation to how most universities around the world turned to Cloud technology and altered practical examination format. Laboratory-based practical, performance-based courses, physical, artefact and inter-personal courses requiring face to face exam or presentation shifted to online based tests. A new virtual learning environment needed to be custom made to support education under such circumstances. On the other hand, such examination methods need collaboration proctoring methods. The Turkish Council of Higher Education (CoHE) conducted a survey regarding the outcomes of online education where 1,255,022 students and 27,820 academic staff from 207 universities responded to the survey. The survey aims to measure the challenges both participants are facing during that shift learning and teaching are facing in Turkish higher education.

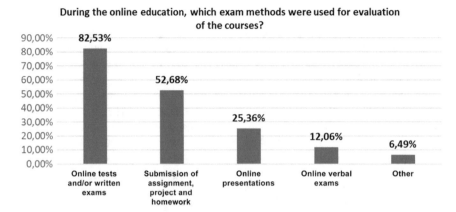

Fig. 3 A chart illustrates different percentages of preferred online exam evaluation methods based on the survey conducted by CoHE

Survey takers were asked their opinion regarding online exams. They were asked about preferred evaluation methods. The percentage of participants preferred to conduct online administered tests and/or written exams compared to other methods (Fig. 3).

After the online experience, during Covid-19 pandemic, the survey presented questions to outlook participants perspective of education after the pandemic. Most of the participants preferred they go back to face to face over digital communication and recommended in class tests over online tests.

Academic staff have both negative and positive responses regarding online teaching experience. On one hand, and based on the survey, 63% of the academic staff agreed on poor student attendance to online classes, where 39% agreed that their research was affected negatively. Seventy-four per cent of the academics indicated that their technological and teaching skills were improved due to online teaching.

Surprisingly, around 49% of faculties supported blended education between the online and face to face education post the pandemic and 44% want life to go back to the way it was and preferred face to face education. The two percentages are close from each other which indicates that educational circumstances need of technology and digital is taking a new curve.

7 Conclusion

It is obvious that digital learning will dominate wider domain of higher education systems in the coming future. Pedagogical methods will be evolved thanks to the effective implementation if Artificial Intelligence especially post Covid-19.

Although e-learning existed before the pandemic, for some fields, the pandemic forced pending digital challenges to explode thus improved with the improvement of technology. There are many new prevalent avenues of intellectual communication and artificial intelligence supported online learning to pursue though the pandemic by recruiting smart software. A new horizon of secured academic environment simulating actual classroom environment is now provided by different designed online educational platforms.

As cheating and plagiarism are most e-learning worst fears that might affect the quality of education, new solutions had to be customized within limited time to overcome online exam limitations. There are many reasons to why students are cheating and many resolutions to morally and ethically encourage students to enjoy the online learning experience virtually. Many educational higher educational institutions are improving different aspect of academic learning quality as they are adapting futuristic solutions. Biometric authentication is only one of these brilliant adapted and continuously updated systems. A revolution in digital literature methods is now available and ready for future educational variables.

This paper is discussing the transformation of ethical online pedagogical methods in general, compared to traditional on campus learning environment, and tackling flawless online examination procedures relying on Artificial Intelligence tools in specific.

Authentication tools and online examinations security browsers are two-winning accomplishments this chapter is negotiating. Many universities and higher educational institutions adapted exams software solutions to manage online examinations remotely and safely.

There are many similarities and differences between online examination systems and the typical exam paper systems. Proctoring and grading methods are discussed in this paper as the findings were supported with many statistics. Both staff and students' opinions had a crucial role in determine suitable situations in different cases.

References

1. James, F. (2020). The challenges and advantages of conducting exams during the COVID-19 crisis. Retrieved 2020, from https://www.qs.com/the-challenges-and-advantages-of-conducting-exams-during-the-covid-19-crisis/.
2. Mitra, S., & Gofman, M. (2016). Towards greater integrity in online exams. Retrieved 2020, from https://aisel.aisnet.org/amcis2016/ISEdu/Presentations/28/.
3. Amigud, A., Arnedo-Moreno, J., Daradoumis, T., & Guerrero-Roldan, A. (2016). A behavioral biometrics based and machine learning aided framework for academic integrity in e-assessment. In *2016 International Conference on Intelligent Networking and Collaborative Systems*. Retrieved 2021, from https://www.researchgate.net/profile/Thanasis-Daradoumis/publication/309496443_A_Behavioral_Biometrics_Based_and_Machine_Learning_Aided_Framework_for_Academic_Integrity_in_E-Assessment/links/5a1fcf610f7e9b9d5e02ceb0/A-Behavioral-Biometrics-Based-and-Machine-Learning-Aided-Framework-for-Academic-Integrity-in-E-Assessment.pdf.
4. National Union of Kuwaiti Students [@NUKS]. (2020, January 20). نظراً لكثرة الشكاوي التي وردتنا في ما يخص استخدام بعض أعضاء هيئة التدريس برنامج respondus و Lock down browser
5. .حيث أن هذه البرامج تسببت بحدوث خلل في الأجهزة الإلكترونية.

6. جامعة_الكويت# #NUKSKU. وعلى ذلك، تم تقديم هذا الكتاب للإدارة الجامعية لإتخاذ اللازم [Tweet].Twitter. https://twitter.com/NUKSku/status/1351937583513821184.

7. Mokadem, W., & Muwafak, B. M. (2021). The difference of the theoretical approach of corporate social responsibility between the European Union and United States of America. *International Journal of Business Ethics and Governance, 4*(1), 124–136. https://doi.org/10.51325/ijbeg.v4i 1.22.

.

An Ethical Discourse on Learning, Communication, and Intersubjectivity in Reference with Digital Technology: A Panacea in the Time of COVID-19 Pandemic

Sooraj Kumar Maurya ⓘ

1 Introduction

The term "intersubjectivity" has been used in a variety of areas of study, including sociology, history, psychology, and anthropology, to describe the intrinsically social essence of human life. Although all intersubjective conceptual frameworks describe the self-other interaction, the literature allows one to distinguish between various focus groups when it comes to the key locus of intersubjective mechanisms. It tends to be a transcendental occurrence, similar to certain conceptual frameworks of intersubjectivity. Intersubjectivity, in relation to subjective perception, refers to the realm of primordial, maternal truth, which has been conceived as an all-encompassing continent [1]; it is the experience of an accommodating, comforting soil, wherein separateness emerges as a component of subjective perceptions, not only through resistance or conflict but also through its primordial incorporation nature.

The absolute differentiation between self and other is the place to start for considering intersubjectivity herein. We are forced into an unconscious view of the universe, according to Heidegger, because we have no alternative. This tacit perception, which is often comprised of our various subjective experiences and interpretive options for the artefacts with which we come into touch, ends up defining us in the sense of culture. In this context, we are still under the influence of the individualistic space of expectations that defines and restricts the circumstances of our perception and the acts we may take. It is the practice that has gone before us and continues to follow us, and it must be recognised as something that has shaped who I am. According to this perspective, intersubjectivity entails a form of involvement in the real and material universe that may not even necessitate an automatically perceptible human existence. Our view of the environment has already been turned into a "cultural world" through human intervention and labour [2]. Merleau-Ponty stressed the primaeval essence of

S. K. Maurya (✉)
Ramanujan College, University of Delhi, New Delhi, India

bodily experiences similarly, explaining how we date back any earliest components of self, finding ourselves engaged in collaborative and extracorporeal mechanisms well before any primeval modes of social knowledge take effect. Although experiences must acquire intersubjectivity with other individuals, the brain development potential and intrinsically social predilections that enable education makes this possible.

Moreover, intersubjectivity has therefore been regarded as an inter-personal feature in classical evolutionary and correlative psychology, as such a newly emerging attribute of the evolving human inner self affection abilities, some of which are even active in neonates. For instance, there is a lot of research into how human neonates are birthed to adapt to faces and communicate with each other by expressions and verbal interactions. Since repetitive encounters feed these impulses, they develop a sense of intersubjective empathy through mutual experiences. As per this physical mimesis claim, these primordial modes of intersubjectivity soil the growth of vocabulary, increasing the production of intersubjectivity to even higher stages. As a result, while initial social interactions are critical for the growth of intersubjectivity, the inter-personal viewpoint on intersubjectivity is focused on certain primaeval human social capabilities that predate social communication systems [3]. In other interpretations of intersubjectivity, the concept is used not as a prerequisite for the emergence of the interpersonal domain but also as the actual interactional mechanisms that socially bring the interpersonal domain into existence.

Intersubjectivity, in this context, applies to the area of connections established in and by encounters between persons, and hence may be called an interactional concept—even though, paradoxically, individuals' assumed expectations regarding an already occurring intersubjectivity perform a key part in assisting them in achieving it. Under this framework, we may differentiate between two approaches to stressing the interactional mechanisms that are most important for achieving intersubjectivity [4]. However, intersubjectivity is achieved through a dynamic systems conversational mechanism of cooperation. The resultant emergent forms of coordination gain significance above and beyond the significance of the person's behavior [4]. Some intercultural communication behaviour in which respondents organise or co-regulate their specific intercultural communication inputs, according to this viewpoint, lends itself to intersubjectivity research.

Furthermore, through a series of widely used normative practises from which behaviour and positions may be interpreted in an understandable manner, intersubjectivity is accomplished in a comprehensible manner, according to ethnomethodology and discourse study. According to this viewpoint, human interaction is rooted in Goff man's communication order described as the reasonably stable regularities and mechanisms of social communication, which should be viewed as a substantial realm in and of itself. As previously said, all conceptual frameworks of intersubjectivity include a description of the partnership between self and Other and an assessment of the types of mechanisms that underpin that partnership in each situation [5]. Such various methods can tend to be incompatible at first glance. We contend, though, that they should not have to be. We would mostly use the intercultural communication conceptual model of intersubjectivity in this article, which includes both procession and procedural emphases present in the literature. We choose this viewpoint because we want to help future scientific studies by explaining tangible markers of intersubjectivity and laying out a potential connection between humans giving priority and growth and reproduction experiences [6]. At the same period, we believe that the interactional mechanisms that underpin intersubjectivity are profoundly embedded in individuals' sentimental capabilities, including such Theory of Mind abilities. We are fascinated with how these abilities express themselves in social contact. As a result, our narrative is heavily influenced by adolescent psychological literature, in which intersubjectivity is often explored in terms of its intrapersonal features. Moreover, certain elements of our model, like the one emphasising the importance of expression in influencing people's intersubjective inclinations, are consistent with the concept of intersubjectivity as a transpersonal concept.

Nevertheless, the people were confronted with a variety of items to cope with during the worldwide holds catastrophe of the 2020s. This age was introduced to new safety rules, such as solitude, protective clothing, lockdowns, and closure times. The sudden shift in the surroundings caused physiological alertness. People across the world were experiencing conflicting emotions. To ensure instructional consistency, academic institutions attempted to react promptly. This broad concept covers a variety of approaches to education. A lack of preparedness characterised the sudden transition of the entire education system. It pushed instructors who had no previous understanding of online education methodologies to participate in online learning and teaching. Several instructors disliked teaching at a distance, but they had little choice [7]. They are required to participate in the process of life-creativity growth. They used their socially defined instructional knowledge and expertise, as well as their emotions about the uniqueness in the setting, to address the unique instructional issue. The goal was to replicate the intersubjective environment that exists in a traditional classroom setting.

This is the reason; intersubjectivity is considered as an integrated activity that places educators in a dialectical arena with pupils. Such physical connection, based on how teacher-students, as well as student-students occupying the very same place and time, interact with each other, maybe favourable to learning

A methodology of online education, according to Sentz, is a collection of specified techniques, tactics, and practises for effectively teaching topics online. The learners are separated from the teaching person or other classmates in a real place. In a school environment, every teaching–learning procedure requires various instruments, methods, practices, and approaches. The notion that the place has been supplanted by expansion, which has replaced emplacement in learning, on the other hand, necessitates the development of other extensions that cross chronological and geographical boundaries. The intrinsic heterotopia of online settings necessitates considerations beyond strategy and tool analysis to comprehend intersubjectivity in this setting.

Even though the classroom has always penetrated a house in many ways, technology-mediated education enters it in various and new ways amid health emergencies. To keep their kids from losing behind during closing schools, parents and carers used a variety of methods. Schools have been closed as a result of the epidemic. As the past of outbreaks demonstrates, it is not a recent concept. Nonetheless, the unpredictability of events has prompted the development of a semiotic procedure in which individuals name the phenomenon individually to make sense of it [8]. As a result, what is seen as a disturbance in the educational environment may be referred to as "corona teaching," "forced switchover," "urgent distant learning and teaching," or "school-run homeschool." People have invented these language alternatives to create understanding of the inevitable instructional process. Participants may recognise and label the item after the first sense of uniqueness even without the feelings it elicits. Their perspective, previous understanding and philosophy are all accounted for by the many terms and meanings. At level 3, the very same topics may generalise and describe the newly established idea, establishing conceptual boundaries that can be expanded or reduced based on the impression of the body of knowledge. Valsiner's development of the 4 layers of semiotic intervention that reflect the processes of higher education is referred to as Gamsakhurdia. They may, nevertheless, destroy their socially created understanding if they believe it is no longer useful, as Valsiner suggests.

The COVID-19 dilemma focuses on our concept of self as well as the psychological concerns that accompany it. Such existential characteristics transcend beyond the true identity, the single member, to the self–other connection. The dread of being contaminated by or spreading another individual's self taps into a fundamental, largely intuitive foundation of the individual in the self–other relationship's social environment. As contrasted to mortality, my own life is now dependent not just on myself but also on the self of all the other individuals. As a result, the COVID-19 issue touches on a fundamental aspect of our being, namely, our intersubjective and social character, as well as our social, economic, and environmental grounding.

The coronavirus outbreak presents a challenge for many civilizations and cultures, as it shows how individuals respond and govern themselves in the face of such a chaotic scenario. Self is profoundly intersubjective, cultural, and social, and it transcends cultural distinctions [9]. What are how individuals of various cultures deal with their self-perception and associated existentialist fears? How do people feel the dread of isolation, psychological class, fear of infection, and mortality that profoundly impact our sense of identity and connectedness with others, in contrast to the restrictions set by regulatory authorities and their administration of the COVID-19 transition?

Let's look at the case of masks. In East Asian cultures like Chinese, Japanese, and Korean, it appears that wearing masks to safeguard other people's identities is not an issue. Wearing a mask, especially if one has a cold, is a question of courtesy and consideration for one of the other people to not transmit another. This is especially true during the COVID-19 crisis. In Western civilizations, like the Anglo-American and European, there appear to be issued with masking a portion of someone's face and with human rights and personal freedoms. According to Tessa Wong's recent report, "Coronavirus: Why Some Countries Wear Face Masks and Others Don't," published in BBC News, Singapore, "the general belief is that anybody may be a carrier of the virus, including healthy individuals." As a result, in the sense of cooperation, one must defend others against oneself.'

Such segment disclosure to more fundamental concerns concerning the immanent characteristics of self; such existential characteristics make contact on the profoundly intersubjective natural world of self, which can be seen in differences in culture such as when trying to identify one's own self in a much more socially autonomous or inter-dependent manner.

Hence, it is required that both intersubjective and cultural aspects of self are intimately linked since they are based on the same neural or, rather, neuro-ecological processes. They represent latent underlying existential levels of identity which are founded on the brain's congruence to its international ecosystem, that is, the world, and supplement the apparent cognitive characteristics of self [10].

2 Intersubjectivity and Digital Technology

A limited fusion of the two i.e., intersubjectivity and technology is essential to fulfil the intersubjective significance purpose. More different communications media are needed, especially to support the signification depth of individual interaction. Both self-control activities and training pathways and mechanisms of intersubjective significance need direction for an educational agenda, without restricting flexibility by too restrictive scripting of behaviour. We have to better grasp how these practises communicated by built objects to make progress in these areas of service. The third big argument of this paper is that the Computer Supported Collaboration and Learning motives technology aspect should rely on the creation and analysis of inherently social technologies that are motivated by their affordances and weaknesses for moderating intersubjective significance [11]. Since interactional and particularly intersubjective epistemologies of education demand it, 'Computer Supported Collaboration and Learning' structures should be widespread today.

To be profoundly social, technology must be created to mediate and encourage actions of intersubjective significance. Being aware of a technology's capabilities and weaknesses ensures that the architecture seeks to exploit the technology's specific benefits rather than replicating help for education that could be accomplished through other means or forcing the innovation to be what it is not well equipped for. There is a growing research agenda in the field of affording costs for intersubjectivity. We must first comprehend the collective techniques people use while interacting through various types of knowledge objects [12]. Human speech and the usage of symbolic instruments in its operation are also flexible: we do not pre-determine their interpretations or communicative purposes. Rather, 'Computer Supported Collaboration and Learning' analysis should look at how associates take advantage of the media's presumed capabilities and how the notational assets of media affect collaboration and coordination. Communicative techniques that can be used on several platforms are probably important. Irrespective of how flammable the system is in terms of such techniques, individuals would want to come up with a way to implement them [13]. As planners, our task is to find more realistic modifier keys, to include sets of affordances that complement attendees' approaches while still offering versatile ways of direction. The rest of this segment addresses some of the interesting possibilities that digital technology offers for intersubjective significance and some potential study topics for the research proposal initiative.

2.1 (Im) Mutable Mobiles

The analytical media is expandable yet reproducible like a notational process. It is simple to modify digital artefacts and duplicate behaviour and objects in other places: space and time may be bridged. The malleability and immutability of virtual engravings allow newly recruited participants to participate throughout the sense-making phase and encourage ongoing participation. How do we use this function of innovation to create modern social orientations and interplay?

2.2 Negotiation Potentials

Any platform has its own set of possibilities for motion. Respondents can feel obligated to achieve consensus on changes to carvings inside the process to the degree that they are socially transmitted. As a result, the medium's capacity for intervention will direct experiences against proposals correlated with the activities available [14]. To start a review, consider the following question: What positive behaviour does the process allow? The options for intervention are the most important? What choices must be taken to choose and execute each of these activities? Will respondents' experiences be fruitful for learning if they resolve these judgments in accordance with the design's epistemology? Such study may also be applied in the opposite direction: if we want consumers of our innovation mechanism to concentrate on specific facets of an issue [15], how can the mechanism be configured to trigger acts that necessitate mediation of certain facets?

2.3 Referential Resource

Since a mediation phase created them, jointly formed symbols become infused with significance for the respondents. These symbolic representatives then allow deictic comparison or specific modification of previous meanings. External interpretations created cooperatively promote future discussions by growing the intellectual ambiguity that can be managed in community encounters and explaining earlier interpretations. A medium's descriptive as well as indexical capabilities can influence its importance as a contextual asset. As a result, people may think about how to emphasise what we want our technology consumers to focus on and apply to new knowledge or thoughts. What are the most common meanings provided to truly representative proxies by participants? How do we achieve the indexicality needed for subsequent interpretive plays in our technology-mediated environments? Likewise, conceptual models like templates [16], models, and visualisations may be used to facilitate mediation and act as discussion starters. Instead of becoming instruments for transmitting expert information, these symbols become subjects about which students have sense-making interactions, and they can be structured to contribute to fruitful communication.

2.4 Integration

Engravings in quantitative media have the potential to last for a long time. It is possible to keep, record, as well as modify a record of operation and its items. This estate may be used to systematically use previous interaction as a learning tool, allowing for the composition of meanings that span time and individuals. We can investigate how a reliable estimate of contact and communication may be used as a framework for intersubjective significance based on previous practice [17]. How should truly representative objects be structured to promote adequate understanding of previous perceptions as well as the ability to interpret them in future experiences such that fresh knowledge and theories can be incorporated?

2.5 Trajectories of Participation

What social advantages do innovations have for habits of engagement across longer time periods and groups of actors? Different interpretations of interpretations scattered through entities and time can be mutually interpreted as a shared assumption mechanism in what forms and scales? Will we promote fruitful interdependence of several human involvement paths through having their reification evident and thus accessible to someone else for analysis?

2.6 Adaptiveness

A computational medium may evaluate workplace conditions as well as communication series, and then reorganise or produce stimuli based on those characteristics. We should investigate the effect of contingent dynamism on the trajectory of intersubjective mechanisms [18]. To reap the benefits of the medium's capacity to prompt, interpret, and strategically react, we do not require humanising it.

2.7   Reflector of Subjectivity

Quantitative media should be used to help people become more conscious of their surroundings. The basic knowledge that everyone else is involved and can judge one's behaviour can impact one's decision-making. People in the collective's activities can be influenced by knowledge regarding their attentional state and perceptions toward already introduced concepts. Visual representations of dispute or cooperation among participants can contribute to further discussion or agreement. Innovation may improve intersubjective sense by projecting self-representations onto a social image or incorporating the physical body in the social simulator [19]. What particular forms should we design technologies to reflect behaviour, subjectivity, and personality to mediate intersubjectivity? Social networking capabilities are dealing with many of these concerns about how the resources of technology should help and even be utilized for intersubjective education. The analysis of technology capabilities should be done with continuous attention to the task that would be assisted by intersubjective significance as well as its educational implications.

3   Framework of Inetrsubjectivity

3.1   Presence

Lombard and Ditton defined six interconnected principles of existence in technology environments in their study of the idea of existence: social diversity, authenticity, movement, absorption, social actor within media, and mode as a social actor. The concept of linked existence is often linked. TI borrows from these ideas of appearance, but it expands the idea in two respects [20]. To begin, TI is presented as a conceptual entity to aid in the layout as well as application of novel modes of connection in HCI, AI, and human–robot interactions. Furthermore, TI emphasises the incredible experience-based qualities of living together.

3.2   Information Subject

Mark Poster's strategic thinking and media studies practice on the partnership of technology and subjectivity has postulated the modern human entity as nothing more than an intelligence subject [21]. Since TI is a technology development based conceptual framework, it varies from Poster's concept.

3.3   Time–Space Compression

David Harvey has expressed the concept of space time contraction as a state of human life in the postmodern period. There is a moment of contraction with the introduction of information and communication technology; social networking platforms like Orkut, My Space, and Facebook; video distribution websites like Flickr and YouTube; and simulated environments like Virtual Worlds. Not only has the physical environment diminished through space and time as a result of technological innovation [22]; intersubjective gaps between mates, relatives, meaningful others, including common outsiders, have also shortened as a result of social innovations.

3.4   Networked Individualism

The concept of interconnected individual freedom, coined by Manuel Castell, is also linked to TI. The distinctions between the two concepts are in their definition and degree. TI is based on philosophy of thought, whereas interconnected individual freedom is based on a broader Cognitive theoretical framework applicable to the Internet. Compared to TI as an emergent concept of locally social relationships, networked individual freedom is focused on macro-level phenomena over long periods. Technological destiny is a latent present of interconnected individual freedom, whereas intercultural communication naturalism may be a potential criticism of TI [23]. Our understanding of communication encompasses not just Human Computer Interaction (HCI)—that is, engaging with technology—but also computational intersubjectivity (TI)—that is, socialising and data. In brief, we aim to use TI as a rich framework to perform a theory-based analytical analysis of technology-mediated human social experiences and relationships, with a particular emphasis on computer-assisted intercultural cooperation.

4   The Construction of Collaborative Intersubjectivity

According to Habermas, culture is a space of intersubjectivity formed by individuals who see themselves as intentional producers of beliefs and definitions by their acts and experiences. Intersubjectivity, according to Habermas, is the collective product of multiple human subjectivities, and it has its independent nature.

Intersubjectivity is viewed as a dynamic and ever-changing aspect of human awareness. Language, expression, and auditory exchanges are technical means of communicating metaphorically mediated behaviour affecting respondents' main objective acts. Language is divided into two categories: material, and social behaviour [24]. Habermas credits mead with developing the radical conceptual interactionism approach to the development of self and community. The relationship between respondents who respond to each other and mutually take a position on one someone else's expressions is defined by Mead as the process of intersubjectivity.

Education, in this opinion, is not about acquiring skills on a personal level but about being a part of skill enhancement training programme of society. Participating as a candidate entails acting similarly to others in terms of beliefs, interpretations, ideals, and desires. Learning is the process of engaging or communicating.

Habermas described the contact as a mutual contact wherein the participants try to reach a consensus on the information recorded and their proposals to organise their activities. People do not gain functional information by studying; instead, it comes through conversational behaviour that is fueled by interaction and its social meaning [25, 26]. On the other hand, instrumental supremacy is structuralist or hermeneutic in nature, in opposition to conversational behaviour. Intersubjectivity is empirical, collective, as well as free, while individual awareness is contextual, egoistic, and closed.

Participative interactionism, as per Habermas, is necessary for the development of analytical awareness and critical reasoning. Participative participation generates the development of a greater form of existence in an optimal conversational culture. Instrumental and conversational reason or intellects are the 2 kinds of reasoning or intellect. The difference between instrumental and conversational awareness is that instrumental awareness is taxonomic, whereas interactional awareness is conceptual and informative [27]. Horizontal natural ecosystems are structural reforms and grouping systems, while vertical ecologies are definitions generated in the conceptual discussion. Collaborative relationships are relational and gradual since they are conversational and produced by discussion conversations.

The current knowledge on environment by Floridi may be thought of as an evolving field of intersubjectivity divided into awareness areas or events. There is a horizontal as well as a vertical aspect of the www. The terms channel and network focus on the physical infrastructure that allows international communication through geographically dispersed computers that are uniquely defined [28]. The actual manifestation of a vertical ecosystem inhabited by mental awareness of human definitions and beliefs is this massive and quickly developing horizontal contact ecosystem.

The principle of environmental structuralism by communication media, as discussed somewhere below, provides an empirical description of how respondents develop the vertical ecosystem of cognitive awareness by communicating with someone on a constant framework in real-life conditions using computer networks and other objects. The term "intersubjectivity" relates to the mutual conceptual awareness generated by individuals who use innovation to participate in constructive experiences in a real-world situation [29]. Each person expresses personal feelings and thoughts related to one another by common community definitions and beliefs during communication. The process that causes intersubjectivity is described in the following principle.

When participants engage in collaboration or other collaborative digital tasks and organised events, mapping their activity is required. Social democratic interactionism believes that communication is essential for the development of effective main objective collective collaboration and the formation of online communities

[30]. It is reasonable to assume that when a community of people work collectively to achieve a task, the chances of success are higher than when single people work individually and alone. Apart from the sheer size of the community, it is the engagement that facilitates and enables the appearance of a response. When they communicate and interpret one another's conversational actions, people respond to one another. Their response, in essence, has an effect on someone in a collective, main objective manner, contributing to the development of alternatives.

The charting approach can make it obvious how encounters in a communications system occur as a result of the introduction of technologies and consumers engaging in innovative experiences. The features of programming restrict or impact user experiences in terms of how the system helps consumers encode their intellectual and emotional responses so that they can be shared with others [31]. According to Gibson and Select, there are two kinds of human–machine implementations: satisfying and maximising conceptions.

Users' perceptual, observational, as well as behavioural engagements in technical frameworks are correctly described by these concepts. When users respond to a presentation or post, they use normal, satisfying techniques to fill in the conceptual sense with significance relevant to the encounter. Users then assess the significance of the show or text, determining if it is meaningful and useful [32].

Technological capabilities instigate a revolutionary sequence of explaining to consumers as well as maximising experiences of online socialising and interactive activities [33, 34]. The design shows that these experiences are made up of three human biological activities together with the abilities provided by the interface.

5   Ethics and Intersubjectivity

Intersubjective links, also known as subject-subject relations or subject-object relations, have long been recognised by sociologists. Five different individuals may be turned into objects, and not only in the benign meaning of "intentional" objects, such as when I remember my mother's smile on a specific day but also in the notion that together we can embrace the phrase "impersonal viewpoint" on some other individual. Whenever one embraces this viewpoint on some other individual, they are not expressly referring to them as someone with a perspective of outlook on the world, like a world-experiencing subject [32]. Instead, they interact with them simply as if they were a complicated tangible object. To be honest, embracing an indifferent viewpoint does not have to be ethically wrong: it might, for instance, aid surgeons in their vital work. Although sociologists would usually stress that such a relationship is not speaking intersubjective, it overlooks the other individual's subjectivity. We meet other individuals even though we experience them as perceiving subjects, which implies subjects with a viewpoint not just on the realm of things but also on ourselves. Jean-Paul Sartre gave these concepts a well-known formulation. It is critical, according to Sartre, to differentiate between the other whom one sees and another who experiences one, i.e., between some of the other as objects and another as subject. And what is special about encountering someone else is that you will be confronted with someone who can observe and objectify you.

Another is the one in front of whom one may behave like an object. Instead of concentrating on the other as a particular object of concern, Sartre contends that the feeling of someone being an individual's object is a much more natural and genuine intersubjective connection. Another one could only be completely or entirely engaged as a topic in this manner. Another individual's initial existence as a whole is, therefore, their appearance as "one who appears at me." Sartre's storey has not moved uncontested. The rigid subject–object distinction that appears to drive Sartre's thoughts on intersubjectivity has been criticised by many [33]. Then according to Sartre, someone is either the subject objectifying some other individual or another objectifying the subject; there really is no middle ground, and one cannot be simultaneously subject and object at the same moment. For starters, there's the notion another individual as a subject "transcends" or completely escapes my objectifying grip.

Therefore, this appears to culminate in a continuous morally corrupt conflict for Sartre: either one exploits, or one shall be exploited. However, in the research of Emmanuel Levinas, a contemporary of Sartre, the issue with aggression and the transcending of another is evolved into unique intersubjectivity ethics. Levinas' works offer a perplexing image from the standpoint of orthodox logical positivism. Levinas' arguments appear to disregard the distinctions between "metaphysics, metaethics, and normative ethics", and his assertions are outrageously extravagant when they appear ethical. It is feasible to insight into Levinas' fundamental assertions if one views him from the perspective of intersubjectivity psychology [34]. Levinas' main goal is to provide a phenomenological account of the intersubjective connection, focusing on its ethical aspect. To put it another way, Levinas believes that ethics has everything to do with the relationship between subjects. A thorough comprehension of intersubjectivity necessitates a study of ethics.

As a result, Levinas argues that I connect to another as a subject when I perceive another as having a genuine hold on myself. At the same time, Sartre said that I experience another as a subject whenever I perceive another as objectifying and disempowering me. As a result, every genuinely intersubjective relationship includes a moral component. Obviously, this is not to imply that one constantly or even most of the time—do whatever is ethically correct, or that he/she always recognise that individuals have a moral responsibility. However, Levinas would argue that humans are already conscious of this responsibility [35]. Levinas constantly refers to the ethical relationship as a "face to face" relationship, in which I connect to the "face" of another. His use of the notion of face is motivated by a desire to highlight that one's ethical duty extends to another individual. Faces are a feature that allows us to differentiate between people. For instance, when people know somebody's face, they are no longer simply "someone" to us, maybe someone specific. This is among the most significant connotations Levinas wishes to convey with the notion of the face: to relate to individuals as nameless and interchangeable homo sapiens samples is not to connect with individuals as humans. However, dealing with people ethically means treating them as distinct, irreplaceable persons. Levinas further emphasises that the face communicates spontaneously, refusing to be limited to any classifications or portrayals as a result. In other terms, another can be limited to any one of this colour, sex, or socioeconomic status as a face; as just a face, the other one is another person simply qua other individuals.

As a result, Levinas characterises this "face-to-face" contact as one in which the other individual "transcends" any conceptions. One can have in an effort to comprehend him or her. Furthermore, he believes that the most basic intersubjective interaction is one where the another stops to be any kind of object, just as Sartre does. Levinas connects the interaction with another's face with not seeing the colour of his eyes, much as Sartre stresses that when one is exposed to another's gaze, then his eyes vanish. But there is an important distinction between Sartre and Levinas: for the earlier, another's gaze objectsifies and "enslaves" me, while for the latter, another's facing basically makes a moral plea or request to me. In the Sartrean perspective, other individuals are basically dangers to one's liberty, whereas they urge me to duty in the Levinasian perspective (It is indeed important to note that neither Levinas nor Sartre disputes that people may seem as both dangerous and ethically required, as well as a variety of other things). Their argument is about how much these individuals are essential to us, before and irrespective of whatever challenges or appeals they may make [36]. But what kind of ethical responsibility does the "face-to-face" contact entail? Whereas Levinas emphasises that the face reflects the biblical commandment "Thou shall not kill," it should be seen not even as a key point to homicide but instead as a broad connection to any kind of aggression and contempt aimed at others. It's also worth noting that Levinasian morality cannot be reduced to a body of norms or concepts.

If norms are to serve any purpose, they should have some level of applicability, which appears to be irreconcilable with the idea that we should only account to the particular, special individual. As a result, Levinas believes that the ethical imperative "must become quiet as fast as one hears its word." No one has the authority to tell others what to do. Rules or standards should never govern my relationship with another individual since if one follows a set of ethical principles, People are just interested in another individual's suffering as processed through the principles, not with another individual's situation as its whole. Levinas' profound distrust of ethical norms and principles corresponds with current moral particularists' basic beliefs. Several of Levinas's other assertions, on the other hand, are probably to be met with scepticism, even from particularists [37]. Because, according to Levinas, the ethical relationship is unequal in the respect that, although one has obligations to whoever he/she am confronted with, one should not anticipate or desire that another individual does the same for him. In other terms, one must behave unselfishly, as though he/she is only answerable for another and not another for him. This may seem to be an outlandish assertion, but Levinas has evidence to back it up.

First and foremost, he is not stating categorically that another has no duty regarding him. He's implying that although the other individual may be accountable for him, it is really his or her concern, not ours as viewed from one viewpoint, those who are more accountable than another. This would not rule out the possibility that another is accountable for him, but it does preclude the possibility that her obligation for each other may be clearly reflected in one's ethical relationship with another.

Second, the individual one is confronting has a perspective, and they are, of obviously, "another" from such a viewpoint. Nevertheless, Levinas goes much farther, claiming that my duty is "infinite." Maybe the simplest way to express Levinas' argument is to say that my duty has no set limit. "The infinite of obligation does not imply its real enormity," Levinas stresses. His argument is not that I have an endlessly long list of responsibilities to manage someplace in a Platonic realm of thoughts, such as it is. Instead, he claims that there will never be a limited set of tasks or duties that will satisfy my commitments toward others. In other terms, his duty is "infinite" in the respect that no specific limit can be placed about what it entails [38, 39]. Levinas' ideas suggest that the conventional difference between mandatory and supererogatory actions is false. There is no such thing as "beyond the call of duty" for him. Examine everything he states about connecting to "the face" once more to understand why that's the case. To connect ethically to someone is to connect to one as an individual, not as a disposable example of a species.

Furthermore, it is to treat one as just a distinct personality whose suffering is my concern, and therefore my duty. To believe that there is still a stage beyond which it is morally acceptable to declare that her/his suffering is not one's business, that does not really have to worry me, is to believe that there is still a level beyond which one obligation to his/her stop and supererogation starts. In Levinas' words, this is equivalent to stating a threshold beyond which somebody's face is no longer required to be reacted to. Stopping replying to someone else's argument or plea may be the wisest course of action in certain cases, but it cannot be morally acceptable [40–42]. As a result, the ethical requirement must be "impossibly demanding"; it must be limitless in the view that a clear limitation is never morally acceptable. However, it is essential to keep in mind that Levinas is not telling others what to do. Another factor that makes Levinas' ideology difficult to locate is that he professes to provide observations instead of recommendations at the end of the day. His representations, however, are not of the ethical issues of any current or historical society, but rather of something which is part of the intersubjective connection as such, the contact with some other individuals as other individuals [43]. This is the reason; intersubjectivity is considered as the integral part of learning and communication. And digital technology can only be relevant if and only, if it serves the applicability and usefulness of doing the inter-personal and intra-personal communication. Moreover, intersubjectivity is responsible for inter-subject communication objectively, which is being done with the help of technologies in today's world of digital technology. Hence, during the pandemic the role intersubjectivity has crossed the expected periphery of serving the world, and has become co-existential with that of technology. This unique amalgamation is now been the best creation of digital technology, which is collectively termed as the technological intersubjectivity. This fusion of two altogether different phenomena shows that humanities is till an integral part of the technology-driven world. That is why, to say that digital world gets the desirable extent of serving the human beings when it gets attached with ethical philosophy. Moreover, technological intersubjectivity is not only relevant in during the pandemic but also useful for the future of the world due to its affection towards the values and virtues of communication and learning.

6 Summary

As many dimensions, universes, or variants on truth and worlds exist as there are people. Each person has the ability to use intersubjectivity resources to create his or her own version of existence, view of the world, or truth. This is clearly a dynamic process that necessitates research into the individuation and creation of various human beings. What, if any, are the connections between individual conceptions of fact and life-world statutes and institutionalised ones shared by groups, nations, sub-cultures, and folk societies? We may realize that some people, communities, and categories of people have more educated and less educated perceptions of nature as well as the universe being constructed [44, 45], as well as more sophisticated and less specialised, almost sophisticated perceptions.

Art, humanities, sciences, philosophy, career, jobs, and other specialized fields may all contribute to the development of more educated truths. The above is referred to as empirical images of truth by Dennett, Sellars, and others. Individual people of everyday, basic logic, existence, universe or fact perceptions, I believe, may have more specialised areas of certain worlds or realms as a result of their participation in those specialised fields, such as science, careers, sport, art, faith, business, politics, culture, and so on. Painters, including an acrylic painting and a watercolour painter, and several paintings, might be conscious of anomalies that some are not.

Chemists, physicists, astronomers, and biochemists may be able to conceptualise events that non-experts in certain disciplines will be unaware of, like apparent and usually unseen beings. Biologists, neurologists, social scientists, sociologists, educators, linguists, zoologists, and other professionals will use jargon that the rest of the audience will not recognize [46]. Many of the experts mentioned above use various intersubjectivities that require them to understand, and communicate with phenomena in the ways that many are unable to do and understand unless they study the specialized discourse of others'.

Nevertheless, Dennett, Searle, and others refer to elements of broader intersubjectivities and concepts, or viewpoints, as imagined representations of life, the universe, and so forth. Hence, it can be summed up that persons may use indistinct regions, realms, or domains of generalized intersubjectivity and regions or aspects, categories of truth that their specialist specialty distinguish and conceptualise in greater depth.

Intersubjectivity has many facets such as philosophical, sociological, and psychological understanding. The experiential academic discipline defines intersubjectivity as the interactional accomplishment between individual subjectivities, such as individuals or personal observations. In psychiatry, the term evolved into a communication philosophy involving a psychoanalyst as well as a client. Intersubjectivity is described in sociology as a complex junction of multiple separate, interacting, subjective processes rather than a static convergence of personal actions. Intersubjectivity originated from its multidisciplinary origins when it was extended to the area of the school in the twenty-first century to become a symbol of information building accomplished by a synergistic evolution through individual donations within a dialogue to a series of interrelated interventions within the dialogue. The importance of inclusion in the framework of intersubjectivity cannot be overstated. Intersubjectivity refers to integrating individual donations throughout a debate, resulting in the sense of development as members draw on one another's ideas. Each respondent begins his or her approach to the conversation by picking up where the preceding respondent left off. Respondents build mutual awareness by connecting one discursive act to the next, drawing on objects generated by the ongoing dialogue to generate new inputs to the debate [47]. These objects are the comments added to the program discussion page, and the respondents in online learning are mainly learners.

It is also interesting to mention that dialogism offers a promising range of analytical methods for studying recursive point of view. A wide range of intertwined experiences characterises contemporary cultures. Such organisations have vastly different perspectives, such as the health system and the people who utilise them. These divergences in interpretation are tractable to study thanks to the suggested approach of dialogical examination [48]. According to Cornish and Gillespie, the importance of every modern methodological methodology is contained not in theoretical or historical claims, but in the opportunities for fruitful study that are opened up.

The goal in writing this article was to make empirical studies on the link between various forms of intersubjectivity as well as the essence of subjectivity easier. The usage of the dialectical technique, according to the speaker, is intended to be another tool in our tool chest for understanding intersubjective and non-subjective modes of subjective subjectivity. The proposed structure may be used to examine an individual's intrapsychological relationships, or subjectivity. The method may also be used to assess the effectiveness of programmes aimed at promoting debate. The platform may be used to code diaries, interviews, life stories, emails, and speak protocol files.

It may be able to determine a person's overt and implied direct, meta, and meta-met perspectives. The dialogical contradictions between overt and implied elements could then be identified. And those who are adamantly dedicated to one point of view are mindful of alternatives, and the challenge is how these various viewpoints communicate within the person [49]. Moreover, empirical analysis is likely to uncover a wide range of potential semiotic relationships between viewpoints within a single person. Hence, the proposed approach of dialogical study needs a rethinking of the individual-society antinomy.

It is interesting to note that the writers' opinions are their own and do not reflect those of the general public mindset. Individuals are seen as intersubjective beings are woven out of social interactions and discourses in dialogical study. People and organizations are re-imagined as well. According to Frida Ghitis, what defines a community or culture is the degree to which its members' views are recognised by one another. Ghitis: Societal discourses and significant others pervade entities, causing social perspectives to be distorted.

She has also claimed that communities may be formed on the basis of differences, but only if a level of shared understanding accompanies such differences. Mead and Schutz's conceptualizations of entities akin to Leibnitz's monad in comparison to the cosmos form the foundation of intersubjectivity theory. Community and culture theories, according to Ghitis, may help one interpret human behaviour, but they are not homogeneous and must be balanced with an awareness of one another's experiences. Hence, it can be summed up that intersubjectivity is an integral part of learning to learn, and communication education, not only in the time of COVID-19 pandemic and for the future too. It is because; the future of learning and imparting knowledge is the age digital technology, where intersubjectivity plays a crucial role for inter-personal and intra-personal communication, essential for dissemination of learning and knowledge.

References

1. Spaulding, S. (2012). Introduction to debates on embodied social cognition. *Phenomenology and the Cognitive Sciences, 11*, 443. https://doi.org/10.1007/s11097-012-9275-x.
2. Friston, K. J., & Frith, C. D. (2015). Active inference, communication and hermeneutics. *Cortex, 68*, 133. https://doi.org/10.1016/j.cortex.2015.03.025.
3. Reddy, V., Liebal, K., Hicks, K., Jonnalagadda, S., & Chintalapuri, B. (2013). The emergent practice of infant compliance: An exploration in two cultures. *Developmental Psychology, 49*, 1754. https://doi.org/10.1037/a0030979.
4. Das Nair, R., Martin, K.-J., and Lincoln, N. B. (2016). Memory rehabilitation for people with multiple sclerosis. Cochr. Database Syst. Rev. 3:CD008754. https://doi.org/10.1002/14651858. CD008754.34.
5. Ohata, W., & Tani, J. (2020). Investigation of the sense of agency in social cognition, based on frameworks of predictive coding and active inference: A simulation study on multimodal imitative interaction. *Front. Neurorob., 14*, 61. https://doi.org/10.3389/fnbot.2020.00061.
6. Newen, A., Bruin, L., & Gallagher, S. (2018). 4E cognition: Historical roots, key concepts and central issues. In A. Newen, S. Gallagher, & L. de Bruin (Eds.), *The Oxford Handbook of 4E Cognition* (Oxford: Oxford University Press), 11.
7. Hodges, C., Moore, S., Lockee, B., Torrey, T., & Bond, A. (2020). The difference between emergency remote teaching and online learning. In *EDUCAUSE Review. Why IT matters in Higher Education*. Stable https://er.educause.edu/articles/2020/3/the-difference-between-emergency-remote-teaching-and-online-learning.
8. Larrieu, M., & Di Gesú, M. G. (2021). Developing educational digital literacies among preservice teachers of Portuguese. An analysis of students' resistance in an Argentinian University. In Di Gesú, M. G., & González, M. F. (Eds.), *Cultural Views on Online Learning in Higher Education*, Cultural Psychology of Education 13.
9. Marková, I. (2003). Constitution of the self: Intersubjectivity and dialogicality. *Culture & Psychology., 9*(3), 249–259. https://doi.org/10.1177/1354067X030093006
10. Marsico, G., Dazzani, V., Ristum, M., & de Sousa Bastos, A. C. (2015). Borders in education-examining contexts. In G. Marsico, V. Dazzani, M. Ristum, & A. C. de Sousa Bastos (Eds.), *Educational Contexts and Borders through a Cultural Lens* (pp. 359–365). Springer.
11. Narodowski, M., & Campatella. (2020). Educación y destrucción creativa en el capitalismo de pospandemia. In I. Dussel, P. Ferrante, & D. Pulfer, (Eds). *Pensar la educación en tiempos de pandemia : entre la emergencia, el compromiso y la espera*. 1a ed (43–52). – Ciudad Autónoma de Buenos Aires. UNIPE: Editorial Universitaria, 2020.ebook, PDF - (Políticas educativas ; 6) Digital file: download and online access. ISBN 978–987–3805–51–6.
12. Pontecorvo, C. (2013). How school enters family's everyday life. In G. Marsico, K. Komatsu, & A. Iannaccone, (Eds.), *Crossing boundaries. Intercontextual dynamics between family and school*. Information Age Publishing. Charlotte, N.C. USA.
13. Ramos, W., Rossato, M., & Boll, C. (2021). Subjective senses of learning in hybrid teaching contexts. In M.G. Di Gesú, & M.F. Gonzalez, (Eds.), *Cultural Views on Online Learning in Higher Education. A Seemingly Borderless Class*. Springer. Switzerland.

14. Rosa, A. (2016). The self rises up from lived experiences: A micro-semiotic analysis of the unfolding of trajectories of experience when performing ethics. In J. Valsiner, G. Marsico, N. Chaudhary, T. Sato, & V. Dazzani, V. (Eds.), *The Yokohama Manifesto. Annals of Theoretical Psychology* (105–128). ISBN 978–3–319–21093–3 ISBN 978–3–319–21094–0 (eBook). https://doi.org/10.1007/978-3-319-21094-0. Springer. Switzerland.

15. Rosa, A., & Gonzalez, F. (2013). Trajectories of experience of real-life events. A semiotic approach to the dynamics of positioning Published online: August 14th 2013. Springer Science & Business Media, New York.

16. Rosenberg, C. (1991). What is an epidemic? AIDS in historical perspective. *Daedalus,* Spring, 1989, Vol. 118, No. 2, Living with AIDS (Spring, 1989), pp. 1–17. The MIT Press on behalf of American Academy of Arts & Sciences. Stable http://www.jstor.com/stable/20025233.

17. Newen, A. (2018). The person model theory and the question of situatedness of social understanding. In A. Newen, S. Gallagher, & L. de Bruin (Eds.), *The Oxford Handbook of 4E Cognition,* (Oxford: Oxford University Press), 483.

18. Murata, S., Namikawa, J., Arie, H., Sugano, S., & Tani, J. (2013). Learning to reproduce fluctuating time series by inferring their time-dependent stochastic properties: Application in robot learning via tutoring. *IEEE Trans. Auton. Mental Dev., 5,* 301. https://doi.org/10.1109/TAMD.2013.2258019.

19. Mathur, M. B., & Reichling, D. B. (2016). Navigating a social world with robot partners: A quantitative cartography of the uncanny valley. *Cognition, 146,* 27. https://doi.org/10.1016/j.cognition.2015.09.008.

20. Maier, M., Ballester, B. R., Bañuelos, N. L., Oller, E. D., & Verschure, P. F. (2020). Adaptive conjunctive cognitive training (ACCT) in virtual reality for chronic stroke patients: A randomized controlled pilot trial. *Journal of Neuroengineering and Rehabilitation, 17,* 15. https://doi.org/10.1186/s12984-020-0652-3.

21. Loetscher, T., Potter, K.-J., Wong, D., & das Nair, R. (2019). Cognitive rehabilitation for attention deficits following stroke. *Cochr. Database Syst. Rev.* https://doi.org/10.1002/14651858.CD002842.

22. Linson, A., Clark, A., Ramamoorthy, S., & Friston, K. (2018). The active inference approach to ecological perception: General information dynamics for natural and artificial embodied cognition. *Front. Robot. AI, 5,* 21. https://doi.org/10.3389/frobt.2018.00021.

23. Lawson, R. P., Rees, G., & Friston, K. J. (2014). An aberrant precision account of autism. *Frontiers in Human Neuroscience, 8,* 302. https://doi.org/10.3389/fnhum.2014.00302.

24. Koster-Hale, J., & Saxe, R. (2013). Theory of mind: A neural prediction problem. *Neuron, 79,* 842. https://doi.org/10.1016/j.neuron.2013.08.020.

25. Kirchhoff, M. D. (2018a). Autopoiesis, free energy, and the life–mind continuity thesis. *Synthese, 195,* 2527. https://doi.org/10.1007/s11229-016-1100-6.

26. Kirchhoff, M. D. (2018b). The body in action: Predictive processing and the embodiment thesis. In A. Newen, S. Gallagher, & L. de Bruin (Eds.) *The Oxford Handbook of 4E Cognition,* (Oxford: Oxford University Press), 248.

27. Kingma, D. P., & Ba, J. (2014). Adam: A method for stochastic optimization. *arXiv preprint* arXiv:1412.6980.

28. Ismail, L. I., Verhoeven, T., Dambre, J., & Wyffels, F. (2019). Leveraging robotics research for children with autism: A review. *International Journal of Social Robotics, 11,* 392. https://doi.org/10.1007/s12369-018-0508-1.

29. Hohwy, J., & Palmer, C. (2014). Social cognition as causal inference: implications for common knowledge and autism. In M. Gallotti & J. Michael (Eds.), *Perspectives on Social Ontology and Social Cognition,* (Dordrecht; Heidelberg; New York, NY; London: Springer), 1672.

30. Granulo, A., Fuchs, C., & Puntoni, S. (2019). Psychological reactions to human versus robotic job replacement. *Nature Human Behaviour, 3,* 1065. https://doi.org/10.1038/s41562-019-0670-y.

31. Goršič, M., Cikajlo, I., & Novak, D. (2017). Competitive and cooperative arm rehabilitation games played by a patient and unimpaired person: Effects on motivation and exercise intensity. *Journal of Neuroengineering and Rehabilitation, 14,* 23. https://doi.org/10.1186/s12984-017-0231-4.

32. Gassert, R., & Dietz, V. (2018). Rehabilitation robots for the treatment of sensorimotor deficits: A neurophysiological perspective. *Journal of Neuroengineering and Rehabilitation, 15*, 46. https://doi.org/10.1186/s12984-018-0383-x.

33. Gallagher, S., & Allen, M. (2018). Active inference, enactivism and the hermeneutics of social cognition. *Synthese, 195*, 2637. https://doi.org/10.1007/s11229-016-1269-8.

34. Fuchs, T. (2013). The phenomenology and development of social perspectives. *Phenomenology and the Cognitive Sciences, 12*, 655–683. https://doi.org/10.1007/s11097-012-9267-x.

35. Froese, T., Iizuka, H., & Ikegami, T. (2014). Embodied social interaction constitutes social cognition in pairs of humans: A minimalist virtual reality experiment. *Science and Reports, 4*, 3672. https://doi.org/10.1038/srep03672.

36. Bruner, J. (2006). Going beyond the information given. In *Search of Pedagogy*. (1st ed., p. 17) Routledge. ISBN 9780203088609.

37. Bruner, J. (2008). Culture and mind: Their fruitful incommensurability. In *Ethos, troubling the boundary between psychology and anthropology: Jerome Bruner and his inspiration 36*(1), 29–45. Wiley on behalf of the American Anthropological Association. Stable URL: http://www.jstor.org/stable/20486559. Accessed 1 Nov 2016

38. De Luca Picione, R. (2021). The dynamic nature of the human psyche and its three tenets: normativity, liminality and resistance. Semiotic advances in direction of modal articulation sensemaking. University Giustino Fortunato Benevento, Italy.

39. Deaton, A. (2020) We may not be all be equal in the eyes of coronavirus. Downloaded from https://www.ft.com/content/0c8bbe82-6dff-11ea-89df-41bea055720b.

40. Demantowsky, M., & Lauer, G. (2020). Teaching between pre- and post-corona. An essay (1). Downloaded from https://public-history-weekly.degruyter.com/issues/.

41. Di Gesú, M. G. (2021). Affect as catalyzer of students' choice of learning environments. In M. G. Di Gesú, & M. F. González, (Eds.), *Cultural views on online learning in higher education* (45–66), Cultural Psychology of Education 13. Springer, Switzerland.

42. Di Gesú, M. G., & Gonzalez, M. F. (2021). The imposed online learning and teaching during COVID-19 times. In M. G. Di Gesú, & M. F. González, (Eds.), *Cultural views on online learning in higher education* (189–202) Cultural Psychology of Education 13. Springer, Switzerland.

43. Foucault, M. (1967). Des Espace Autres. *Architecture /Mouvement/ Continuité*-October, 1984; Translated from the French by Jay Miskowiec.

44. Gamsakhurdia, V., & Javakhishvili,. (2021). The dynamic nature of the human psyche and its three tenets: Normativity, liminality and resistance. Semiotic advances in direction of modal articulation sensemaking. Tbilisi State University.

45. Hermans, H. (2001). The dialogical self: Toward a theory of personal and cultural positioning. *Culture and Psychology., 7*(3), 243–281. https://doi.org/10.1177/1354067X0173001.

46. Sentz, J. (2020). The balancing act. Interpersonal aspects of instructional designers as change agents in higher education. In J. Stefaniek, (Ed.), Cases on learning design and Human performance technology. IGI- Global.

47. Trevarthen, C. (2005). Action and emotion in development of the human self, its sociability and cultural intelligence: Why infants have feelings like ours. In J. Nadel & D. Muir (Eds.), *Emotional Development* (pp. 61–91). Oxford University Press.

48. Trevarthen, C. (2012). Embodied human intersubjectivity: Imaginative agency, to share meaning. *Journal of Cognitive Semiotics., 4*(1), 6–56. https://doi.org/10.1515/cogsem.2012.4.1.6.

49. UNESCO. (2020). No dejar a nadie atrás en tiempos de la pandemia del COVID-19 ODS Educación 2030. *Seminario web regional n°11: "El COVID-19 y la Educación Superior: Impacto y recomendaciones".*

50. Pineau, P., & Ayuso, M. L. (2020). De saneamientos, trancazos, bolsitas de alcanfor y continuidades educativas: brotes, pestes, epidemias y pandemias en la historia de la escuela argentina. In I. Dussel, P. Ferrante, & D. Pulfer, (Eds.), *Pensar la educación en tiempos de pandemia: entre la emergencia, el compromiso y la espera*. 1a ed (19–32). Ciudad Autónoma de Buenos Aires. UNIPE: Editorial Universitaria, 2020. ebook, PDF - (Políticas educativas ; 6) Digital file: download and online access. ISBN 978–987–3805–51–6.

51. Terigi, F. (2020). Aprendizaje en el hogar comandado por la escuela: cuestiones de descontextualización y sentido. In Dussel, I., Ferrante, P, & Pulfer, D. (Eds.), *Pensar la educación en tiempos de pandemia : entre la emergencia, el compromiso y la espera.* 1a ed (243–251) . – Ciudad Autónoma de Buenos Aires. UNIPE: Editorial Universitaria, 2020.ebook, PDF - (Políticas educativas ; 6) Digital file: download and online access. ISBN 978–987–3805–51–6.
52. Trevarthen, C. (2004). Stepping away from the mirror: Pride and shame in adventures of companionship. Reflections on the Nature and Emotional Needs of Infant Intersubjectivity Department of Psychology, The University of Edinburgh, Edinburgh, Scotland, U.K.

The Impact of Artificial Intelligence (AI) on the Development of Accounting and Auditing Profession

Manal Abdulameer, Mahmood Mohamed Mansoor, Mohammed Alchuban, Abdulrahman Rashed, Faisal Al-Showaikh, and Allam Hamdan

1 Introduction

Artificial Intelligence (AI) can be defined as the creation of software that executes a set of procedures mimicking human behaviour [13]. Artificial intelligence is categorized into 4 types, Reactive Machines, Limited Memory, Theory of Mind, and Self-Awareness [14].

In recent years, AI has developed massively, and all industries have benefited enormously from these developments [6]. The accounting and auditing industry is one of these industries, it has benefited from the application of AI in many ways [23]. However, as AI highly benefits the accounting and auditing field it also comes with some risks. These risks touch different points and one of the major concerns is the replacement of human capital with machines and software [25].

The application of AI in the field of accounting and auditing is mainly found in performing the routine tasks, because of the large number of errors in the traditional methods the AI practices was proven to be more efficient [20].

Literatures argued that even though AI will replace the routine tasks, it will also lead to the emerge of new employment opportunities. Therefore, AI is an opportunity for the development of the accounting and auditing industry as well as the organizations that adapt AI-based accounting [19].

The adoption of AI-based accounting leads to the development and innovation of the field [22]. The utilization of AI increases the efficiency and effectiveness via improving the accuracy, speed and decreasing the errors [9].

M. Abdulameer · M. M. Mansoor · M. Alchuban · A. Rashed
College of Business and Finance, Manama, Bahrain

F. Al-Showaikh (✉) · A. Hamdan
Ahlia University, Manama, Bahrain
e-mail: falshowaikh@ahlia.edu.bh

© The Author(s), under exclusive license to Springer Nature Switzerland AG 2022
A. Hamdan et al. (eds.), *Technologies, Artificial Intelligence and the Future of Learning Post-COVID-19*, Studies in Computational Intelligence 1019,
https://doi.org/10.1007/978-3-030-93921-2_12

2 Research Problem

Many studies highlighted the impact on AI on the accounting and auditing profession as a threat to the human recourse. However, this study aims to address the AI-based accounting as an opportunity for development of both the profession and the industry.

Therefore, this study aims to answer the following questions:

1. What is the impact of AI on the accounting and auditing industry?
2. How does the adoption of AI effects the development of the accounting and auditing field?

3 Literature Review

3.1 What is Artificial Intelligence?

Artificial Intelligence (AI) name was first introduced by a math professor in the year 1955. It did not have many advances at the beginning instead experts' predictions had many upsets. An economist named Herbert Simon in 1957 said that AI can be used by computers to play chess against humans and it will AI will win in 10 years but that happened only after 40 years. In addition, Marvin Minsky a scientist predicted in 1967 that in one generation someone will figure how to develop an application with AI, but his prediction also did not come true [8]. Artificial Intelligence simply is creating an application or software to execute a set of procedures mimicking human behaviour. These pieces of code can be executed very fast and can have more accuracy than when it is done by a human being [13].

3.2 Types of Artificial Intelligence

Artificial intelligence is categorized into 4 types, Reactive Machines, Limited Memory, Theory of Mind, and Self-Awareness.

3.2.1 Reactive Machines

The lowest level type of AI is limited because it comes up with decisions without using the history of occurred events. Some scientists suggest that we should only develop software that has this type of AI only as the other types can harm human beings when they advance their knowledge [14].

3.2.2 Limited Memory

This type of AI has a limited amount of memory that stores information about the past. Such applications can observe the surrounding environment and save these observations in memory and use them as inputs before coming up with a decision [14].

3.2.3 Theory of Mind

The third type of AI will take an extra step than the limited memory, this type has the theory of mind where it understands anything surrounding it either living beings or objects.

3.2.4 Self-Awareness

This type is the most dangerous of the four types, it extends the features of the theory of mind by adding consciousness which allows them to act the same as human beings and maybe even better [14].

3.3 AI in Real Life

In the latest years, AI has developed massively, and all industries and sectors have benefited enormously from these developments. Websites and mobile applications have been using AI to provide a personalized experience for each customer when using them. For example, online shops recommend items based on what the customer has been purchasing or searching for. In addition, it has been used in cars auto-driving, where the AI observer the surrounding and uses some hardware like sensors and cameras and applications like GPS to drive the car without any interaction from the driver. Moreover, chatbots have been used in customer service which uses AI to reply to and answer customers, and machine learning to enhance the quality of the answers [6].

Health has been using AI very much especially in the latest years, for example, an organization has developed an AI system that can analyze a patient and warn the doctor if it noticed he/she has symptoms of a heart stroke [18]. The health care sector is already using AI in areas like oncology and radiology which support the decision-making when examining patients and AI also has a huge role on E-Health platforms which are systems used to exchange patient information between hospitals [1].

Also, AI has been used during the research to find a cure and a vaccine for the Covid-19 virus and for tracking and forecasting the spread of the virus so they can such information to prevent the spread [1]. Accounting and Auditing have their share of these developments, as AI and other advanced technologies are being used to enhance the process and automate others [26].

4   AI in Accounting and Auditing

As it has in every other business, artificial intelligence is having a big influence in the world of accounting and auditing. AI-enabled technologies for accounting and auditing are the way accounting professionals and their businesses will stay competitive and attract the next generation as workers and customers, saving time and money and offering insights. Would you be interested if you could save expenses by 80% and cut the time it takes to complete activities by 90%? Robotic process automation (RPA), according to Accenture Consulting, will produce these outcomes for the financial services business. Accounting companies and financial experts must begin to embrace artificial intelligence to provide services that their clients will need and compete for business with other professionals [23].

In every industry, new technology is altering the way people work. It is also altering clients' expectations while working with businesses. The same may be said about accounting. Accountants may benefit from artificial intelligence to become more productive and efficient. Human accountants would be able to focus more on offering advice to their customers if job times are reduced by 80–90 percent. Adding artificial intelligence to accounting procedures will improve quality by reducing mistakes [23].

Moreover, accounting companies that incorporate artificial intelligence into their practices become more appealing as an employer and service provider to millennials and Gen Z workers. This generation grew up with technology, so they will expect potential employers to have the most up-to-date technology and innovation to support not only their working preferences of flexible schedules and remote locations but also to let them focus on creative works instead of daily repetitive tasks that can be automated [23].

"Many internal business, municipal, state, and international standards must be observed in accounting. AI-enabled solutions facilitate audits and assure compliance by comparing documents to norms and legislation and flagging those that violate them. Fraud costs businesses billions of dollars each year, with financial services companies incurring $2.92 in expenses for every $1 of fraud. Machine learning algorithms can swiftly filter through massive quantities of data to identify possible fraud concerns or suspicious behaviour that people might have missed and flag it for further investigation" [23].

Auditing has historically trailed behind business in terms of technological adoption, although it is ripe for automation due to its labour-intensive nature and variety of decision structures. Furthermore, various technologies are being developed that can act as automation motivators as well as change auditing methods. This is because of new technical capabilities or changes in the cost/benefit of doing specific activities. The phenomena we shall refer to as Technological Process Reframing may be characterized as the reassessment of techniques and processes in a field of endeavour because of the introduction of disruptive technology. Standards and rules tend to cause delays, which is especially evident in the auditing field [17].

When used in auditing, a successful expert system provides several advantages, including automated comprehension of audit job procedures, improved knowledge, and knowledge transferability. The use of expert systems in accounting, auditing, and taxation began in the 1980s. Early expert system research was conceptual and mostly intended for demonstration. As a result of the early study, accounting practitioners investigated the feasibility and potential of expert systems and expressed a strong interest in this field. Public accounting companies have made significant expenditures in developing expert systems to assist audit planning, compliance testing, substantive testing, risk assessment, and decision-making [17].

Contract reviews, such as leases contracts, are the most prevalent use case for AI in audit procedures. Organizations may use AI to continually examine a bigger number of contracts in real-time. Auditors may utilize AI to automatically extract data from contracts using NLP techniques and identify important terms for accounting treatment such as lease start date, payment amount, renewal, and termination options, and so on. As a result of these results, auditors will be able to examine and assess contract risks more properly. Another common use is identifying any material presentation for statements and general ledgers [11].

Auditors have been working with the "reasonable assurance" slogan due to issues such as the huge number of data, inadequate time, and intrinsic constraints of internal control systems and accounting. Auditors analyse a subset of data rather than the whole data set, which raises the risk of material misstatement. Machine learning enhances ledger data testing by evaluating the full dataset in a short period of time to discover major misstatements based on risk analysis rather than traditional audit standards. This implies that AI-powered tools can highlight transactional data depending on how much it deviates from a typical set. Expenses are another example of how AI may be used to automate an audit process. Artificial intelligence-powered technologies may assist firms in detecting duplication, out-of-policy expenditure, erroneous quantities, suspicious merchants or attendees, and excessive spending [11]. The benefits of using AI are many and some of them are:

- The auditor burden is decreased since there is no need to go back and forth with the client to ask questions.
- Cost savings: The use of AI decreases the costs associated with manual hours of research and analysis.
- Audit Quality: AI systems learn and adapt to datasets on a continual basis, allowing them to increase anomaly detection accuracy as more data is analysed. As a result, the application of AI/machine learning improves audit quality.

5   Risks of AI

If artificial system intelligence considerably exceeds human intellect, it would pose a major risk to human beings [24]. AI became sophisticated and the voices of warning against its current and potential hazards are now more audible than before. The risk is impacting directly human beings in many areas including the automation of some jobs, gender and racial prejudice concerns originating from obsolete information sources, and controlled weapons that work without human control such as autopilot drones. We are still at the very beginning of AI and more risks will raise in the future. The risks of AI touch different points and one of the major concerns is human jobs. Job automation is usually considered to be the promptest issue. Whether AI will replace some sort of employment is not an issue anymore, but to what degree. Disruption is well underway in several industries [24].

A survey by Brookings Institution in 2019 showed that 36 million people are employed in employment with elevated exposure to automation, indicating that soon, they will be employed by AI at least 70% of their activities—from retail sales and market analysis to hospitality and warehouse labour. In an even more recent Brookings' research, white-collar occupations might well be the most threatened. The other major concern is privacy, security, and fake communications transactions over the internet. AI might endanger digital security by criminals training computers to hack or socially engineer victims at superhuman levels of competence. The same is true for fake audio and video generated by altering voices and likenesses. An audio recording of any given politician may be altered using machine learning, a subset of AI engaged in natural language processing, to make it appear as though that person spouted racist or sexist ideas when they did not. The video quality is very good, and it could mislead the public and evade discovery, it could destroy a political campaign [25].

Mitigating the risks of AI can be achieved by regulating the AI implementations and applications. Regulations can determine where AI can be used and where it cannot. Where it is appropriate and where it is not. Different countries will make different decisions. AI researchers Fei-Fei Li and John Etchemendy, of Stanford University's Institute for Human-Centred Artificial Intelligence, proposed "Our future depends on the ability of social—and computer scientists to work side-by-side with people from multiple backgrounds—a significant shift from today's computer science-centric model [25].

The creators of AI must seek the insights, experiences, and concerns of people across ethnicities, genders, cultures, and socio-economic groups, as well as those from other fields, such as economics, law, medicine, philosophy, history, sociology, communications, human–computer interaction, psychology, and Science and Technology Studies (STS). This collaboration should run throughout an application's lifecycle—from the earliest stages of inception through to market introduction and as its usage scales" [25]. As mentioned, there are many risks involved in AI but at the same time, AI will provide humans with the capability to overcome the toughest challenges in the world.

6   Applications of AI in Accounting and Auditing

Artificial Intelligence is not a concept for Accounting and Auditing, it has been used for some time in this field. It has been used to conduct analysis and it is also is used to automate procedures and processes within banks and financial institutions [27]. Robotic Process Automation (RPA) is part of the AI, it is the use of a virtual robot to automate the routine tasks performed inside a firm. As well RPA can be used in non-routine tasks which require decisions base on complex inputs [20].

6.1   Accounting and Auditing Tasks Using AI

AI has been used by accounting and auditing to automate routine tasks and processes, and it is predicted to have more other tasks which the AI will replace the process that is done currently by humans and even introduce new ones. Previously Financial transactions have been recorded manually by the accountant which can result in some human errors but now such transactions can be recorded automatically by the systems [20].

In addition, accountants and auditors had been using lots of files and papers to analyse firms' data, but now we have tools like Business Intelligence (BI) that can perform such analysis and produce results with just a click of a button. Moreover, accountants and auditors before were using software like Excel and PowerPoint to present their analysis and finding, now interactive dashboards can be developed to present such findings in user-friendly manners. Furthermore, using AI in financial institutions will introduce new jobs and tasks, for example, such institutions will need someone expert in AI to maintain and change these AI tools if a change is required [20].

6.2   AI Accounting and Auditing Applications

Financial firms and institutions have been developing software and tools to ease the work and process performed by accountants and auditors. The big four accounting companies have introduced many tools in this area and here in Bahrain, there are many financial institutions that have developed and upgraded their systems to use AI.

6.2.1   Ernst and Young

Ernst and Young have introduced an online software called Canvas which enables the connections between auditors and their customers and provides tools that can be used to monitors milestones and allow a wide range of customizations [20].

6.2.2   PwC

Cash.ai was developed by PwC which uses AI and machine learning to perform an automatic audit on the transaction and process conducted by financial firms which involve cash [20].

6.2.3   Deloitte

Cortex is a cloud application by Deloitte which was initially developed to perform audit and taxes for the customers. But currently, it uses advanced technologies and algorithms to perform analysis and provide financial consultation and advisory [20].

6.2.4   KPMG

KPMG has developed a software called Clara which uses AI and Machine Learning to conduct a smart audit by analyzing data and detect risks [20].

6.2.5   Ila Bank

Ila is a new bank in Bahrain that has all of its customers' operations online using its mobile application. Customers register, open one or more accounts and request debit and credit cards by only using the application without the need for human interaction. Also, the bank has transformed some traditional financial techniques for saving money into a digital function within the application. They have introduced Hassala which is like a piggy bank where the customer can save money occasionally. Also, they have added a function called Jamiya where a set of customers initializes a saving community [15].

6.2.6   Benefit

Benefit Pay has been developed by Benefit which is an application used by anyone in Bahrain who has an account in a Bahraini Bank. It allows customers to transfer money by using only the recipient's mobile number within a maximum of 30 s without any interaction from an employee from the banks [7].

6.3   The Impact of AI on the Performance

In the recent years, the rapid enhancements in technologies, automations, cloud computing and interactions with artificial intelligence have absorbed noticeable attentions globally. As a result, technologies made its attributable interactions to all aspects of life ranging from the lowest involvement and participation in our life's activities to potentially becoming the main and core component of our daily life's activities. Accounting and auditing professions were not an exception in adopting and tailoring such enhanced technologies in their conducts and practices.

"Nowadays, Artificial Intelligence (AI) is democratized in our everyday life. To put this phenomenon into numbers, International Data Corporation (IDC) forecasts that global investment on AI will grow from 12 billion U.S. dollars in 2017 to 52.2 billion U.S. dollars by 2021" [16]. "Meanwhile, the statistics portal Statista, expects that revenues from the AI market worldwide will grow from 480 billion U.S. dollars in 2017 to 2.59 trillion U.S. dollars by 2021" [2, 21].

So, it can be concluded out of the aforementioned and the referenced articles that artificial intelligence will make a solid difference in the way of conducting accounting and auditing activities in today's world since huge reliance is being devoted to systems and technologies while handling and performing the routine accounting activities. Accordingly, it can be generalized that at least one system, if not many are being deployed today in firm's setup that meant mainly to handle the accounting and auditing processes with minimal reliance on manual interventions other than feeding the system. Such systems were improved rapidly once compared to the past era and used to perform further complex activities. Due to that fact that accounting and bookkeeping activities have changed their focus from paperwork era to technological era especially with the introduction of systems, automation and cloud computing services (Fig. 1).

Fig. 1 Artificial Intelligence Interlinked Technologies

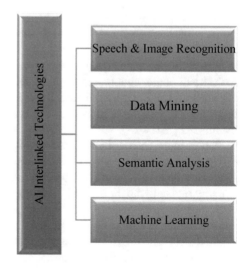

"Accounting today is handled by a system called, Robotic Process Automation (RPA). According to AIIM (2018), Robotic process automation (RPA) is a term that denotes software tools that partially or fully systemize human activities that are repetitive, manual, and rule based. Boulton (2018) sees RPA as an application of technology, directed by business logic and arranged inputs, aimed at automating business processes".

AI in accounting and auditing fields are not yet fully practiced by businesses due to its implementation costs which seems relatively high for many businesses and the level of the required expertise and organizational setup within businesses that is marked in some aspect as a prerequisite requirements for success implementation of AI. Despite the fact of the concerns about the implementation costs, the fruitful aspects of the AI can be noticed in so many aspects of our daily life and accounting industry is not an exception for such popularity. This resulted in complete transformation from the traditional accounting systems to the most technological advanced accounting systems [5, 4, 28, 3].

A study on 2018 revealed that AI has a positive impact on the performance. It highlighted that the application of AI brings about a massive increase in the performance and encourages the firms to utilize its resources for the adoption of AI-based accounting. This can allow the organization to enhance its accounting practices, reduce the time, effort and increase the quality and therefore the performance [9].

6.4 The Impact of AI on the Development of Accounting

Literatures suggested that the implementation of AI in the field of accounting and auditing will lead to the development and innovation of the industry. Also, it has the ability to create a sustainable competitive advantage to the organization [22].

Previous studies highlighted the impact of AI on the accounting and auditing field. Some of these impacts are speed, accuracy, improvement in the reporting, less paper usage, high flexibility and efficiency, reduction of errors and improvement of the competitive advantage [9].

Moreover, researchers argued that AI has the ability to reduce the financial fraud. In the traditional accounting method, the financial personnel can disorder some of the accounts for his own personal gain. However, the implementation of the AI can limit this manipulation, but it cannot resolve it because these systems are still managed by humans [29].

Furthermore, the traditional accounting and auditing method is time consuming, and in order to finish some tasks on schedule workers might need to work overtime. Therefore, the utilization of AI improves the quality and efficiency of the accounting work by reducing the time, effort, and cost [29, 4, 3].

Many Literatures suggested that in the upcoming years AI will takeover many jobs, and the human recourses will be replaced with machines [12]. However, some authors claimed that AI cannot entirely takeover the accounting and auditing industry because of the need for managerial decisions, that's why it should be seen as an opportunity not a threat. Therefore, accountants should try to improve their professional and managerial skills to keep up with the changes in the industry that resulted from the development and utilization of AI [29].

Nonetheless, other authors argued that in the long run AI could take over even the decision-making roles in the organization. They suggested that the availability of large figures could allow the company to use the AI in the analysis of information to predict the future [19].

In 2018, a study about the impact of AI on the performance revealed that the application of AI in the accounting and auditing industry influence the performance positively. The authors also argued that the AI is a very powerful tool in the accounting profession that can enhance the organizations and create a sustainable competitive advantage [9].

Moreover, literatures anticipated that the continuous development of new technologies will result in the rapid change in the methods used in accounting and auditing [19].

The usage of AI start by utilizing the technology in the daily activities, this practice alters throughout the process till it gradually create new tasks and routines and finally takeover the activity. This process forces the firm to make new adjustments in the roles and activities that will eventually result in creating new positions. Moreover, the new positions created by the utilization of AI requires new qualifications and tasks [19].

A study in 2021 found that the application of AI will lead to the replacement of human resources in the routine tasks such as the collection of data with machines. It was anticipated that by 2030 the human capital will be monitoring the AI- based accounting instead of performing the daily tasks [19].

However, even though literatures anticipated the replacement of human resources with machines as a result from the adoption of AI-based accounting, they also claimed that it will result in creating new jobs, tasks and roles that don't exist yet. Therefore, AI showed not be seen as a threat, but it is an opportunity to shift the practices used in the field of accounting and auditing into a new form that is more efficient and effective that will generate new employment opportunities [10].

7   Conclusion and Recommendation

The accounting and auditing field is highly affected by the usage and utilization of AI. And in order for the firms to cope and deal with these changes, the existing skills and qualifications should be upgraded. In the short-term daily and routine activities will be replaced with machines. As a result, the number of human capitals will be reduced, and new roles and tasks will emerge. Therefore, the survival of human capital in this field requires enhancement and development of skills. However, in the long-run new roles will emerge and better employment positions will be created.

Moreover, literatures argued that AI is changing the accounting and auditing industry. These changes are allowing the firm to be more efficient, and flexible via reducing the cost and providing more efficient figures in the decision-making process. Also, AI has a positive impact on the performance, productivity, and effectiveness.

Therefore, the adoption of AI should be encouraged. As a result, the accounting and auditing profession will not be eliminated as some fears, but it will flourish. AI- based accounting is the revolution that will alter the accounting and auditing profession as we know it. It will only eliminate the routine work and will create more creative and innovative roles based on decision-making and the AI monitoring.

Furthermore, the industry will witness a shift that will lead to a new era. The large data that can easily predict the future will help in the long-run survival of the firm, the reduction of errors will help is making better judgments, and the reduction of time will increase the quality and efficiency of the work.

AI-based accounting will open the floor for improvements on the profession level, organizational level and the whole industry of accounting and auditing.

7.1 Limitations of Research

Although this research summarizes and explains the AI- based accounting and its impact on the profession, it is a study based on literature review, a more advanced quantitative research that measures the impact of using AI in the firm on the performance and the development of the profession should provide more accurate data for further understanding and more detailed analysis.

References

1. AL-Hashimi, M., & Hamdan, A. (2021). Artificial intelligence and coronavirus COVID-19: Applications, impact and future implications. *The Importance of New Technologies and Entrepreneurship in Business Development: In The Context of Economic Diversity in Developing Countries.* , 830–843.
2. Adadi, A., & Berrada, M. (2018). Peeking inside the black-box: A survey on explainable artificial intelligence (XAI). *IEEE Access*, 52138–52160.
3. Al Shehab, N., & Hamdan, A. (2021). Artificial intelligence and women empowerment in Bahrain. *Studies in Computational Intelligence, 2021*(954), 101–121.
4. Ali Saad, A. Z., & Mohd. Noor, A. B., & Sharofiddin, A. (2020). Effect of applying total quality management in improving the performance of Al-Waqf of Albr societies in Saudi Arabia: A theoretical framework for "Deming's Model." *International Journal of Business Ethics and Governance, 3*(2), 12–32. https://doi.org/10.51325/ijbeg.v3i2.24.
5. Alshurafat, H., Al Shbail, M. O., & Mansour, E. (2021). Strengths and weaknesses of forensic accounting: An implication on the socio-economic development. *Journal of Business and Socio-economic Development, 1*(2), 85–105. https://doi.org/10.1108/JBSED-03-2021-0026.
6. Bansal, S. (2020). *15 Real World Applications of Artificial Intelligence.* Retrieved from analytixlabs: https://www.analytixlabs.co.in/blog/applications-of-artificial-intelligence/.
7. Benefit. (2021). *Introducing BenefitPay.* Retrieved from benefit.bh: https://www.benefit.bh/Services/BenefitPay/BenefitPay.aspx.
8. Brynjolfsson, E., & Mcafee, A. (2017). The business of artificial intelligence.

9. Chukwudi, O. L., Boniface, U. U., Victoria, C. C., & Echefu, S. C. (2018). Effect of Artificial Intelligence on the Performance of Accounting Operations among Accounting Firms in South East Nigeria. *Effect of Artificial Intelligence on the Performance of Accounting Operations among Accounting Firms in South East Nigeria*, 1–11.

10. Davenport, T. H., & Barro, S. (2019). People and machines: Partners in innovation. *MIT Sloan Management Review*, 22–28.

11. Dilmegani, C. (2021). *AI Audit in 2021: Guide to faster & more accurate audits*. Retrieved from AI Multiple: https://research.aimultiple.com/ai-audit/.

12. Frank, M. R., Autor, D., Bessen, J. E., Brynjolfsson, E., Cebrian, M., Deming, J. D., . . . Youn , H. (2019). PERSPECTIVE Toward understanding the impact of artificial intelligence on labor. *Proceedings of the National Academy of Sciences*, 6531–6539.

13. Frankenfield, J. (2021). *Artificial Intelligence (AI)*. Retrieved from investopedia: https://www.investopedia.com/terms/a/artificial-intelligence-ai.asp.

14. Hintze, A. (2016). *Understanding the Four Types of Artificial Intelligence*. Retrieved from govtech: https://www.govtech.com/computing/understanding-the-four-types-of-artificial-intelligence.html.

15. ilabank. (2021). *About Us*. Retrieved from ilabank.com: https://www.ilabank.com/AboutUs.

16. International Data Corporation IDC. (2018). *Worldwide Semiannual Cognitive Arti_cial Intelligence Systems Spending Guide*. Retrieved from International Data Corporation IDC: https://www.idc.com/getdoc.jsp?containerId=prUS43662418.

17. Issa, H., Sun, T., & Vasarhelyi, M. A. (2016). Research ideas for artificial intelligence in auditing: The formalization of audit and workforce supplementation. *Emerging Technologies in Accounting*, 1–20.

18. Johari, A. (2020). *AI Applications: Top 10 Real World Artificial Intelligence Applications*. Retrieved from edureka: https://www.edureka.co/blog/artificial-intelligence-applications.

19. Lehner, O. M., Eisl, C., Forstenlechner, C., & Leitner-Hanetseder, S. (2021). A profession in transition: actors, tasks and roles in AI-based accounting. *Journal of Applied Accounting Research*, 539–556.

20. Leitner-Hanetseder, S., Lehner, O. M., Eisl, C., & Forstenlechner, C. (2021). A profession in transition: actors, tasks and roles in AI-based accounting. *Journal of Applied Accounting Research*.

21. Liu, S. (2020, December 7). *Statista*. Retrieved from Revenues From the Arti_cial Intelligence (AI) Market: https://www.statista.com/statistics/607716/worldwide-arti_cialintelligence

22. Luo, J., Meng, Q., & Cai, Y. (2018). Analysis of the impact of artificial intelligence application on the development of accounting industry. *Open Journal of Business and Management*, 850–856.

23. Marr, B. (2021). *Artificial Intelligence In Accounting And Finance*. Retrieved from Bernard Marr & Co: https://bernardmarr.com/default.asp?contentID=1929.

24. Müller, V. C. (2016). *Editorial: Risks of Artificial Intelligence*. Retrieved from philarchive: https://philarchive.org/archive/MLLERO-2.

25. Thomas, M. (2021). *6 DANGEROUS RISKS OF ARTIFICIAL INTELLIGENCE*. Retrieved from Bulletin: https://builtin.com/artificial-intelligence/risks-of-artificial-intelligence.

26. Ucoglu, D. (2020). Current Machine Learning Applications in Accounting and Auditing. *Press Academia Procedia (PAP)*, *12*, 1–7.

27. Ukpong, E. G., Udoh, I. I., & Essien, I. T. (2019). Artificial intelligence: Opportunities, issues and applications in banking accounting, and auditing in Nigeria. *Asian Journal of Economics, Business and Accounting*, 1–6.

28. Zainal, M. M., & Hamdan, A. (2022). Artificial intelligence in healthcare and medical imaging: Role in fighting the spread of COVID-19. https://doi.org/10.1007/978-3-030-77302-1_10.

29. Zheng, L., & Li, Z. (2018). The impact of artificial intelligence on accounting. *Advances in Social Science, Education and Humanities Research*, 813–816.

Higher Education a Pillar of FinTech Industry Development in MENA Region

Amira Kaddour, Nadia Labidi, Saida Gtifa, and Adel Sarea

1 Introduction

The term FinTech is a mixture of the terms "finance" and "technology" and is meant to indicate using technology to deliver a financial solution. Interest in FinTech business has been growing. According to [14] the value of global funding in FinTech grew 75% in 2015. Despite the enormous boom, instructional studies approximately FinTech is still scarce, which motivates this work.

Many industries are experiencing large transformational business model adjustments due to improved digitization [25]. Energy, healthcare, schooling, and hospitality are only a few prominent examples. Perhaps one of the maximum profound alterations is happening in the financial sector. The infusion of digital technology into monetary services, typically called monetary generation or FinTech, has brought about an explosive boom in university program in new startups and exceptional change within the finance area [7]. It has been argued that each banking carrier is being focused through FinTech companies globally, both to lessen costs and serve customers better, while in the long run disrupting the financial incumbents.

A. Kaddour (✉)
Lab Life, ENSTAB, Carthage University, Tunis, Tunisia
e-mail: amira.kaddour@ensta.u-carthage.tn

N. Labidi
Sup'Com, Carthage University, Tunis, Tunisia
e-mail: nadia.labidi@supcom.tn

S. Gtifa
FSEGN, Carthage University, Tunis, Tunisia
e-mail: saida.gtifa@fsegn.u-carthage.tn

A. Sarea
Ahlia University, Manama, Kingdom of Bahrain

© The Author(s), under exclusive license to Springer Nature Switzerland AG 2022　　215
A. Hamdan et al. (eds.), *Technologies, Artificial Intelligence and the Future of Learning Post-COVID-19*, Studies in Computational Intelligence 1019,
https://doi.org/10.1007/978-3-030-93921-2_13

Traditionally, consumers accessed economic offerings through one or more huge institutions. In this "standard" version, incumbents usually offer an extensive product portfolio along with retail, non-public, business, investment, and transaction banking, at the side of wealth, asset control, and insurance. In today's platform and app-centric world, customers are much less involved about receiving all their services from one provider [25]. Instead, purchasers count on a seamless enjoy across diverse offerings, responsive and personalised to their expectancies, and available anywhere and at every time [19, 26]. FinTech startups and computer science universities are consequently focusing on improving precise elements of the conventional model; they design, build, and execute individual parts of the traditional value chain, while being better, cheaper, and faster than existing incumbent offerings. With this strategy, startups and computer science students are often able to establish a niche market position.

Enabled with the aid of the proliferation of cloud, mobile, and social computing, computer science universities and startups are knowing this new value expectation and feature began to "unbundle" most of the traditional financial services [8]. Overall, this would now not be a problem to the massive incumbents who've always handled technological modifications [7]. However, given the sheer quantity of FinTech players, the tempo of improvements is highly excessive and exposes incumbents to a large scale and scope of disruption. Since a single agency cannot suit this rapid rate of disruption, incumbents must shift their techniques for that reason.

New collaboration and opposition models have to be embraced [25]. The emergence of application programming interfaces, digital manage points that enable broader get right of entry to information and offerings are allowing such novel organizational arrangements [25].

In a totally short time, the FinTech surroundings have grown remarkably. To call FinTech a technology fad could be dismissive of the tectonic activities shaping the panorama. According to a recent industry file, 13.8 billion USD had been invested into organizations by means of each challenge capitalists and corporate banks in 2016 [25].

Despite the prominence of this emerging research, there is confined understanding inside the converging structure, dynamics, and evolution of the FinTech atmosphere. Earlier work on FinTech is predominantly practitioner oriented, specializing in understanding individual segments or international locations or presenting anecdotal qualitative evidence. Rigorous empirical studies presenting systemic perception into the entire global atmosphere are lacking. Building on earlier work of provider ecosystems and digital disruption, the overarching goal of this study is to analyze the characteristics of the FinTech surroundings, higher education role and technology transfer especially in MENA region. To gain this objective, we need to look at the unbundling of services that using and shaping the FinTech ecosystem. Specifically, we pick out and describe fintech exclusion and inclusion and the role of universities in the structural evolution of the FinTech ecosystem. We take a descriptive method to offer a foundation for destiny predictive or prescriptive analyses. In doing so, we propose similarly our overall understanding of digital innovations made by way of financial services and infrastructures.

We focus our analyses on four parts; the first part investigates financial inclusion with its opportunity and perspectives for MENA region, the second part focuses on FinTech industry. In the third part we highlight the gap between needed profile and the actual higher education curricula, toward a new student profile. Since FinTech is a disruptive innovation, tackling effectively this challenge must be done through an effective technology transfer policy in our universities; element discussed in the fourth part.

2 Financial Inclusion and FinTech in MENA Region

The level of financial inclusion of MENA Region is determined by looking at factors that are seen to affect the awareness and accessibility to financial services such as banking, savings, credit, and digital payments. The factors used include financial literacy, initiatives targeting underserved population, micro-financing for SMEs (Small and Medium Entreprises), and the push for FinTech and digitalization.

For 1.7 billion unbanked individuals through the world in 2017, the global financial inclusion rate is 76%, but the financial inclusion in the MENA Region is significantly weaker 20% [4]. Hence, it's crucial for countries as well as relevant stakeholders in the region to be aware of adopt specific measures to increase financial inclusion.

2.1 Financial exclusion

According to Demirguc-Kunt and Klapper [11], effective and inclusive financial systems are appropriate to poor people and other disadvantaged groups because «without inclusive financial systems, poor people must rely on their own limited savings to invest in their education or become entrepreneurs and small enterprises must rely on their limited earnings to pursue promising growth opportunities. This can contribute to persistent income inequality and slower economic growth».

One question arises, why so many poor people are still financially excluded while financial inclusion is improving worldwide? Financial exclusion is driven by several reasons: high costs of banking services, operation hours may not be suitable for all people, high documentation requirements to open a bank account, commercial bank products are not designed for low-income clients and banks don't want low-income clients.

The terms "unbanked" and "underbanked" are associated with the financial exclusion/inclusion but they are distinct. Unbanked refers to individuals who do not have access or make use of any form of banking services including not having a bank account, debit card, checking account, or savings account. While underbanked individuals represent people, who do not take advantage of all banking services but use few ones. They can also use alternative financial services.

The financial exclusion of the unbanked and underbanked can be voluntary or involuntary. In the case of voluntary exclusion, individuals can access to financial services but don't use them due to personal reasons or beliefs. However, involuntary exclusion is when barriers like low income or physical proximity of banks/ATMs (Automated Teller Machines) preclude individuals to access financial services [4].

Financial services contribute to economic growth. They allow people to avoid poverty by helping them to invest in their health, education, and businesses. Poor people around the world don't access to financial services such as bank accounts and digital payments, they rely on cash which is unsafe and hard to manage. That's why the promotion of financial inclusion (access to and use of formal financial services) represents a key priority of the World Bank.

2.2   Financial Inclusion

The definition of the World Bank is: «Financial inclusion means that individuals and businesses have access to useful and affordable financial products and services that meet their needs—transactions, payments, savings, credit and insurance—delivered in a responsible and sustainable way. Being able to have access to a transaction account is a first step toward broader financial inclusion since a transaction account allows people to store money and send and receive payments. A transaction account serves as a gateway to other financial services, which is why ensuring that people worldwide can have access to a transaction account is the focus of the World Bank Group's Universal Financial Access 2020 initiative» (The World Bank, Financial inclusion overview).

The Center for Financial Inclusion defines financial inclusion as «A state in which everyone who can use them has access to a full suite of quality financial services, provided at affordable prices, in a convenient manner, with respect and dignity. Financial services are delivered by a range of providers, in a stable, competitive market to financially capable clients».

According to Chehade and Navarro [6], «Financial inclusion refers to a state where individuals, including low-income people, and companies, including the smallest ones, have access to and make use of a full range of formal quality financial services (payments, transfers, savings, credit, and insurance) offered in a responsible and sustainable way by a variety of providers operating in a suitable legal and regulatory environment».

The financial inclusion has been adopted globally by central banks and international standard-setting bodies because of its impact on country's economy. In fact, these institutions support financial growth via the improvement of the economy, insure the monetary stability, reduce Anti-Money Laundering through well formalized financial system. Various financial institutions, policymakers, and government entities are interested in financial inclusion because of it offers more economic growth opportunities and less poverty. The most important organizations devoted to improve financial inclusion globally are the G20 (an international forum for the governments and central bank governors from 19 countries and the European Union), the United Nations and the World Bank.

Furthermore, the goal of the Council of Arab Central Banks is to improve the financial inclusion indicators in the Arab countries. The latest statistics reflect the efforts made by the Arab countries to upgrade access to financial services. In fact, the percentage of the adult population in the Arab countries who have access to formal finance and financial services has risen, on average, to 37%, and achieve 265 for women, and to 28% for low-income segments. Although these statistics still indicate the great potential opportunities—especially for private financial and banking institutions—allowing more access to financial services in the Arab communities [10].

In addition, the Council appreciates the efforts of Arab central banks and monetary authorities, commercial banks and official financial institutions in the past years for the Arab Day of Financial Inclusion as well as the related activities and events on this day. These efforts have contributed to reveal awareness and financial education among all members of the community.

2.3   FinTech, Financial Inclusion and Economic Growth

In the case of voluntary financial exclusion, people don't want to deal with banks because of the lack of awareness of the digital financials' advantages or the absence of education about the use of digital finance [23]. Moreover, the world's poor are excluded from formal financial services because they are expensive once provided by traditional banking system, thus advantageous digital financial channels are suggested [15].

The concept of FinTech, includes technological innovation in the financial sector. It emerged from start-ups and technology firms developing user-friendly, cheap and well-suited financial products based on digital channels. FinTech solutions reduce operating costs, serve more clients and making firms profitable. For financial-service providers, Manyika et al. [20] forecast that technology can reduce costs to 90 percent, consequently and the expected gain for companies and their clients increase significantly, that's why developing economies focus on moving from traditional banking to digital finance.

According to the Board of Governors of the Federal Reserve System, «FinTech is an industry composed of companies that use technology to make financial systems and the delivery of financial services more efficient». Technologies, from artificial intelligence and mobile applications provide new solutions to increase the efficiency, the accessibility and the security of financial services. New solutions based on cloud, digital platforms and distributed ledger technologies, covering mobile payments and peer-to-peer applications, fill the gaps. FinTech firms represent a small share of the total revenue of the financial services industry but their growth and contribution to innovation is apparent.

2.4 Measure of Financial Inclusion

The degree of financial inclusion can be measured by assessing the availability of key financial services. The evaluation focus on four key areas that act as a gateway to financial inclusion which are:

- Banking: easy access to traditional and digital transactions.
- Savings: facilitation of investments and reduction of poverty.
- Credit: easy access to credit protect households and businesses to avoid income shocks.
- Payments: increase the circulation of money and boost the economy.

G20, the partner for Financial Inclusion, determined the following "Basic Set" of indicators to measure financial inclusion:

- Access: ability to access available financial services considering barriers (costs, proximity of bank branches and ATMs),
- Usage: permanence and frequency of financial services usage,
- Quality: quality of financial services determined by attitudes and opinions of its relevance, ease of use and added value.
- Welfare: impact of financial services on the client's lives, businesses activities and wellness.

2.5 FinTech Beyond Covid-19

(a) **The need for resilience**

The development of the FinTech industry represents a goal that meets the aspirations of the Sustainable Development Goals. Based on the possibilities of inclusion and poverty eradication, FinTech solutions are increasingly finding takers for their usefulness in bringing financial services closer to previously unbanked populations. Resilience is an important quality that is becoming increasingly important in the era of climate change. The Covid-19 pandemic that shook the usual standards and norms of the economic circuits imposed the resilience factor as a fundamental need in order to respond to specific conditions and ensure the continuity of socio-economic services for the populations.

(b) **FinTech market performance in general during Covid-19**

Based on the World Bank's report, FinTech has seen a plethora of development during the pandemic explained by an adaptability and access to affordable financial services based on innovation. The pandemic has accelerated people's transition to digitalized services, digital payment and remittances were the most important requested services, they increased by 65% in emerging economies). Funding and payment solution have experienced an important growth (tripled to $4 billion, and to over than $6 billion). In general FinTech filed has experienced a positive impact; the retention of existing customers/users improved by 29%, the pandemic created a promising condition to merge to FinTech solution.

To meet the need of customers, two thirds of FinTech firms made two or more changes to their existing products or services especially by deploying additional payment challenges (30–45%). The need for substantial governmental intervention was confirmed. Based on the Global FinTech Regulatory Rapid Assessment Study (2020), there is an urgent need for tax relief by 31% with a further need to down the line (30%). The access to liquidity facility was mentioned (38% as urgent action, with additional needed measures in the future 26%).

(c) **Confirmed challenges**

The hard core of FinTech success is innovation, hence there is an important need to promote education that can improve this capability, Gomber et al. [18] stipulates that the maturation process is still challenging and needs mutual efforts. Lack of access to information about FinTech services especially in developing countries (people with less socioeconomic resources) challenge its development and limit the potential of inclusion and disparities eradication, in addition cybersecurity, operational risk and customer protection present important risk that FinTech industry must master regarding this important growth. Regarding these perspectives the university must play an active role toward talking these challenges and promote curriculum that can offer a substantial core of capabilities ranging from technical side to legal and financial aspects.

2.6 Impact of Financial Inclusion in MENA Region

Financial inclusion facilitates development of households, businesses and the economy in general. According to the «MENA financial inclusion report 2020», 6 problems were solved by financial inclusion in MENA region:

- Eliminating poverty: by accessing to financial services, poor people can control their economic life. Hence, the savings increase as well as investment, consumption and lending which reduces poverty and income inequality,
- Fostering quality education: human capital and education are important for economic growth; hence education helps individuals to make better decisions about their finances and more usage of financial services,
- Achieving gender equality: inequity gender in financial inclusion generates income loss for women about 27% in MENA region. To achieve gender quality, women must have more control for their finances, to have credit and invest,
- Promoting sustainable Economic Growth: accessing people and businesses to more financial services generate economic growth, more income and employment opportunities,
- Promoting innovation and sustainable industrialization: MSMEs (Micro, Small and Medium Enterprises) have 5,2 trillion $ of financial needs non satisfied every

year. Digital payment via innovative mechanisms helps them to tracking funds, reducing risk of fraud, consequently creating investment opportunities,

– Reducing inequality: the gap between rich and poor people in term of education, health and finance increase income inequality about 11%. For decreasing this gap, financial inclusion must reduce barriers to accessing financial services for poor people by offering cheaper services, relaxing credit constraints. Hence, poor people can access to savings which help them to absorb financial shocks, accumulate assets and ease consumption.

The effort of different countries reduces the unbanked global population from 51% in 2011 to 69% in 2017.

The Table 1 demonstrates that Europe and Oceania have the higher financial inclusion rates worldwide.

The MENA region financial inclusion rate is 20% less than the worldwide rate. Men are more dominant; they have a rate of 19% while the rate of women is 9% during 2016. Moreover, 40% of adults of poor households have no account.

The following Table 2 indicates financial inclusion rate of MENA region countries.

Table 3 indicates in details various levels of financial inclusion of MENA region countries.

The first three places are for UAE, Bahrein and Saudi Arabia, they have the best financial inclusion rates in MENA region. However, Morocco is at the end of the list.

Table 1 Financial inclusion rate by region

Region	Financial inclusion rate
North america	28
South america	23
Europe	13
Africa	16
Asia	28
Oceania	65

Source World Bank report 2017

Table 2 Financial inclusion rate by country—MENA region

Countries	Financial inclusion rate
Morocco	9
Algeria	13
Tunisia	17
Egypt	11
Bahrain	39
Jordan	31
KSA	18

Source «MENA financial inclusion report 2020», pp 18

Table 3 Financial inclusion rates of MENA Region countries, 2017

	Access to banking	Access to credit	Access to savings	Digital payment
Algeria	24	3	11	14
Bahrain	59	17	31	51
Egypt	18	6	6	13
Jordan	26	17	10	20
Morocco	16	3	6	9
KSA	54	11	14	43
Tunisia	23	9	18	17

Source «MENA financial inclusion report 2020», pp 20–30

In comparison to the financial inclusion rates of the worldwide, MENA countries must ameliorate its rates. Therefore, with the current state of financial inclusion in the MENA region, a generic strategy must be implemented to eliminate the barriers to financial inclusion from enabling access, increasing usage, maximize usage, and improving quality and range. This strategy can be realized by structuring recommendations in addressing financial inclusion including:

- Regulations: A catalyst for financial inclusion, they concern; transparency and consumer protection, convenient and economical financial services, initiation of agent banking services and activation of bank accounts.
- Digitalization: It represents FinTech for the unbanked. It includes establishing a strong technological infrastructure and increasing the use of FinTech and digital innovation.
- And Financial Literacy: It consists of understanding financials through; encouraging a national agenda for financial inclusion, programming collaborative efforts to extend financial services and promoting financial knowledge via educational institutions.

3 FinTech Industry: From Innovation to Possibilities for MENA Region

Students in postsecondary education need a flexible ecosystem, integrated, green and less expensive. Institutions, teachers, and directors should recall policies and practices that expect and adapt to students wishes over the course of their lives, and may include both conventional and new structures, applications, and institutional practices.

What may be wanted for the new regular postsecondary pupil is broader environment possibilities to examine within both conventional institutions and new vendors, underpinned through a digital infrastructure that permits students to create, apprehend, and fee nice gaining knowledge of studies anywhere and each time they are most convenient, and that rewards the understanding they broaden inside and outside of formal establishments over their lifetimes. This imaginative and prescient of the higher education would in addition permit students to transport plenty greater fluidly inside and out of various styles of institutions, depending on their desires, and transfer as they relocate or pursue increasingly more disturbing education and career paths.

In an effort to meet the needs of those varieties of learners, new packages and providers of training have started to emerge within and in partnership with institutions, providing new fashions of getting to know possibilities which includes industry aligned, job-based schooling applications; FinTech; blockchain; on line gaining knowledge of; quick-time period boot camps; and competency-primarily based education.

3.1 Getting Business and Finance College Students on Track for FinTech

FinTech is a term implemented to diverse digital era adjustments impacting banking, insurance, and other sectors of the finance enterprise. As a new and rising area, technologically diverse FinTech groups are gaining market percentage and converting the way business is performed in traditional economic institutions.

FinTech is a brand-new form of finance which must be understood before it may be carried out. If you don't recognize these developments and new kinds of finance, you're at a disadvantage. There are some exemplars within better training, with some first movers starting to deal with the want for FinTech education and profession improvement. These models can assist universities locate progressive approaches to evolve educational programs to better put together business and laptop technology students for careers in FinTech, or in existing economic corporations redefined with the aid of new generation.

Universities round the sector are beginning to examine FinTech preparedness and a few try to create coursework to fill the skills hole. Characteristically, the majority of college FinTech programs have targeted on regulation, safety issues and information the impact of era in the marketplace. This high-level technique does now not pass into intensity about character technologies, leaving college students to analyze on the job or pursue extra optionally available coursework. To avoid teaching students the incorrect model of FinTech, which could depart them without-of-date competencies for a dynamic and hastily converting field, universities in MENA region and especially in Saudi Arabia should seek chronic insertion to the industry, in which the change is taking place.

3.2   Nurturing FinTech Talents in PC Technology Students

While business students may additionally sense the impacts of FinTech boom maximum immediately of their selected discipline, computer science college students at Saudi Arabia universities may additionally find themselves operating for groups presenting FinTech software program or within present financial services companies doing software program layout, improvement and configuration. However, as FinTech will become a developing consciousness for a few laptop technology college students, the skill sets required can also start to resemble the soft abilities already extensively taught in enterprise packages. Whether working in tightly integrated groups at small organizations or massive firms, pc technological know-how graduates will need to transcend the technical elements of their fields for success collaboration, or to pitch their FinTech innovation thoughts to investors and shoppers. Only a few Saudi Arabia universities require computer science students to take communications training. Fewer nevertheless focus on interpersonal in addition to written and oral communique.

Computer science departments at Saudi Arabia universities can work toward greater integration with business departments. Like business schools, these programs could create specific tracks for FinTech and count some business coursework toward the computer science degree, such as introductory courses in finance and accounting.

Because of the centrality of the economic services and era industries in Arabic word, our universities are in a unique position to play a lead function in developing packages that provide graduates an aggressive area inside the new age of FinTech. Some Saudi Arabia universities have already made development closer to preparing their business and computer science students for careers in FinTech organizations, as well as in traditional financial businesses redefined by using changing era. However, others are just getting started out. Overall, additional training is wanted all through the kingdom's college device. Academic leaders can begin a technique of assessing modern coursework and degree paths, potentially developing new prerequisites or totally new avenues to degree of completion for enterprise and pc technological know-how students. By running to apprehend shifts within the marketplace added on by way of FinTech and their implications for process competencies call for, universities can enterprise to better put together undergraduate and graduate students for the future and improve one of the middle industries of Saudi Arabia.

3.3   Recommendations

A.   To make certain the green and effective transmission of information from enterprise, public universities should create a statewide advisory board to advise at the evolution of FinTech that consists of enterprise experts, professionals and public college and community university personnel to study applicable direction offerings, expert certifications, tracks inside relevant diploma packages and aligning the ones services. MENA region public schools and universities need to:

- Expand coursework and associated electives to include FinTech subjects, together with in gadget mastering, records analysis, and FinTech in conventional commercial enterprise/finance disciplines.
- Establish professional certification programs for people wishing to hold their schooling and adapt to the converting position of FinTech in finance and commercial enterprise.
- Create new tracks within established business and computer finance diploma programs to combine FinTech subjects and talents into students' instructional experience.
- Plan opportunities for interdisciplinary studies for business, laptop technological know-how, engineering, and other diploma candidates to expand a strong set of competencies wished for FinTech—associated jobs.

B. Any better schooling institution, public or private, must are looking for enter from industry representatives on talents wanted for the body of workers and the way FinTech is changing the finance enterprise. Finally, even though outside the limits of this work, Arabic vocational high schools would do nicely to explore the development of course providing and packages on FinTech so as to make certain that Arabic college students are well-prepared for the future.

4 Higher Education Role: Which Actions Toward a New Student Profile

Explained the advent of FinTech is doubly reasoned, on the one hand it is a response to the failure of financial institutions following the subprime crisis, on the other hand it is a trend well rooted in the digital transition, seeking better solutions in terms of cost, time and proximity towards sustainable development and an inclusive economy. As explained above, inclusive finance is about solving poverty, economic discrimination and underdevelopment, and has great potential for developing countries on the one hand, but also for countries with a young population in great need of employability on the other hand. In this regard, it is important to review the training of agents working in the financial sector. In this paper we try to relate a vision of a transitional adaptation that our universities in the Arab countries specifically in MENA region can follow in the perspective of an effective integration in this emerging trend of FinTech professions (Fig. 1).

Indeed, the basic question we can ask ourselves is the following: do we need simply additional training? Or do we need to create new courses totally dedicated to FinTech?

In this sense, which student profile is best placed to study FinTech, students in Finance or engineering students in technology?

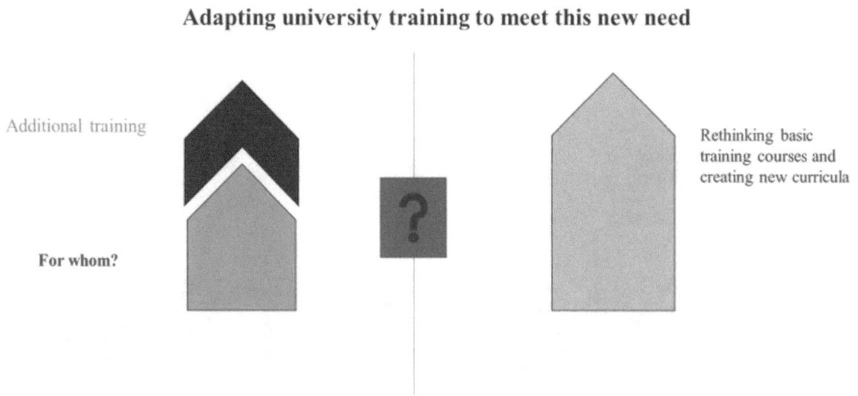

Adapting university training to meet this new need

Additional training

For whom?

Rethinking basic training courses and creating new curricula

Fig. 1 Which strategy to adopt?

To answer these questions, let's first look at the core engine of a FinTech training course:

The current pillars in the world in FinTech are China and India, as a benchmark we will take the teaching units offered by the University of Hong Kong as well as prospective for technologies that FinTech will develop in the future proposed by "FinTech Trend 2020/2021: Top prediction according to Experts".

The distribution of startups shows the current lack in Arab countries (except Egypt), this virgin area could constitute a development field with a high potential of employability for young graduates (Table 4).

Counting the training needs, shows the great component of advanced technologies and data analytics that the professional must have, besides strong capabilities in innovation and a basic knowledge in finance, but also an ethical component and

Table 4 Benchmark of core engine on FinTech	Core engine on FinTech training Hong Kong University	FinTech Trend 2020/2021 Experts
	– Digitization of Financial Services	– Artificial Intelligence
	– Financial infrastructure	– Blockchain
	– Artificial Intelligence	– Digital capabilities
	– RegTech (Regulatory Technologies)	– Legacy system
	– Blockchain	– Security
	– Regulatory Sandboxes	– Identity
	– Cryptocurrencies and crypto assets	– Privacy
	– Data analytics	
	– FinTech ethics and risks	
	– Identity	
	– Research, Development and interactions with industries	

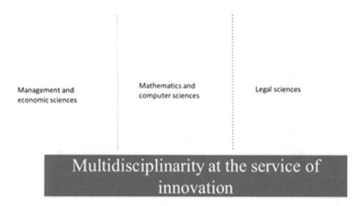

Fig. 2 Multidisciplinarity conjunction

an adequate legal framework. This suggests that FinTech is a multi-dimensional subject, at the heart of which is financial services, with digitization tools based on advanced technologies that are undergoing strong development. The current training in Finance, does not offer the learner the mastery of these advanced technologies, or data analytics based on Bigdata and Machine Learning. Similarly, the training of engineers in advanced technologies enables them to master these tools but orienting them towards the needs of financial services remains critical, given the lack of a theoretical basis in this area.

RegTech capabilities are basically a transformative knowledge linked to the legal sciences, which can always be integrated into the ordinary curriculum in form of a specific master's degree.

Taking these considerations into account, a review of financial training with the aim of leading a career in FinTech is necessary, with actions at different temporalities (Fig. 2).

4.1 Short Term Actions: Tackling the Trend

In order to align with this new need, professional master's programs in FinTech or RegTech or InsurTech (Insurance Technology), can be set up as short-term actions.

This type of training could be offered to students with a basic education in finance, engineering or legal sciences.

A first phase of harmonization is crucial as it will fill the lack of basic knowledge in technology for students in finance and legal sciences and will also fill the lack of basic knowledge for financial engineers and the wheels that FinTech will be able to serve.

Limit: The harmonization phase can't summarize all these basic knowledges (Fig. 3).

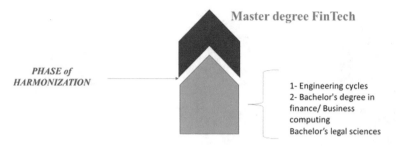

Master degree FinTech

PHASE of
HARMONIZATION

1- Engineering cycles
2- Bachelor's degree in
finance/ Business
computing
Bachelor's legal sciences

Fig. 3 Short term actions

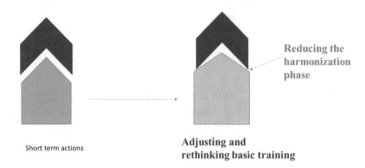

Reducing the
harmonization
phase

Short term actions

**Adjusting and
rethinking basic training**

Fig. 4 Medium term actions

4.2 Medium Term Actions: Gradually Towards a Full Curriculum in FinTech

The main aim of this phase is to setup a long-term vision for what do we need as finance curriculum, what will be the need of future financial institutions and customers. Through the assessment of short-term actions and training utility, gaps and shortcomings, the basics of a bachelor's degree in finance can be adjusted by adding incrementally computer sciences capabilities that FinTech needs. This phase is to be aligned with SDGs 2020–2030 (Fig. 4).

4.3 Long Term Actions: A Full Curriculum in FinTech

The important aspect of innovation, as well as the acceleration of the application of artificial intelligence requires the implementation of this new curriculum which will be more effective in the form of engineering studies (5 years) to master the basics of economics and management and also master the use of technology (Fig. 5).

All these actions require a good coupling between university courses and the needs of the ever-changing economy, on the one hand, but also a good capacity to develop

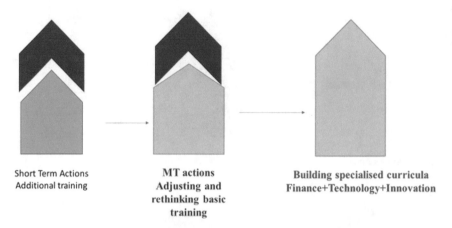

Short Term Actions	**MT actions**	**Building specialised curricula**
Additional training	**Adjusting and**	**Finance+Technology+Innovation**
	rethinking basic	
	training	

Fig. 5 Long term actions

the spirit of innovation that FinTech targets, on the other hand. This need leads us to look at the notion of technology transfer existing in the region, with the current innovation and competitiveness indicators that higher education has a large impact on.

5 Technology Transfer and Innovation in the Era of Challenging Conditions: A Promising Orientation in Such Environment

It is an undeniable fact today that the green and digital transition marks the beginning of a new economic cycle, which opens horizons towards a new impetus of innovation and needs. Contemplating the proponents of Kondratieff waves (1929), a new cycle is looming at the end of 2020 characterized by a new era of globalization marked by artificial intelligence, massive data sharing and a new relationship between man and nature under the aegis of climate change and health challenges. The Covid-19 pandemic is the asset of such need of transition where digitizing and greening the economic model stipulate a new relationship between Man and Nature, but also strengthening the connection between the economic, the social and the environmental.

In this type of situation, the organization faces an alarming need to adapt on the one hand, but also to optimize and regain competitiveness on the other hand. Hence, scientific research is in this sense a real development pillar that can provide solutions to these concerns. The university and research centers must play their role as a think tank for reflection and improvement of social welfare, towards the resolution of problems of poverty, resource scarcity and optimization.

Research, valorization, transfer and innovation seem to be the process that a scientific result must convey as added value. In this way, the coupling between university and business will be strengthened in order to boost competitiveness on the one hand and to direct university education towards the real needs of the economic system on the other hand. Goal 9 of *UN Agenda—2030* meets these orientations.

Understanding Technology Transfer

"The next frontier for university technology transfer will likely be in the transformation of data-rich sectors using artificial intelligence and machine learning technologies. One area largely accumulating data is the healthcare sector. Medical knowledge is doubling every 73 days, yet we are barely scratching the surface of utilizing this data." [21]. For Penny [23] six linked steps are defined in order to ensure the transfer process. Hence, the current technological development makes it possible to proceed to tangible gains for the economic structure on several levels; from an optimization point of view, effectively ensuring the transfer of a new technology adapted to the specific context of the company will be able to compensate for deficiencies in the consumption of resources (raw materials, energy) or also contribute to the improvement of the productivity and quality of the production process in general.

The basic step is Technology innovation: this is the basic step, where the researcher will try to formulate theoretical ideas with a potential valorizable impact (solving a problem, new service or product). At this level, orienting scientific research towards the real needs of humanity by developing strategic research plans. However, the potential for innovation remains critical in developing and underdeveloped countries, where improving organizational performance, solving public service problems and improving social welfare are very important based on the need for development. Switzerland, Sweden, USA and Netherland are in the top countries according to the Global Innovation Index 2019.

Clearly these top countries in terms of innovation are also the leaders of the global competitiveness rankings (Fig. 6).

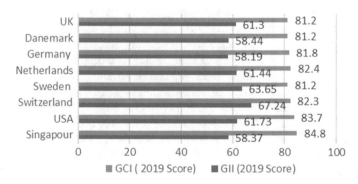

Fig. 6 Global innovation index versus Global competitiveness index (2019). *Data source* World Economic Forum

Based on the *ESCWA Critical Analysis of Innovation Landscape of Arabic Countries (AC) (2017)* which places it within the framework of sustainable development, the *National Innovation System* in AC faces a complex context, because of the number of young people, the existence of the informal economy and the social troubles since 2011, the link between innovation and employment is not yet welded. Expenditure on research and development are very important, statistics stipulates that *Saudi Arabia* is ranked thirtieth in terms of *Global Research and Development Companies* and twenty-fifth in terms of the *Quacquarelli Symonds University Ranking* average score of the top three universities.

Graduates in science and technology as a percentage of total tertiary graduates are very promising. For *Morocco* and *Tunisia,* they rank highly in the index of *Human Capital and Research. Tunisia* ranks third worldwide in terms of graduates in science and engineering, and *Morocco* ranks fourth. However, these two countries show weaknesses in *Global Research and Development Companies* and in the *Quacquarelli Symonds University Ranking.*

The report concludes that there are several weaknesses among the different pillar of the Global Innovation Index, especially in term of policy, limited financial resources, lack of incentives and the absence of an overall promising vision gathering university, economic institution and policy framework.

Vac and Fitiu [30] demonstrate the important role of technology transfer offices TTOs in matching the needs of companies with research work. In order to contribute to sustainable development efforts, TTOs need to mobilize the right human skills. From the second step "The invention disclosure" to "the transfer evaluation step", patent licensing and intellectual property seems to be the most critical issues that researchers must master. Indeed, according to the World Organization of Intellectual property WIPO, from 1980–2018 Egypt is in the first rank with 34.1% of total patent, Morocco and KSA are in the second and third rank (Fig. 7).

This level of patent and intellectual property contract remains by far the desired level that will ensure sustainable development, and the objectives of a better competitiveness of companies. Boosting the *National System of Innovation* on the one hand, alongside the orientation of academic training towards the real needs of the economy on the other hand, are decisive. The TTO offices will be able to perform this dual function of transfer and feedback toward necessary adaptation that the university needs to have to be more entrepreneurial and innovative (Fig. 8).

Drawing up a strategic vision of the National Innovation System, the objectives of scientific research as well as the mobilization of sufficient financial resources, will be able to give the university the ability to create the right profiles worthy of this transitional period, such as FinTech. The large number of engineers and students in science and technology in these countries, stipulates their decisive role in this transition period in general, but also their potential as a driver for the desired FinTech industry development.

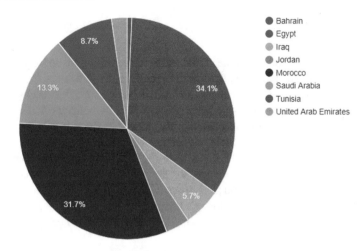

Fig. 7 Total patent grants (direct and PCT national phase entries)—AC (1980–2018). *Data* WIPO statistics

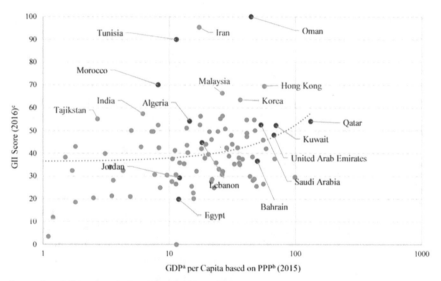

Sources: Cornell University and others, 2016; World Bank, 2017.

Notes: [a] In thousands of United States Dollars.
    [b] PPP – Purchasing Power Parity.
    [c] Each country is scored from 0 to 100 for every pillar in the Global Innovation Index.

Fig. 8 Science and engineering graduates, 2016

6   Conclusion

It is an undeniable fact today that the digital and green transition are no longer choices, but obligations in view of the multiple challenges. In order to achieve the objectives of sustainable development and meet the United Nations 2020–2030 agenda, we need to think across all sectors as well as policies. The tremendous change that our society faces in terms of climate vulnerability, addresses the role of education and culture at the heart of the key factors of change.

The FinTech industry came to address the limitations of the traditional functioning of financial institutions, which could not meet the aspirations of the global community in terms of proximity, security and convenience. While the banking sector represents the basic source of financing for today's economies, the disparities in wealth and the poverty rate suffered by countries stipulate the limits of the existing financing mechanisms.

Crowdfunding, peer to peer lending, FinTech assets and many others are the future of the sector, which requires a new understanding of what a banker's and insurer's job should entail. Advanced technologies are at the core of these changes which require a capacity for mastery and innovation. To respond to these considerations, higher education, whose role is to prepare profiles of graduates who can meet the needs of the market, is at the crossroads of a new evolution similar to the one it underwent during the era of the integration of computer sciences in all fields. Through this paper we have tried to outline the potential of FinTech in the case of Arab countries specifically in MENA region. Catching up with the trend towards sustainable economic inclusion requires effective financial inclusion. At this level, by analyzing the indicators of innovation and education in science and technology alongside the evolution of arab universities, higher education must support this trend through a long-term strategy that can merge the current profiles of graduates in the field, towards new types of profiles mastering the technology side and the financial side of FinTech. The paper set out the short, medium, and long-term actions that can gradually upgrade university training and nuances the lines of thought on the nature of training. Without a sustained policy of technology transfer and a good coupling between the university and the economy, benefiting from the potential for innovation will remain far from its possibilities. For the case of MENA region, engineering, science, and technology students have the best profile to support this change in the short and medium term.

References

1. Aita, S. (2017).The innovation landscape in Arab countries. ESCWA-UN, Retrived on https://www.researchgate.net/publication/330656282_The_Innovation_Landscape_in_Arab_Countries_A_Critical_Analysis.
2. Alt, R., & Puschmann, T. (2016). *Digitalisierung der Finanzindustrie : Grundlagen der Fintech-Evolution.* Springer.
3. Alt, R., & Ehrenberg, D. (2016). FinTech–umbruch der Finanzbranche durch IT. *Wirtschafts informatik Management, 8*(3), 8–17.

4. Bahrain Fintech Bay, FinTech Consortium and Jordan Fintech Bay. «MENA financial inclusion report 2020», Retrived on https://www.bahrainfintechbay.com/financial-inclusion.
5. Center of Financial Inclusion at Accion. «Microfinance vs financial inclusion : what's the difference ?», Retrived on https://www.centerforfinancialinclusion.org/microfinance-vs-financial-inclusion-whats-the-difference.
6. Chehade, N., & Navarro, A. (2017). Financial inclusion measurement in the Arab World. Working paper, CGAP and the Arab Monetary Fund's Financial inclusion Task Force. Retrieved on https://www.cgap.org/sites/default/files/Working-Paper-Financial-Inclusion-Measurement-in-the-Arab-World_1.pdf.
7. Chishti, S., & Barberis, J. (2016). *The FinTech Book: The Financial Technology Handbook for Investors, Entrepreneurs and Visionaries.* John Wiley & Sons.
8. Christensen, C. M., Bartman, T., & Van Bever, D. (2016). The hard truth about business model innovation. *MIT Sloan Management Revue, 58*(1), 31–43.
9. Clement Ancri, the Board of Governors of the Federal Reserve System, Washington D.C. (2016). Fintech Innovation : An overview, Retrieved on http://pubdocs.worldbank.org/en/767751477065124612/11-Fintech.pdf.
10. Council of Arab Central Banks and Monetary Authorities' Governors Arab Financial Inclusion-April 27, 2020. «Arab financial inclusion day 2020» under the theme of «Towards building financial literacy that promotes financial inclusion». Retrieved on http://www.rtb.iq/news/arab-financial-inclusion-day2020/.
11. Demirguc-Kunt, A., & Klapper, L. (2012). Measurin financial inclusion: The global findex database. *Policy Research Working Paper, The World Bank.* Retrieved on https://openknowledge.worldbank.org/bitstream/handle/10986/6042/WPS6025.pdf?sequence=1&isAllowed=y.
12. Evans, P. C., & Basole, R. C. (2016). Revealing the API ecosystem and enterprise strategy via visual analytics. *Communication of the association for Computing Machinery (ACM), 59*(2), 26–28.
13. FinTech Trends 2020/2021: Top prediction according to Experts. Retrived on https://financeonline.com/fintech-trends/.
14. Gagliardi, L., Dickerson, J., & Skan J. (2016). FinTech and the evolving landscape: Landing points for the industry. *Accenture,* 1–12.
15. Grossman, J., & Nelson, P.K. (2019). Digital finance for development. *A Handbook for USAID STAFF.* Retrived on https://www.marketlinks.org/sites/marketlinks.org/files/resource/files/Digital_Finance_Handbook_-_Complete_Vesion_-_v11.pdf.
16. Gomber, P., Kauffman, R.J., Parker, C., & Weber, B.W. (2018). On the Fintech revolution: Interpreting the forces of innovation, disruption, and transformation in financial services. *Journal of Management Information Systems, 35*, 220–265.
17. Mackenzie, A. (2015). The FinTech revolution. *London Business School Review, 26*(3), 50–53.
18. Maglio, P. P., Kwan, S. K., & Spohrer, J. (2015). Commentary—Toward a research agenda for human-centered service system innovation. *Service Science, 7*(1), 1–10.
19. Maglio, P. P., & Spohrer, J. (2013). A service science perspective on business model innovation. *Indust. Marketing Management, 42*(5), 665–670.
20. Manyika, J., Lund, S., Singer, M., White, O., & Berry C. (2016). Digital finance for all: Powering Inclusive growth in emerging economies. *McKinsey Global Institute.* Retrieved on https://www.mckinsey.com/~/media/McKinsey/Featured%20Insights/Employment%20and%20Growth/How%20digital%20finance%20could%20boost%20growth%20in%20emerging%20economies/MG-Digital-Finance-For-All-Full-report-September-2016.pdf.
21. Nag, D. G., Gupta, A., & Turo, A. (2020). The evolution of university technology transfer: by the number. Retrived on https://www.researchgate.net/publication/340766806_The_Evolution_of_University_Technology_Transfer_By_the_Numbers.
22. Nienaber, R. (2016). Banks need to think collaboration rather than competition. In S. Chishti, J. Barberis (Eds.) *The FinTech book: The financial technology handbook for investors, entrepreneurs and visionaries* (Wiley, Hoboken, NJ), 20–21.
23. Ozili, P.K. (2018). Impact of digital finance on financial inclusion and stability. *Borsa Istanbul Review, 18*(4), 329–340. https://doi.org/10.1016/j.bir.2017.12.003.

24. Penny, R. (1992). *Understanding the Technology Transfer Process.* University of Pennsylvania African studies center. Retrived on http://www.africa.upenn.edu/Comp_Articles/Technology_Transfer_12764.html.
25. Basole, R. C., & Patel, S. S. (2018). Transformation through unbundling: Visualizing the global FinTech ecosystem. *Service Science., 10*(4), 1–18.
26. De Reuver, M., Sorensen, C., & Basole, R. C. (2017). The digital platform. *Journal of Information Technology.* https://doi.org/10.1057/s41265-016-0033-3.
27. Rogers, D. L. (2016). *The Digital Transformation Playbook: Rethink Your Business for the Digital Age.* Columbia University Press.
28. Smedlund, A. (2012). Value cocreation in service platform business models. *Service Science, 4*(1), 79–88.
29. The World Bank. «Financial inclusion overview», Retrieved on https://www.worldbank.org/en/topic/financialinclusion/overview.
30. Vac, C. S., & Fitiu, A. (2017). Building sustainable development through technology transfer in a Romanian university. *Sustainability, 9*(11), 2042. https://doi.org/10.3390/su9112042.

Willingness to Communicate in Face-To-Face and Online Language Classroom and the Future of Learning

G. Bhuvaneswari[ⓘ]**, Rashmi Rekha Borah**[ⓘ]**, and Moon Moon Hussain**

1 Introduction

With the advancement of technology, it is becoming increasingly vital to track and evaluate the communication patterns that emerge from these changes. Individuals can employ a variety of communication styles because of technological advancements [1].

Several approaches have waxed and waned in the area of L2 learning and teaching over the last century. Each of these approaches has attempted to discover the best and most successful methods for achieving their objectives. Since the 1980s, communicative approaches have dominated the market, radically altering how L2s are taught. Consequently, contact in the target language is fundamental to language learning and acts as the base for language instruction. Even though contemporary language pedagogy emphasizes communication, learning a second language and communicating, it does not follow a series of steps. It is, on the contrary, a dynamic phenomenon affected by affective, personal, and linguistic influences. The ability to interact is one of the variables that have been observed.

Second language acquisition (SLA) research shows that contact in the target language and linguistic production contributes to language growth [2]. Gender, grade level, self-efficacy, self-concept, apprehension, perceived importance of writing, self-efficacy for self-regulation, and previous writing achievement were used by Pajares and Valiante to predict the writing capacity of 742 L1 middle-school students [3]. MacIntyre and Charos extended the limits of the construct and applied it to L2

G. Bhuvaneswari (✉) · R. R. Borah
School of Social Sciences and Languages, Vellore Institute of Technology, Chennai, India

M. M. Hussain
B.S. Abdur Rahman Crescent Institute of Science and Technology, Chennai, India
e-mail: moon@crescent.education

© The Author(s), under exclusive license to Springer Nature Switzerland AG 2022
A. Hamdan et al. (eds.), *Technologies, Artificial Intelligence and the Future of Learning Post-COVID-19*, Studies in Computational Intelligence 1019,
https://doi.org/10.1007/978-3-030-93921-2_14

contexts, based on research in one's native language (L1) [4]. L2, described as "readiness to enter into discourse with a specific individual or per-sons, using an L2 at a specific time" [5], is thought to have a direct effect on L2 use [6].

According to research, females also outperform males in L1 learning, reading, and verbal capacity, according to research [7]. In this regard, the literature indicates that variables such as self-perceived communication competence (SPCC) [8], foreign language anxiety (FLA) [9], age, and gender have affected a learner's and thus his/her involvement in communication [10].

In learning and teaching, a classroom is a social setting in which various interconnected factors such as teacher support, student cohesiveness, and task orientation play critical roles. These components can affect students' thoughts and behaviour. This line of research has indicated that a high-quality classroom atmosphere facilitates student learning. As a consequence, the absolute importance of a healthy classroom climate is undeniable.

Personal and situational interests are the two types of interest [11]. Personal interest is a common phenomenon that typically lasts a long time. Individual involvement can spur learners' participation in a task independently, and such engagement is assumed to be a deep-seated rather than a surface-level phenome-non. In addition, personal interest is often less affected by external factors, resulting in greater stability than situation interest [11].

2 Research Questions

Due to the crucial role that a classroom interaction plays in learning and analysis of willingness to communicate is crucial. Therefore, this study sets intent on un-veil the difference between online and face-to-face class environment. Besides, it attempts to seek out answers to the subsequent research questions:

1. How comfortable are the students to communicate with strangers in an online classroom?
2. How willing are the students in initiating a talk with a group of strangers in an online classroom?
3. How comfortable are the students in making presentations in front of their peers in the online classroom?
4. Do body language/cues affect willingness to communicate through online communication?

3 Literature Review

Walther presents a three-part hierarchy of communication that starts at the lowest level, impersonal and ascends to the hyper-personal in his theoretical analyses of how humans engage with one another in a computer-mediated world [12].

Despite a large body of research on and variables related to it, little research has been done on L2 following the classroom climate. Students, especially in countries where English is learned as a foreign language (EFL), spend a considerable amount of time in class to learn a language [13]. In these classes, it is of utmost importance that students can speak and participate in activities because they are mostly limited to classrooms in terms of interaction and communication.

Consequently, this personality trait explains why one person would interact while others would not fall under the same circumstances [14]. Following the conception, a variety of experiments were performed to bring it to the test. This line of inquiry into L1 showed that a strong desire to communicate is likely to contribute to increased language use [15–17].

After discovering WTC as a variable that influences communication frequency, MacIntye and Charos [3] tested the concept in L2 situations to see if WTC in L2 predicted target language use frequency. They discovered that WTC was a measure of language performance and was influenced by perceived competence and language anxiety by administering questionnaires to language learners. MacIntyre et al. [5] proposed the heuristic model of variables affecting WTC in their seminal work. According to the model, WTC is the most immediate factor that leads to L2 usage. It is characterized as "a readiness to enter into discourse at a specific time with a specific individual or persons, using an L2" [5]. In addition, a range of interconnected personal, motivational, affective, and cognitive factors have been suggested to influence WTC and communication behaviour.

"Using the language strongly implies a preexisting behavioural intention, a desire to interact in the L2," according to MacIntyre, Baker, Clément, and Conrod [18]. Since "talking to learn" [19] is vital to the L2 learning process, students must speak. As a consequence, they must first be able to communicate. Kang hypothesizes that developing L2 WTC makes learners become more independent learners who are more likely to attempt to learn the language without the assistance of teachers and lead learners to increase their learning opportunities [20] (Fig. 1).

People differ a lot in terms of how much they speak, according to McCroskey and Baer [21]. Some people talk a lot, while others tend to remain quiet, and still, others speak only under certain circumstances. McCroskey and Baer suggested the definition of in one's native language as a personality variable, based on differences in the amount and frequency of language use among people [21].

4 Procedure

A total of 326 universities (undergraduate, post-graduate and PhD scholars) students took part in this study, with 111 females (34%), 215 males (66%). Students of two Universities in a city in South India were chosen as participants. Many of the participants had passed the extremely competitive university entrance test, and they had English as one of the compulsory courses. All the students had English as the first language in the curriculum even though English is not their regional language. The

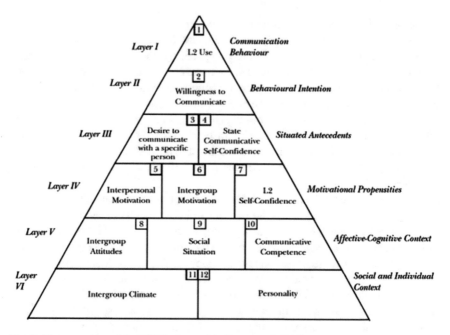

Fig. 1 The pyramid model of WTC. *Source* MacIntyre et al. (1998)

Table 1 Mean and standard deviation	Age group	Mean age	SD of age
	17–45	20.75	11.56

participants ranged in age from 17–45 with an average age of 20.75 (SD = 11.56). Non–English major university students were not chosen because they do not acquire a functional English proficiency, cannot speak English in classes and have limited English class time.

4.1 Age Group Distribution

The participants who took part in answering the questionnaire fall in the 17 to 45 age group category, as shown in Table 1.

4.2 Data Collection and Data Analysis

To perform the Data analysis, two sets of questions were given. 10 questions to assess online communication level and 10 to assess offline communication regarding willingness, comfort and motivation. The questionnaires were given to 326 students and used for validity and reliability analyses. The responses collected were qualitative nominal measurement responses, and they were converted to quantifiable numbers to analyze the statistical data we collected.

The knowledge obtained through the questionnaires was computerized and analyzed using SPSS. To begin, descriptive statistics were used to get a clear picture of the students' WTC levels and their classroom experiences. Second, the chi-square method was used to see a substantial difference between online and face-to-face communications in classroom environments.

5 Methodology

5.1 Level of Comfort to Communicate with Strangers in an Online Classroom When Compared to Face-To-Face Communication in the Classroom

There have been varied factors that play a vital role in affecting the willingness to communicate, such as not being confident enough, fear of being ridiculed, not a native speaker, stage fear, not having sufficient language skills etc., the factor that was intended to be analyzed in our research was analyzing how comfortable the participants were in communicating with strangers.

Hypothesis

H0: There is no significant difference in the comfort level in communicating to a group of strangers between online and face-to-face communication.

H1: There is a significant difference in the comfort level in communicating to a group of strangers between online and face-to face communication.

To analyze the student's (Table 2) willingness to communicate with strangers, there were 20 questions composed of statements concerning online and offline communication concerning communicating to a group of strangers to determine how comfortable students are communicating in the classroom. On a four-point Likert scale, responses to the items were anchored at one end by "never" and at the other end by "always".

When the effect of "being comfortable online" was examined, there was a significant difference between the means of variance between online and face to face

Table 2 Distribution of data on comfort in communication in classroom with strangers

Mode of communication	Never	Sometimes	Frequently	Always	Total
Online	40	76	53	157	326
Face-to-Face	146	45	100	35	326

communication. Students tend to open up more to talking to strangers online. In face-to-face communication, people become familiar; students do not want to lose face if they face any issues when communicating, such as exhibiting their body cues to everyone present in the face-to-face communication. It gives them a way to hide their "being uncomfortable" in online mode. They were able to open up more with people though they were strangers.

Surprisingly, nearly half of the people have said that they were never comfortable talking to strangers face to face. In face-to-face communication, people didn't feel comfortable talking to strangers even though they looked at the people they were talking to. While 48% of the students always felt comfortable in online communication, 44% felt uncomfortable in face-to-face communication.

In terms of speaking with strangers, the higher the score, the better the conversation in the classroom. Thus, the table value of Chi-square $= 7.819$ at $\alpha = 0.05$. Degrees of freedom $= 3$. Calculated Chi-square $= 160.30$, which is more than its critical value.

Therefore, the null hypothesis is rejected. Hence we conclude that there is a significant difference in feeling comfortable talking to a group of strangers in between online and face-to-face communication.

5.2   Willingness to Initiate a Talk with a Group of Strangers in an Online Classroom When Compared to Face-To-Face Communication

Students' willingness to communicate when they are not aware of who they are conversing with in their online mode of communication is profoundly influenced as a result of this. When students are asked to strike up a conversation with a group of strangers, their hesitancy is evident. The purpose of this study is to see how comfortable students are with initiating a conversation with a group of classmates they have never met before. We investigate if the act of initiating a conversation is influenced by the online mode of communication.

Hypothesis

H0: There is no significant difference in the willingness to communicate in communicating to a group of strangers between online and face-to-face communication.

H1: There is a significant difference in the willingness to communicate to a group of strangers between online and face-to-face communication, (Table 3).

Table 3 Distribution of data on willingness to initiate a talk with a group of strangers

Mode of communication	Never	Sometimes	Frequently	Always	Total
Online	25	78	56	167	326
Face-to-Face	76	43	75	132	326

There are 20 statements on online and offline communication regarding readiness to speak with a group of strangers to determine the willingness to communicate in the classroom. Among the 20 questions, the questions to determine the willingness were, "I can start a conversation with a group of strangers online in the classroom," and "I prefer to start a group discussion in face-to-face contact." Responses to the items were anchored at one end by "never" and at the other end by "always" on a four-point Likert scale. Communication was rated positively by 51% of students, and face-to-face communication was rated positively by 40% of students. However, 7 percent said they would never initiate a conversation via internet communication, while 23% said they would never initiate contact using face-to-face conversation. The reasons for this could be the non-requirement of approval of others and the non-existence of fear of being judged by others.

Thus, the table value of Chi-square = 7.819 at α = 0.05. Degrees of freedom = 3. Calculated Chi square = 42.72 which is more than its critical value.

Therefore, the null hypothesis is rejected. Hence we conclude that there is a significant difference in feeling comfortable talking to a group of strangers in between online and face-to-face communication. The variance between the two samples taken was, however, small.

The results obtained correlates highly towards initiating the talk to strangers irrespective of communicating online or face-to-face. Students were able to initiate the conversation with strangers though they did not feel comfortable talking to them.

5.3 Comfort Level in Participating in Group Discussions Online When Compared to Face-To-Face Communication

Students learn how to hold a conversation in small and big groups when they participate in group discussions. We gathered information to see how eager students were to participate in online and face-to-face group discussions. The primary analysis showed that the correlations obtained were substantially overlapping with others. The results showed good stability in the way people responded when it comes to group discussion. Individual student traits closely behaved in group situation context, as shown in Table 4.

Hypothesis

H0: There is no significant difference in feeling comfortable in participating in group discussions in online and face-to-face classroom communication.

Table 4 Distributions of data of willingness to initiate in group discussion

Mode of communication	Never	Sometimes	Frequently	Always	Total
Online	71	33	73	149	326
Face-to-Face	113	14	81	118	326

H1: There is a significant difference in feeling comfortable in participating in group discussions in online and face-to-face classroom communication.

In the questionnaire, among the 20 questions asked, the questions asked to determine the comfort level of the participants were, "I am willing to imitate in online Group Discussions in the online classroom" and "I like to initiate in group discussion in Face to Face communication". While 45% of the participants responded that they are always willing to initiate in the online classroom, 36% responded positively to face-to-face classroom communication. Initiating in group discussion seems to be lesser in face-face communication than online communication as 21% are never ready to initiate in the online classroom, and 34% are never ready to initiate group discussions in a face-to-face classroom. This could be that the students need not face stage fear, mockery by classmates, or the initial hiccups. They can even hide their videos for a while to feel the initial comfort, which is impossible in a face-to-face classroom.

Thus, the table value of Chi-square $= 7.819$ at $\alpha = 0.05$. Degrees of freedom $= 3$. Calculated Chi square $= 11.28$ which is more than its critical value.

Therefore, the null hypothesis is rejected. Hence we conclude that there is a significant difference in feeling comfortable talking to a group of strangers between online and face-to-face communication. People showed more willingness to communicate and initiate discussion when the platform was online. The mean difference between online and face-to-face communication variance falls very close, indicating that though there was a significant difference in the way people acted when they communicated to strangers, they all correlated towards showing their willingness to communicate in talking to strangers irrespective of the platform.

5.4   Comfort Level in Presenting a Speech to a Group of Strangers in Online and Face-To-Face Communication

When examining the results obtained on giving a presentation in online and face-to-face communication, results revealed that Students showed more comfortable-ness. Each of their responses was evaluated using the Chi-square method. Though there were significant differences in the way students felt comfortableness when they delivered a speech among strangers, there was a positive correlation.

Furthermore, the difference in the mean–variance showed that the analysis produced very slight variations and showed a positive correlation in the results, indicating that students showed more comfortableness towards delivering a speech in front of strangers irrespective of online or face-to-face communication platform, as shown in Table 5.

Hypothesis

H0: There is no significant difference in feeling comfortable giving a presentation to a group of strangers in online and face-to-face classroom communication.

Table 5 Distributions of data of comfort in presenting a speech to strangers

Mode of communication	Never	Sometimes	Frequently	Always	Total
Online	51	69	64	142	326
Face-to-Face	65	40	80	141	326

H1: There is a significant difference in feeling comfortable giving a presentation to a group of strangers in online and face-to-face classroom communication.

To determine the willingness to initiate in a group, questions "I can present any speech online comfortably with among group of strangers"—(online and face to face) were asked. While 43.5% of the participants feel comfortable giving a presentation among strangers in online classrooms, 43.2% of the participants feel comfortable in face-to-face classroom communication. 15% of the participants are never comfortable presenting a speech in an online classroom, and 19% of the participants are never comfortable presenting a face-to-face classroom speech.

Thus, the table value of Chi-square $= 7.819$ at $\alpha = 0.05$. Degrees of freedom $= 3$. Calculated Chi square $= 11.28$ which is more than its critical value.

Therefore, the null hypothesis is rejected. Hence, we conclude that there is a significant difference in feeling comfortable presenting a talk among strangers between online and face-to-face communication.

Being in a group gives them additional motivation to eliminate their fear and shyness towards initiating their talk, making them feel comfortable.

5.5 Willingness to Communicate—Talking to Teacher Online Without Making Eye Contact and Talking to the Teacher in Face-To-Face Communication

Hypothesis

H0: There is no significant difference between comfortable talking to teachers without making eye contact in online communication and listening to teachers in face-to-face communication.

H1: There is a significant difference between feeling comfortable talking to teachers without making eye contact in online communication and listening to teachers in face-to-face communication.

To analyze the body language impact on the level of willingness to communicate (Table 6), the responses to questions, "I feel comfortable talking to teacher online without the need to look at the teacher and "I can talk to my teacher comfortably in Face to Face communication" were analyzed. 43% of the participants feel comfortable without eye contact for communication, while 15% never feel comfortable without eye contact in online communication. In face-to-face communication,

Table 6 Distribution of Data—Communication with respect to Eye-contact

Mode of communication	Never	Sometimes	Frequently	Always	Total
Online	52	91	41	142	326
Face-to-Face	141	10	99	76	326

while 23% always feel comfortable communicating with eye contact, 43% never feel comfortable doing so.

Thus, the table value of Chi-square $= 7.819$ at $\alpha = 0.05$. Degrees of freedom $= 3$. Calculated Chi square $= 150.01$ which is more than its critical value.

Therefore, the null hypothesis is rejected. Hence, we conclude that there is a significant difference in feeling comfortable talking to a group of strangers between online and offline communication modes.

5.6 Willingness Towards Listening to a Teacher in Online and Face-To-Face Communication

Listening is a much-needed skill in a classroom, and students' willingness to listen to a teacher is a needed skill for a student to learn better. Though the student's ability to listen gets affected mainly because of how well the student can maintain the eye-contact, asking relevant questions on the subject being taught, ability to recollect or repeat what teach has taught etc., Having eye-contact signals the teacher that the student is actively listening and that the student is focused towards what the teacher is teaching. For a teacher, this becomes much more difficult when the teaching platform changes to online because the student no longer has eye contact with the teacher. The power of connection that needs to be present between the student and the teacher for a healthy environment for learning is nearly lost and it reflects in the students' willingness to communicate as well. When the responses were evaluated for listening, there were more significant differences in feeling comfortable listening to a teacher face-to-face than listening to a teacher online, as shown in Table 7.

Hypothesis

H0: There is no significant difference between feeling comfortable listening to teachers without making eye contact online and listening to teachers in face-to-face communication.

Table 7 Distribution of Data—Ability to listen better in online versus face to face communication

Mode of communication	Never	Sometimes	Frequently	Always	Total
Online	67	88	49	122	326
Face-to-Face	196	8	110	183	326

H1: There is a significant difference between feeling comfortable listening to teachers without making eye contact online and listening to teachers in face-to-face communication.

Result: The table value of Chi square $= 7.819$ at alpha $= 0.05$. Degrees of freedom $= 3$. Calculated Chi square $= 151.58$which is more than its critical value. Therefore, the null hypothesis is rejected. Hence we conclude that there is a significant difference in feeling comfortable when talking to a group of strangers in between online and offline mode of communication.

5.7 Willingness Towards Talking to Acquaintances Versus Talking to Strangers—A Correlation Analysis

The correlation research of students' propensity to talk to strangers and their willingness to talk to acquaintances revealed that regardless of the form of communication, students displayed comparable qualities in talking with individuals. A almost equal percentage of participants gave comparable responses. This study demonstrates that people's tenacity does not alter greatly when different forums are provided for them to open up and discuss. It is their behavioral quality that neither online nor face-to-face communication platforms have much influence over.

Analyzing why students show comparable attributes based on their confidence level, language proficiency, and fear factor would provide considerably more interesting study results; nevertheless, such an analysis is outside the scope of our research (Table 8).

Hypothesis

H0: There is no significant difference between feeling comfortable talking to strangers and talking to acquaintances irrespective of online or face to face communication.

H1: There is significant difference between feeling comfortable talking to strangers and talking to acquaintances irrespective of online or face to face communication.

Result: The table value of Chi square $= 7.819$ at alpha $= 0.05$. Degrees of freedom $= 3$. Calculated Chi square $= 6.982$ which is less than its critical value. Therefore, the null hypothesis is accepted. Hence we conclude that there is no substantial difference in feeling comfortable when talking to a group of strangers versus talking to acquaintances irrespective of online or face to face communication.

Table 8 Distribution of Data—Strangers versus Acquaintances correlation study of WTC

Strangers/Acquaintances	Never	Sometimes	Frequently	Always	Total
Acquaintances	54	154	63	55	326
Strangers	76	157	53	40	326

Table 9 Distribution of number of students—Influencing factors—WTC

Factors Influencing WTC	Good rapport	Emotional connection	Interaction	Comfortable platform
Online	28	20	34	52
Face to face	298	306	292	274
Total	326	326	326	326

5.8   Willingness to Communicate—Influencing Factors

Better communication between students and teachers is ensured by different factors and we analyzed the below list of aspects from the students perspective. The word "platform" would mean either online or face-to-face communication platforms (Table 9).

Factors Taken for Analysis

1. Which platform provides good chances for students in building good rapport with his/her teacher?
2. Which platform provides good changes for students in making better emotional connection with his/her teacher?
3. Which platform do the students find suitable for better communication?
4. Which platform do the students find more comfortable for their communication?

Good Rapport Between Teacher and Student

A positive and respectful relationship between students and teachers is essential for effective communication. It can boost student engagement and willingness to speak, promote social interaction, and create a positive learning environment, all of which lead to improved learning. From the above data, 91% of the students believe that face-to-face communication improves their chances of developing a good rapport with their teacher.

Positive Emotional Connection Between Teacher and Student

Positive emotional connection is crucial for students to learn better and communicate better with teacher. Establishing a positive emotional connection between teacher and student greatly aids in their learning and motivation. 93% of the students believe that face to face communication helps them to build positive emotional connection with their teacher.

Interaction Between Teachers and Students

Relationships between teachers and students develop as a result of real-time interactions between teachers and students. Interpersonal content, structure, and complementarity can all be used to characterize these real-time encounters. 84% of the students feel that face-to-face communication is more comfortable for them to communicate.

6 Challenges of WTC in Online Communication

Students struggle to adjust to an online learning environment immediately following traditional classroom instruction. They are unable to adapt to computer-based learning due to the abrupt change. Students who have been studying in a traditional classroom setting their entire lives are unable to concentrate on online platforms. It is critical for them to approach the new learning environment with a positive attitude. Many students may not have access to a high-speed internet connection, which is necessary for online learning. As a result, they are having difficulty launching virtual learning and other platforms that require an internet connection. They have technical problems since they are unfamiliar with technology and computer programmes.

Factors	Readiness to communicate	
	Online	Offline
Individual differences	High	Low
Rapid synchronous exchanges of information	High	Low
Teachers' Personality	Low	High
Teachers' ability to communicate	Low	High
Teachers' ability to lead	Low	High
Teachers' prompts for responses	Low	High
Opportunity to Interact	Low	High
Technical glitches	High	Low

A fast or slow internet connection can affect how quickly you can join the class and avoid missing any live sessions. During online learning, students lack appropriate communication skills. Teachers make tasks to help students improve their reading and writing skills, but it's possible that they won't be able to write eloquently enough for educators to comprehend what their assignments are about. Due to the new learning approach, some kids are hesitant to communicate with their professors and peers. It could be due to a lack of interest, a lack of technological abilities with apps and video calls, or an inability to communicate via live chats, emails, or text messages.

7 Implications and Conclusions

According to the findings presented in this research, students had little preference for online communication over the face-to-face conversation. Individual differences in communication may have an impact on the challenges of online communication. For example, one student may love online learning while another has a strong dislike for it. Therefore, the challenge becomes a matter of determining who communicates online and who communicates better in person and tailoring the needs to improve a student's readiness to communicate.

The online classroom has its advantages like rapid synchronous exchanges of information sharing such as text, files, chats, visual clips and videos that makes online classrooms easy accessible source that suits promotion of "Electronic Collaborative Learning Groups" [22]. Adding to this, a student can review their presentations and self-reflect on their abilities to assess how tone, rhythm, questioning style, tempo, and other factors influence the quality of their delivery. However, even though the online classroom has its advantages, such as the ability to record sessions, which allows academics to reflect on online learning episodes, and the objective nature of online communications, which allows for easier coding and decoding of interactions, there is still a lack of motivation to communicate.

A teacher's strong and pleasant personality, communication abilities, and ability to lead can be just as significant as the knowledge and skills of a foreign language imparted. They aid in creating positive relationships with pupils, which is critical for effective teaching [23]. Teachers' prompts for responses, more opportunities to interact, recognizing friends' voices, and the willingness of shy students to contribute because they don't have to show their video, unlike face-to-face communication, where feeling self-conscious in front of others is unavoidable, are some of the factors that support willingness in online classrooms. Lack of opportunities to interact and co-create with friends, technical glitches, a lack of cues from peers, difficulty determining who speaks, not knowing who listens to the speaker, and not understanding when and how to communicate are all factors that may hinder students' willingness to communicate [24].

Students' communication skills must be emphasized during the learning process. The ability to recognize and use effective and appropriate communication patterns in various contexts is characterized as communication competency [25]. In addition, willingness to communicate depends on factors like grasping communication ethics, acquiring cultural awareness, using computer-mediated communication, and thinking critically; communication competency is required.

Conflict of Interest   No conflict of interest exists.

Appendix

Questionnaire

Directions: We would like to get your inputs for the below questions to understand, analyze and evaluate how willingly people communicated concerning online and face-to-face communication. People respond differently when the environment changes from online to a face-to-face environment. Therefore, a questionnaire for online communication and face-to-face communication has been given.

Online Communication

Select any one of the values from "Never"," Sometimes", "Frequently", and "Always" for each of the questions. There will be a corresponding question in the face to face communication section for analysis.

1. I feel comfortable talking to a group of strangers without looking at their face in the classroom.
2. I can initiate talking to a group of strangers online in the classroom.
3. I can participate in online Group Discussions in the classroom.
4. In an online classroom, I can talk to my classmates, who I don't know before.
5. I can present any speech online comfortably.
6. I feel comfortable talking without looking at people.
7. Communications are easy online.
8. I have felt miscommunication in online classes.
9. I feel comfortable talking to the teacher online without the need to look at the teacher.
10. I feel comfortable listening to teachers online.

Offline Communication—Face to Face communication.

Select any one of the values from "Never"," Sometimes", "Frequently", and "Always" for each of the questions.

11. I am shy about talking to a group of strangers.
12. I like to initiate group discussion.
13. I enjoy talking to new people.
14. I can present a talk to a group of strangers.
15. I can talk in a large meeting of acquaintances.
16. I can initiate asking for doubts/questions in the classroom.
17. Communications are easy in face-to-face communication.
18. I can listen comfortably face-to-face.
19. Eye contact is important in communicating with the teacher.
20. I feel miscommunication in the classroom.

References

1. Tomas, C. L., & Carlson, C. L. (2015). How do Facebook users believe they come across in their profiles?: A meta-perception approach to investigating Facebook self-presentation. *Communication Research Reports, 32*(1), 93–101.
2. Ellis, R. (1991) Grammar teaching—practice or consciousness-raising?" In R. Ellis (Ed.), Second language acquisition and second language pedagogy, 232–241. Clevedon, Avon: *Multilingual Matters*.

3. Pajares, F., & Valiante, G. (1999). Grade level and gender differences in the writing self-beliefs of middle school students. *Contemporary Educational Psychology, 24*, 3907405.
4. MacIntyre, P. D., & Charos, C. (1996) Personality, attitudes, and affect as predictors of second language communication. *Journal of Language and Social Psychology, 15*, 3–26. https://doi.org/10.1177/0261927X960151001
5. MacIntyre, P. D., Clément, R., Dörnyei, Z., & Noels, K. A. (1998). Conceptualizing willingness to communicate in a L2: A situational model of L2 confidence and affiliation. *The Modern Language Journal, 82*(4), 545–562. (1998).
6. Hashimoto, Y. (2002). Motivation and willingness to communicate as predictor Predictors of reported L2 use: The Japanese context. *Second Language Studies, 20*(2), 29–70.
7. Cao, Y., & Philp, J. (2006). Interactional context and willingness to communicate: A comparison of behavior in whole class, group and dyadic interaction. *System, 34*(4), 480–493.
8. Oz, H., & Demirezen, M., & Pourfeiz, J. (2015) Emotional intelligence and attitudes towards foreign language learning: Pursuit of relevance and implications. *Procedia—Social and Behavioral Sciences, 186*, 416–423. https://doi.org/10.1016/j.sbspro.2015.04.118.
9. Alemi, M., Parisa, D., & Pashmforoosh, R. (2011) The impact of language anxiety and language proficiency on WTC in EFL context. *Cross-Cultural Communication, 7*, 150–166. https://doi.org/10.3968/j.ccc.1923670020110703.152.
10. Donovan, L. A., & MacIntyre, P. D. (2004). Age and sex differences in willingness to communicate, communication apprehension, and self-perceived competence. *Communication Research Reports, 21*(4), 420–427. https://doi.org/10.1080/08824090409360006.
11. Hidi, S., & Renninger, K. A. (2006). The four-phase model of interest development. *Educational Psychologist, 41*(2), 111–127. https://doi.org/10.1207/s15326985ep4102_4.
12. Walther, J. B. (1996). Computer-mediated communication impersonal, interpersonal, and hyperpersonal interaction. *Communication Research, 23*(1), 3–43.
13. Peng, J.-E., & Woodrow, L. (2010) Willingness to communicate in English: A model in the Chinese EFL classroom context. *Language Learning, 60*, 834–876. https://doi.org/10.1111/j.1467-9922.2010.00576.x.
14. Bhuvaneswari, G., & Vijayakumar, S. (2021). Emotional intelligence and values in digital world through emoticons among indian students and faculty. *International Journal of Asian Education, 2*(2), 267–276. https://doi.org/10.46966/ijae.v2i2.142.
15. Chan, B., & McCroskey, J. C. (1987). The WTC scale as a predictor of classroom participation. *Communication Research Reports, 4*(2), 47–50.
16. MacIntyre, P. D., Babin, P. A., & Clément, R. (1999). Willingness to communicate: Antecedents & consequences. *Communication Quarterly, 47*(2), 215–229.
17. Zakahi, W. R., & McCroskey, J. C. (1989). Willingness to communicate: A potential confounding variable in communication research. *Communication Reports, 2*(2), 96–104.
18. MacIntyre, P. D., Baker, S. C., Clément, R., & Conrod, S. (2001). Willingness to communicate, social support, and language-learning orientations of immersion students. *Studies in Second Language Acquisition, 23*(03), 369–388.
19. Skehan, P. (1991). Individual differences in second language learning. *Studies in Second Language Acquisition, 13*, 275–298. https://doi.org/10.1017/S0272263100009979.
20. Kang, S. J. (2005). Dynamic emergence of situational willingness to communicate in a second language. *System, 33*(2), 277–292.
21. McCroskey, J. C., & Elaine, B. J. (1985). Willingness to communicate: The construct and its measurement.
22. Fåhræus, E. R., Bridgeman, N., Rugelj, J., Chamberlain, B., & Fuller, U. (1999). Teaching with electronic collaborative learning groups: Report of the ITiCSE'99 Working Group on Creative Teaching of Electronic Collaborative Learning Groups. In *Annual Joint Conference Integrating Technology into Computer Science Education.* (Cracow, Poland). ACM, NY, USA, 121–128. (1999)
23. Bhuvaneswari, G. (2016). 'Teacher Cognition' *International Journal of Economic Research, 13*(3), 693–695.

24. Bhuvaneswari, G., Swami, M., Jayakumar, P. (2020). Online classroom pedagogy: Perspectives of undergraduate students towards digital learning. *International Journal of Advanced Science and Technology, 29*(04), 6680–6687. Retrieved from http://sersc.org/journals/index.php/IJAST/article/view/28069.
25. Cooley, R. E., Roach, D. A. (1984). "A Conceptual Framework," In R. N. Bostrom (Ed.) *Competence in Communication: A Multidisciplinary Approach*, (Beverly Hills, CA: Sage) 25.

Mental Health, E-learning, and Future of Education in Palestine After the COVID-19 Pandemic

Fayez Azez Mahamid, Dana Bdier, and Abdulnaser Ibrahim Nour

1 Introduction

1.1 The COVID-19 in Palestine

The first case of COVID-19 in Palestine was diagnosed on March 5th 2020. The Palestinian Authority (PA) immediately declared a state of emergency and initiated containment measures including a lockdown in the governorate of Bethlehem, followed by lockdowns in other governorates of the West Bank on March 22nd by restricting movement and closing all non-essential facilities [1]. As of 8 June 2021, there have been 338, 694confirmed total cases of coronavirus and 3 784 deaths in the Gaza Strip and the West Bank including East Jerusalem [2].

When the pandemic reached the Israeli occupied Palestinian West Bank, the mental health status of Palestinians was already compromised because of the political context [3]. Since 1967, and the fall of the West Bank (including East Jerusalem) and the Gaza Strip, under Israeli military rule, Palestinians have endured chronic exposure to political violence, oppression, subjugation, and lack of freedom [4]. Generations of Palestinians have suffered human rights violations, including land confiscation, displacement, control by the Israeli army of the movement of people and goods from one area to another—and outside the country—death, injury, disability, imprisonment, the lack of freedom, and injustice. The Palestinian Authority (PA) has little

F. A. Mahamid (✉) · D. Bdier
Psychology and Counseling Department, An-Najah National University, Nablus, Palestine
e-mail: mahamid@najah.edu

D. Bdier
University of Milano-Bicocca, Milan, Italy

A. I. Nour
Accounting department, An-Najah National University, Nablus, Palestine

A. Hamdan et al. (eds.), *Technologies, Artificial Intelligence and the Future of Learning Post-COVID-19*, Studies in Computational Intelligence 1019,
https://doi.org/10.1007/978-3-030-93921-2_15

authority in practice, is burdened by starvation for funds and dependence on foreign aid, and was unable to fulfill the basic needs of the population even before the pandemic reached the West Bank [5].

The Palestinian Ministries of Health in both Gaza and Ramallah have acknowledged that their capacity to contain the spread of COVID-19 is limited by ongoing and pre-existing shortages in healthcare equipment, including medications and disposable equipment. Poor public health conditions compound the insufficient amount of equipment needed to treat COVID-19 in the occupied Palestinian territory (e.g., 87 intensive care unit beds with ventilators for nearly 2 million people and a paucity of personal protective equipment): a water and electricity crisis, rampant poverty, and a high population density [6, 7].

In this scenario, neither individuals nor the state itself has the power to control/secure their borders, the ability to create a comprehensive country strategy for prevention; the Palestinian Authority (PA) currently administers only 39% of the West Bank. 61% of the West Bank remains under direct Israeli military and civilian control [8].

2   Theoretical Background

2.1   COVID-19 Pandemic and E-learning

On January 6, 2020, a novel coronavirus emerged and was termed COVID-19 [9]. Generally, COVID-19 is an infectious acute respiratory disease that could also be deadly, and It is categorized by several symptoms including fever, chills, cough, fatigue, and shortness of breath [10, 11].

Since it is an infectious disease individuals around the world needed to be isolated or in quarantine, as a result of this many aspects of life have been affected as daily routine has been changed, in which most companies, public places, institutions (e.g. educational institutions) were closed [12].

As closing the educational institutions was one of the protective measures during COVID-19 pandemic, and since education considered one of the key factors in building a good nation, most countries around the world have migrated from the traditional methods of learning to imparting education through online means as a solution to protect education [13, 14].

Oxford presents e-learning as a type of "learning conducted via electronic media, typically on the Internet", also it could be defined as the development of knowledge and skills through the use of information and communication technologies (ICTs), particularly to support interactions for learning—interactions with content, with learning activities and tools, and with other people [15]. These definitions highlights the importance and the role of social interaction as it is connected on the one hand with the teacher's feedback and on the other hand with the students' communication and collaborativeness with classmates [16].

There are two categories in which the modes of e-learning can be delivered: synchronous or asynchronous. Synchronous delivery is similar to a traditional classroom environment, except the interaction is virtual instead of physical, and the activities that could be conducted through this mode are lectures, tutorials, quizzes and discussions through the using of video-conferencing platforms, such as; Microsoft Teams and, Google Meet, Zoom, Cisco WebEx [17]. Asynchronous e-learning is similar to synchronous e-learning which is a learner-centered process which uses online learning resources to facilitate information sharing regardless of the constraints of time and place among a network of people, in which sharing information can be done through prerecorded video lectures, lecture slides, external resources such as YouTube videos or educational websites, and the students' feedback can be sent through email or messaging apps like WhatsApp [18].

2.2 E-learning in Palestine During COVID-19 Pandemic

On March 5, 2020 Palestine was declared a state of emergency as a result of the outbreak of COVID-19 pandemic, and to help contain the spread of COVID-19 and prevent a possible large-scale outbreak among students, the Ministry of Education and Higher Education has decided to close all educational institutions temporarily from 6 March 2020, until further notice in order to combat and contain the spread of COVID-19, as well as to protect students, teachers and lecturers [19]. So as most countries around the world the Palestinian Ministry of Education (MoE) immediately launched its National Response Plan for COVID-19 where distance learning (e-learning) was highlighted as an alternative solution to ensure the continuation of learning to their students [20].

Despite e-learning is considered as the best solution to continue learning in the time of crisis, it was found that there are a lot of obstacles and challenges that faced both teachers and students in Palestine as a result of implementing e-learning. Subaih et al. [21] found that the most important obstacles and challenges that associated with applying e-learning during COVID-19 pandemic in Palestine are, the lack of commitment of students to attend the full lessons, the limited number of ways for evaluating students; the weakness of the Internet or its continuous interruption; the absence of technical support from the school; problems in the operation of the educational system; and the difficulty of learning on e-learning programs. As well as, [20] found that students faced a number of challenges while using the e-learning system, as they had problems accessing the technology such as computer availability and permanent internet connection, and lack of skills for the use of e-learning system.

Students engagement in e-learning found to be influenced by several factors during remotely teaching in Palestine. These factors include cultural factors, namely parental concerns which related to religious issues as well as local culture and norms and traditions; the digital inequality among students as family did not have an equal chance to enjoy online learning; the digital privacy as some parents considered online learning as a threat to their child's digital space; the quality of the digital content as both teachers and students reported that the quality of the content in the e-learning was lower than that in face-to-face learning; the weakness of Internet as many participants complained about the weakness of the Internet signal which prevented them from attending synchronous sessions and communicating with their peers and teachers; technical support as most students reported that they faced many technical challenges while trying to access Google classroom or to attend the synchronous sessions which prevented them from continuing online learning; Availability of the technological devices such as desktop computers, laptops, smartphones, and tablets [22, 23].

Despite social media and cloud computing found to be useful for design and delivery of educational materials as well as raising safety awareness, and communication during the COVID-19 pandemic in Palestine, the findings also identify various challenges including the widening of the education's digital divide and the negative attitude towards online education, are identified as challenges that correlated with applying e-learning [24].

While e-learning found to provide students with the opportunity to self-learn, enhance the ability to solve problems, think critically and communicate more easily with a larger group of social groups, it may have a negative side if it is not used in a scientific and pedagogical way, and it may require additional effort from the teacher; what may distract the teacher from his basic mission in education [25].

A study aimed at revealing teaching English as a foreign language (TEFL) professors' e-learning experiences during the Covid 19 pandemic in the Gaza Strip higher education institutions. The results revealed that Palestinian TEFL professors spent (4–7) hours daily preparing, meeting, guiding, helping and facilitating learning by using different eLearning programs and network social media such as Moodle, Google classroom, Zoom, WhatsApp and Facebook, and the most challenges for both TEFL professors and learners were insufficient knowledge about eLearning and technical problems [26].

Students in Gaza Strip faced several challenges while using e-learning including electricity cuts and the instability of the Internet.; the opportunity to have discussions with the teachers declined, so it was very difficult to discuss about unclear points in the recorded lectures; the process of communication between professors and students is not as it should be; students believe that e-learning is unfair with regard to student's levels, and first-level students believe that e-learning is somewhat weak in terms of the extent of their interaction and benefit [27].

Bashitialshaaer et al. [28] explored the obstacles to achieving good quality distance learning and electronic exams from the point of view of the university professors and students in different Arabian countries including Palestine. the results revealed that the most obstacles and challenges characterized by the weak motivation of students to distance learning and that some professors are not convinced of the usefulness of distance learning, the unsuitable home environment, and technical difficulties (e.g., inadequate technological compatibility) that hinder the teaching and learning system, such as students and learners being dissimilar and increased frustration and confusion, while challenges regarding electronic exams characterized by the lack of financial and technical capabilities of some students and the need for power (electricity) to complete the electronic test, and the possibility of using the internet regularly to facilitate the task was considered as one of the most important obstacles to applying simultaneous electronic exams in higher education institutions.

Zboun and Farrah [29] explored students' perspectives towards benefits and challenges of fully online classes at Hebron University, and they found that students preferred face to face classes more than online classes, as online classes found to be related with several challenges from the students' perceptions such as the weak internet connectivity, poor interaction, less motivation, less participation and less understanding. While, [30] analyzed the reality of e-learning at Palestine Technical University-Khudouri/Tulkarem, and identified the most important challenges facing students when using the education system, the results indicated that (63.136%) of the researched believe that the reality of e-learning at the university suffers from different problems, and the infrastructure considered as one of the most barriers in e-learning.

A study that examined the perceptions of EFL students in Palestinian universities about online learning advantages, challenges, and solutions during Corona pandemic, it was found that e-learning has several advantages in which it turning students to be researchers, making students more confidence and have self-reliance, improving their technological skills, and offering them valuable experience, but at the same time students found to suffer from different challenges while using e-learning such as the lack of technical support that the universities have; the need for training lecturers to improve their technological skills; the unreliable evaluation system; and the poor technological infrastructure [31].

Moreover, the Ministry of Education's plan that was designed to manage school education during the pandemic in Palestine, found to suffer from several challenges characterized by the lack of clarity of the plan; technical related issues of the e-learning system; financial and logistical support issues; fear and confusion about the plan; and the lack of confidence in the procedures followed [32].

2.3   Mental Health and E-learning Among Palestinians

Despite e-learning found to be the best solution to continue education during COVID-19, it was found that e-learning had a negative impact on mental health of students around the world, as they found to be psychologically distressed as a result of being afraid from losing the academic year, and the ineffective e-Learning systems [33].

The movement restrictions, school closures and stay-at-home during the COVID-19 pandemic were expected to lead to a rise in the rates of domestic violence, loneliness, depression, fear, panic and anxiety, and substance use among Palestinian school students [34]. Moreover, the obstacles and challenges that found to be associated with implementing e-learning in Palestine during COVID-19, teachers, students, and parents' mental health found to be affected negatively. For example, in the Gaza Strip school and university students found to feel that they are losers as they were not able to cope with e-learning procedures or able to access to the Internet, so they felt as if they are losers because they are not able to continue their education which is considered especially for girls as a key for their self-esteem, autonomy and future career development, also as school is not only a place to learn, but also a space to establish social contacts, students started to have feelings of loneliness [35].

In a study that explored the psychological impact of educational process disturbance upon the COVID-19 pandemic on the parents and primary school students in the Gaza Strip, it was found that the mental health of both parents and children was negatively affected due to the educational process disturbance upon COVID-19 pandemic in the Gaza Strip as they reported symptoms of anxiety, stress, depression, fear, loneliness, and behavioral problems [36].

During COVID-19 pandemic school and university students showed an increase in weight gain and food intake, and this could be due to the fact that the physical activities decreased as a result of the lockdown and that students have to stay for a long time in front of laptops, computers, and mobiles screens in order to study which led for gaining weight as they ate were eating food without doing any physical activity [37].

Moreover, [38] assessed the psychological distress (stress, anxiety, depression) among a convenience sample of Palestinian dental students after few weeks of closing universities, switching to e-learning and shutting down all aspects of life in response to the COVID-19 pandemic, the results of this study showed that half of the dental students in this sample had sever or extremely severe anxiety and one third had severe or extremely severe depression, as dental students bear an extra burden in the lockdown, which is the complete suspension of clinical training that is considered a core curriculum requirement to pass any course or to proceed to the year after, and online education couldn't completely address the challenges and the requirements of dental education.

During the closure of schools, students are using social media to continue their learning, which made it more easy for the students to access the Internet and social media applications which exposed them to more information about the COVID-19 pandemic, so as a result of this the fear and panic episodes increased among primary and secondary school students in Palestine [39]. Furthermore, found that Palestinian young adults especially university students, more vulnerable for depression than older age in which they are under a huge pressure as a result of online educational activities were implemented to continue the ongoing academic semester, and they had to deal with the dramatic changes in requirements to pass the academic semester [40].

Palestinian students are expected to suffer from the negative consequences for a long time due to the lack of mental health care institutions in Palestine since there is a lack of qualified specialist staff, insufficient funding, the political conflict, and poor community awareness of psychosocial services, as well as low salaries and unemployment among psychosocial practitioners, which means that Palestinians in general especially students are unable to get the needed mental health care, despite they are living under very difficult circumstances [41, 42]

2.4 Future of Education in Palestine After the COVID-19 Pandemic

Coronavirus outbreak has significantly accelerated development of online education in Palestinian higher education. Internet, big data, Artificial Intelligence, and cloud-based platforms, among other technologies, have been put into service of education. However, a more flexible way of teaching and learning does not end up with infrastructure. Rather, infrastructure is only the first step towards a new paradigm of teaching and learning in post-pandemic time. This paradigm could represent a shift from traditional, teacher-centered, and lecture based activities towards more student-centered activities including group activities, discussions, hands-on learning activities, and limited use of traditional lectures. This requires conceptual and philosophical rethinking of nature of teaching and learning, roles, and connections among teachers, learners, and teaching materials, in post digital learning communities [43].

Although this online delivery can present barriers to the teachers since they need to acquire online-driven competencies in planning, implementing, and assessing the performance of their students, providing teachers with adequate training courses can assist them to effectively implement the courses through electronic delivery. There are various devices available with innovative tools for the teachers to access to promote learning for the students with diverse educational needs. The technology devices, design of the program, choices of instructors, responsive curriculum, and supportive stakeholders are necessary and significant for the successful delivery of the lessons in an online environment [44]. In that case, this brings an opportunity for higher education institutions to scale up the training of the teachers for online learning instruction. The training for the teachers can improve student learning in educational programming for the instructors to facilitate the goals aligned to the learning goals of higher education institutions [45].

Although there have been overwhelming challenges for educators, schools, institutes and the government regarding online education from a different angle, there are several opportunities created by the COVID-19 pandemic for the unprepared and the distant plans of implementing e- learning system. It has forged a strong connection between teachers and parents than ever before. The homeschooling requires parents to support the students' learning academically and economically [46].

The use of online platforms such as Google Classroom, Zoom, virtual learning environment and social media and various group forums like Telegram, Messenger, WhatsApp and WeChat are explored and tried for teaching and learning for education in the future. This can be explored further even after face to-face teaching resumes, and these platforms can provide additional resources and coaching to the learners. Teachers are obliged to develop creative initiatives that assist to overcome the limitations of virtual teaching. Teachers are actively collaborating with one another at a local level to improve online teaching methods. There are incomparable opportunities for cooperation, creative solutions and willingness to learn from others and try new tools as educators, parents and students share similar experiences [47].

Many educational organizations in Palestine are offering their tools and solutions for free to help and support teaching and learning in a more interactive and engaging environment. Online learning has provided the opportunity to teach and learn in innovative ways unlike the teaching and learning experiences in the normal classroom setting [23].

3 Conclusion

The global outbreak of the COVID-19 pandemic has spread worldwide, affecting almost all countries and territories. The outbreak was first identified in December 2019 in Wuhan, China. Bhutan first declared closing of schools and institutions and reduction of business hours during the second week of March 2020 (Kuensel, 2020, 6 March).

The COVID-19 pandemic has disrupted the lives of students in different ways, depending not only on their level and course of study but also on the point they have reached in their programmers. Those coming to the end of one phase of their education and moving on to another, such as those transitioning from school to tertiary education, or from tertiary education to employment, face particular challenges. They will not be able to complete their school curriculum and assessment in the normal way and, in many cases, they have been torn away from their social group almost overnight [48].

The COVID-19 pandemic has brought into focus the mental health of various affected populations. It is known that the prevalence of epidemics accentuates or creates new stressors including fear and worry for oneself or loved ones, constraints on physical movement and social activities due to quarantine, and sudden and radical lifestyle changes. Several studies indicated the negative effect of COVID-19 on students' mental health outcomes, such as in the occupied territories of Palestine, who are under the unique circumstances of having the risk of infection be controlled by a hostile foreign occupier. The situation of the occupied territories of Palestine is fraught with environmental stressors militarization, poverty, lack of employment opportunities, cultural pressures, etc.) and few positive social outlets due to the restrictions on movement between communities, a lack of recreational facilities, and a lack of external support [49–51].

Israeli military occupation of the West Bank and Gaza Strip has lasted over 50 years, with lack of access to land, water, borders and the freedom of movement of people and goods from one part to another and even within parts of the country. This has resulted in stunted development, also known as dedevelopment, weak and underfunded health and social services. In the wake of the COVID-19 outbreak, this structural predicament has heightened political, economic and social instability [52].

In Palestine, It's the right time for teachers, students, and administrators to learn from this critical situation and to overcome these challenges. Online learning could be a greater opportunity as a result of this crisis [53]. Students are young and energetic, and they are capable of learning through the online platform. Schools can motivate the younger minds and draw them into active participation. Ministry of Education should encourage students and teachers to stay connected through the online or any social media platform and move forward together during this extremely difficult time. Students should be provided with course instruction and other services in an online format to support academic continuity [54]. The training program should be organized as quickly as possible for the teachers to tackle the online learning platform, this force experimentation will guide academic institutions in Palestine to upgrade their technical infrastructure and make online a core aspect of teaching and learning [55].

References

1. AlKhaldi, M., Kaloti, R., Shella, D., Al Basuoni, A., & Meghari, H. (2020). Health system's response to the COVID-19 pandemic in conflict settings: Policy reflections from Palestine. *Global Public Health, 1–13*. https://doi.org/10.1080/17441692.2020.1781914.
2. World Health Organization. (2021). Weekly epidemiological update on COVID-19 - 20 July. Retrieved July 20, 2021 from https://www.who.int/publications/m/item/weekly-epidemiolo gical-update-on–covid-19---20-july-2021
3. Ghandour, R., Ghanayem, R., Alkhanafsa, F., Alsharif, A., Asfour, H., Hoshiya, A., & Giacaman, R. (2020). Double burden of COVID-19 pandemic and military occupation: mental health among a Palestinian university community in the West Bank. *Annals of global health, 86*(1), 131. https://doi.org/10.5334/aogh.3007
4. Giacaman, R., Khatib, R., Shabaneh, L., Ramlawi, A., Sabri, B., Sabatinelli, G., Khawaja, M. & Laurance, T. (2009). Health status and health services in the occupied Palestinian territory, *The Lancet, 373*(9666), 837–849. https://doi.org/10.1016/s0140-6736(09)60107-0 .
5. Mataria, A., Khatib, R., Donaldson, C., Bossert, T., Hunter, D. J., Alsayed, F., & Moatti, J.-P. (2009). The health-care system: An assessment and reform agenda. *The Lancet, 373*(9670), 1207–1217. https://doi.org/10.1016/s0140-6736(09)60111-2.
6. Mahamid, F.A., Veronese, G. &; Bdier, D. (2021a). Fear of coronavirus (COVID-19) and mental health outcomes in Palestine: The mediating role of social support. *Current Psychology*. Advance online publication. https://doi.org/10.1007/s12144-021-02395-y .
7. Mahamid, F.A., Veronese, G., Bdier, D., & Pancake, R. (2021b). Psychometric properties of the COVID stress scales (CSS) within Arabic language in a Palestinian context. *Current Psychology*. Advance online publication. https://doi.org/10.1007/s12144-021-01794-5
8. Mahamid, F. A., Berte, D. Z. & Bdier, D. (2021). Problematic internet use and its association with sleep disturbance and life satisfaction among Palestinians during the COVID-19 pandemic. *Current Psychology*. Advance online publication. https://doi.org/10.1007/s12144-021-02124-5 .
9. Pan, F., Ye, T., Sun, P., Gui, S., Liang, B., Li, L., et al. (2020). Time course of lung changes on chest CT during recovery from 2019 novel coronavirus (COVID-19) pneumonia. *Radiology, 295*(3), 715–721. https://doi.org/10.1148/radiol.2020200370.
10. Khatib, S. F., & Nour, A. N. I. (2021). The impact of corporate governance on firm performance during the COVID-19 pandemic: Evidence from Malaysia, *Journal of Asian Finance, Economics and Business, 8*(2), 0943–0952. https://doi.org/10.13106/jafeb.2021.vol8.no2.0943.
11. Xu, Z., Shi, L., Wang, Y., Zhang, J., Huang, L., Zhang, C., et al. (2020). Pathological fndings of COVID19 associated with acute respiratory distress syndrome. *The Lancet Respiratory Medicine, 8*(4), 420–422. https://doi.org/10.1016/S2213-2600(20)30076-X.

12. Mahamid, F. A., & Bdier, D. (2021). The association between positive religious coping, perceived stress, and depressive symptoms during the spread of coronavirus (covid-19) among a sample of adults in palestine: Across sectional study. *Journal of Religion and Health, 60*(1), 34–49. https://doi.org/10.1007/s10943-020-01121-5.

13. Baiyere, A., & Li, H. (2016). *Application of a Virtual Collaborative Environment in a Teaching Case.* Paper presented in AMCIS: Surfing the IT Innovation Wave - 22nd Americas Conference on Information Systems. https://aisel.aisnet.org/amcis2016/ISEdu/Presentations/33/.

14. Soni, V. D. (2020). Global Impact of e-learning during COVID 19. *Advanced online publication.* https://doi.org/10.2139/ssrn.3630073.

15. Tirziu, A. M., & Vrabie, C. (2015). Education 2.0: e-learning methods. *Procedia-Social and Behavioral Sciences, 186*, 376–380. https://doi.org/10.1016/j.sbspro.2015.04.213.

16. Bylieva, D., Lobatyuk, V., Safonova, A., & Rubtsova, A. (2019). Correlation between the practical aspect of the course and the e-learning progress. *Education Sciences, 9*(3), 167. https://doi.org/10.3390/educsci9030167.

17. Azlan, C. A., Wong, J. H. D., Tan, L. K., Huri, M. S. N. A., Ung, N. M., Pallath, V., & Ng, K. H. (2020). Teaching and learning of postgraduate medical physics using Internet-based e-learning during the COVID-19 pandemic–A case study from Malaysia. *Physica Medica, 80*, 10–16. https://doi.org/10.1016/j.ejmp.2020.10.002.

18. Shahabadi, M. M., & Uplane, M. (2015). Synchronous and asynchronous e-learning styles and academic performance of e-learners. *Procedia-Social and Behavioral Sciences, 176*, 129–138. https://doi.org/10.1016/j.sbspro.2015.01.453

19. Marbán, J. M., Radwan, E., Radwan, A., & Radwan, W. (2021). Primary and secondary students' usage of digital platforms for mathematics learning during the COVID-19 outbreak: The case of the Gaza Strip. *Mathematics, 9*(2), 110–131. https://doi.org/10.3390/math9020110.

20. Khalilia, W. (2020). Attitudes and challenges towards e-learning system in time of Covid-19 from the perspective of AlIstiqlal university students. *Elixir Educational Technology, 145*, 54676–54682.

21. Subaih, R. H. A., Sabbah, S. S., & Al-Duais, R. N. E. (2021). Obstacles facing teachers in Palestine while implementing e-learning during the COVID-19 pandemic. *Asian Social Science, 17*(4), 44–45. https://doi.org/10.5539/ass.v17n4p44.

22. Khlaif, Z. N., & Salha, S. (2020). The unanticipated educational challenges of developing countries in Covid-19 crisis: A brief report. *Interdisciplinary Journal of Virtual Learning in Medical Sciences, 11*(2), 130–134. https://doi.org/10.30476/IJVLMS.2020.86119.1034.

23. Khlaif, Z. N., Salha, S., & Kouraichi, B. (2021). Emergency remote learning during COVID-19 crisis: Students' engagement. *Education and Information Technologies*, 1–23. Advanced online publication. https://doi.org/10.1007/s10639-021-10566-4.

24. Shraim, K., & Crompton, H. (2020). The use of technology to continue learning in Palestine disrupted with COVID-19. *Asian Journal of Distance Education, 15*(2), 1–20. https://doi.org/10.5281/zenodo.4292589

25. Bsharat, T. R., & Behak, F. (2021). The impact of microsoft teams' app in enhancing teaching-learning english during the Coronavirus (COVID-19) from the English teachers' perspectives' in Jenin city. *Malaysian Journal of Science Health & Technology, 7*(Special Issue), 102–109. https://doi.org/10.33102/mjosht.v7i.116.

26. Abou Shaaban, S. S. (2020). TEFL professors' e-learning experiences during the COVID 19 pandemic. *European Journal of Foreign Language Teaching, 5*(1), 82–97. https://doi.org/10.46827/ejfl.v5i1.3202.

27. Shehab, A., Alnajar, T. M., Marni, N. B., & Hamdia, M. H. (2020). A study of the effectiveness of e-learning in Gaza strip during COVID-19 pandemic: The Islamic University of Gaza "case study". *Elementary Education Online, 19*(4), 2627–2643. https://doi.org/10.17051/ilkonline.2020.04.764626.

28. Bashitialshaaer, R., Alhendawi, M., & Lassoued, Z. (2021). Obstacle comparisons to achieving distance learning and applying electronic exams during COVID-19 pandemic. *Symmetry, 13*(1), 99–114. https://doi.org/10.3390/sym13010099.

29. Zboun, J.S. & Farrah, M.(2021). Students' perspectives of online language learning during corona pandemic: Benefits and challenges. *Indonesian EFL Journal, 7*(1), 13–20. https://doi. org/10.25134/ieflj.v7i1.3986.
30. Almbayed, H. (2020). Analsis e-learning status in Palestinian universities, a case study of Palestine Technical University Kadoorie Tulkarm. *Palestine Technical University Kadoorie Research Journal, 8*(3), 154–178. https://rj.ptuk.edu.ps/index.php/ptukrj/article/view/118.
31. Farrah, M., & al-Bakry, G. H. (2020). Online learning for EFL students in Palestinian universities during corona pandemic: Advantages, challenges and solutions. *Indonesian Journal of Learning and Instruction, 3*(2), 65–78. https://doi.org/10.25134/ijli.v3i2.3677.
32. AL-Rub, I. O. I. A. (2020). The procedures of education administration and the exploration of challenges during COVID-19 pandemic in Palestine. *PalArch's Journal of Archaeology of Egypt/Egyptology, 17*(6), 6195–6212. https://archives.palarch.nl/index.php/jae/article/view/1905.
33. Hasan, N., & Bao, Y. (2020). Impact of "e-Learning crack-up" perception on psychological distress among college students during COVID-19 pandemic: A mediating role of "fear of academic year loss." *Children and Youth Services Review, 118*. https://doi.org/10.1016/j.childy outh.2020.105355.
34. Radwan, E., Radwan, A., & Radwan, W. (2020). The mental health of school students and the COVID-19 pandemic. *Aquademia, 4*(2), ep20020. https://doi.org/10.29333/aquademia/8394.
35. Hamad, S., Abu Hamra, E., Diab, R., Abu Hamad, B., Jones, N., & Małachowska, A. (2020). Exploring the impacts of Covid-19 on adolescents in the Gaza Strip. *ODI, June*.
36. Saleh, S., Habib, A. A., Hamam, R., Aita, S. A., Alzir, M., Jourany, H., & Abu Jamei, Y. (2021). The psychological impact of educational process disturbance upon COVID-19 Pandemic among primary school students and their parents in the Gaza Strip. *Journal of Education, Society and Behavioural Science, 1–10*. https://doi.org/10.9734/jesbs/2021/v34i130286.
37. Allabadi, H., Dabis, J., Aghabekian, V., Khader, A., & Khammash, U. (2020). Impact of COVID-19 lockdown on dietary and lifestyle behaviours among adolescents in Palestine. *Dynam Human Health, 7*, 2170. https://journalofhealth.co.nz/?page_id=2170.
38. Abu Kwaik, A., Saleh, R., Danadneh, M., & Kateeb, E. (2021). Stress, anxiety and depression among dental students in times of Covid-19 lockdown. *International Journal of Dentistry and Oral Science, 8*(2), 1560–1564. https://doi.org/10.19070/2377-8075-21000310.
39. Radwan, E., Radwan, A., & Radwan, W. (2020). The role of social media in spreading panic among primary and secondary school students during the COVID-19 pandemic: An online questionnaire study from the Gaza Strip, Palestine. *Heliyon, 6*(12),e05807.https://doi.org/10. 1016/j.heliyon.2020.e05807.
40. Al Zabadi, H., Al-Hrouh, T., Yaseen, N., & Haj-Yahya, M. (2020). Assessment of depression severity during COVID-19 pandemic among the Palestinian population: A growing concern and an immediate consideration. *Frontiers in Psychiatry, 11*, 1486. https://doi.org/10.3389/ fpsyt.2020.570065.
41. Mahamid, F., & Veronese, G. (2020). Psychosocial interventions for third-generation Palestinian refugee children: Current challenges and hope for the future. *International Journal of Mental Health and Addiction, 1–18*. Advanced online publication. https://doi.org/10.1007/s11 469-020-00300-5.
42. Marie, M., Hannigan, B., & Jones, A. (2016). Mental health needs and services in the West Bank, Palestine. *International Journal of Mental Health Systems, 10*(1), 1–8. https://doi.org/ 10.1186/s13033-016-0056-8.
43. Zhu, X., & Liu, J. (2020). *Education in and after Covid-19: Immediate responses and long-term visions*. Advance online publication. https://doi.org/10.1007/s42438-020-00126-3.
44. Barr, B., & Miller, S. (2013). Higher Education: The Online Teaching and Learning Experience. Retrieved from https://files.eric.ed.gov/fulltext/ED543912.pdf
45. Toquero, C. M. (2020). Challenges and opportunities for higher education amid the COVID-19 Pandemic: The philippine context. *Pedagogical Research, 5*(4), em0063. https://doi.org/10. 29333/pr/7947.

46. Kogan, M., Klein, S. E., Hannon, C. P., & Nolte, M. T. (2020). Orthopaedic education during the COVID-19 pandemic. *The Journal of the American Academy of Orthopaedic Surgeons, 28*(11), e456–e464. https://doi.org/10.5435/JAAOS-D-20-00292.
47. Doucet, A., Netolicky, D., Timmers, K., and Tuscano, F. J. (2020). Thinking about pedagogy in an unfolding pandemic: an independent report on approaches to distance learning during COVID19 school closures. Available at: https://issuu.com/educationinternational/docs/2020_r esearch_covid-19_eng
48. Daniel, S. J. (2020). Education and the COVID-19 pandemic. *Prospects, 49*, 91–96. https://doi.org/10.1007/s11125-020-09464-3.
49. Berte, D. Z., Mahamid, F. A., & Affouneh, S. (2021). Internet addiction and perceived self-efficacy among university students. *International Journal of Mental Health and Addiction, 19*, 162–176. https://doi.org/10.1007/s11469-019-00160-8.
50. Mahamid, F. A., & Berte, D. Z. (2019). Social media addiction in geopolitically at-risk youth. *International Journal of Mental Health and Addiction, 17*(1), 102–111. https://doi.org/10.1007/s11469-017-9870-8.
51. Mahamid, F. A., & Berte, D. Z. (2020). Portrayals of violence and at-risk populations: Symptoms of trauma in adolescents with high utilization of social media. *International Journal of Mental Health and Addiction, 18*, 980–992. https://doi.org/10.1007/s11469-018-9999-0.
52. Hammoudeh, W., Kienzler, H., Meagher, K., & Giacaman, R. (2020). Social and political determinants of health in the occupied Palestine territory (oPt) during the COVID-19 pandemic: who is responsible? *BMJ Global Health, 5*(9), e003683. https://doi.org/10.1136/bmjgh-2020-003683
53. Chick, R. C., Clifton, G. T., Peace, K. M., Propper, B. W., Hale, D. F., Alseidi, A. A., & Vreeland, T. J. (2020). Using technology to maintain the education of residents during the COVID-19 Pandemic. *Journal of Surgical Education*. Advance online publication. https://doi.org/10.1016/j.jsurg.2020.03.018.
54. Sahu, P. (2020). *Closure of universities due to coronavirus disease 2019 (COVID-19): Impact on education and mental health of students and academic staff*. Advance online publication. https://doi.org/10.7759/cureus.7541.
55. Chang, T.-Y., Hong, G., Paganelli, C., Phantumvanit, P., Chang, W.-J., Shieh, Y.-S., & Hsu, M.-L. (2020). Innovation of dental education during COVID-19 pandemic. *Journal of Dental Sciences*. Advance online publication. https://doi.org/10.1016/j.jds.2020.07.011.
56. Khlaif, Z. N., Salha, S., Affouneh, S., Rashed, H., & ElKimishy, L. A. (2020). The Covid-19 epidemic: Teachers' responses to school closure in developing countries. *Technology, Pedagogy and Education, 30*(1), 95–109. https://doi.org/10.1080/1475939X.2020.1851752.
57. Mahamid, F. A., Bdier, D., & Berte, D. Z. (2021). Psychometric properties of the Fear of COVID-19 Scale (FCV-19S) in a Palestinian context. Manuscript submitted for publication.
58. Mahamid, F. Veronese, G. Bdier, D. (2021). The Palestinian health care providers perceptions, challenges and human rights related concerns during the COVID-19 pandemic. Manuscript submitted for publication.

Cheating in Online Exams: Motives, Methods and Ways of Preventing from the Perceptions of Business Students in Bahrain

Zahera Baniamer and Bishr Muhamed[ID]

1 Introduction

Many governments have established stringent measures to decrease crowding forms in attempt to restrain the spread of the covid-19. Education, which is a university's initial goal, has obviously been harmed by the lockdown. therefore, they had to decrease class sizes or close entirely and shifted to online learning.

YERUN [1] earlier expected that higher education will be strongly affected, Hybrid teaching (online for some students, face-to-face for others) and blended education (online content plus rotation system for on-site activities) will be in place at least until the end of 2020, possibly until the middle of 2021. This means that classes will revert to a reduced physical format, with some protections in place for a while longer.

Academic dishonesty, that can be explained in a variety of different ways like cheating in tests and plagiarism, is one of the critical issues that gained researchers academic attention. Regarding the fact that cheating has been reported at many universities, the conversation around online education has expanded to how to prevent academic dishonesty while providing online exams.

Z. Baniamer
Islamic Economics and Finance Pedia, Irbid, Jordan

B. Muhamed (✉)
Applied Science University, Sitra, Bahrain
e-mail: Bishr.mohd@asu.edu.bh

© The Author(s), under exclusive license to Springer Nature Switzerland AG 2022
A. Hamdan et al. (eds.), *Technologies, Artificial Intelligence and the Future of Learning Post-COVID-19*, Studies in Computational Intelligence 1019,
https://doi.org/10.1007/978-3-030-93921-2_16

1.1   The Study Problem

Academic dishonesty, which severely affects educational quality, is becoming a main consideration for most academics. Particularly after covid-19 pandemic that imposed different ways of assessment through online exams. A growing number of students and lecturers are worrying about cheating in online exams. As a result, it is best to examine the causes, changes in practice especially the perspectives of students toward the motives and best ways to thwart online cheating.

1.2   Study Questions

1.   What are the main perceptions of cheating in online exams?
2.   What are the methods that business student use to cheat in online exams?
3.   What are the most effective ways to impede cheating in online exams from business students' views?

1.3   Importance of the Study

The following points highlight the importance of the study:

1.   The growing phenomenon of cheating in online exams and the persistent need to eliminate this phenomenon that threatens the quality of higher education.
2.   Exploring the reasons and the motives that stands behind this phenomenon from the perspective of students themselves.
3.   Help the academics and educational institutions to determining the best methods to eliminate cheating in online exams.

1.4   Objectives of the Study

(1)   Determining the motives of cheating in online exams from the perspective of business students in private universities in Bahrain.
(2)   Identifying the method that are used in cheating in online exams by business students in private universities in Bahrain.
(3)   Assessing the most effective ways to eliminate cheating in online exams from the perspective of business students in private universities in Bahrain.

1.5   Research Structure

1.   Introduction: Includes

   1.1   Study problem.
   1.2   Study questions.
   1.3   Study objectives.
   1.4   Importance of the study.

2.   Results

   2.1   Respondent demographics.
   2.2   Student perceptions.

3.   Conclusion
4.   References.

2   Review of Related Literature

2.1   Relation Between Academic Dishonesty and Online Exams

According to researchers, Students had a broad awareness of many sorts of academic dishonesty, as well they believed that academic dishonesty decreases the value of academic qualifications, [2]. Cheating, as a salient form of academic dishonesty, is defined as a violation of academic integrity that involves taking an unfair advantage of a student's abilities and knowledge. This involves getting inappropriate help from an internet source or adjutant, plagiarism, and fraudulent self-representation in the current online environment [3]. However, the question raised here is that is there a relation between cheating and online exams?

King & Case's study, 2014 showed that students cheat in a variety of ways, regardless of how the course is delivered. In 2013, for example, 15 percent of students acknowledged to cheating on tests, 10 percent stated they utilized technology to cheat on exams, and 15 percent let other students copy from their exams.

On the other hand, Number of studies revealed that Students believe that cheating to be more widespread in online classes [3, 4]. Academics as well think that online scores are likely exaggerated by cheating. Such arguments are based on the premise that academic dishonesty is more common in online courses than in face-to-face courses [5, 6].

On the other hand, [7]'s study showed that there is a link, not a causation, between online testing and cheating. Depending on the parameters, online testing allows for cheating. Cheating behavior is predicated on the combination of opportunity, need and rationalization.

In a different study, 73.6 percent of the surveyed students believed it is easier to cheat in an online course than in a traditional course. Responding students revealed that they were more than four times more likely to cheat in an online class than in a traditional in-class format [3].

There are several reasons why online classes can be more vulnerable to academic dishonesty. One is that it is difficult to establish the identity of the test taker because assessments are frequently conducted in unsupervised or unproctored circumstances. [8]. During assessments, online test takers can use unlawful resources (such as cheat sheets, books, or internet materials). Furthermore, the absence of a close contact and engagement with an instructor in the online setting can stimulate collaborative (group) work with other students [9–11].

2.2 Methods of Cheating

Number of scholars identified at least four methods in which students cheat on online exams. Students accessing other websites, talking with others via instant messaging programs and email during the exam, seeding test machines with answers, and bringing in non-exam disks containing exam solutions. [12, 13].

2.3 Motives of Cheating in Online Exams

Why the students cheat in online exams also is one of the main questions has been raised lately. According to [14] academic cheating is a function of three broad factors: incentive (motivation to cheat derived from internal or external pressures), opportunity (ability to cheat because the environment permits it), and rationalization (ability to consider acts of cheating as not contrived). Others believe that neuroticism, extraversion, openness to experience, agreeableness, and conscientiousness are five characteristics that influence academic dishonesty [15–17].

Bachore [18] determined the most popular reasons for participating in such activities (academic corruption). The level of difficulty of the test is one of the factors that forces people to engage in such activities, according to the majority of students. Time scarcity was assigned the second factor. Irrelevance of course material and pressure to get good grades were ranked third and fourth, respectively. and having little clarity on policy was ranked fifth.

McCabe et al. [19] identified numerous causes in a basic model. Lack of interest and/or preparation, pressures to perform, misunderstandings about what constitutes academic dishonesty, a lack of understanding in course and exam policy/expectations, and a belief that others are participating in similar behavior are just a few of them. Other reasons such as the student's desire "to get ahead", the desire to help others, procrastination, the need to pass the class are other motives for such behavior. [20, 21].

According to [22] Students were asked to express their views for the most important cause of cheating by placing numbers from 1 to 12 on a scale of 1 to 12. Grade competition eventually came in first. In addition, "parental expectations" was ranked as the second most common reason for cheating good invigilation as the third explanation. Lack of motivation received the fourth. Additionally, the ease of cheating in college and the pressure of good grades were placed fifth and sixth, respectively. Furthermore, kids are motivated to achieve good grades in order to succeed. Paradoxically, despite being ranked eighth, "the meaninglessness and difficulty of tests" was not regarded as a major contributor. Pressure from peers was ranked ninth. Furthermore, the nature of the test and the lack of penalty were ranked ninth and tenth, respectively. More intriguingly, the level of punishment may have a significant impact on students' cheating trials. Heavy workload and insufficient study time were ranked ahead of the last item. Students cheat for various reasons that are more important than not having enough time to revise, the lack of teacher competency comes in last, implying that instructors' effectiveness and skills have less of an impact on students' cheating than the other factors.

According to Raines et al. [23] the majority of students did not view cheating as a technique of achieving success. However, slightly more than a third of respondents said cheating was a good method to get better grades as a measurable result.

2.4 Techniques to Prevent Cheating in Online Exams

There are numerous methods to dcrease cheating in exams depending on the type of the exam, Chirumamilla et al. [24] investigated three types of written examination: paper exams, Bring Your Own Device e-exams (BYOD e-exams), and e-exams. BYOD e-tests are said to be easier to cheat on than paper exams or university computer exams. Teachers regard BYOD as permitting easier cheating as compared to university PCs. Proctoring is thought to be more effective with paper tests than with e-exams in terms of countermeasure effectiveness.

Despite that, researchers think that proctoring e-exams is very important. Therefore, they suggested colleges should create a uniform online exam policy requiring the use of a camera to capture each student's computer screen and room. This will allow to verify a student's identification and eliminate the chance of someone else taking the test [25].

However, there is still a significant issue with online exam proctoring, which is the high cost for both students and teachers.

Therefore, to avoid the high cost of proctoring online exams, Jr. et al. identified number of Online Exam Control Procedures to prevent and detect cheating. These include Limiting exam time, limiting student access, changing test characteristics, and Verifying Student ID. Others suggested that the use of random question generation to make each exam unique is the most successful approach for preventing cheating in online exams, Watters et al.

Besides, instructors could utilize different techniques such as, Policy dissemination, Strict Test Taking Timeline, Cheating Trap, Surveillance, Randomized exam questions and responses and Statistical Analysis to detect Common Errors or proctoring [26, 27].

3 Method

The researchers surveyed business students in number of public and private universities in Bahrain during the summer 2021 semester. Students were asked to complete a soft version of the questionnaire which was distributed via link in a google-formatted questionaire.103 students answered the questionnaire, 86 students were from private universities, a 83.5 percent response rate, while 17 students were from public university,16.5 percent response rate.

The questionnaire consisted of four sections. Section one was designed to gather demographic data about the respondent. Section two gathered data regarding the respondent's perceptions about cheating in online exams. In section three, the respondent was asked to identify the possible ways of cheating, Finally, section four gathered data about the student's perceptions of techniques that may be used to prevent cheating in online exams.

4 Results

4.1 Respondent Demographics

Table 1 summarizes the demographic characteristics of the 103 students responding to the survey. Of the 103 students responding to the survey 70.9 percent were between 18–24 years old, 21.3 percent were between 25–35 years old, and 8.7 were above 35. Of those responding, 30.1 percent were of the first and second year students of bachelor degree, 58.3 percent were third and fourth year students and 11.7 percent were of master degree. Regarding type of university, 30.1 percent of respondents were from private universities and 16.5 percent were from public universities. Concerning the level of monthly income, 19.4 percent of respondents indicated that their income was Less than BD 500, 50.5 percent were between BD 500–1500, and 30.1 percent their income was more than BD 1500.

Table 1 Demographics data

Characteristics		Frequency	Percent
Age	From 18–24	73	70.9
	From 25–35	22	21.3
	Above 35	8	7.8
	Total	103	100
Education	BSc Level 1 & 2	31	30.1
	BSc Level 3 & 4	60	58.3
	MSc	12	11.7
	Total	103	100.0
University	Private university	86	83.5
	Public university	17	16.5
	Total	103	100.0
Income	Less than BD 500	20	19.4
	500–1500	52	50.5
	More than 1500	31	30.1
	Total	103	100.0

4.2   Student Perceptions

4.2.1   Perceptions Regarding Cheating in Online Exams

Students were asked to respond to many questions about their perceptions of cheating in online exams (Table 2). In response to a question weather students practiced cheating in online exam, 45.6 percent of students strongly disagreed that they had practiced cheating, 13.6 percent disagreed, 22.300 percent were nuetral, while 2.9 percent strongly agreed that they practiced cheating and 15.5 percent reoprted that they agreed.. The findings show that majority of students have a fairly tendency of rejecting cheating in online exams.

In response to point (2), 22.3 percent indicated that cheating in online exams is easier than cheating in paper exams, as well 19.4 agreed with that. 25.2 percent on the opposite side strongly disagreed with this point. 5.8 percent also disagreed with that.

In response to point (3), Is cheating in electronic exams more expensive than cheating in paper exams? 35.9 percent indicated that they strongly disagreed, and 23.3 percent indicated that they disagree, and the percentage for who were neutral about this point was 19.4 percent. on the other hand the percent of who agreed was 12.6, also the percentage for the option (strongly agree was 8.7).

According to respondents, 27.2 percent strongly disagreed with the statement that cheating in online examinations requires more equipment and devices than cheating in paper exams point (4), 24.3 percent did not agree, 12 percent highly agreed, and

Table 2 Students' perceptions about cheating in online exams

Perceptions about cheating in online exams	Strongly disagree	Disagree	Neutral	Agree	Strongly agree	Mean	Std. deviation	General percent (%)
Q1: I did practice cheat in online exams	45.600	13.600	22.300	15.500	2.900	2.165	1.245	43
Q2: Cheating in online exams is easier than cheating in paper exams	25.200	5.800	27.200	19.400	22.300	3.078	1.473	62
Q3: Cheating in online exams is more expensive than cheating in paper exams	35.900	23.300	19.400	12.600	8.700	2.350	1.319	47
Q4: Cheating in online exams requires more equipment and devices than cheating in paper exams	27.200	24.300	20.400	15.500	12.600	2.621	1.366	52
Q5: Cheating in online exams is more difficult to detect than cheating in paper exams	27.200	17.500	28.200	13.600	13.600	2.689	1.365	54

15.5 actually agreed.20.4 percent were not affiliated with any of these opposing views.

Is cheating in online exams more difficult to detect than cheating in paper exams? Was the last question in this section. 42.8 percent indicated that they disagreed with this point, 27.2 percent strongly disagreed, and 14.6 percent agreed. On the contrary, 27.2 percent agreed with this statement. While 28.2 percent indicated that they are neutral about it.

4.2.2 Methods of Cheating in Online Exams

The first question: Do you believe that cheating in online tests is widespread by keeping the test material, notes, and expected answers on a computer or mobile phone?

The answers to this question in percentages were as follows: the percentage for the option (strongly disagree was 28.2%, the percentage for the option (I disagree was 17.5%, the percentage for the option (neutral was 25.2%, the percentage for the option (agree was 17.5%, and the percentage for the option (Strongly agree was 11.7%.

The second question: Do you think using traditional means such as writing notes and sticking them on the screen, installing a plastic panel behind the screen, that can be moved during the test are prevalent in online exams?

The answers to this question in percentages were as follows: 42.7percent was for who strongly disagree and those who disagreeing were 18.4, who indicated that they are neutral were 19.4%). Those who indicated that they are agreeing 15.5percent and 3.9percent who strongly agree.

The third question: Do you think searching for answers through multiple browsers on the Internet and cheating from them is widespread among students?

The answers to this question in percentages were as follows: 28.2%, percent was for who strongly disagree and those who disagreeing were 17.5%), who indicated that they are neutral were 19.4%). Those who indicated that they are agreeing 15.5percent and 3.9percent who strongly agreeing.

The fourth questions: Do you believe that cheating in online tests is widespread by receiving answers from other students who have already taken it?

Students' perception about this question in percentages were as follows: 42.70 percent was for who strongly disagree and those who disagreeing were17.5%), who indicated that they are neutral were 19.4%). Those who indicated that they are agreeing 14.6percent and 5.8percent who strongly agreeing.

The fifth question: do you think that cheating through participation in groups for cheating through various social media such as WhatsApp, Facebook and others are prevalent among students?

Students' perception about this question in percentages were as follows: 40.800 percent was for who strongly disagree and those who disagreeing were 14.600%, who indicated that they are neutral were 26.200%. Those who indicated that they are agreeing 8.700 percent and 9.70 percent who strongly agreeing.

The sixth question: do you think that having someone answered the exams on the student's behalf is widespread among students in online exams?

The answers to this question in percentages were as follows: 51.5 percent was for who strongly disagree and those who disagreeing were 11.7, who indicated that they are neutral were 20.4. Those who indicated that they are agreeing 10.7 percent and 5.8 percent who strongly agree.

4.2.3   Methods of Preventing Cheating in Online Exams

To collect evidence on student opinions of methods that could be used to combat cheating in online exams. The researchers investigated students' perceptions about the effective techniques could be applied to prevent cheating in online exams. In Table 3, students were asked to indicate whether they strongly agreed, agreed, disagreed, strongly disagreed or had no opinion with regard to several statements regarding the effectivness of selected techniques to combact cheating (Table 4).

Statement 1:   receiving proctored exams in the university using its devices, contributes in reducing cheating?

Students' perception about this question in percentages were as follow: 33 percent was for who strongly disagree and those who disagreeing were 13.6%), who indicated that they are neutral were 24.3%). Those who indicated that they are agreeing 15.5 percent and 13. 6 percent who strongly agreeing.

Statement 2:   requiring students to use "Lockdown" browser while examining contributes to preventing cheating?

Students' perception about this question in percentages were as follow: 41.7% percent was for who strongly disagree and those who disagreeing were 13.6%), who indicated that they are neutral were 24.3%). Those who indicated that they are agreeing 8.7 percent and 11.7 percent who strongly agreeing.

Statement 3:   requiring students to operate the camera or monitoring devices during the exam reduce the possibility of cheating?

The answers to this question in percentages were as follows: 34 percent was for who strongly disagree and those who disagreeing were 13.6, who indicated that they are neutral were 20.4. Those who indicated that they are agreeing17.5 percent and 14.6 percent who strongly agree.

Statement 4:   changing the nature of questions, lengthening them and presenting them in a random and variable ways, so that there is no time for cheating or surfing, reduce the ability to cheat in online exams?

The answers to this question in percentages were as follows: 36.9 percent was for who strongly disagree and those who disagreeing were 15.5 percent who indicated that they are neutral were 23.3

Table 3 Methods of cheating in online exams

Methods of cheating in online exams	Strongly disagree	Disagree	Neutral	Agree	Strongly agree	Mean	Std. deviation	General percent (%)
Q7: Save the test material, notes, expected answers, on your computer or mobile device	28.200	17.500	25.200	17.500	11.700	2.670	1.360	53
Q8: Using traditional methods such as writing notes and sticking them on the screen, installing a plastic sheet behind the screen, which can be moved during the exam	42.700	18.400	19.400	15.500	3.900	2.194	1.253	44
Q9: Find answers across multiple browsers on the internet	28.200	17.500	23.300	22.300	8.700	2.660	1.333	53
Q10: Get answers from previous tests from those who took the test previously	42.700	17.500	19.400	14.600	5.800	2.233	1.300	45
V	40.800	14.600	26.200	8.700	9.700	2.320	1.345	46
Q12: Hiring another person (friend, relative) to take the test on the student's behalf	51.500	11.700	20.400	10.700	5.800	2.078	1.296	42

Table 4 Methods of preventing cheating in online exams

Methods of cheating in online exams	Strongly disagree	Disagree	Neutral	Agree	Strongly agree	Mean	Std. deviation	General percent (%)
Q1: receiving proctored exams in the university using its devices, contributes in reducing cheating?	33.000	13.600	24.300	15.500	13.600	2.631	1.428	53
Q2: requiring students to use "Lockdown" browser while examining contributes to preventing cheating?	41.700	13.600	24.300	8.700	11.700	2.350	1.398	47
Q3 requiring students to operate the camera or monitoring devices during the exam reduce the possibility of cheating?	34.000	13.600	20.400	17.500	14.600	2.650	1.467	53
Q4: changing the nature of questions, lengthening them and presenting them in a random and variable ways, so that there is no time for cheating or surfing, reduce the ability to cheat in online exams?	36.900	15.500	23.300	13.600	10.700	2.456	1.385	49

(continued)

Table 4 (continued)

Methods of cheating in online exams	Strongly disagree	Disagree	Neutral	Agree	Strongly agree	Mean	Std. deviation	General percent (%)
Q5: reducing the duration of the exam, so that the student engages in the answering questions instead of cheating, reduce cheating in exams?	46.600	21.400	14.600	6.800	10.700	2.136	1.358	43
Q6: doing exam at a specific time for all students reduces the possibility of conspiracy and cheating?	32.000	16.500	21.400	15.500	14.600	2.641	1.441	53
Q7: strict sanctions considered a strong deterrent and keep student away from cheating?	25.200	10.700	29.100	19.400	15.500	2.893	1.393	58
Q8: religious beliefs that prohibit cheating considered an effective deterrent that prevents cheating?	19.400	6.800	22.300	16.500	35.000	3.408	1.504	68

percent. Those who indicated that they are agreeing 13.6 percent and 10.7 percent who strongly agree.

Statement 5: reducing the duration of the exam, so that the student engages in the answering questions instead of cheating, reduce cheating in exams?

Students' perception about this question in percentages were as follow: 46.6% percent was for who strongly disagree and those who disagreeing were 21.4%), who indicated that they are neutral were 14.6%. Those who indicated that they are agreeing 6.8 percent and 10.7 percent who strongly agreeing.

Statement 6: doing exam at a specific time for all students reduces the possibility of conspiracy and cheating?

Students' perception about this question in percentages were as follow: 32% percent was for who strongly disagree and those who disagreeing were 16.5%), who indicated that they are neutral were 4%. Those who indicated that they are agreeing 15.5 percent and 14.6 percent who strongly agreeing.

Statement 7: strict sanctions considered a strong deterrent and keep student away from cheating?

Students' perception about this question in percentages were as follow: 22.5% percent was for who strongly disagree and those who disagreeing were 10.7%), who indicated that they are neutral were 29.1%. Those who indicated that they are agreeing 19.4 percent and 15.5 percent who strongly agreeing.

Statement 8: religious beliefs that prohibit cheating considered an effective deterrent that prevents cheating?

Students' perception about this question in percentages were as follow: 19.4% percent was for who strongly disagree and those who disagreeing were 6.8%, who indicated that they are neutral were 22.3%. Those who indicated that they are agreeing 16.5 percent and 35 percent who strongly agreeing.

5   Conclusion

Generally, results indicate that many students surveyed in this study reported they did not practice cheating in online exams. about 41 percent believed that cheating in online exams is more easier than cheating in paper exams. the majority of respondents disagreed that cheating in online exams is more expensive or it requires more equipment or devices than paper exams.

Concerning the methods used in online exams, respondents believed that Saving the test material, notes, expected answers, on computer or mobile device and surfing answers across multiple browsers on the internet are the most widespread among students.

With regard to the best techniques that could be applied to prevent cheating in online exams, the majority of the respondents agreed that religious beliefs that prohibit cheating is considered an effective way to prevent cheating. 66 percent disagreed that reducing the duration of the exam would prevent cheating in online exams. Also, they disagreed that using lockdown browser or doing exams in the labs of the university or operating proctoring devices during exams would prevent cheating. As well, changing the nature of questions or reducing the duration of the exam were not effective way to prevent cheating in online exams according to the majority of respondents in this study.

References

1. YERUN. (2020). THE WORLD OF HIGHER EDUCATIONAFTER COVID-19. https://www.yerun.eu/wp-content/uploads/2020/07/YERUN-Covid-VFinal-OnlineSpread.pdf.
2. Akakandelwa, A., Jain, P., & Wamundila, S. (2013). Academic dishonesty: A comparative study of students of library and information science in Botswana and Zambia. *Journal of Information Ethics, 22,* 141–154.
3. King, Chula G., Guyette, Roger W., Piotrowski, Chris, (2009), Online exams and cheating: An empirical analysis of business students' views. *The Journal of Educators Online, 6*(1).
4. Watson, G., & Sottile, J. (2010). Cheating in the digital age: Do students cheat more in online courses? *Online Journal of Distance Learning Administration, 13*(1).
5. Kennedy, K., Nowak, S., Raghuraman, R., Thomas, J., & Davis, S. F. (2000). Academic dishonesty and distance learning: Student and faculty views. *College Student Journal, 34*(2), 309–314.
6. Young, J. R. (2012). Online classes see cheating go high-tech. *The Chronicle of Higher Education, 58,* A24–A26.
7. Varble, D. (2014). "Reducing Cheating Opportunities in Online Test," *Atlantic Marketing Journal, 3*(3), 9. https://digitalcommons.kennesaw.edu/amj/vol3/iss3/9.
8. Kraglund-Gauthier, W. L., & Young, D. C. (2012). Will the real 'John Doe' stand up? Verifying the identity of online students. In L. A. Wankel & C. Wankel (Eds.), *Misbehavior online in higher education* (Vol. 5, pp. 355–377). Bingley.
9. Hearn Moore, P., Head, J. D., & Griffin, R. B. (2017). Impeding students' efforts to cheat in online classes. *Journal of Learning in Higher Education, 13*(1), 9–23.
10. McGee, P. (2013). Supporting academic honesty in online courses. *Journal of Educators Online, 10*(1).
11. Sendag, S., Duran, M., & Fraser, M. R. (2012). Surveying the extent of involvement in online academic dishonesty (e-dishonesty) related practices among university students and the rationale students provide: One university experience. *Computers in Human Behavior, 28,* 849–860.
12. Baker, R., Papp, R. (2003). "Academic Integrity Violations in the Digital Real." In *Proceedings of the Southern Association for Information Systems Conference,* Savannah, G.A
13. Jones, K. O., Reid, J., & Bartlett, R. (2006). E-Learning and E-CHEATING. 3rd E-Learning Conference, Coimbra, Portugal.
14. Becker, D., Connolly, J., Lentz, P., & Morrison, J. (2006). Using the business fraud triangle to predict academic dishonesty among business students. *Academy of Educational Leadership Journal, 10*(1), 37–54.
15. Nathanson, C., Paulhus, D. L., & Williams, K. M. (2006). Predictors of a behavioral measure of scholastic cheating: Personality and competence but not demographics. *Contemporary Educational Psychology, 31*(1), 97–122.

16. Williams, K. M., Natanson, C., & Paulhus, D. L. (2010). Identifying and profiling scholastic cheaters: Their personality, cognitive ability, and motivation. *Journal of Experimental Psychology: Applied, 16*(3), 293–307.
17. Alqooti, A. A. (2020). Public governance in the public sector: Literature review. *International Journal of Business Ethics and Governance, 3*(3), 14–25. https://doi.org/10.51325/ijbeg.v3i 3.47.
18. Bachore, M. M. (2016). The nature, causes and practices of academic dishonesty/cheating in higher education: The case of Hawassa University. *Journal of Education and Practice, 7*(19)
19. McCabe, D. L., Trevino, L. K., & Butterfield, K. D. (2002). Honor codes and other contextual influences on academic integrity: A replication and extension to modified honor code settings. *Research in Higher Education, 43*(3), 357–378.
20. Owunwanne, D., Rustagi, N., & Dada, R. (2010). Students' perceptions of cheating and plagiarism in higher institutions. *Journal of College Teaching and Learning, 7*(11), 59–68.
21. Simkin, M. G., & McLeod, A. (2009). Why Do College Students Cheat? *Journal of Business Ethics, 94*, 441–453.
22. Abdaoui, M. (2018). Strategies for avoiding cheating and preserving academic integrity in tests. *Alkhitab w el-Tawassol Journal, 4*(1)
23. Raines, D. A., Ricci, P., Brown, S. L., Eggenberger, T., Hindle, T., & Schiff, M. (2011). Cheating in online courses: The student definition. *Journal of Effective Teaching, 11*(1), 80–89.
24. Chirumamilla, A., Sindre, G., & Nguyen-Duc, A. (2020). Cheating in e-exams and paper exams: the perceptions of engineering students and teachers in Norway. *Assessment & Evaluation in Higher Education, 45*(7).
25. Bilen, Eren & Matros, A. (2020). Online cheating amid COVID-19. *Journal of Economic Behavior and Organization, 182*, 196–211. www.elsevier.com/locate/jebo.
26. Moten Jr, J., Fitterer, A., Brazier, E., Leonard, J., & Brown, A. (2013) Examining online college cyber cheating methods and prevention measures. *Electronic Journal of E-Learning, 11*(2), 139–146.
27. Al Kurdi, O. F. (2021). A critical comparative review of emergency and disaster management in the Arab world. *Journal of Business and Socio-economic Development, 1*(1), 24–46. https://doi.org/10.1108/JBSED-02-2021-0021.
28. Cluskey Jr, G. R., Ehlen, C. R., & Raiborn, M. H. (2011). Thwarting online exam cheating without proctor supervision. *Journal of Academic and Business Ethics.*
29. King, Darwin L. & Case Carl J. (2014). E-CHEATING: INCIDENCE AND TRENDS AMONG COLLEGE STUDENTS. *Issues in Information Systems, 15*(I), 20–27. https://doi.org/10.48009/1_iis_2014_20-27.
30. Watters, M. P., Robertson, P. J., & Clark, R. K. (2011). Student perceptions of cheating in online business courses. *Journal of Instructional Pedagogies.*
31. https://dolphinuz.com/blog/single/41/social-media-marketing-statistics   Citation   Date: 04/14/2020

Foreign Language Learning Then, Now and After COVID-19: An Exploration of Digital Tools to Augment the Receptive and Productive Skills of Language Learners

Subhasri Vijayakumar⊙

1 Introduction

In a multilingual country like India, where every citizen is fluent in more than one language, colonization brought in the need for learning a foreign language—English. Over the years after English, the need to learn another foreign language like German or French or Japanese or Chinese has been fortified by globalization, internationalization of higher education, the advent of Multi-National Companies, overseas projects and global clientele. Educational institutions offer foreign language classes as part of their curriculum to build global citizens. Communication is the central nerve of language classes and both receptive and productive skills are its core. Reading and Listening aid in receiving the information and these receptive skills are built in tandem with the productive skills of writing and speaking. A foreign language classroom is an amalgamation of innovation and creativity, a pot pourri of cultures with a blend of varied activities for fortifying target language mastery.

If we analyze the evolution of the foreign language teaching/learning process from the twentieth century to the twenty-first century, change has been the only constant factor. Right from the grammar translation method to the current Artificial Intelligence enabled language learning, foreign language classrooms have been the epitome of educational process transformations. Learners' motivation to learn languages impact the teaching methodology and the evolving learning styles of the learners dictate the pedagogical tools used to teach the target language. The advent of Information and Communication Technology (ICT) in education has redefined the teaching/learning process at large. Viewed as the building block of the modern society, ICT's impact is felt in every walk of life from governance to business to education and has reduced the world into a global village. The influence of IT

S. Vijayakumar (✉)
School of Social Sciences and Languages, VIT Chennai, Chennai, India
e-mail: subhasri.v@vit.ac.in

© The Author(s), under exclusive license to Springer Nature Switzerland AG 2022
A. Hamdan et al. (eds.), *Technologies, Artificial Intelligence and the Future of Learning Post-COVID-19*, Studies in Computational Intelligence 1019,
https://doi.org/10.1007/978-3-030-93921-2_17

in education is phenomenal and technology integration has become inevitable to academia. ICT in education has been playing a vibrant role in both formal and informal settings and has also brought in a paradigm shift from the traditional instructive approach to the modern constructive approach.

1.1 *Foreign Language Classrooms till 2020*

Indian classrooms have always been characterized as teacher centric with the sage on stage portrayal and traditional chalk and talk method of teaching. A typical Indian classroom comprises of fixed furniture and a compactly packed class of 40 to 70 students. The facilities like projectors, smart boards, audio players etc. are available only in a few schools and colleges. In a physical classroom text books are the widely used resources along with the chalk and board method. Oral drills for pronunciation practice, role plays, solving worksheets are few of the learning activities that are common and feasible in such a rigid setting. Although teachers strive hard to strike a balance between practicing all the four language skills—Reading, Writing, Speaking and Listening, many a times doing speaking and listening activities is either not possible or is not given adequate importance given the factors like huge batch size, or the unavailability of time or the facilities just not being conducive for the activity. The foreign language courses are also examination centered and tests only their knowledge rather than the skill. Speaking and listening skills are not tested directly in formative or summative assessments in many institutions. Pen and Paper method of testing is what is largely carried out and unlike English, the time period allotted for foreign language classes in schools or colleges is limited and more emphasis is laid on completion of syllabus and conduct of exams. This results in speaking and listening skills being left out by many teachers as reading, writing and grammar take precedence.

Set against this context, Vijayakumar et al. in their paper explored the motivations of the Indian learners of German as a foreign language at tertiary level and to bridge the gap identified between learners' needs and the teaching/learning process, proposed a "context-based, indigenous curriculum for tertiary level" that emphasized more on speaking and listening skills than the reading and writing skills as the former two skills were the most attributed skills by the learners [1]. The study also suggested a more learner-centered approach that made use of Information and Communication Technology to teach learning strategies for building autonomous learning and to integrate cultural components for fortified intercultural communication.

Literature is ripe with studies that analyzed the role of ICT in education and it is concurred that ICT facilitates digital learning and any technique becomes a good technique especially in a foreign language classroom, as the exposure to the target language is minimal outside the classroom and limited inside the classroom given the constraints. To cater to the growing needs of the tech savvy digital native learners, ICT tools are being extensively integrated to augment teaching/learning of foreign languages. The World Wide Web offers a plethora of authentic information and accessing resources in target language is easier. Websites, online magazines, videos and podcasts are a few among the limitless supply of materials that are available online. They provide real life language inputs and cultural information to the learners. To bring in a constant touch with the target materials apart from the text books, teachers have been employing these tools in their foreign language classes. It's a well situated argument that "authentic materials help students acquire an effective communication competence in the target language" [2]. But many teachers also opine that authentic materials are easier to use with advanced learners but with beginner level learners a different set of learning materials is what is required [3].

Creating additional learning materials is a herculean task for any teacher for it includes the time to plan and create the resource, the cost of printing and sharing it with the students and the effort to correct and give feedbacks. Often combined with minimal or no support from the department/institution, teachers resort to text books to teach the target language. Amidst these challenges, whenever possible teachers do bring their personal laptops or CD-Players to do audio/video exercises and copy additional resources in pen drive and distribute them to students or share them through common drives. Despite its potential and availability, the usage of ICT tools was minimal and its advantages were not fully leveraged.

1.2   The Pandemic and Its Effects on FL Classes in 2020

Due to the outbreak of COVID-19 pandemic in India in March 2020, educational institutions were forced to close and emergency remote teaching was adopted to continue classes. The emergency remote teaching that was adopted by every educational institution initially tried replicating the physical classroom virtually. With the academic year nearing its end by May 2020, for the intermittent two months, it seemed to be fine as many anticipated going back to normalcy for the next academic year. But when the pandemic spread and the next academic year had to be carried out in the online mode, academia explored online platforms, different Learning Management Systems (LMS), digital tools and the optimum methods to conduct classes, share materials or homework and assessments. Digital divide among learners and teachers became evident and many institutions couldn't continue online classes also. The lack of proper training to use tech-tools by the teachers loomed as a problem to be solved. Webinars and Workshops in Zoom / Meet became frequent and every teacher had to sharpen his/her digital acumen in a short span of time and get ready for the online classes. Setting up virtual classrooms through Google, Microsoft or Zoom was learned and E-Resource creation and sharing was learnt. Audio/video editing softwares were learnt and lesson videos were developed and shared. Various digital tools were explored and other language learning applications that provide lessons, activities, simulations, game tasks, and assessments were studied. The best advantage of these online tools is that they facilitate effortless generation of instructional materials and enable differentiation in the activities created based on learners' needs and learning styles. Various web tools act as content creating platforms as well as assessment kits. Based on the teachers' interest and implementation, many such online tools were used in virtual classes. Although initially online classes were mere virtual substitution of the physical class room, ICT tools were employed gradually from material sharing to assessments. Slowly teachers started to explore the many facets of digital learning and more digital tools are being employed to garner the attention of the students and to bring in student engagement in online classes.

2 Background

While these concepts may be completely new for many subject teachers and language teachers, literature has various studies that showcase the various tools available for language teaching/learning [4], studies that assess the teacher/learner perspective of using digital tools for language learning from an ecological perspective [5], the benefits individual tools bring in to the teaching/learning process [6] and the engagement and motivation these tools boost in the learners [7].

Kubler [8] in his study discussed the principles encompassed in developing course materials for a technology-mediated language learning program. Audio drills designed with an intention to assist independent and distance learning enabled the learners to obtain clarity in pronunciation and achieve confidence while communicating in the target language.

Khoiriyah [9] in her study analyzed two ESL based websites for their listening content based on the Computer-Assisted Language Learning (CALL) and Second Language Acquisition (SLA) theories and proposed a framework for selecting the appropriate materials and tools for fostering the listening skills.

Vijayakumar [10] in the study on Gamification through Moodle explored various game based tools that the open source learning management system—Moodle provided for enhancing vocabulary learning in a foreign language classroom.

Podcasts, digital storytelling, role-plays etc. continue to be some of the techniques to assist speaking activities in the virtual mode and many web-based tools are available for both synchronous and asynchronous evaluation of speaking activities. Syahrizal and Rahayu [11] in their study analyzed Padlet, a web tool for conducting speaking activities in their class and observed the various advantages like collaboration, peer feedback, flexibility and autonomous learning that it brought in.

The benefits of creating digital portfolios using self-made videos were explored by Cabrera-Solano [12] in her EFL classroom. It was reported that such recording of speaking activities in different situations brought in authenticity, increased the fluency and enhanced the pronunciation skill of the learners.

A digital taxonomy based mapping of web tools for creating grammar activities was proposed by Vijayakumar [13]. The study propositioned a descriptive digital pedagogical model that could be used as a framework by any language teacher to develop learning activities and for the learners to progress cognitively from lower order skills to higher order skills.

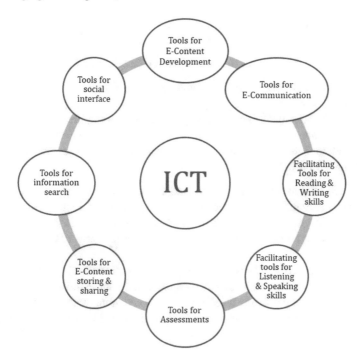

Fig. 1 ICT tools and the various purposes they facilitate

Among the emerging technologies, Augmented Reality [AR] and Virtual Reality (VR) are gaining popularity and their application is being seen in educational contexts too. The three dimensional (3D) technology that AR presents or the visual simulations of the VR bring in contextual visualization, authentic learning opportunities and interactivity to learning [14].

From this we know that there are a plenty of E-Tools and Fig. 1 illustrates the various purposes they facilitate.

2.1 Objectives of This Study

Backed by these earlier studies and based on the Indian context of teaching/learning a foreign language, through the current work, this study illustrates the status quo of foreign language classes in general and during the pandemic period along with suggestions for a post pandemic learning scenario.

2.2   Research Questions

The study also aims to answer the below given research questions:

- What are the web tools or ICT enabled tools available for creating activities for different language skills?
- What are the advantages and challenges in implementing these tools?
- Can these tools be used post pandemic in a physical class also?
- How should foreign language teaching/learning process evolve in the post pandemic scenario?

Such a list of web tools available to create receptive and productive activities in both physical and online classes along with their merits and demerits in using them in foreign language classrooms adds magnitude to the research in the field of foreign language teaching and learning in general and to the teaching fraternity in India in particular.

The following section details these tools for implementation in any language classroom. Discussion on the merits and challenges in using these tools is also presented. Suggestions for post pandemic teaching are explored along with the limitations and scope for further research in this area. Finally, concluding remarks are outlined.

3   Web Tools for Developing Receptive Activities—Reading and Listening

3.1   Web Tools for Developing Listening Activities

Self-Assessed quizzes through Google/MS Forms

As the foreign language classroom predominantly follows the textbook, these textbooks have audio CD's included in them. Students are encouraged to listen to these CDS but many a time it goes in vain. So to make the students listen and do activities based on the audio, self-assessed quizzes can be created using Google forms or MS forms. The audios can be uploaded in Google drive and the link can be shared in the form. Reading texts, Images etc. can also be included as questions to test comprehension. Both Google forms and MS forms allow creation of different types of questions like MCQs, Fill in the blanks etc.... and suitable questions can be designed. Figure 2 shows how a video can be integrated in Google form and related questions based on the video can be asked.

These forms can then be easily shared with the students through WhatsApp groups or Google Classrooms or Teams Class groups or even through any Learning Management System (LMS) that the institution uses. The quiz so created can be made self-assessable, so that at the end of the quiz, learners know their scores and the correct answers as shown in Fig. 3.

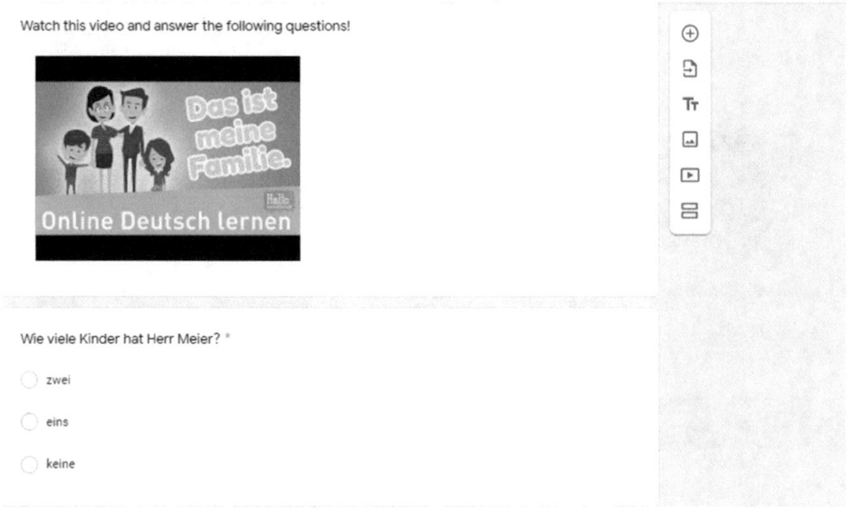
Watch this video and answer the following questions!
Das ist meine Familie.
Online Deutsch lernen
Wie viele Kinder hat Herr Meier? *
zwei
eins
keine

Fig. 2 Integrating video based questions in Google Forms

Pros:

- Creating questions along with the correct answers is a onetime activity. Link to the audio file in the drive and questions below it organizes the entire listening activity at one place and hassle free.
- Ability to check the answers and know the scores immediately after the quiz makes it easy for learners to rectify their mistakes and re-learn quickly.

Cons:

- From the teachers' perspective it becomes cumbersome a process to upload the files in drive, share the link in the form and create questions in the form. Despite the time involved, this can become a one-time activity and can always be re-used n number of times if it is well planned and designed.

Immersive Reader in MS Teams

Microsoft's free inbuilt tool called immersive reader provides a full screen reading experience of the given text. Immersive reader can be used with Word, Outlook, PowerPoint and OneNote application of Microsoft. It gives an option to choose either a male voice or a female voice and supports many languages also. This is highlighted in Fig. 4.

The main advantage of immersive reader is that it assists people who have reading difficulties to listen to the texts being read out loud. This assistive feature is a boon in the foreign language classroom as texts or dialogues can be shown to the learners and if they listen to the immersive reader, they can also listen and learn the intonation, pronunciation and melody of reading/saying the text in the given target language.

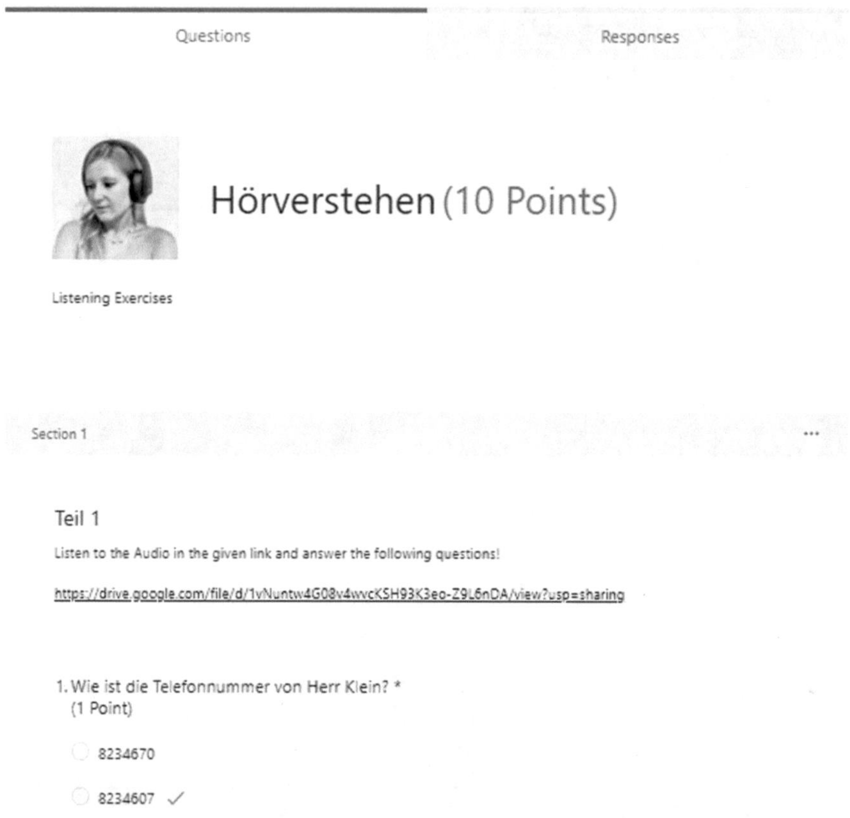

Questions   Responses

Hörverstehen (10 Points)

Listening Exercises

Section 1   ...

Teil 1

Listen to the Audio in the given link and answer the following questions!

https://drive.google.com/file/d/1vNuntw4G08v4wvcKSH93K3eo-Z9L6nDA/view?usp=sharing

1. Wie ist die Telefonnummer von Herr Klein? *
   (1 Point)

   ○ 8234670

   ○ 8234607 ✓

Fig. 3 A self-assessable listening quiz created through MS Forms

Fig. 4 Immersive reader in microsoft one note

Pros:

- Learners can read and listen to the texts in real-time and practice the pronunciation of individual words in the target language
- It facilitates pronunciation drills that the learners can do at their own pace and place
- This feature can be used free of cost with MS office products in MS Teams

Cons:

- The institutions must have institutional access to Microsoft 365 and MS teams should be the common online platform for virtual classes
- Teachers need some orientation as to how to set up the OneNote feature and activities using this tool.

Immersive Reader in MS Teams

Quizlet is an online American application that enables development of flash cards for to study any subject. The flash cards so developed can be used for self-learning, self-assessments as well as for conducting live games in foreign language classrooms. The main feature here is the ability to listen to the flash card content created through the text-to-speech facility incorporated. Quizlet is the virtual flash card that has two sides: the terms and the definitions. By clicking on the audio icon in the learn mode, learners can listen to the words/sentences typed in the terms/definitions side of the flash card. This is shown in Fig. 5.

At a click of the links given in the left side menu bar, different activities like writing, spelling and easy tests can be created. The Gamification feature found in Quizlet makes it an engaging tool for individual playing as well as conducting live

Fig. 5 Listening in Quizlet flashcards

games in class. Options to include images, audio files etc. facilitates listening and comprehension. These features are highlighted in Fig. 6.

Pros:

- It is very user-friendly and the free version itself has various features for the teachers to develop their content
- Pre- created flash cards can also be searched and customized to suit ones requirements
- Provides a play way platform to learners to learn the content
- Tests can be created from the sets in one-click and shared as online test or also as a.Pdf document to the students

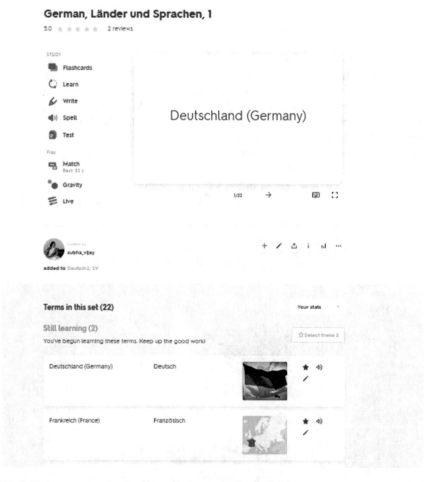

Fig. 6 Various game options and inserting images/audios in Quizlet

- Can be integrated with many platforms like Moodle, Google Classroom and MS Teams

Cons:

- Quizlet live cannot be played from the mobile app. Web app is required to start the live game.
- Free version of quizlet allows the usage of existing pictures only. This could be limiting in some occasions. Our own images can be uploaded only in the paid version.

Podcast

Podcast as an innovative tool has garnered the interest of teachers and learners in the tertiary sector in the recent years. As a tool it has been seen effective in developing personalized learning, tailored to the learners' needs. A Podcast is a "digital audio/video file that is created and uploaded to an online platform to share with others" [15]. It can be then downloaded by anyone and can be played on any computer/portable device. Since they are the paradigm of Mobile Assisted Language Learning (MALL) they are congruent with the learners' tech-savvy, mobile addicted as well as always networked life style. Audio Podcasts are unidirectional (teacher-to student) and foster authentic listening by supplementing the textbooks. As Phillips in his 2005 study analyzed, students developed Podcast has also been gaining momentum recently and brings with it various learning advantages to the students [16]. To produce a Podcast, a learner has to research the content, draft and prepare the speech, rehearse to deliver the content appropriately with the correct pronunciation, intonation and fluency. While Podcast developed by the teachers can be a good listening activity material, students' podcast can be a good source of speaking material too.

Pros:

- It offers learners a variety of listening information to be used both in-outside the class room
- This can bridge the gap between the formal usage of the target language in the class room and the language used in the real-life communication
- It can also aid in reducing the foreign language learning/speaking anxiety among learners
- It is a useful collaborative tool and can be used to reach larger audiences.

Cons:

- Learners and teachers need time to explore and learn how to create and set up podcasts for their classes.
- It can be a cumbersome process if the required infrastructural facilities are not available.

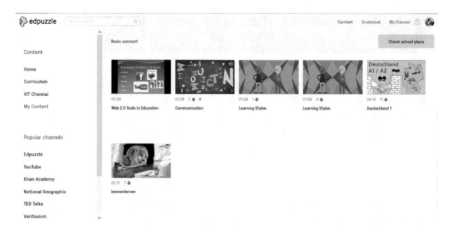

Fig. 7 Edpuzzle and adding videos from the channels

Edpuzzle

Edpuzzle is a tool that can be used to integrate interactive content to the pre-existing video. Edpuzzle has a vast in-built video database from sources like Ted, YouTube, Khan Academy etc., along with the option to use one's own video also (through Youtube). The starting page in Edpuzzle is as shown in Fig. 7.

The required video can be selected and interactive MCQ questions, True/False types or Notes, additional Audio tracks and comments can then be integrated into the video. This is a newfangled app that facilitates listening and comprehension of the text. Figure 8 illustrates how interactive questions pop up at the various stages they are added.

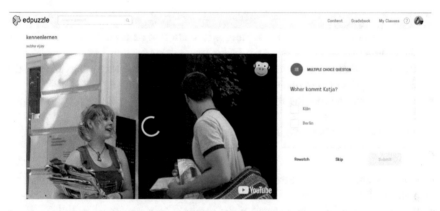

Fig. 8 Interactive questions integrated into the video

Pros:

- It is a free application that can be used with all Learning Management Systems
- You can use the LIVE mode option to play the videos live in real time in both physical and virtual class and gamify the experience of listening and doing the activities
- Can be effectively used for formative assessments

Cons:

- Apart from the initial orientation of how to use the Web app, this tool is simple and easy to use
- There is a limitation to the number of interactive videos that can be created in the basic free plan.

3.2 Web Tools for Developing Speaking Activities:

Flipgrid

Flipgrid is a Microsoft platform for video discussions. It allows sharing of user developed videos in a social learning community created by the teacher. Teachers create a Grid and invite/add their students to that class grid. Within the Grid, various topics can be created and assigned for the students. The students prepare the speech content as per the topics assigned to them, record them using the in-built video recording features of Flip grid and upload them. Such videos can be viewed by their peers in the class, they can be liked, shared and commented and the teacher can also assess the students' performance based on the Rubrics. This is shoown below in Fig. 9.

Pros:

- Setting up personalized rubrics for evaluation of the videos created enhances the feedback system
- Peer feedback brings in responsibility to the learners to critically analyze the performances of their class mates
- It provides an infinite practice scope to the students via a non-threatening learning environment [17].

Cons:

- In a large classroom, evaluating the individual videos of all the learners will be a challenge and a time consuming process for the teachers. This can be overcome by the peer reviewed feedback option that Flip Grid provides.

Video-Mediated Communication (VMC) Tools:

Meaningful and effective oral production has been easy and instantaneous with the advent of VMC tools like YouTube, Padlet and WhatsApp. Learning Management

Fig. 9 Flipgrid for empowering oral presentations

Systems (LMS) like Moodle, Google Classrooms and Ms Teams also have in built audio and video creating tools, which can be used at the click of the mouse to develop video presentations. Figure 10 illustrates how audios are incorporated in Moodle for a listening activity.

Any video production requires planning of the content, practice of the text and collaboration among people for efficient audio/visual outputs. Vlogs are catchy and trending and can be used as a Project-Based-Learning (PBL) activity in the foreign language classroom to instill real-time communication.

Pros:

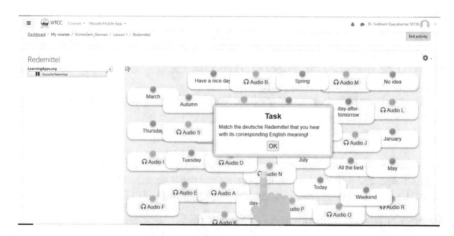

Fig. 10 Integrating audios in Moodle for a listening activity

- VMC tools are simple and easy to use
- They are available free of cost and can be used/accessed across devices
- Brings in learner autonomy as well as collaboration
- Emphasizes on body language and other non-verbal techniques in communication also

Cons:

- VMC can be time consuming and construed as intruding privacy of people
- It can never substitute the benefits of face-to-face interactions.

In a nutshell Table 1 summarizes the various tools discussed above.

Table 1 Summary of digital tools for augmenting receptive and productive skills

Name of the web tool	A brief note
Google/MS forms	Creation of Self assessed quizzes using texts, images, audio/video links/files Creation of different types of questions like MCQs, Fill in the blanks etc Easily shareable with students
Immersive reader in MS teams	Full screen reading experience of the given text in any language Assists people having reading difficulties Helps in improving the pronunciation and intonation of the learners
Quizlet	Listen to flash cards Play and learn vocabulary sets Self-Tests and Live Games in Online classrooms
Podcast	Fosters authentic listening activities custom made for specific learning Helps bridge the gap between the formal usage of the target language in the class room and the language used in the real-life communication
Edpuzzle	Video based assessment tool with options to include additional audio tracks and comments Facilitates listening and comprehension of the text
Power points, Prezi	Facilitates Digital Storytelling, creative writing
Flipgrid	Learner developed videos on any given speaking topics Peer evaluation and personalized rubrics for evaluation Scope to practice in a non-threatening learning environment
Video mediated communication tools	Youtube, Padlet, WhatsApp, Screencastify are few of the VMC tools that can be used to create video presentations/Vlogs Effective for PBL and collaborative learning Promotes creativity and learner autonomy Body Language and Non Verbal Communication can also be fortified

4 Discussion

In the triangulation of pedagogy, methodology and technology, technology needs to be the "mediating instrument for cognitive simulation, skills development, ad-hoc learning and knowledge construction/re-construction/co-construction" [18]. Needless to reiterate, technology strengthens the foreign language class rooms and offers myriad tools and applications to fortify LSRW skills in the target language.

Learners can learn at their own place, anytime and anywhere and practice the target language. They can collaborate and communicate with their teachers and peers and also with natives to learn how to creatively use the target language in the real world. In essence, learners can learn with and learn from technology proficiently. With information at everyone's fingertip, teachers can't be expected to always control their students' learning experiences. They now become facilitators, advisors or guides by the side.

Teachers on the contrary face innumerable challenges when it comes to teaching with technology. They endure this clash between giving in to learner-centric methods while executing within the defined timings, pre-set syllabus and assessment patterns. To balance learner autonomy, individual technical knowhow and the affordances of digital tools, teachers are expected to master the triangulation with little or no support from the other stakeholders of education. Teachers may be digital literates with the ability to curate the information from the varied sources and create meaningful content but they also need training and technical know-how to develop effective activities for the online classes, to support blended/hybrid learning. Starting from access to technological tools and technology driven pedagogy, teachers have to be trained on continuous transformations of learning resources, assessment strategies and teaching styles. Ability to use and integrate games, social networking activities etc. by the teachers has to be encouraged to increase learner's motivation and effective engagement with the content in online classes.

4.1 Post Pandemic Teaching/Learning

Post Pandemic teaching/learning scenario should focus on sustainability in language education. Blended or hybrid learning what is being tried and tested now should be made an integral part of the language class. Be it synchronous or asynchronous learning, emphasis should be given to impart learning strategies for self-learning using the myriad tools available. Understanding the heavy reliance on web tools for translations and quick learning, activities have to be redesigned to enable learners' continuous learning, their ability to reflect on their learning and promoting connections between theories to practice. Research skills are to be given importance and students have to be taught to distinguish between reliable and unreliable sources of information in this era of Google and YouTube. Instituting a change is arduous but within their means teachers are discretely advocating the change they want to see in the language classrooms. This article is towards snow balling these independent efforts of many teachers to the mass to re-engineer the teaching/learning process holistically to attain its goal of fruitful and meaningful communication in real-time.

From easy and simple tools to complex and dynamic varieties, every day there is a tool or app introduced in the market. There's also a surge in ed-tech sellers' market and many commercial learning solutions are not always based on pedagogical techniques. The debate over Ed-Tech reducing or redefining the teaching/learning concepts has been there for a while now and the narrative that "Technology can fix broken parts of the education" is highly contested [19]. To ensure continuation in studies, employing quick fixes with readymade/free Ed-Tech app is not the solution. Many free tools might cease to be 'free' post pandemic and that should not be the cause to relapse back to conventional methods of teaching also. Education cannot be situated on readymade contents made by someone else. Rather it has to be co-created by the teachers and the students based on their needs and requirements. A major attitude change seen among learners amidst the ongoing pandemic is that any ways the information required is in the digital form, so why should I understand it or commit it to my memory. This is also not a healthy attitude to be fostered in the long run. For the assessment purpose or for just clearing the exams, if the learners resort to copy, paste mode that will also not be beneficial when the physical classes and normal classes begin. Probably now is the best time to make some choice decisions on the implementation of quality standards for assessments and assessment patterns. If flipped classrooms, blended or hybrid learning has to be effectively integrated with face-to-face instruction, then design of activities, evaluation rubrics and modes of assessments have to be re-engineered to maintain the quality of teaching/learning process and to sustain learner motivation. The tools discussed above are not a comprehensive list but is only a bird's eye view of them. These tools are tried and tested by teachers especially in the ongoing pandemic period, which has for once transformed the teaching/learning scenario all over the world.

Some suggestions for language activities in post pandemic era:

- Developing personalized learning sets using web tools for constant practice
- Communicating with natives through social media and reporting/describing them
- Collaborative projects to understand the target culture and system
- Watching videos/MOOC courses and creating appropriate dialogues for everyday situations
- Book marking authentic websites, resources etc. and sharing with the class
- Student teaching sessions
- Student videos for grammar concepts; Peer evaluations and feedbacks using web tools
- Searching, assessing and collating activities based on topics learnt
- Strategy training like using online dictionaries/thesaurus to understand forms and meanings
- Training to translate from and to target language using translation software
- Error detections in written or oral communication

- Culture specific responses and behaviors or inter cultural communication based case studies
- Comparing and Contrasting customs and beliefs
- AR or VR augmented trips to target countries for better understanding of facts and information

5 Limitations and Scope for Further Research

This paper significantly discusses within the Indian foreign language classroom perspective but many of the tools and activities discussed can be implemented in any other subject class also. The paper also does not quantitatively measure the effectiveness of the tools used or the frequency in which tools are used in FL classes; it only elucidates a general narrative of the tools available.

This concept can be developed further and the suggestions put forward above can be developed using these tools. Such a course can also be tried in a foreign language classroom and quantitative data can be collected to analyze the effectiveness of such a model in fortifying the receptive and productive skills of the learners. More inquiry can also be done on the practical challenges encountered by the teachers in implementing such a design in the Indian class room and the motivation of the learners in adapting to such a learner centered model.

6 Conclusion

Technology is integral to our classrooms and methods of language teaching have to be diversified to include digital tools [20]. With the technology, teachers' activity design should also evolve, new experiences to practice/master a skill has to be constantly designed to empower students' learning. ICT offers the ways and means to develop differentiated instructional materials for the learners and it becomes the responsibility of the concerned teacher to explore what suits their class and teaching requirements. It is certainly not about latest tools, it is about the innovative activities with the known tools too. It might be arduous for the teachers initially to design and develop activities using these tools and to get their students acclimatize to the hybrid learning process. But once they get on board with the process, the benefits it brings to the learners is only manifold. This paper thus highlights few of the tools that can be easily used for designing differentiated activities for learners' self-paced, anywhere, anytime learning and offers recommendations for a robust and functional post pandemic foreign language pedagogy.

References

1. Vijayakumar, S., Kumar, G., & Sathiaraj, J. (2017). The role of foreign languages at tertiary level – towards developing a contextual curriculum for a German Course in Indian technical universities. *German as a Foreign Language (GFL), 1*(3), 100–126. http://www.gfl-journal.de/3-2017/vijayakumar.pdf.
2. Akbari, O., & Razavi, A. (2015). Using authentic materials in the foreign language classrooms: Teachers' perspectives in EFL classes. *International Journal of Research Studies in Education, 4*(5), 105–116. https://doi.org/10.5861/ijrse.2015.1189. Retrieved 21 March, 2021.
3. Polio, C. (2014). Using authentic materials in the beginning language classroom. *CLEAR: Centre for Language Education and Research, 18*(1). Retrieved 21 March, 2021, from http://clear.web.cal.msu.edu/wp-content/uploads/sites/22/2018/10/2014-Spring.pdf..
4. Yunina, O. (2019). Digital tools in foreign language teaching. *Education Innovation Practice, 1*(5), 17–22. Retrieved 25 March, 2021, from http://www.eip-journal.in.ua/index.php/eip/article/view/64.
5. Liu, Q., & Chao, C. (2017). CALL from an ecological perspective: How a teacher perceives affordance and fosters learner agency in a technology-mediated language classroom. *ReCALL, 30*(1), 68–87. Retrieved 21 March, 2021, from https://doi.org/10.1017/s0958344017000222.
6. Castillo-Cuesta, L. (2020). Using digital games for enhancing EFL grammar and vocabulary in higher education. *International Journal of Emerging Technologies in Learning IJET, 15*(20), 116–129.
7. Panagiotidis, P. (2021). Technology as a motivational factor in foreign language learning. *European Journal of Education, 1*(3), 43–52. Retrieved 17 March, 2021, from http://journals.euser.org/index.php/ejed/article/view/3980.
8. Kubler, C. (2021). Developing course materials for technology-mediated Chinese language learning. *Innovation in Language Learning and Teaching, 12*(1), 47–55. Retrieved 31 March, 2021, from https://doi.org/10.1080/17501229.2018.1418626.
9. Khoiriyah, K. (2021). CALL and SLA theory: Developing a framework to analyze web-based materials for teaching listening skills. *IDEAS: Journal on English Language Teaching and Learning, Linguistics and Literature, 8*(1), 80–92. (2020). Retrieved 18 March, 2021, from https://doi.org/10.24256/ideas.v8i1.1296.
10. Vijayakumar, S. (2020). Gamification through Moodle to enhance vocabulary learning. *Journal of Critical Reviews, 7*(13), 3108–3111. Retrieved 17 March, 2021, from http://www.jcreview.com/index.php?mno=129888.
11. Syahrizal, T., & Rahayu, S. (2021). Padlet for english speaking activity: A case study of Pros And Cons on ICT. *Indonesian EFL Journal, 6*(2), 149–156. Retrieved 17 March 2021, from https://doi.org/10.25134/ieflj.v6i2.3383.
12. Cabrera-Solano, P. (2020). The use of digital portfolios to enhance english as a foreign language speaking skills in higher education. *International Journal of Emerging Technologies in Learning (iJET), 15*(24), 159–175. Retrieved 18 March, 2021, from https://www.learntechlib.org/p/218569/.
13. Vijayakumar, S. (2021). A digital taxonomy based instructional design to teach and learn German Grammar. *Journal of Information and Computational Science, 11*(1), 77–86. (2021). Retrieved 18 March, 2021, from http://joics.org/VOL-11-ISSUE-1-2021/.
14. Huang, X., Zou, D., Cheng, G., Xie, H. (2021). A systematic review of AR and VR enhanced language learning. *Sustainability, 13*(4639). https://doi.org/10.3390/su13094639. Retrieved from https://www.mdpi.com.
15. Phillips, B. (2017) Student-produced podcasts in language learning – exploring student perceptions of podcast activities. *IAFOR Journal of Education, 5*(3), 157–171. Retrieved 18 March, 2021, from https://iafor.org/journal/iafor-journal-of-education/volume-5-issue-3/article-8/.
16. Phillips, B. (2015). Empowering students: Using technology-enhanced learning to foster learner autonomy. In: A. Trink (Ed.), *Pannonia research award: Regionale und europäische Zukunftsfragen* (pp. 83–94). Graz: Leykam, Hdl.handle.net. Retrieved 19 March, 2021, from http://hdl.handle.net/20.500.11790/57.

17. Nadjwa Miskam, N. (2019). The use of flipgrid for teaching oral presentation skills to engineering students. *International Journal of Recent Technology and Engineering (IJRTE)*, 8(1), 536–541. Retrieved 19 March, 2021, from https://www.ijrte.org/wp-content/uploads/papers/v8i1C2/A10880581C219.pdf.
18. Gaballo, V. (2019). Digital language teaching and learning: A case study. In *Innovation in language learning, conference 2019* (pp. 1–6). Retrieved 26 March, 2021, from https://conference.pixel-online.net/ICT4LL/files/ict4ll/ed0012/FP/4156-QIL4217-FP-ICT4LL12.pdf.
19. Teräs, M., Suoranta, J., Teräs, H., et al. (2020). Post-Covid-19 education and education technology 'Solutionism': A seller's market. *Postdigital Science and Education, 2*, 863–878. https://doi.org/10.1007/s42438-020-00164-x. Retrieved 10 June, 2021,from https://link.springer.com/article/https://doi.org/10.1007/s42438-020-00164-x.
20. Badran, M. (2017). Digital tools and language learning. *Master's Degree Project, Malmö Högskola.* https://www.diva-portal.org/smash/get/diva2:1490406/FULLTEXT01.pdf.

Language and Electronic Media Competencies Required for the Arabic Language Teacher for Non-native Speakers in the Time of Covid 19

Ali Omran

1 Introduction

Perhaps it is obvious for education practician to say that it is not possible to apply any educational curriculum—no matter how prepared it is without a teacher prepared in based on psychological-educational techniques, that is, he has a set of competencies and necessary skills that achieve his goals, and thus enable him to motivate the student to reach the stage of perfection that God has prepared for him. This is according to his capabilities, capabilities, preparations…etc. It is obvious that the type pf preparation process is differ between the teacher of children and the teacher of adults. The teacher of the child should have specific competencies that help him to manipulate the children in order motivate those children to fulfill their duties.

The educators emphasized that a good teacher for childhood must be prepared and trained to acquire three sets of skills: personal, social, and cognitive skills, in addition to professional skills in the age of Covid-19 virus. The qualifications and components of a good teacher are described in three dimensions in the personality. The first: full knowledge of the specialization subjects that children teach, and his control over the various arts and skills (specialization dimension), second: full knowledge of the nature of learners, their psychological, scientific and technical characteristics, and the appropriate sources of knowledge for them (the professional dimension), and third: the availability of a knowledge of student culture that help him to understand the world in which his students live. (cultural dimension); Therefore, the preparation and training of any teacher requires attention to many basics, including: the specialized or cognitive basis, the professional basis, and the cultural basis.

A. Omran (✉)
Rhetoric and Discourse Analysis, Ahlia University, Manama, Kingdom of Bahrain
e-mail: aomran@ahlia.edu.bh

© The Author(s), under exclusive license to Springer Nature Switzerland AG 2022   303
A. Hamdan et al. (eds.), *Technologies, Artificial Intelligence and the Future of Learning Post-COVID-19*, Studies in Computational Intelligence 1019,
https://doi.org/10.1007/978-3-030-93921-2_18

The study aims to identify the linguistic, professional and cultural competencies necessary for an Arabic language teacher for non-native children in age of e Covid-19 virus.

Characteristics of the Arabic language teacher for non-native speakers of children in light of the spread of the Covid 19 virus [1]:

A person's behavior reveals a personality, and a teacher is a human, so his behavior is a consequence of his personality, and this behavior appears in varying degrees from one teacher to another. These characteristics have been categorized to facilitate their handling under the following headings:

A- Enthusiasm: The enthusiast shows interest in the subject and believes that the subject is an interesting value. It also shows his activity and strength in all his performance and the behavioral patterns of this teacher:

He seems confident.

- Highlights the importance of learning Arabic for children, and show it.
- Uses vivid and clear gestures to emphasize and reinforce certain points.

Innovative and diversified in his teaching method.

- Maintains close contact with the eyes of all children.
- Uses a diverse sound tune.

Uses movement to control the attention of his students.

b- Warmth and humor are two important factors in providing a supportive, satisfying, and productive environment for students, especially for foreign children.

T- Credibility: It means the students' trust in the teacher, and this confidence does not exist between short time, but it requires a long time, intense effort, and clever emotional behavoir from teacher.

2- Orientation towards success [2]:

A good teacher has high expectations of success for himself and his students, and his behavior can be as follows:

Clearly informs children of the objectives of the lesson.

Uses waiting time to allow children to think before responding.

He rarely disrupts his children while they work.

Helps children modify incorrect or inaccurate responses.

Uses intensive and frequent feedback.

3- Professional Conduct [1]:

Many features fall into this category, including:

a. Seriousness at work.

B. goal orientation.

T. serious.

d. deliberate.

c. organized.

h. He bases his work on scientific theories, believes in children's imagination, and introduces tangible, material objects, and scientific details.

x. Adaptable and Flexible.

Dr. Knowledgeable.

Some studies have added other features to the children's teacher in light of the spread of the Covid-19 virus, including:

Honesty, straightforwardness, spontaneity, sensitivity to others, acceptance of individual differences, tenderness, activeness, vitality, self-confidence, self-acceptance, emotional stability, endurance, and the ability to learn from others [1].

The Arabic language teacher for non-native speakers can set a good example for children, through his personality, and through all his behavior. This becomes an honorable interface for the Arabic language with its Arab and Islamic culture.

The teacher requires educational preparation after the academic preparation of which purpose to convey teaching strategies, instructional methods and practical application under specialists and educational professors, and he requires to study the following very important subjects [3]:

1. Developmental Psychology: In Which the Student/teacher Studies the Developmental Characteristics of the Learners and the Characteristics of Each Stage of Development.
2. Educational psychology: in which the student/teacher is acquainted with some theories related to this science, including: learning theories, and how to formulate educational goals ... and others.
3. Educational Measurement and evaluation: in which the student/teacher studies how to prepare tests in the correct educational manner and how to evaluate the student achievement.
4. And other important educational materials that if we were not afraid of prolonging, we would have listed them.

(F) The necessary professional competencies for an Arabic language teacher for non-native speakers in light of the spread of the Covid-19 virus [4]:

In light of above discussion, it is possible to come up with a set of professional competencies necessary for an Arabic language teacher for non-native speakers, as follows:

1. To formulate the general and specific objectives of teaching Arabic as a foreign language accurately, with the setting of the appropriate time to attain them.
2. To determine the appropriate content to attain the objectives.

3. To use the appropriate methods for educational situation, with diversification in these methods.
4. To prepare the means and techniques used in each activity.
5. To plan a variety of activities: individual and group, to meet the individual differences between the children.
6. To prepare suitable and diverse tools for assessment.

Competencies of implementing activities in light of the spread of the Covid-19 virus [5]:

1. To prepare interesting introduction for the subject of the activity.
2. To diversify the methods of introduction to lessons in the light of the child interests.
3. To teach Arabic language in an integrative manner.
4. To diversify the use of activities.
5. To use the sign language to help the child understand the language, taking into account the issuance of instructions in clear and simple manner.

1. To move from easy to difficult, from abstract to tangible, and from known experience to unknown experience gradually.
2. To use the appropriate educational means for each activity, with diversification in them.
3. To take into account safety of children in the means used.
4. Diversity in the implementation of the activity.
5. To encourage children to use the Arabic language as a means of fun.
6. To correct common mistakes that affects the meaning.
7. To deal with individual differences, and take into account the child's concentration period.
8. To use feedback when executing each step of the activity.

Competencies of interacting with children in light of the spread of the Covid-19 virus [6]:

1. To establish a friendly relationship with children, based on trust and affection.
2. It provides opportunities to establish friendship between children each other, by providing opportunities for teamwork.
3. To use a variety of tools and procedures to control the class.
4. To distribute his attention to the whole class.
5. To encourage children to discuss and be initiative.
6. To create an atmosphere of fun.

Comptencies of evaluating activities in light of the spread of the Covid-19 virus [7]:

1. That the activity is carried out in accordance with its objectives.
2. To use the diagnostic, structural, formative and final evaluation.

3. To measure the different cognitive levels: remembering, understanding, and applying.
4. To observe the children's work on an ongoing basis, taking into account the individual differences between the children.
5. To use the appropriate means for each activity, and for each skill: listening, speaking, reading, writing.

Competencies of teaching listening in light of the spread of the Covid-19 virus [8]:

1. To clearly define the objectives of listening education, formulate them empirically.
2. To identify the steps for teaching listening.
3. To help children acquire the etiquette of listening: to listen attentively to others, not to interrupt others...etc.
4. To identify the problems that educated children face when they learn to listen, and to address these problems.
5. To diversify the use of listening activities, with the use of appropriate means for each activity.
6. To employ natural situations in teaching listening.

Speech teaching competencies in light of the spread of the COVID-19 virus [9]:

1. To clearly define the objectives of teaching speaking, and formulate them empirically.
2. To identify the steps of teaching speaking.
3. To employ opportunities for oral communication between children.
4. To use speaking skills in a proper manner, and help children acquire them.
5. To diversify the use of speaking education activities.
6. To identify the problems that children encounter when they learn to speak, and try to treat them.
7. To employ natural situations in teaching speaking.

Competencies of teaching reading in the age of Covid 19 virus [10]:

1. To clearly define the objectives of teaching reading, and formulate them empirically.
2. To specify the steps for teaching reading.
3. To be able to master the skills of silent reading, and aloud, maintaining what is in Arabic in terms of stress and intonation, and helps children to acquire it.
4. To use different approaches to teach reading.
5. To choose the appropriate materials for reading. Taking into account individual differences, and the different interests.
6. To identify the problems that children face when they learn to read, and try to treat them.

7. To employ the rules of Arabic grammar in teaching reading.
8. To provide opportunities for reading through natural situations.
9. To develop the inclination to read in Arabic.

Competencies of teaching writing in light of the spread of the Covid 19 virus [4]:

1.   To clearly define the objectives of teaching reading, and formulate them empirically.
2.   To determine the steps of teaching writing.
3.   To use different approaches to teach writing.
4.   To train children to write from right to left.
5.   To help children writing Arabic letters with various forms.
6.   To train children to use punctuation marks.
7.   To employ Arabic grammar in writing.
8.   To provide opportunities for children to choose the activities they like, through natural situations.
9.   To identify the problems that children face when they learn to write, and to try to remedy them.
10. To develop the inclination to write Arabic.

Third: Cultural Competencies in era of Covid-19 virus [11]:

In those competencies, teachers differ, and by it we distinguish between a teacher and a teacher, it is the competencies that means what the teacher gains from the general culture, it is neither reasonable nor acceptable for a teacher to teach a students from China, and he does not know anything about the nature of their country, language and religion. The teacher of Arabic for non-native speakers must be familiar with everything new in the field of teaching Arabic to non-native speakers, and be a good follower of conferences and seminars that are held, research published, and books taught in this field, knowing the most famous and best, and choosing the most appropriate for their students.

This sufficiency is one of the most important competencies from my point of view, as it shows the culture of one teacher from another, and base in those competencies, respect and understanding between the teacher and his foreign students is occurred.

(C) The necessary cultural competencies for an Arabic language teacher for non-native speakers in light of the spread of the Covid-19 virus [12]:

One of the sources of deriving the necessary cultural competencies for a teacher of Arabic for non-native speakers is the nature of Arab-Islamic society and culture. God has created man to meet. On this basis, the human society consists of four elements:

1. The individuals who make up the group.
2. The connections that arise, necessarily, between individuals.

3.　The system that controls and directs the behavior of the group.
4.　Doctrine, which is the greatest component of society.

And the most dangerous of them" is that it controls society, sets its rules, and establishes its systems [13], and if the human society consists of individuals who make up the community, connections, system, and doctrine, then that complex whole that includes knowledge, beliefs, art, morals, law, and habits acquired by man as a member of a society are called culture [9].

The concept of culture and its aspects:

It is well known that culture plays an effective role in building an individual's personality. Culture is the main reference for understanding the behavior of peoples, present and future, as it contains all aspects of human life [14]. Culture is the overall style of community life, which is consistent with its general conception of divinity, the universe, human, and life. According to this definition, the aspects of culture can be summarized as follows [15]:

A- The normative aspect: the society's perception of divinity, the universe, human, and this is the special aspect, as it distinguishes each society from the other, and it is the one that preserves the distinctive characteristics of culture, and prevents its liquefaction [16].

A- The behavioral aspect or the applied aspect: it includes the individual and collective behaviors of the society such as political, economic, and social behaviors. It also includes the relationships, connections, the way of work and building systems and institutions. It also contains the education system, the family system, the road traffic and driving regulations. Moreover, it comprises language, its arts, skills, methods of use and linguistic practices in general, and includes arts, literature, media, advertising, ways of thinking and creativity....etc. [17].

C- The civilized aspect: which represents the fruits of culture. When the realistic behavioral aspect of culture is consistent with its normative aspect, the civilizational, spiritual, intellectual, scientific, artistic fruits are ripen [18].

And the way in which cultural aspects and concepts are presented will highlight the importance of culture, and work to form positive attitudes towards it. Language is the main tool for transmitting culture, through which the child acquires culture in an automatic manner. Thus, language becomes the link of the contract for the cultural and educational system [19].

In the field of teaching second languages, culture and its teaching represent the fifth dimension as educational skill. Culture complements the four well-known dimensions. It connects them in the form of an integrated system, as the four dimensions cannot be imagined practicing them in a vacuum. Hence, it became one of the most important principles of teaching the language [20].

The necessary cultural competencies for the Arabic language teacher for non-native speakers in light of the spread of the Covid 19 virus:

From the above it is possible to extract some of the cultural competencies necessary for the Arabic language teacher for non-native speakers:

1. Memorizing some parts of Qur'an, memorizing a number of Prophetic hadiths, and memorizing a number of Arab speeches.
2. Defining the concept of culture and its characteristics.
3. Differentiation between Arab and Islamic culture [19]
4. Knowing the aspects of culture, and presenting the appropriate ones to the foreign child.
5. Knowing the nature of society and its elements and understanding the nature of interaction between these elements.
6. Mastery of English language skills: listening, speaking, reading, writing.
7. Mastering computer-driving skills, using the Internet to support its information, and developing its teaching methods [21].

For extracting the competencies of teaching culture that are necessary for the Arabic language teacher to non-native speakers in light of the spread of the Covid 19 virus, they are as follows:-

1. To start with the cultural aspects that the child is interested in.
2. To use cultural vocabulary that is close to the child's perception.
3. To present the values, customs and behaviors that represent the Islamic culture.
4. To diversify the use of methods and activities related to the presentation of culture.
5. To link between the cultural and Islamic aspects and the aspects of the mother's culture of the child.
6. To understand the goals and culture of child parents and work to cooperate with them in order to develop the linguistic and cultural skills of the child.
7. To avoid making intolerant judgments about other cultures [22].
8. To avoid issuing fanatic judgement about Islamic culture.

The way in which cultural aspects and concepts are presented would highlight the importance of the Arab-Islamic culture and work to form positive attitudes towards it. Language is the womb that generates thought that expresses culture, and it is the system through which a child acquires culture automatically [23].

The role of the teacher in interacting with children in light of the spread of the Covid-19 virus:

It is normal that when the teacher implements the activities, to deal with the children in one way or another, and interaction occurs, so there may be an overlap between the steps of implementation and interaction, and what the teacher should take into account when interacting with the children, taking into account the following:-

a- Establishing a trust relationship with children, and encouraging them to do similar work with friends.

1. Encouraging group work. The teacher should prepare activities that encourage group work and support it with fun atmosphere. He can adapt these activities to match the subject of language and children's interests, so that children feel comfortable [7]

2. Providing opportunities for children to use their bodies, through activities, to express and present their experiences in a second language community. This creates an atmosphere of fun, and helps in the process of communication between children [24].

3. Developing the spirit of initiative, asking questions, and discussion, as this supports self-confidence and gives the opportunity for more activities [14].

4. Giving the opportunity for parents to participate in the activities, which benefits the child, and feels that he is the focus of attention and care by both the teacher and the parents, and the interaction between the teacher and child parents should occur through the implementation of activities in the classroom with the presence and participation of the parents if possible [15].

5. Distributing attention to the children of the whole class, so the teacher does not let one "child" monopolize the attention, rather all children must be encouraged, so that the children feel equality [25], and the teacher can choose a child in each lesson to help him in a task in the lesson, as this makes the child feel confident, and his use of language develops.

2- The teacher's role in strengthening in light of the spread of the Covid 19 virus:

Effective evaluation begins with taking into account the objectives, actions, experiences, and attitudes of children. Hence, many approaches can be used in the evaluation process. The results of these approaches lead to a deeper understanding of children, and to make sound decisions about their education [26], and although there are many approaches to assess children in Learning a foreign language, but this role is very difficult when implementing it, for the following reasons [27]:

A- Children practice language activities among other activities.

2- Children are in constant motion, so notes should be stored in various situations, perhaps many times.

3- The difference and diversity of the cultural environment for children according to the mother tongue.

Hence, the second language teacher for children, when performing the evaluation process, should do the following [9]:

1. To understand the process of individual differences between children and use this understanding in determining evaluation methods.

2. Using the method of observation to improve the educational process, and to reach the maximum level of education [28].

3. To expand observational methods and models to include cognitive, emotional and motor abilities in linguistic situations [27].

4. To know the child's level in the first language, because this helps in acquiring the second language [38].

5. The teacher's knowledge of the child's level in the first language necessarily requires exchanging information with both the language teacher and the family, so the process of exchanging information is considered as a part of the evaluation process [30].

2- The teacher should use the "work sampling system" where this system helps the teacher in comparing the objectives of the curriculum, the level of achievement of the student, in addition to helping him to strengthen the student's motivation, and enhance the learning process, and it also helps the student to know the relationship Between what to learn to do and how to do it. Three elements are associated with this system [5]:

A - Preparing lists and guidelines for the process of developing the child's skills.

B - Preparing the Portfolio of assessment bag.

c - Preparing reports.

A - Preparation of lists: in which the teacher prepares lists of the activities carried out in the classroom and the extent to which the student is expected to achieve in this activity, so that these lists include all aspects of the curriculum and all the skills that the teacher wishes to develop in the student, as well as the cognitive, emotional, and motor areas that is intended to apply. The teacher has a number of points for each child in a gradual manner, consistent with his level of skill or achievement. These lists can be applied by more than one teacher, and for greater accuracy. Then the data and results obtained are documented, and then the teacher can obtain the level of child's progress in language acquisition, control skill, academic achievement, and social interaction.

B - Preparing the Portfolio: It means a follow-up file for all the activities that the student performs in a specific period of time in Arabic as a second language. Preparing Portfolio is required to determine pivotal titems and Individualized Items.

The pivotal items represents how the child performs during the school year in the dimensions of language, mathematical thinking, science, social studies, and arts, and these items clarify certain parts of learning within each dimension.

For the individual items, it reflects the child's goals, interests, and abilities in various fields. For example, it shows the significance of what is happening: the child may write, draw, or try to write a short story...etc. These activities suggest many aspects of education, thinking, and performance. The child's journal entries and his ability to express his ideas, how he organizes the context of writing, the extent of errors and misspellings, and the scope of vocabulary use. The newspaper article also presents daily events, interests and trends, by shielding the light on the daily events of the child [31].

In addition, preparing the portfolio is an important step for children, as it not only clarifies the process of the differences between them, but also enables them to understand the effective system in evaluating their own actions, by discussing it together, and then both the teacher and the child can make educational decisions, and the portfolio reflects Class activities, as it leads to new activities aimed at the progress of the child, taking into account his interests, as it draws a "cross-section" of the child's life in the classroom, including its contradictions and patterns ... etc. The Portfolio not only collects the child's work, but clearly displays progress, achievement, and participation in the class.

It appears in the form of a report on the student's performance in all aspects of the curriculum, and it can be in the form of a short story book, brief judgments, completion of specific phrases, or fill out previously defined criteria. The preparer can make these reports monthly, or three times annually, or At the end of each semester.

The role of the teacher in teaching language skills in light of the spread of the Covid-19 virus [32]:

The communicative approach in teaching languages around the world aims to enable learners to have communicative competencies, which is one of the most important goals of teaching a foreign or second language, and to develop procedures for teaching the four language skills, through the overlap between language and communication [33], and the basic language skills in communication: listening and reading. Reception, speaking and writing skills: the two skills of sending, where the communication approach depends on the skills of reception and transmission together." These skills take place in a framework of active interaction between the sender and the receiver, and the role of the receiver (the meeting) is no less than a positive role towards what he hears or reads [34], And it receives knowledge in order to build knowledge and make meaning [35] Hence, the relationship of language reception and transmission is a relationship of cause and effect. The efficiency of reception affects the efficiency of transmission, in other words: the quality of inputs affects the presence of outputs [6].

The teacher must meditate with these skills in an integrated manner, the integration of language skills is one of the most important modern trends in language education, and one of the justifications for this trend is that the language arts are nothing but the language itself.

As a whole, its functions are fully clear, and similarly, the grammatical, morphological, or rhetoric rule when studied in a natural linguistic situation leads to the speed of learning, and to the learner's awareness of its function [10]. Also, teaching the Arabic language is more effective if all language arts are dealt with as basic units, a means of communication and systems to these skills in a synergistic manner [12], so the teacher must link these skills as a "strong" support for other skills, or an "extended" activity that can be used in other skills integration.the language arts allows the formation of a cognitive background, enabling children to interact with their peers and adults, by reading stories, listening to them, speaking and writing about them [36], and using integration in real materials develops second language skills and increases the motivation of learners [37]. Therefore, the division of skills into: listening, speaking, reading, writing in this research is only abstraction that helps the teacher in the process of teaching Arabic to non-native speakers. Here are to follow many rules, including [38]:

- The rules should not be studied separately from the rest of the skills, especially in the childhood stage. Rather, they must be take care of skill to be familiar in child's ear, and tongue, and writes it with his pen.
- Provide language skills through the curriculum activities, so that these activities take into account the integration between diverse skills.

The role of the teacher in teaching listening in light of the spread of the Covid-19 virus:

The teacher's ability to listen to the Arabic language is one of the most important ingredients for success when teaching non-native speakers. The listening skill contributes to the development of language skills: speaking, reading, and writing, and developing linguistic wealth. Listening is one of the most commonly used language skills, as it represents About 45% of the total linguistic activity, while speaking represents 30%, reading 16% and writing 9% [12]. The students who are trained in good listening at the primary stage are better able to listen well in the following stages [36] and the student's superiority in the whole study is due to that he excelled in listening skills [11].

Therefore, the teacher should observe the following when teaching the child to listen in the second language in light of the spread of the Covid-19 virus [9]:

- Develops his listening skill, so that he is a good listener and thus becomes a role model in front of his students.
- He listens carefully to every child every day, especially children who have some problems of understanding or listening.
- Encourage children to use as many words as possible by smiling, nodding, holding eyes, or paying attention to what the child says.
- The teacher asks the child in a friendly manner, and listens carefully to the child's answers.
- Helps children acquire the etiquette of listening, as studies by Wadha Ali Suwaidi and Fayza Al-Sayyid Muhammad confirmed that the listeners' keenness to practice listening etiquette contributes to acquiring correct pronunciation and expressive performance in reading aloud [8].
- The teacher employs natural attitudes in teaching listening, perhaps through the media and programs with sound language. Several studies have indicated this [9].
- Presents activities, games and listening in a varied and fun way, for example, "At the time of the children sitting in a circle, the children listen to the sounds around them: the sound of a car, a siren...etc. The teacher asks them to mention this sound, and to whom, then mention a different sound,and so on [28].
- He uses diverse audible intructional means such as a child's voice, his voice, a recorded voice...etc. [11].

The role of the teacher in teaching speaking in light of the spread of the Covid-19 virus:

Speaking skill is one of the most frequently used skills by the teacher in the teaching process. Children are able to express their feelings, communicate, pay attention, make reactions, discover the language used in the classroom, and even create verbal funny situations. This applies to the foreign language, as the skills learned in one language are transferred to the other, and thus if the children speak their mother language in a proper manner, this enables them to learn to speak in the new language in a more specialized manner [2].

The speaking process is not a simple movement that occurs suddenly." Rather, it takes place in several steps: stimulation - thinking - formulation - pronunciation [39], so the second language teacher deduces the content of speech in the non-linguistic expressions and impressions of children, as the second language is just another way to express what they want to Express it [5], so the teacher should do the following: -

- Plans well for the new language used, in order to reach the meaning effectively, taking into account to follow the rules of phonology of the language.
- Enhancing the meaning of the spoken language by: body language (gestures, representation of movements, facial expression), visual keys (signs, symbols, pictures, cards, schematic drawing), diversity in the speed of performance, taking into account the tone, paraphrasing in other words while preserving the meaning, the use of similarities and structures, and written reinforcement such as common words and sentences wall hangings [32].
- It provides opportunities for emotional communication between children, through various fields: play, hobbies, religion … etc., and through normal situations which has been confirmed by many studies [26].
- Encouraging students to have conversations between him and them and each other and motivate them to ask questions and comments [33].

The role of the teacher in teaching reading in light of the spread of the Covid-19 virus:

There are many definitions of the reading process in light of the spread of the Covid-19 virus, and the definition that the researcher adopt: reading is a process of thinking and insight, as this concept includes the following skills:

1. Seeing with the eye with reflection and contemplation.
2. Understanding.

In light of the previous definition, the characteristics of a good reader includes the following:

1. Remember the parts of the reading material and understand its meaning
2. Analysis and interpretation of the article.
3. Criticizing the reading material in the light of objective criteria.
4. Evaluation of the material read [35].

Since the main objective of reading is to develop the student's ability to read, understand and respond to written language, the teacher of Arabic for non-native speakers should have the following abilities and skills:

1. The ability to quickly know the meaning of the written symbols of the language.
2. The ability to adjust the reading speed to suit the nature and purpose of the material being read.
3. The ability to control the basic skills of reading, so that he uses the convenient skills for the activities.
4. The ability to remember the above.
5. The ability to distinguish between linguistic material that requires meditative and analytical reading and those that do not require more than passing attention [22].

6. The ability to clearly define goals.

7. The ability to support previous information.

8. Use one language phare either Arabic or mother language, and not using an multi-language phrases. The teacher has to explain the meanings by different means, including using objects (pen, book, puppets ... etc.) to clarify the meaning, including role playing, as confirmed by the study of Christina Maxwell and Rebecca BrentIt also helps to develop confidence in the student and the use of facial and hand signs, and colored pictures [5].

9. Mastering the skills of reading aloud and silent, and knowing the problems that non-Arabic speakers face in this field.

10. The ability to choose the appropriate materials for reading, taking into account some conditions, including:

    a. It should be in the correct Arabic language, taking into account that it does not contain words from a particular dialect, or a specific vernacular.

    b. It matches the interests and tendencies of the children.

    c. The text should contain vocabulary related to the field for which they want to learn Arabic.

    d. To develop in the children a certain moral values or present an Arab and Islamic cultural pattern without transgressing the culture of others.

    e. The text should be graded in terms of vocabulary, structure, and type. He begins with what he has studied orally, and what they can use in actual communication situations [29].

    f. The ability to provide opportunities for reading in normal situations. There are many studies [14] that confirm the effectiveness of using practicing Naturalistically in language learning, whether it is through reading newspapers, advertisements, or signs…etc. Those chances require encouraging motivation of students, to focus and pay attention, and to develop their abilities to predict, anticipate what they will read, and determine the purpose of reading. It also requires a real linguistic environment in which the child exercises the language as reception and "transmission".

There are other general, sensory and psychological abilities including: general health, visual strength, emotional stability, intelligence, ability to focus, strong attention to the practice of reading, and interest in the content of the material.

The role of the teacher in teaching writing in light of the spread of the Covid-19 virus:

Writing is a means of linguistic communication such as listening, speaking, and reading. The education of writing is focused on three matters [28]:

1. Pupils' ability to write correctly.
2. The clarity and beauty of hand writing.
3. The ability to express ideas clearly and concisely.

Since the main objective of the writing process in the second language is to develop the student's ability to express and communicate through the written language, the teacher of Arabic for non-native speakers should [29]:

1. Define clearly the goals in the writing process and to familiarize the students with these goals as much as possible.

2. Defines the steps to be followed in teaching writing.

3. It employs what is presented to the students, in the sense that the student is not asked to write something unless he is familiar with it by hearing, distinguishes it by speaking, and recognizes it in writing.

4. He chooses the appropriate time to teach writing, so that it is neither early nor late. Rather, it stands in a "middle" position. It is sufficient for the student to have a simple balance of a group of words and sentences that he uses in situations of meaningful communication, understanding and comprehensive.

5. Leaving children free to choose their favorite writing activities. This is better than memorizing a set of activities that include specific structures and sentences, which then become repetitive and traditional methods (clichés).

6. He trains children to write in a good handwriting, by following the rules of Arabic calligraphy, taking into account the use of grammar rules for writing.

7. The principle of gradual practice is taken into account when teaching writing, whether in terms of choosing the linguistic material, or in terms of the method of teaching.

8. Recognize the problems that children face when teaching them to write Arabic [37], and how to deal with them. Such as:

    - Practice writing from right to left.
    - Training on the forms of Arabic letters, as they differ according to their location in the word.
    - Training to write some letters that are written and not pronounced.
    - Training in punctuation, as it differs from the English language.
    - Training to distinguish between similar letters.
    - Training in writing through texts that children have already read and understood.

2   Conclusion

After this wandering through the sources and dimensions of the subject, we can refer to its most important results, which are as follows:

1. A good teacher for childhood must be prepared and trained, whereby he acquires three sets of skills: personal, scientific, and professional in light of the spread of the Covid-19 virus, through a multidimensional and angular diagnosis of the reality of teaching Arabic to non-native children and through ambitious and promising visions of development and positive change. Towards a better future for the Arabic language teacher at the local and global levels; Therefore, it is necessary to prepare the childhood teacher with preparation that transcends the traditional boundaries that he used to deal with crises, epidemics and pandemics with an educational spirit armed with the latest advances in electronic and technological sciences.

2. The Teacher of Arabic Language for Non-Native Speakers Must Focus in His Personality on Three Dimensions:

   a.   Specialized dimension
   b.   Occupational dimension
   c.   The cultural dimension

3.   The Characteristics of the Arabic Language Teacher for Non-Native Speakers Are Attracted by Three Axes:
     Motivation axis: An effective teacher has a motivated personality.
     Orientation to success: A successful teacher has high expectations for himself and his children.
     The focus of professional behavior: seriousness at work, goal-orientation, careful, regular, tidy, adaptive, flexible, and knowledgeable. He bases his work on scientific theories.
4.   The teacher of Arabic for non-native speakers is a good role model for children through his personality, and all his behaviors, thus becoming an honorable interface for the Arabic language and Islamic culture.
5.   The Arabic language teacher for non-Arab requires educational preparation after the academic preparation, which is to know the learning strategy and teaching methods, and to apply this matter in practice in front of the educational specialists.
6.   The Teacher of Arabic for Non-Native Speakers is Required to Study Very Important Subjects Such as:

     • Educational psychology, developmental psychology, educational assessment and measurement and others, so that this matter can set a forward-looking vision that will help it develop and update programs for teaching Arabic to non-native speakers in light of Covid 19.

7.   Preparing our specialized and non-specialized governmental and private institutions to update and develop their programs for teaching Arabic to non-native speakers has greatly helped the teacher of the non-native language stage to overcome this pandemic so that distance education remains the most appropriate option for this difficult stage in the history of education in the world.

References

1.   Abd al-Latif al-Qazzaz. (1991). *The needs and skills of oral communication among elementary school students in Egypt and Kuwait, PhD missionaries, unpublished*. Mansoura University.
2.   Richard, J. Thioror Rogers, doctrines and methods in teaching languages, translated by: Mahmoud Ismail Chinese and others, Al-Riyadh, Book World House.
3.   Seei, H. N. (1975). Teaching culture strategies for foreign language educations, U.S.A. National Text Book company.
4.   Richard, P. (1996). Making it happen: Interaction in the second language classroom from theory to practice, California. Long man
5.   Tegano, D. W., Moran, J. D., & Sawyers, J. K. (1991). *Creativity in early childhood classrooms* (P102). Washington, DC: National Education Association.

6. Taameya, R. A. A working guide for preparing educational materials for Arabic teaching programs.
7. Younes, F. A. (1999). *Strategies for teaching the Arabic language at the secondary stage.* Faculty of Education, Ain Shams University.
8. Abdel-Hamid, J. Teacher of the Twenty-first Century, Skills and Professional Development, House of Arab Thought.
9. Al-Naqa, M. K., & Tameya, R. The basic book for teaching the Arabic language to speakers of other languages, Makkah, Umm Al-Qura University.
10. Coleman, M., & Wallinga, C. (2000). Connecting gamilies and classrooms using family involvement webs. *Childhood Education, 76*(4).
11. Tabors, P. (1997). One child two languages: A guide for preschool educators of Children learning English as a second language, Baltiomre: Paul.H. Broods.
12. Adrian blackleg: we can't tell our stories in English: Language, Story, and Culture in the Primary School, Language, Culture, and Curriculum, Printed in Great Britain, Short Run, vol.6, No.2, 1993.
13. Al-Sayed Muhammad, F. (1995). The effect of the Noble Qur'an on the development of reading and writing skills for basic education students. *Journal of Social Educational Studies, Faculty of Education, Helwan University, 1.*
14. Taameya, R. A. (1990). Cultural content in arabic learning programs at a second cost in Islamic Societies: A proposed framework, conference toward contemporary Islamic educational theory, Amman, Jordan July 24–27, 1990.
15. Mesiels, S. J., Marseden, J. D. B., Dichtelmiler, M. M. L., & Doefman, A. (1994). The work sampling system, planning associates.
16. Couchenour, D., & Dimino, B. (1999). Teacher power, who has it, how to get it, and what to do with it. *Childhood Education, 75*(4).
17. Younes, F. A., Al-Kandari, A. (1995). *Teaching Arabic for beginners, Kuwait house of translation.*
18. Brent, R., & Anderson, P. (1993). Developing Children's classroom listening strategies, the Reading Teacher. October 1993. Vol 47, Bo. 2.
19. Caruso; Teacher Enthusiasm (1892). Behaviors reported by teachers and students paper Presented at the annual meeting of the American Research association, New York.
20. Carton, C. E., & Allen, J. (1993). Early childhood curriculum. Maccillan Publishing Company, U.S.A.
21. Ali Ahmed Madkour: Education and Technology Culture, Cairo, Arab Thought House.
22. Maxwell, C. (1997). Role play and foreign language Learning. Paper presented at the annual meeting of the Japan association of language Teachers October
23. Rubinm, D. (1980). *Teaching elementary language arts,* 2nd ed. New York Holt and Winston.
24. Robert de Bojraidek Text, Discourse and Procedure, translated by Tamam Hassan, Cairo, Alam Al-Kitab.
25. Salameh, W. Environmental education for kindergarten children, Cairo, House of Arab thought, Reference series in education and psychology.
26. Walczynski, W. (1993). Listening to authentic Polish EEIC: Ed 353831.
27. Ali Al-Suwaidi, W. (1994). The relationship between memorizing the Noble Qur'an and the level of performance of basic reading and writing skills among a sample of students of the fourth grade of primary school in the State of Qatar, Contemporary Education Magazine, Issue 22, Year 9, December 1994.
28. Beaty, J. J. (1992). *Skills for Preschool Teachers,* 4th Edition Macmillan Publishing Company printed in the U.S.A.
29. Al-Arabi, S. A.-M. Learning and teaching live languages, Beirut, Lebanon Library.
30. Samuel, J. (1996). *Mesiels : Performance in context: Assessing children, achievement at the outset of School in the five to seven years shift: The age of reason and responsibility.* University of Chicago Press.
31. Taameya, R. A. Muhammad al-Sayed Manna: Teaching Arabic in General Education: Theories and Experiments, Cairo, the House of Arab Thought.

32. Johnson, K., & Morrow, K. (1981). *Communication in the Classroom Oxford University Press.*
33. Hyson, M. (2000). Professional development, opcit, young children, November.
34. Porcaro, J. W. (2001). Integrating authentic materials and language skills in English for science and technology English Teaching Forum 2001, Vol 39, No 2.
35. Gunt, R. (1999). Making positive multicultural early childhood education happen.
36. Halliday, M. A. (1989). *Spoken and Written Language Oxford University Press.*
37. Patterson, G. B. (1999). Using newspaper articles to teach English Language skills, English teaching forum 1999 Vol 29. No. 3
38. Al-Naqa, M. K. Teaching the Arabic language and the cultural challenges facing our curricula, Faculty of Education, Ain Shams University.
39. Fathi Ali Younes, Abdullah Al-Kandari: Teaching Arabic to Beginners, Kuwait House of Translation, 1995.
40. Meisels, S. J., Dorfman, A., & Steel, D. (1885). Equity and excellence in group administered and performance based assessment. In M. T. Nettles, & A. L. Boston (Eds.), *Equity educational assessment and testing.* Kulwer Academic Publishers.

Controlling the Implementation of Precautionary Measures in Conducting Exams Within a University During the COVID-19 Pandemic

Abobakr Aljuwaiber⬤, Abdelkader Derbali⬤, Ahmed Elnagar⬤, and Roaa Aldhahri⬤

1 Introduction

From time to time, countries and organizations from local to global levels face challenges that require concerted efforts and an exchange of experiences and information to alleviate the potential risks. In December 2019, SARS-CoV-2 first appeared in the city of Wuhan's seafood market in China. The virus skyrocketed and spread throughout China and then throughout the world [26]. Consequently, many countries took precautions and suspended activities (e.g., workplaces, schools, restaurants, and commercial shops); moreover, some countries imposed a curfew. Under these circumstances, collaboration on efforts and the exchange of information between governmental organizations is necessary to achieve common goals. Furthermore, improving effective responses to COVID-19 outbreaks requires comprehending how different organizations can harness their potential to mitigate risk and limit the virus's spread by sharing needed information, data, and practices. The sharing of knowledge from different entities can assist policymakers in making the best possible decisions.

The newly emerging SARS-CoV-2 virus became a global pandemic due to its spread worldwide. Although the number of individuals who recover increases daily, the number of cases continues to rise due to this virus's characteristics, enabling it to spread rapidly. Health institutions in all countries have shared ways to curb the spread. Likewise, the education sector has not been isolated from this pandemic's impact, which forced schools and universities to transfer teaching to online learning.

However, most research has mainly focused on studying the psychological factors of COVID-19 on university students [19] or health facets (e.g., anxiety and mental;

A. Aljuwaiber · A. Derbali (✉) · R. Aldhahri
Taibah University, Madinah, Saudi Arabia

A. Elnagar
Saudi Arabia and Suez Canal University, Ismailia, Egypt

© The Author(s), under exclusive license to Springer Nature Switzerland AG 2022 321
A. Hamdan et al. (eds.), *Technologies, Artificial Intelligence and the Future of Learning Post-COVID-19*, Studies in Computational Intelligence 1019,
https://doi.org/10.1007/978-3-030-93921-2_19

[25]) or e-learning challenges [1]. Notably, research that investigates implementing exams in person during the COVID-19 pandemic remains insufficient. According to [20], "There are no best practices for universities and higher educational institutions to mimic and no known models to follow." Hence, this current study presents an original and novel contribution to the knowledge of the field of higher education by sharing empirical practices regarding implementing in-person exams for students attending college.

Considering these motivations, the present research explores students' perceptions regarding attending the first-term exams during the first semester of 2020's academic year. It elucidates the precautionary and preventive measures the Department of Administrative and Financial Sciences Technologies in Community College at Taibah University followed. Precisely, this paper aims to shed light on the following objectives:

1. To explore the department's students' views regarding the precautionary and preventive measures that have been employed for conducting in-person exams in light of COVID-19.
2. To examine the framework that the department has applied to implementing in-person exams.
3. To share empirical practices for implementing in-person exams among universities during the pandemic.

This research presents a case study of the department's efforts in executing in-person exams during the COVID-19 pandemic. The study proposes a practical framework that can assist universities and colleges in pursuing in-person exams during human-made crises. Hence, the research question is as follows: To what extent do students feel safe when taking exams through the department implemented precautionary-measure phases during the first mid-term in-person exams?

2   Literature Review

2.1   The Impact of COVID-19 on Higher Education Institutions

The impact of COVID-19 has brought lifestyle changes on individual and organizational levels due to the pandemic's continuation. Adapting to these changes has been inevitable in various circumstances, such as work or education. Universities around the world transferred their teaching systems to online learning due to lockdowns. But not all universities worldwide had adequate infrastructure or resources to facilitate online teaching due to the immediate action that was required [23]. Regardless, teachers and students had no choice but to adapt to the online mode of learning. According to Aristovnik et al. [5], students were able to deal with technology effectively, which helped them considerably when overcoming the difficulties of transitioning to online learning.

Several studies (e.g., [15, 16]) identified factors that impacted higher education institutions during the COVID-19 pandemic. These factors include the following: (a) a reduction in international enrollment due to the difficulty of international students' mobility; (b) a disruption of the academic calendar; (c) cancellations of local and international conferences; (d) a destabilization of all educational activities, such as examinations and admission; (e) budget cuts; and (f) technological learning gap between lecturers and students. Additionally, COVID-19 has negatively influenced researchers by eliminating travel or opportunities to work with others—especially for scientific groups and laboratory research. However, from a positive standpoint, academics obtained sufficient time for theoretical research [16, 20].

The pandemic has also impacted some academic programs more than others. Particularly concerning was when the university lockdowns interrupted medical and clinical subjects. Cairney-Hill et al. [6] asserted that although remote learning was imperative during this time, it could not replace clinical experience for medical students. Several universities also suspended semesters and final examinations, while others utilized continuous assessments via online classes. Hence, educators were forced to alter the types of assessment to suit the online mode [22].

In Saudi Arabia and around the world, adjusting to distance learning to prevent the spread of COVID-19 remains a challenge for all educators. Since March 8, 2020, the Saudi Arabian Ministry of Education has directed all schools, colleges, and universities to indefinitely suspend in-person classes and activate distance learning as an alternative [4]. Consequently, traditional teaching in universities was replaced by virtual classrooms. Some learning management systems were applied in Saudi universities, such as Blackboard and Microsoft Teams.

Furthermore, academic departments across all Saudi Arabian universities accepted the responsibility of handling this unusual situation to ensure the quality of teaching and course learning outcomes by using appropriate means of remote learning [4]. In Saudi universities, the semester consists of 15 academic weeks, excluding the final examination period. Based on the decision to suspend in-person study in Saudi Arabia, the first semester of the 2019–2020 academic year was held remotely for the remaining eight academic weeks. The Ministry of Education directed assessment modification to all Saudi universities through university admissions by assigning 80% of the total course grade to coursework. The remaining 20% was allocated to final assessments.

The acquired experience in remote teaching during the first semester due to the pandemic has helped many academic departments manage their learning process in subsequent semesters, as the pandemic is ongoing [4]. Thus, remote teaching continued during the second semester of the 2019–2020 academic year. With a decrease in COVID-19 cases in many Saudi Arabian regions (according to the Ministry of Health's daily reports), Saudi universities continued giving lectures remotely, especially for courses that did not require in-person attendance. This present study, therefore, reflects on the university department experience while performing the first mid-term exams in person and the precautionary measures that the academic staff and students followed.

2.2   Teaching and Learning in Universities During COVID-19 Pandemic

The COVID-19 pandemic has swept across most countries. In an effort to contain the virus, educational institutions ceased in-person education, which allows physical closeness and an opportunity for infections to spread. The solution to the shutdown was adopting e-learning or distance education: Consequently, approximately 1.5 billion children and youth in 188 countries around the world have had to stay at home and learn online [2].

Aljaser [3] identified among Saudi Arabian fifth-grade students the e-learning environment's effectiveness in developing academic achievement of English-language learning. The e-learning environment was designed to use a test and a scale to assess the trend toward learning the English language. The quasi-experimental curriculum was applied to a sample of fifth-grade students. A control group was taught using the traditional method, and an experimental group was taught in the e-learning environment. The study results showed statistically significant differences in favor of the experimental group in both the post-achievement test and the measure of the trend toward learning the English language.

Both [27, 28] argued that e-learning could be effective if teachers do the following: organize the educational content, choose the appropriate educational means, define measurement tools, individualize learning, meet different learning needs and styles, and participate in professional development.

Draissi and Yong [9] examined the COVID-19 response plan and implementation of distance education in Moroccan universities. The study used a content analysis approach. The results indicated that the COVID-19 pandemic is presenting universities with an ongoing challenge to continue overcoming the difficulties facing both students and professors and to invest in scientific research to continue their vaccine-development efforts.

Sahu [22] reviewed the impact of university closures due to the COVID-19 pandemic on the education and mental health of university students. As COVID-19 spread rapidly around the world, many universities postponed or canceled activities, and universities have taken extensive measures to protect all students and staff from the highly contagious disease. Faculty members have moved to an electronic teaching system, and the research highlighted the potential impact of COVID-19's spread on education and mental health. For students, the study results showed that universities should enforce strict rules to slow the virus's spread.

Yulia [27] presented a descriptive study aimed at clarifying the ways the COVID-19 pandemic has reshaped education in Indonesia. The researcher explained the types and strategies of Internet learning that teachers worldwide have used due to the closure of universities. The study showed the advantages and effectiveness of Internet learning and concluded that the COVID-19 epidemic had an almost immediate impact on the education system, as the traditional method of education was significantly revised to stop the virus's spread. Thus, learning through the Internet was vital, as it supported learning from home and reduced the spread of the virus by eliminating face-to-face contact.

Hodges et al. [13] researched the difference between distance teaching in emergencies and online education. Their study concluded that online learning experiences differ from learning in emergencies in terms of the quality of planning and the online courses provided in response to a crisis or disaster. Colleges and universities have focused on preserving education during the COVID-19 pandemic.

Favale et al. [11] analyzed the impact of the lockdown on the campus computer network traffic and e-learning during the COVID-19 pandemic and how the epidemic changed the traffic on the campus of a university in Turin, Italy. Their results indicated that the Internet enables individuals to deal with sudden needs and that remote work, e-learning, and online collaboration platforms are viable solutions to deal with the social distancing policy during the COVID-19 pandemic. These options also helped manage network traffic while instituting e-learning.

2.3   Factors and Challenges Influencing Students Taking Exams During COVID-19

The examination system has many factors, which directly or indirectly affects students' performance on examinations. Singha [24] indicated that the main factors that affect a student's performance on exams are (a) inadequate question structure, (b) type of questions, (c) subjective scores, (d) individual variations in the evaluation of answers, (e) the direction of dishonest staff, and (f) scoring errors. These factors create hurdles for measuring actual student performance. As a result, many students fail exams. However, a student's failure is not their fate; instead, some problems become an obstacle to their success. Sometimes, deserving students are denied opportunities to display their knowledge despite their excellent IQs. Rasul and Bukhsh [21] classified these factors into (a) extrinsic factors, (b) intrinsic factors, (c) personal factors, and (d) various factors. Extrinsic factors comprise eight subfactors: temperature, the distance between rows and lines of students, lighting, lack of airflow, the sound inside and outside the examination hall, seating positions, invigilation staff, and lack of discipline. Intrinsic factors include the style or pattern of questions, strict grading, and inappropriately sequenced questions. Personal factors include the selection of questions, tension, family problems, overconfidence, or a lack of confidence. Various factors include handwriting, exam phobia, presentation of material, and selective study.

Mushtaq and Khan [18] classified the factors that affect students' exam performance into four categories:

1. Communication: Student performance is affected by communication skills; thus, communication can be considered a variable that can be positively related to student performance in open learning.
2. Learning facilities: Students' performance corresponds with an appropriate scholastic climate such as the library, the PC lab, and the overall establishment.
3. Proper guidance: Students face many issues in creating positive study attitudes and habits. The focus is on how a student can improve these attitudes and habits to improve scholastic accomplishment. The students who receive appropriate parental or guardian guidance have been found to perform well on tests. Direction from the educator additionally influences the student's performance. Thus, direction from guardians and educators seems to influence student success.
4. Family stress: The student's performance relies upon various elements that include their class participation, family income, parents' schooling, teacher–student ratio, educator qualifications, gender, and the school's distance. Thus, student achievement is improved if the ideal wellbeing-related hindrances are low.

Jabor and Hamid [14] showed that two types of factors affect student performance during studying and test-taking: (a) internal study environment factors incorporate the climate of the class, class size, class plans, learning offices, class test results, the unpredictability of the course material, schoolwork, innovations used in the class, test frameworks, technology, and instructor's role inside the class; (b) external study environment factors incorporate extracurricular exercises, work, finances, family problems, and social or other issues.

The COVID-19 pandemic has constrained numerous proactive tasks around the world, including instructive exercises. The present circumstance has left instructive organizations no choice but to move to web-based learning. Although web learning is anything but a novel wonder, this abrupt change into online learning has presented significant difficulties for instructive exercises worldwide, and instructive organizations, instructors, and students have generally been unprepared for this surprising interruption to the customary instructing and learning strategies; this digital transformation in the provision of education has come with many logistical challenges and attitudinal adjustments [12].

According to Denial [8], many other challenges to online learning have emerged during the COVID-19 crisis. For instance, online or distance learning amid a pandemic creates more tension and frustration, such as the isolation of students who have lost the opportunity for peer interaction. An unprecedented shift to online learning has also occurred, increasing concerns about cybersecurity, cyberbullying, online violence and exploitation, difficulties, and other psychological issues; thus, the uncertainties associated with online learning during the COVID-19 pandemic are numerous.

3   Study Context

In the context of this present study, Saudi Arabia was not alone in facing the effects of the COVID-19 pandemic. On March 2, 2020, the first case of COVID-19 appeared in Saudi Arabia—brought by a citizen who arrived from abroad. The tests' results confirmed that social contact with the initial patient infected other citizens. Accordingly, the government employed sequential measures to control the virus's spread across the country.

Taibah University is one of the universities in Saudi Arabia that issued a manual listing precautionary and preventive measures for dealing with COVID-19, including measures related to students' attendance to laboratories and the semester-exam periods. The Department of Administrative and Financial Sciences and Technologies is one of the community college departments at Taibah University. This section provides the measures and instructions that the department has taken to implement the first mid-term exams for the 2020 academic year during the ongoing pandemic. The department board approved the implementation of the first mid-term exams in person, based on the information issued by the University Vice-Presidency for Educational Affairs regarding the arrangements necessary to conduct periodic, mid-term, and final exams in person for all remote theoretical courses and according to the scope in which the university operates (orange scope, plan B: partial attendance of male and female university students). The goal of preparing an integrated plan was to ensure successful exam implementation: Thus, apply the plan's measures and precautions to return to in-person university studies for the first semester of the 2020 academic year. Figure 1 illustrates the pre-steps that the department took to prepare for the first mid-term exam.

The department's Schedules and Exams Committee recommendations: These recommendations were recorded in the meeting minutes and concerned holding the

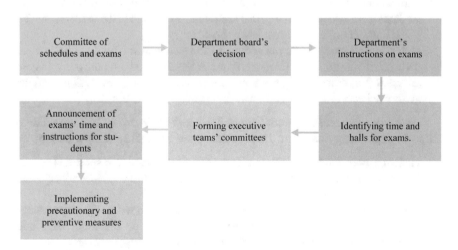

Fig. 1 Preparation steps for attendance exams within the college

first mid-term examination and applying precautionary and preventive protocols in conducting the tests in person.

The department board's decision: The department board decided to implement most of the department's courses' first exams to males in person. They committed to following the distribution of grades mentioned in the course descriptions for the semester work and mid-term and final exams in the department's programs. These descriptions were as follows:

1. Perform the test within one week.
2. Cancel all lectures during the exams' week.
3. Take one test daily for each two-academic program (office management and automated accounting).
4. Conduct one remote test per course.

The department's instructions on exam measures: Two primary directives were given to students:

1. Academic instructions were provided for the exam.
2. Preventive precautionary instructions were specified for in-person exams.

Identification of times and classrooms for exams: This step consisted of four key elements:

1. **Exam classrooms**: Three large classrooms were allocated (accommodating up to 50 students in the normal teaching mode). Additionally, a small hall was designated for students with health problems or chronic conditions.
2. **Exam times**: The first period (9:00–10:00 a.m.) was dedicated to the office management program. The second period (11:00 a.m.–12:00 p.m.) was dedicated to the automated accounting program.
3. **Student distribution**: The students were directed to the exam halls so that the number of students would not exceed 20 students per hall. The seats allocated to students were also numbered so that each student sat in the seat that corresponded to his number on the test attendance sheet.
4. **Determining exam halls according to the programs**: The three halls were dedicated to the third-level students based on the allocated time and the academic program. The office management program was designated two sections, and the automated accounting program was designated three sections.

Formation of executive exams' committees and their tasks: Two committees within the department were formed to provide support during the exam period (Fig. 2).

1. **Health Supervision Committee**: This committee aimed to inspect and ensure that the department's staff and students followed the precautionary and preventive protocols.
2. **Examination Control Committee**: This committee aimed to supervise implementing precautionary and preventive protocols inside the exam halls and regularly excused students from the halls after completing the exam.

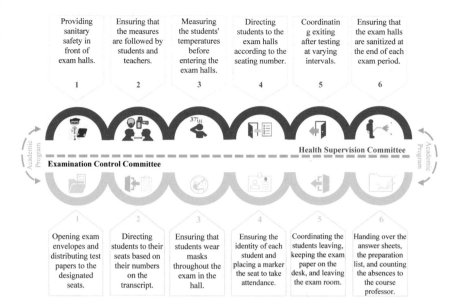

Providing sanitary safety in front of exam halls.	Ensuring that the measures are followed by students and teachers.	Measuring the students' temperatures before entering the exam halls.	Directing students to the exam halls according to the seating number.	Coordinating exiting after testing at varying intervals.	Ensuring that the exam halls are sanitized at the end of each exam period.
1	2	3	4	5	6

Opening exam envelopes and distributing test papers to the designated seats.	Directing students to their seats based on their numbers on the transcript.	Ensuring that students wear masks throughout the exam in the hall.	Ensuring the identity of each student and placing a marker the seat to take attendance.	Coordinating the students leaving, keeping the exam paper on the desk, and leaving the exam room.	Handing over the answer sheets, the preparation list, and counting the absences to the course professor.
1	2	3	4	5	6

Fig. 2 Executive exams' committees and their tasks during the exam period

Announcement of exam times and instructions for students: Each lecturer was required to make an announcement for students on the Blackboard platform that included the following:

1. Each course's exam date and the means to take it were specified (remotely or in person).
2. Instructions and guidelines for students to take the first mid-term in-person exam at the college specified academic instructions for the exam and preventive, precautionary COVID-19 procedures.

Implementing phases of the precautionary and preventive COVID-19 measures during the exam period: The department implemented several steps to ensure that its academic staff and students were fully prepared during the exam period. Figure 3 shows these measures' phases that were employed during the exam period.

4 Methodology

4.1 Sample and Data Collection Process

The target population was all third-level students in the Department of Administrative and Financial Sciences and Techniques in the Community College of Taibah University. According to the college's registration records, a total of 88 students

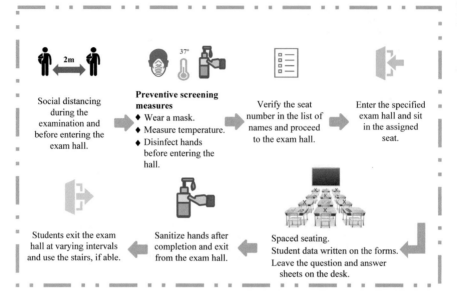

Fig. 3 Phases of the precautionary and preventive measures during the exam period

were registered in both academic programs. The total number of participants was 75, amounting to a response rate of 85.2%. A survey research design was adopted to gather information from students, using a web-based questionnaire to assess the department's experience in implementing the framework that was applied during the first-period examinations at the college. Students were not asked to provide any identifying background or personal information. Therefore, participation in the survey was entirely voluntary and anonymous to reduce bias. The questionnaire was designed to assess the department's precautionary and preventive measures during the first mid-term exam at the college.

The descriptive-analytical method was used for its suitability for such types of studies. The study population consisted of all third-level students enrolled in the major of Office Management and Automated Accounting in the Department of Administrative and Financial Sciences and Techniques of the Community College of Taibah University. The researchers distributed the study tool electronically to the student sample, retrieving 75 questionnaires. Table 1 shows the sample members' distribution.

4.2 Measures

This study aimed to measure the effectiveness of precautionary measures taken during the first period of testing in the Department of Administrative and Financial Sciences

Table 1 Demographic profile of the respondents

Variables		Frequency	Percentage (%)
Specialization	Office management	30	40
	Automated accounting	45	60
Place of residence	Inside Medina	69	92
	Outside Medina	6	8
Total		75	100

and Techniques of the Community College of Taibah University during the COVID-19 pandemic. The questionnaire was divided into five main sections: The first section consisted of basic information, and the other four sections consisted of questions with answers that used a Likert-type scale: *Strongly agree, agree, neutral, disagree, strongly disagree.*

The questionnaire was composed of five sections: Sect. 1 (S1) contained the data the student initially provided on the questionnaire data; Sect. 2 (S2) measured the student's willingness to take the first mid-term exam in person during the COVID-19 pandemic; Sect. 3 (S3) measured the effectiveness of the department's precautionary measures before the student entered the exam hall; Sect. 4 (S4) measured the precautionary measures' effectiveness after the student entered the exam hall; Sect. 5 (S5) measured the effectiveness of conducting the first mid-term in-person exam during the COVID-19 pandemic at the Department of Administrative and Financial Sciences and Techniques in the Community College of Taibah University.

4.3 Statistical Analysis

A tool's validity is based on whether the tool measures what it is intended to measure. The researchers presented the questionnaire to two third-level students in the two majors. Four students from the two majors were participated in the pilot study to evaluate the clearness and feasibility of questionnaire. The questionnaire's paragraphs were modified according to the students' feedback via notes and proposed amendments. The questionnaire was reformulated in its final form accordingly to include 23 questions.

To verify the measurement instrument's stability, the researcher examined the internal consistency and stability of the questionnaire items by calculating Cronbach's alpha coefficient, shown in Table 2.

Table 2 illustrates that the alpha value to measure the effectiveness of the precautionary measures while performing the department's first mid-term exam in person

Table 2 Coefficient of reliability and validity of the scale constructs

No	Constructs	No. of items	Cronbach's Alpha	Validity
1	Student willingness to take the exams in person (S2)	5	0.969	0.984
2	Effectiveness of the precautionary measures inside the college and before entering the examination halls (S3)	5	0.892	0.944
3	Effectiveness of the precautionary measures after students enter the test rooms (S4)	6	0.887	0.942
4	Effectiveness of conducting examinations in person (S5)	5	0.898	0.948
Overall Total Scale		21	0.934	0.966

(during the COVID-19 pandemic) ranged between 0.887 and 0.969. The measurement of students' readiness to take the exam after receiving information from the department had the highest reliability coefficient. In contrast, the measurement of the precautionary measures' effectiveness after entering the exam hall received the lowest reliability coefficient. The α value over the total score was 0.934, which indicates the study's measuring instrument obtained high accuracy.

After collecting the study data, the researcher reviewed it before entering the data into the computer. Verbal answers were converted numerically. For example, a 5-point Likert scale was implemented, with the answers ranging from 1 (strongly disagree) to 5 (strongly agree). Thus, the questionnaire became a measure of the precautionary measures' effectiveness during the first mid-term in-person exam implementation at the community college during the COVID-19 pandemic. The data's statistical processing was done by extracting numbers, arithmetic averages, standard deviations, the Cronbach-alpha reliability equation, and regression analysis using the SPSS (Statistical Packages for Social Sciences).

5 Findings

This researcher's findings are presented through the study sample members' responses to questions about the effectiveness of the precautionary measures while performing the first mid-term exam in person in the Department of Administrative and Financial Sciences and Techniques in the Community College of Taibah University during the COVID-19 pandemic. The mean value of the phrases in the study tool can be interpreted in Table 3 as follows:

Table 4 illustrates the students' willingness to take the department's first mid-term exam in person during the COVID-19 pandemic. The mean was 3.53, with a standard

Table 3 The significance of the arithmetic mean

Arithmetic mean	Significance
1.00–1.79	Very low
1.80–2.59	Low
2.60–3.39	Average
3.40–4.19	High
4.20–5.00	Very high

Table 4 Descriptive analysis for student willingness to take the exams in person (S2)

No	Items	Mean	Std. div
1	The department issued educational instructions directed to students to take the exam in person	3.44	1.50
2	The department issued precautionary and preventive instructions for students to take the exam in person	3.39	1.40
3	An announcement to the student about the exam date was posted in a timely manner	4.03	1.21
4	The advertisement method for the test was effective	3.96	1.16
5	Remote study via the Blackboard system helped the student prepare to take the test in person	2.87	1.53
Total		3.53	1.05

deviation of 1.05. The highest indications of students' readiness to take the first mid-term test regarded the department's guidelines. The highest scoring statement said, *An announcement to the student about the test date was posted in a timely manner* and averaged 4.03 with a high standard deviation of 1.21. The lowest indication of students' readiness to take the first mid-term exam in person in the department during the COVID-19 pandemic was in the phrase that stated *remote study via the Blackboard system helped the student to prepare to take the test in person*, which had a mean of 2.87 and a standard deviation of 1.53. This statement was preceded by the phrase that stated that *the department issued precautionary and preventive instructions for students to take the exam in person*, which had a mean of 3.39 and a standard deviation of 1.40. Next was the department issued educational instructions directed to students to perform the exam in person, which had a mean of 3.44 and a standard deviation of 1.50. Last was the statement expressing that *the declaration method for the test was effective* earned a mean of 3.96 and a standard deviation of 1.16.

Table 5 shows that the measurement of the department's precautionary measures' effectiveness before the students entered the examination room was generally average

Table 5 Descriptive analysis for the effectiveness of the precautionary measures inside the college and before entering the exam halls (S3)

No	Items	Mean	Std. div
1	The department is committed to following the precautionary measures in place at the Ministry of Health and Taibah University (such as social distancing, temperature measurement, and mask-wearing)	4.03	1.18
2	The department provided health-safety supplies in front of the exam hall (mask, sanitizer, gloves)	3.95	1.21
3	The student adhered to the precautionary and preventive measures before entering the exam hall	3.92	1.28
4	The student's temperature was taken before entering the exam hall	3.97	1.28
5	Directing the student to the exam hall according to the seat number was appropriate	4.17	1.08
Total		4.01	1.07

with a mean of 4.01 and a standard deviation of 1.07. The statement that scored the highest for measuring the effectiveness of the department's precautionary measures referred to the period before the student entered the room. The test indicated the phrase that stated *directing the student to the exam room according to the seat number was appropriate earned* a mean of 4.17 and a high standard deviation of 1.08. The statement receiving the lowest measurement of the effectiveness of the department's precautionary measures before the student entered the examination room stated, *The student committed to the precautionary and preventive measures before entering the examination room*; it obtained a mean of 3.92 and a standard deviation of 1.2. The statement that preceded this said *the department provided health-safety supplies in front of the exam halls (mask, sanitizer, gloves)*; it obtained a mean of 3.95 and a standard deviation of 1.21. The next statement read, *The student's temperature was taken before entering the exam hall*; it received an average of 3.97 and a standard deviation of 1.28. Finally, the next statement read, *The department is committed to following the precautionary measures instituted by the Ministry of Health and Taibah University (such as social distancing, temperature measurement, mask-wearing)* and had a mean of 4.03 and a standard deviation of 1.18.

Table 6 shows that the measurement of the precautionary measures' effectiveness after entering the exam hall was average, with a mean of 4.11 and a standard deviation of 1.06. The statement earning the highest score measuring the precautionary measures' effectiveness came after the student entered the examination room: it stated, *The student's identity was verified with the number assigned to the seat in a way that ensured the safety of the student and the supervising professor* with a mean of 4.29 and a high standard deviation of 1.24. Whereas the statement measuring the effectiveness of the precautionary measures receiving the lowest score came after the

Table 6 Descriptive analysis for the effectiveness of the precautionary measures after students enter the exam halls (S4)

No	Items	Mean	Std. div
1	Directing the student to the seat assigned in the examination room was appropriate	4.21	1.10
2	Distributing test questions to the seats in advance achieved the principle of health safety	4.15	1.07
3	The student was agreeable about wearing a mask throughout the test period	3.76	1.46
4	The student's identity was verified with the number assigned to the seat in a way that ensured the safety of the student and the supervising professor	4.29	1.24
5	The testing room was equipped and organized in a manner consistent with the established precautionary measures	4.17	1.20
6	The coordination of the students' exit (leaving the exam paper on the seat and leaving the exam hall) was organized	4.09	1.16
Total		4.11	1.06

student entered the exam hall: it stated, *The student was agreeable about wearing a mask throughout the test period* with a mean of 3.76 and a standard deviation of 1.46. The statement that preceded expressed, *The coordination of the students' exit (leaving the examination paper on the seat and leaving the exam hall) was organized*, with a mean of 4.09 and a standard deviation of 1.16. The next statement referred to distributing test question papers on the seats in advance to achieve the principle of health safety with a mean of 4.15 and a standard deviation of 1.07. Next was the statement, *The exam hall was equipped and organized in a manner consistent with the precautionary measures*, which had a mean of 4.17 and a standard deviation of 1.20. Finally, the next statement said, *Directing the student to the seat assigned to him in the exam hall was adequate and appropriate* and had a mean of 4.21 and a standard deviation of 1.10.

Table 7 illustrates that effectiveness measurements of conducting the first mid-term exam in person during the COVID-19 pandemic at the Department of Administrative and Financial Sciences and Techniques in the Community College of Taibah University were average, with a mean of 3.77 and a standard deviation of 1.03. The statement measuring the highest for the effectiveness of conducting the mid-term test stated that *precautionary and preventive measures were applied at the same level of quality throughout the testing period*; it had a mean of 3.97 and a high standard deviation of 1.19. The lowest scoring statement measured the effectiveness of conducting the first mid-term test and stated that *Performing the first mid-term test onsite and in person in light of the department's commitment to health conditions was a successful experience*; it had a mean of 3.44 and a standard deviation of 1.42. The preceding

Table 7 Descriptive analysis for the effectiveness of conducting in-person examinations (S5)

No	Items	Mean	Std. div
1	The precautionary and preventive measures were applied equally throughout the testing period	3.97	1.19
2	The allocation of one seat for each student throughout the exam period contributed to ensuring the student's health and safety	3.85	1.25
3	The time available for the test was the same as the number of questions asked	3.69	1.37
4	Performing the first mid-term test onsite and in person in light of the department's commitment to health conditions was a successful experience	3.44	1.42
5	There is a difference between the first mid-term in-person test score and the second mid-term test score for the same course	3.87	1.29
Total		3.77	1.03

statement expressed that *The time available for the test was appropriate for the number of questions asked*, which had a mean of 3.69 and a standard deviation of 1.37. The next statement said that *Allocating one seat to each student throughout the exam period contributed to ensuring the student's health and safety*; it had a mean of 3.85 and a standard deviation of 1.25. The final statement read, *There is a difference between the first in-person mid-term test score and the second mid-term test for the same course*, which had a mean of 3.87 and a standard deviation of 1.29.

Table 8 shows that the highest levels of correlation according to the Pearson correlation coefficient were between the second and third se sections (0.932), indicating a relationship between the part devoted to measuring the effectiveness of the precautionary measures in the department before the student entered the exam room and the part devoted to measuring the effectiveness of the precautionary measures after the student entered the exam hall. Additionally, the degree of correlation was high

Table 8 Correlation of Pearson between sections

Variable	S2	S3	S4	S5
S2	1.000			
S3	0.647**	1.000		
S4	0.658**	0.932**	1.000	
S5	0.632**	0.893**	0.911**	1.000

Note (**) Correlation is significant at 0.01 level (2-tailed)
S2 = Section 2, S3 = Section 3, and S4 = Section 4
S1 is not included purposefully

between the fourth and fifth sections (0.911). This result indicated a relationship between the part devoted to measuring the effectiveness of precautionary measures after entering the exam hall and the part devoted to measuring the effectiveness of conducting the first mid-term test in person during the COVID-19 pandemic in the Department of Administrative and Financial Sciences and Techniques in the Community College of Taibah University. The lowest degree of correlation between the second and fifth sections was 0.632 (i.e., a weak correlation between the students' readiness to take the first mid-term test in person and the effectiveness of conducting the first mid-term test in person during the COVID-19 pandemic).

6 Discussion

This study aimed to explore to what extent the students in the Department of Administrative and Financial Sciences and Techniques in the Community College of Taibah University perceived the effectiveness of the precautionary-measure phases applied while taking the first mid-term exam in person during the first semester of the 2020 academic year, amid the COVID-19 pandemic. The results revealed that the students positively perceived the department's practical phases to guide their attention to following the precautionary measures' instructions during COVID-19. This study showed that a majority of the sampled students had an overall positive experience regarding their safety during the first mid-term onsite and in-person exam. Several studies (i.e., [23]) have argued that some test types such as lab tests, practicums, and performance tests may not be conducive for online formats. This study, therefore, appears to present a well-substantiated, viable, and practical framework to conduct in-person exams.

Regarding the students' readiness to take the first mid-term exam in person during the COVID-19 pandemic, the empirical study's analysis revealed that the department's precautionary measures established significant positive perceptions of readiness to take the exams among students. The results confirm that participants' perceptions of the announcement of the exam's dates in a timely manner and the method used for exam announcement (i.e., Blackboard) were effective for students to prepare to take the exam in person. Notable to highlight is that the student perceptions that clear announcement instructions posted in advance via Blackboard (such as the exam date, time, location, and precautionary-measure information) can help prepare the student mentally and psychologically to take the exam in person. The university's precautionary guide (based on the Ministry of Health's precautionary instructions on how to deal with the COVID-19 pandemic) likely increased the students' confidence in attending the college for in-person testing. It can be concluded that taking in-person exams can occur amid the COVID-19 pandemic, provided that officials use a clear system to announce guidelines and follow the precautionary instructions. Scientific departments should emphasize these findings to faculty members, so they can inform students about the precautionary instructions that will be applied throughout the exam period.

Students reacted positively when asked about the effectiveness of the department's precautionary instructions inside the college and before entering the exam halls. Regarding the assigned seat number throughout the exam period, the students indicated that directing them to the exam hall was a helpful factor. A total of 80.6% also agreed that the department followed the precautionary instructions in place, such as social distancing, measuring temperature, and wearing masks. Such results and opinions based on the students' real experiences give the department confidence in the possibility of implementing in-person exams during crises (e.g., the COVID-19 pandemic). In a similar vein, a study by Mukherjee et al. [17] asserted that preventive measures, such as mask-wearing, social distancing, and reducing contact among individuals, are indispensable precautions to consider when reopening universities.

Additionally, the department's strict implementation of precautionary instructions before entering the examination hall contributed to ensuring students' safety. It helped them take the test with confidence. Precautionary measures that the department followed before allowing students to enter the exam halls may play a role in reassuring students and helping them feel comfortable before taking the exam. Limiting person-to-person contact also helps mitigate the spread of infections [10, 22].

The results revealed that the department's procedure of assigning a specific seat for the student during the exam period was considered significant, concerning the effectiveness of the precautionary measures implemented while students took the exam in person. Similarly, placing the students' school ID on the seat where the exam proctor can verify their identities was considered essential to ensuring a contactless procedure, assuring both the exam proctor's and the student's safety. Assigning a seat to each student for the entire exam period might have also reassured the student about attending the college for an in-person exam. Therefore, students' anxiety might have lessened regarding the possibility of infection from surface contact. Conversely, despite the department's emphasis on the importance of adhering to the exam's precautionary measures, the results showed that several students tended to not be careful about wearing a mask during the exam. Considering that the exam duration was an hour, and the windows were opened for ventilation in the exam hall, the students might have found wearing a mask throughout the exam period as a nuisance.

In the current context, the students expressed that the experience of implementing the exam in person was successful. This finding implies that the department's precautionary and preventive measures' phases during the exam were successfully managed. However, the measures were successful in one particular department within the college, likely because of a low number of cases in the city and country. Each college may have different situations, depending on the prevalence of the virus in a college's location [7]. Although the framework applied in this study was executed in a single scientific department, the findings provide positive indications that in-person exams can be organized in educational circles during an ongoing pandemic. Nevertheless, this study emphasizes the importance of pursuing the precautionary measures and instructions issued by the university and other relevant bodies.

7 Conclusion

To the best of the authors' knowledge, this study is the first to offer a practical framework for performing in-person exams. The research investigated students' perceptions in the Department of Administrative and Financial Sciences and Techniques in the Community College of Taibah University regarding the effectiveness of precautionary-measure phases while conducting the first mid-term exam in person during the COVID-19 pandemic. The results showed the difficulty of avoiding a complete disruption of the institution's activities and how managing teaching activities at a distance is possible to some extent. Nonetheless, a crisis (such as the COVID-19 pandemic) might negatively affect the education quality and increase the inequality of learning opportunities. The efforts put in place to prevent a void in the academic year, ensure future planning despite the high degree of uncertainty, and with a risk of decreasing private and potentially also public funding, demonstrate the incredible amount of pressure higher education institutions have endured during the current crisis—and at the same time their resilience and creativity.

This study makes both theoretical and practical contributions. First, theoretically, existing research regarding conducting in-person and onsite examinations during COVID-19 is limited; hence, this empirical study could enhance the theoretical perception of conducting in-person exams at higher education institutions. Second, this study appears to be one of the few studies that have investigated students' views regarding the effectiveness of performing onsite and in-person exams. Empirically, this study's results provide some evidence that conducting in-person exams is possible, provided that the precautionary measures and instructions issued by the authorities concerned are followed strictly.

For practical implementation, employing a framework of onsite examinations during the COVID-19 pandemic can be utilized as a guide during various circumstances and used as an opportunity to promote attitudes toward sustainable development of teaching in higher education. Hence, it aims to offer insights to be developed and facilitated in future scientific works with different scopes intending to understand this potential phenomenon while simultaneously providing contributions for policymakers so that the process could be achieved effectively and efficiently.

The limitations of this paper need to be acknowledged. First, the study is based on the case of a single department in a Saudi university to examine the effectiveness of precautionary-measure phases, which cannot give a full picture or be generalized across universities. Further studies could investigate these phases in different contexts to verify their feasibility and share experiences. Second, the survey sample size was based on perceptions of the department's male students while female students took the exam online. Hence, any potential indicator for the gender difference is not reflected in the results. Furthermore, the perceptions of academic staff are not covered by this study, so future studies could provide a more comprehensive picture from both views. Third, the results might not be generalized due to the location features of the study, which was conducted in Madinah, a medium-sized Saudi Arabian city. The health conditions and the governmental restrictions regarding the COVID-19 pandemic in the city were suitable for conducting onsite and in-person examinations compared to other locations. In this case, more research needs to be done to compare the effectiveness of the framework in other locations and under different health conditions. Last and most important, the study only examined one instance during the first-term exams in the first academic semester of 2020. The remaining examinations were implemented online. As a result of this issue, the opportunity to reassess the framework and examine the stability of the findings was not attainable. Further study is needed to increase the validity level of the framework and assure the stability of its results.

References

1. Adedoyin, O. B., & Soykan, E. (2020). COVID-19 pandemic and online learning: The challenges and opportunities. *Interactive Learning Environments*, 1–13.
2. Affouneh, S., Salha, S., & Khlaif, Z. N. (2020). Designing quality e-learning environments for emergency remote teaching in coronavirus crisis. *Interdisciplinary Journal of Virtual Learning in Medical Sciences, 11*(2), 1–3.
3. Aljaser, A. M. (2019). The effectiveness of e-learning in developing academic achievement and the attitude to learn English among primary students. *Turkish Online Journal of Distance Education-TOJDE, 20*(2), 176–194.
4. Alqurshi, A. (2020). Investigating the impact of COVID-19 lockdown on pharmaceutical education in Saudi Arabia–A call for a remote teaching contingency strategy. *Saudi Pharmaceutical Journal, 28*(9), 1075–1083.
5. Aristovnik, A., Keržič, D., Ravšelj, D., Tomaževič, N., & Umek, L. (2020). Impacts of the COVID-19 pandemic on life of higher education students: A global perspective. *Sustainability, 12*(20), 1–34.
6. Cairney-Hill, J., Edwards, A. E., Jaafar, N., Gunganah, K., Macavei, V. M., & Khanji, M. Y. (2021). Challenges and opportunities for undergraduate clinical teaching during and beyond the COVID-19 pandemic. *Journal of the Royal Society of Medicine, 114*(3), 113–116.
7. Cheng, S. Y., Wang, C. J., Shen, A. C. T., & Chang, S. C. (2020). How to safely reopen colleges and universities during COVID-19: Experiences from Taiwan. *Annals of Internal Medicine, 173*(8), 638–641.
8. Daniel, S. J. (2020). Education and the COVID-19 pandemic. *Prospects, 49*, 91–96.
9. Draissi, Z., & Yong, Q. Z. (2020). COVID-19 Outbreak response plan: Implementing distance education in Moroccan universities. Retrieved February 19, 2021, from https://papers.ssrn.com/sol3/papers.cfm?abstract_id=3586783.
10. Ezeonu, C. T., Uneke, C. J., & Ezeonu, P. O. (2021). A rapid review of the reopening of schools in this COVID-19 pandemic? How ready are We in Nigeria? *Nigerian Journal of Medicine, 30*(1), 8–16.
11. Favale, T., Soro, F., Trevisan, M., Drago, I., & Mellia, M. (2020). Campus traffic and e-Learning during COVID-19 pandemic. *Computer Networks, 176*(20), 1–9.
12. Heng, K., & Sol, K. (2020) Online learning during COVID-19: Key challenges and suggestions to enhance effectiveness. In *Cambodian Education Form, Short Article Series* (pp. 1–16).
13. Hodges, C., Moore, S., Lockee, B., Trust, T., & Bond, A. (2020). The difference between emergency remote teaching and online learning. *Educause Review, 27*, 1–12.
14. Jabor, I. A., & Hamid, A. A. (2019). External and internal factors affecting student's academic performance. *The Social Science, 14*(4), 155–168.
15. Jacob, O. N., Abigeal, I., & Lydia, A. E. (2020). Impact of COVID-19 on the higher institutions development in Nigeria. *Electronic Research Journal of Social Sciences and Humanities, 2*(2), 126–135.

16. Jena, P. K. (2020). Impact of COVID-19 on higher education in India. *International Journal of Advanced Education and Research, 5*(3), 77–81.
17. Mukherjee, U. K., Bose, S., Ivanov, A., Souyris, S., Seshadri, S., Sridhar, P., & Xu, Y. (2021). Evaluation of reopening strategies for educational institutions during COVID-19 through agent based simulation. *Scientific Reports, 11*(1), 1–24.
18. Mushtaq, I., & Khan, N. (2012). Factors affecting student's academic performance. *Global Journal of Management and Business Research, 12*(9), 16–22.
19. Odriozola-González, P., Planchuelo-Gómez, Á., Irurtia, M. J., & de Luis-García, R. (2020). Psychological effects of the COVID-19 outbreak and lockdown among students and workers of a Spanish university. *Psychiatry Research, 290*, 113108.
20. Rashid, S., & Yadav, S. S. (2020). Impact of Covid-19 pandemic on higher education and research. *Indian Journal of Human Development, 14*(2), 340–343.
21. Rasul, S., & Bukhsh, Q. (2011). A study of factors affecting students' performance in examination at university level. *Procedia Social and Behavioral Sciences, 15*, 2042–2047.
22. Saad, A. Z., Mohd. Noor, A. B., & Sharofiddin, A. (2020). Effect of applying total quality management in improving the performance of Al-Waqf of Albr societies in Saudi Arabia: A theoretical framework for "Deming's Model." *International Journal of Business Ethics and Governance, 3*(2), 12–32. https://doi.org/10.51325/ijbeg.v3i2.24
23. Sahu, P. (2020). Closure of universities due to coronavirus disease (COVID-19): Impact on education and mental health of students and academic staff. *Cureus, 12*(4), 1–6.
24 Singha, H. S. (1997). Anatomy of unfair means in university examination. In A. Singh, A. S. Singh (Eds.), *The management of examination*. Association of Indian Universities, Rouse Avenue.
25. Tang, W., Hu, T., Yang, L., & Xu, J. (2020). The role of alexithymia in the mental health problems of home-quarantined university students during the COVID-19 pandemic in China. *Personality and Individual Differences, 165*, 110131.
26. Wang, C., Cheng, Z., Yue, X. G., & McAleer, M. (2020). Risk management of COVID-19 by universities in China. *Journal of Risk Financial Management, 13*(2), 1–6.
27. Yulia, H. (2020). Online learning to prevent the spread of pandemic coronavirus in Indonesia. *ETERNAL (English Teaching Journal), 11*(1), 48–56.
28. Basilaia, G., & Kvavadze, D. (2020). Transition to online education in schools during a SARS-CoV-2 coronavirus (COVID-19) pandemic in Georgia. *Pedagogical Research, 5*(4), 1–9.

Experiences and Practices of Technology Application in Times of the COVID-19 Pandemic

The Experience of the Kingdom of Saudi Arabia in the Field of E-learning During the Coronavirus Pandemic

Abdulsadek Hassan

1 Introduction

The information revolution and modern communication technologies have been able to develop the method of teaching and learning, especially with the expansion and spread of communication technology, which is no longer the issue of delivering information to anywhere, or exchanging it between vast geographical areas, is a difficult matter, but has become available within seconds or minutes [1]. E-learning has added to the educational process the exchange of educational experiences in various fields, and the benefit from the scientific development that developed countries have reached, in addition to saving money, time and effort, as some educational institutions in the world have begun to actually apply this modern educational method, which It will help transfer everything new in various fields of science to universities [2].

Cultural openness has brought about the latest strong tendency towards a philosophy of education to be more open, and from this point of view e-learning came as a mechanism of action and a philosophy at the same time to keep pace with developments and benefit from modern technologies, there are several characteristics that helped the spread of e-learning, the most important of which is availability, which means that opportunities education, especially higher education [3], are available to all, regardless of temporal, spatial and age constraints, in addition to flexibility in overcoming bureaucratic obstacles, routine and time, as well as learner control, and it means that learners can set priorities in their arrangement of curriculum topics to adapt it according to their situations and abilities [4].

In the context of the educational deterioration witnessed by most Arab countries, from deficiencies in curricula and poor teaching performance, e-learning appeared as one of the means that students resorted to compensate for this deficiency in learning,

A. Hassan (✉)
Ahlia University, Manama, Bahrain
e-mail: aelshaker@ahlia.edu.bh

as e-learning represented the link that gives the student everything he needs with the least time and effort for all different age groups and spectrums.

E-learning means the use of modern electronic means, represented in the computer and the Internet [5], in investigating and searching for information and learning away from traditional means of indoctrination, and this type of education broadens the perceptions of the student, so that he searches for the information himself without hearing it directly from the lecturer, and thus it is more entrenched in the minds, provided the safe use of e-learning [6].

With deficiencies and gaps in most of the curricula in Arab countries, the emergence of modern technologies, technological development and the communications revolution, countries began looking to move out of the scope of the traditional educational system to non-traditional educational patterns, which were represented in programmed or electronic education and other forms of education methods. E-learning [7], especially with the development of remote communication systems to connect remote classrooms and enrich education through integration with various means of communication [8].

Among the most important advantages of e-learning is that it provides flexibility and effectiveness in learning, as it provides an opportunity for real education through perception and comprehension, in addition to consolidating and fixing information, as well as saving time and money, and it also continuously develops and updates information, within the quality of educational materials [9].

Numerous studies conducted by international universities have confirmed that e-learning is of great effectiveness whenever the methods, means and techniques of teaching, methods of evaluation and follow-up are continuous, and direct contact and feedback between lecturer and student is appropriate [10], and that it is more effective than the traditional method and direct education, especially with regard to the performance of examinations that are usually taken in a stress-free atmosphere [11].

Specialists point out that e-learning is an opportunity for students to learn real, by taking advantage of modern technology and the information revolution [12], in light of the inadequacy of the curriculum, its poor quality, and its reliance on indoctrination, and achievement in order to obtain certificates, while others explain that there are many challenges that prevent the application of the concept of e-learning in Arab countries [13]. Despite the necessity and importance of e-learning in Arab countries, there are undeniable challenges that make the application of this method in education difficult, and even impossible in some countries, explaining that one of the most important obstacles is Arab e-learning is the lack of familiarity with the concept of this type of education in a specialized way, as there are no competencies capable of adopting the idea of e-learning, to accommodate huge numbers of students [14], in addition to the technical challenge, and there are no schools or universities prepared electronically for this purpose, in some countries Its schools do not basically have computers [15], in addition to poor coverage of communication networks and the Internet, which are two basic requirements in the e-learning process. All this requires financial costs that many Arab countries may not be able to provide [16].

Despite the many challenges that hinder the application of e-learning in the Arab world, however, there are some models in the Arab countries that were able to overcome these challenges, and began to apply this method in education, and perhaps the most prominent of those countries, the Kingdom of Saudi Arabia [17], which uses e-learning methods at the University King Abdul-Aziz has been for a long time and owns a large electronic library containing 16,000 e-books. It also established with the Malaysian Meteor Company the National Center for E-learning and Distance Education, which aims to create a nucleus for a central e-learning incubator for university education institutions and unify the efforts of institutions seeking to adopt the technologies of this The type of education, as well as there is also Egypt [18], which launched the e-learning initiative with the aim of eradicating computer and internet literacy for students, in addition to that hundreds of schools have been linked to a free internet service, and a new e-learning model has been created, and the establishment of the Egyptian University for E-learning has been approved [19].

The world's education systems have witnessed unprecedented turmoil this year due to the Corona pandemic, so most of the world's schools and universities closed their doors to more than 1.5 billion students, or more than 90% of the total number of students, according to recent figures issued by the UNESCO Institute for Statistics [20]. Education experts have agreed that post-corona education will not be the same as before it, especially with the emergence of a highly automated infrastructure using the data of the Fourth Industrial Revolution and artificial intelligence systems [21], and that there are expected changes that will be large and structural in education patterns, methods, directions, and policies, whether at the level of public or university education, and the signs of these transformations are already emerging [22].

Distance education: This pattern has been used, in many countries of the world, as an alternative to traditional education, since the outbreak of the pandemic. In terms of its positive results, the possibility of giving lectures, student response, and faculty members' adaptation to this type of education, even student attendance was better than attendance in classrooms." A UNESCO expert notes that efforts have been made by many governments to reach to the students during the lockdown, it was large and fruitful, and we had evidence that education reached students who were not being reached, when the schools were open [23].

E-learning: which combines distance education in the classroom, through modern means and mechanisms of communication, including computers, networks and multiple media, combining sound, image, graphics, search mechanisms, and digital libraries, with the aim of reaching the student in the shortest time and at least effort [24]. It is expected that this educational pattern will prevail in most educational institutions around the world during the foreseeable future, and one of its most important forms is what has become known as blended education, which combines technology-enhanced education with direct education (face to face) [25].

In an American study that analyzed the nature of the changes in the education system in 213 educational institutions, it was found that blended education has become the dominant method in these institutions, being effective, and preferred by most students, as a stimulating teaching method [26]. On the East Bank of the Atlantic, the European association of distance Teaching Association (EADTU) issued a report on the future of education in the world, in which it was stated that there is a significant increase in the number of educational institutions- at all levels of education - that have already adopted the blended education model [27]. It attributed this to the importance of this model in raising the level of skills, whether among students or lecturers, and that it is ideal to face the increase in the number of students, and it also pushes towards increasing the quality of the educational process [25]. And the United Nations Educational, Scientific and Cultural Organization (UNESCO) emphasized the importance of blended education, as an approach that enhances learning and pushes towards achieving the fourth goal of the sustainable development goals contained in the United Nations (UN) report, known as "Education 2030" [28].

Artificial Intelligence: The trend is increasing towards adopting artificial intelligence techniques, in order to promote online education, adaptive learning software which aims to transform the learner from the position of the negative receiver of information to the collaborator in the educational process, and research tools that allow students to quickly interact, benefit from information, and acquire skills [29].

And the results of many academic studies have indicated that the use of adaptive learning benefits the student's progress in his educational path, promotes active education for students, and assesses the factors affecting student success [24]. However, the effective integration of these new technologies into the university curriculum requires good planning and the provision of the necessary resources [30].

The pattern of artificial intelligence is also linked to the use of robots in the field of education, as educational institutions' adoption of robots in teaching is increasing day after day [31], especially after the success of the experiment with robots that teach languages, as well as the teaching of some basic subjects, as is the case in China and some Scandinavian countries [32].

2   E-learning in the Kingdom of Saudi Arabia in the Light of Corona Pandemic

As a result of the repercussions of Corona, education in the Kingdom is heading towards expansion in digitalization, through the use of advanced education applications and programs, based on artificial intelligence data [33]. The repercussions of the pandemic imposed maximizing this aspect in order to continue the progress of educational processes, while at the same time achieving protection and safety for students [34].

The Ministry of Education has introduced on its smart learning portal 13 global educational platforms based on artificial intelligence techniques, to provide multiple educational options for students during the distance learning process [35], and the platforms include all academic materials and curricula applied by the ministry, in addition to other educational systems applied in the Kingdom. In addition to the advanced educational solutions offered by the ministry's platform, through its electronic gate, which includes thousands of interactive educational clips [36].

The educational platforms that support the Ministry's gate for smart learning are distinguished by advanced educational features and are able to identify the various levels of students for their reliance on artificial intelligence techniques adapted to the educational and educational needs of students [37], so they are able to bridge the knowledge gaps for students and provide a pioneering educational experience that they can rely on in the future [38] to develop the student levels, and all platforms are available to all students regardless of their educational systems, provided that they register them through "Al Manhal", which is concerned with student data, so that they can benefit from the various educational content provided by the Ministry of Education in various platforms [39].

E-learning platforms have been activated for public education and private education, in addition to taking many urgent measures as follows: Completion of the virtual school building in one week, during which the school headquarters were furnished and satellite broadcasting, 20 smart boards were supplied and installed, and the teaching staff was trained to photograph educational clips [40].

Daily lessons were prepared to explain the curriculum, with the participation of 276 lecturers, and 73 supervisors. 3368 lessons were explained, and the number of teaching hours reached 1684 h [41].

In addition to lessons for reviewing the curriculum, with the participation of 123 lecturers, and 73 supervisors, 1107 lessons were reviewed, and the number of teaching hours reached 554 h [39].

The Ministry followed the application of distance education according to the methods of simultaneous interaction and asynchronous interaction as follows:

Simultaneous interaction was implemented through the application of the virtual school [42], which included the unified education system, the Ain gate, the gate of the future, and the application of the virtual kindergarten [43].

Asynchronous interaction was applied through Ain satellite channels, and Lessons Ain channels via YouTube [39].

The previous channels and platforms provided many options for students to continue education and learning remotely, via the Internet or via satellite channels, for those who do not have access to the Internet [44], and these channels witnessed a great interaction by male and female students, and contributed greatly to the successful continuation of the educational process [40]. A brief introduction to these channels follows:

Future Gate: An integrated e-learning platform for the various educational levels. It has had more than 20 million visits, and more than 700,000 students have benefited from its services [45].

Ain Satellite Channels: A group of 20 satellite channels covering all academic levels [43].

Ain Channels on YouTube: These are live channels on YouTube, with more than 60 million views [46].

National Education Portal (Ain): The portal provides digital content that contains more than 45,000 virtual educational materials, and more than 2000 digital books [47].

Virtual Kindergarten: an integrated educational system for early childhood stage [48].

Unified education system: The system contains digital educational lessons and activities in an interactive educational environment between the teacher and the student, and the Ain platform provides daily explanations of the curricula for all educational stages [49], and the Kingdom's decision to activate the "Blackboard" system came after the spread of the Corona virus [50], and the system provides many advantages for teachers, including:

- Recording the attendance and absence of students.
- Add assignments, tests, and courses for students [51].
- Set up virtual lessons.
- Set up educational paths for a dedicated group of students [52].
- Answer student inquiries sent via teacher [49].

The unified education system has become one of the most available prominent solutions for distance education in the Kingdom of Saudi Arabia in order to communicate with students and includes all services related to education in Saudi Arabia for all students [53], in addition to many means that made it easier for all students to complete their educational journey in an easy way [43].

It is evident from this previous presentation of the experience of the Kingdom of Saudi Arabia that there are many options available to all students in the Kingdom to follow the educational process [54], and to provide distance education with multiple alternatives via the Internet or via satellite channels, as well as the immediate initiative of the Ministry of Education to take the necessary measures and not stop educational services for all levels of education [55]. The Saudi experience agreed with its counterpart in the United Arab Emirates, as the UAE has great experience and expertise in the field of using technology in education [56], and this experience differed from the experiences previously reviewed, which witnessed a slowdown in the application of distance education procedures, or a greater focus on the use of Internet in this area [57].

The Saudi Ministry of Education's initiative to deal with the Corona epidemic and the threat it poses to the educational process can be explained in light of the Kingdom's previous experience in managing educational crises [58], as the Saudi Ministry of Education had previous rich experience in providing distance education services to students on the southern borders of the Kingdom, and the plan included persevering the student's basic rights in obtaining education through direct contact with his teachers through the school twinning project [15]. Other electronic educational alternatives were also activated, such as the educational channels and the satellite channel, and benefiting from the online platforms education [43].

The educational platforms (the 13 platforms) provide for the smart learning portal, namely, Mc Graw Hill School, Oxford University Press, College Board, Code.org cde.org, Matific and Alef, Twig, Ynmo, Nahla and Nahl, Booklip, Learntech, and

Microsoft Teams, "advanced educational solutions supported by modern educational technologies that create an interactive educational atmosphere that attracts the attention of the student and provides him with information in innovative ways. Students can benefit from these platforms through the smart learning portal of the Ministry of Education." Education is through access to the Learning Management System (LMS) [50].

3   School Platform

The school platform was launched in 2018 and includes 5,000 educational video lessons, including physics, chemistry, biology, mathematics and general sciences, and covers various school curricula, from kindergarten to grade 12 [59].

In addition to educational videos, a school includes exercises and applications in various scientific subjects, in order to supplement the educational process in an integrated framework [60].

These educational videos were prepared and produced based on the latest international education curricula, as part of a deliberate Arabization and production plan in which students' educational needs were taken into account in the various educational stages, and the highest standards and technical controls were applied in selecting scientific materials [36], Arabizing them, and harmonizing them according to the curricula adopted in the Arab countries through the translation challenge, the school provides educational services to more than 50 million students [55].

The figures recorded by the "My School" platform are 92% of students entering the platform, 97% of male and female teachers, and 37% of parents, and training more than (389) thousand trainees who hold educational positions on the "My School" platform, through (2500) training programs that reflect the size of the achievement that has been achieved in a short period [59].

Source   https://edu.moe.gov.sa/Mokhwah/Departments/mdir/it/Pages/schools. aspxMc

Graw Hill

As for "McGraw Hill", it is a platform for designing learning solutions that covers mathematics and science curricula from kindergarten to the twelfth grade. It includes approximately 1334 lessons in science and 2131 lessons in mathematics. Under this platform there are several other supporting platforms that take into account students' levels and address their interests [41].

Source https://www.mheducation.com/

Alef Platform

While the "Alef" platform, applied in 150 schools, provides supportive educational content of school curricula. The platform contains more than 2000 digital lessons. The platform allows access to any child anywhere and at any time. It includes several characteristics, including the possibility of early intervention to address any weakness in performance [61].

Source https://www.alefeducation.com/ar/

Nahla and Nahl

In turn, Nahla and Nahil platform offers more than 1000 Arabic e-books according to several levels, with a set of exercises associated with them, and provides the opportunity for the learner to advance in different reading levels according to what he accomplishes from reading the books it provides. The platform enables its supervisors to create specialized assignments for one or several students [62].

Source https://nahlawanahil.com/

Code.org

Code.org encourages students to learn computer science, as the website includes free programming lessons from kindergarten to high school and about 15% of the world's students participate in it. The site also provides specialized lessons in concepts related to digital citizenship, computing and programming,logarithms, artificial intelligence for all school levels [48].

Source https://code.org/

Matific

The Matific platform provides more than 2000 educational videos in both Arabic and English languages targeting the mathematics curriculum and providing a learning experience through games. This platform can, according to its characteristics, automatically assign courses to students based on a set of information that depends on advanced data and technologies through which the platform can learn all levels of students. The platform also offers workshops through interactive laboratories that help teachers and students explore mathematics concepts and methods. Matific provides many adaptive worksheets that surpass traditional mathematics training papers [52].

Source https://www.matific.com/au/en-au/home/

TWIG

The Twig platform provides more than 1700 videos in English, including 300 in simplified English, as well as 750 videos in Arabic, in addition to more than 150 educational resource packages and more than 130 plans for math lessons, and it includes several platforms such as the Tigtag platform that provides more From 800 videos in English, as well as 140 films in simplified English, the platform Tigtag offers 100 English videos for kids, 49 lesson plans, and more than 50 educational games [63].

Source https://www.twig-world.com/

Microsoft Teams

Microsoft has made Microsoft Teams available free of charge through the Ministry of Education network, and the network includes more than 32,000teachers and 600,000 students, as this application provides all the necessary features to enable students to complete their studies remotely without any difficulties that prevent the educational process from progressing to the fullest, by providing tools that help students and teachers to communicate and collaborate with each other - whether this communication is visual or audio through the options in the program [43].

Microsoft Teams falls under the list of Office 365 available by the Ministry of Education for both teachers and students in public schools in all the kingdom and which has been trained for all educational staff over a period of no less than four years in order to enable all elements of the educational community to make optimal use for this technique [46].

Source https://www.microsoft.com/en-us/microsoft-teams/group-chat-software

Shifts in universities … and curricula

The high education sector will witness new situations in many countries of the world in general after Coronavirus pandemic, including:

- **Social distancing:** where access to the classroom will be graded, and the principle of social distancing will be carefully observed. No handshakes, no physical closeness. Friendships, social networks and much more will remain suspended [64].
- **Multiple working hours within the same university building:** The need for social distancing among students will impose fewer of them in the classroom. Then it will become necessary for educational institutions to operate all the day [65]. This will undoubtedly put more pressure on the faculty and administrative staff [66].
- **Review of study abroad:** All forms of international education have been affected by the pandemic, and this will continue for some time at least. This effect extends to study abroad plans, training programs, and exchange of experiences [67].

  Acquiring new skills: After spending months in home learning, during the lockdown period, students have become more familiar with the tools and means of educational technology [68], while having sufficient control over their own lessons. Not only will they learn the lessons directed according to the curriculum, but they will also acquire experiences in the many new available applications, which they can use for study and learning [69].
- **Redefining the role of the lecturer:** The concept of the lecturer 's role as the owner of knowledge, who imparts information to his students, will change, especially with the expansion of students' access to knowledge resources through

digital education systems, in which the traditional lecturer's role is shrinking [70].

- **The curriculum level,** important transformations will also impose to keep pace with the post-pandemic reality, especially with the emergence of new educational programs that adopt smart strategies in developing educational contents, by using the latest applications developed by emerging companies and major institutions in the education sector [71].

It is well known that in traditional education systems, students of one stage learn almost the same curriculum, without regard to each student's interests or individual skills [72]. However, in the new curricula, thanks to the creation of a global digital infrastructure, students will have the opportunity to choose and learn at the pace that suits them and pay more attention to the things they enjoy doing [73].

Studies indicated that the shortcomings in the effectiveness of the traditional learning will push the direction of using big data analytics and artificial intelligence properly, and perhaps create personal learning experiences through a customized learning experience [74], make for each student a completely unique curriculum, fully designed according to his individual abilities and needs. This would increase the motivation of students, and it could also provide lecturer s with a better understanding of the learning process for each student [75].

The examination and correction will be subject to change, as the data of the modern technology revolution will help lecturer s to deal with the evaluation, track the performance of each student and take score fairly. These tasks will become simple, which will allow lecturer more time and effort to focus on improving the course Educational, teaching quality and competency development [76].

According to a report issued by the Global Market Insight, it is expected that by the year 2025 AD, the market size for these educational applications at the global level will exceed 300 billion US dollars … and at the Arab level, and according to the Dubai Future Foundation, the market size of Educational technologies in the United Arab Emirates will exceed 40 billion dollars by 2022 [18].

4   Studies Conducted on the Experience of Public and Higher Education in the Kingdom During the Corona Pandemic

Two comprehensive studies were conducted on the experience of public and higher education in the Kingdom during the Corona pandemic, with the aim of documenting and studying the reality of the experience and coming out with initiatives to develop and improve e-learning practices in accordance with the latest international practices and standards in this field [57].

These two studies were conducted with the participation of more than 342,000 students, school members, teachers, parents and school leaders. The number of participants in the public education study reached 318,000, while the number of participants in the higher education study reached 24,000 [40].

The first study was prepared by the online Learning Consortium (OLC), with the participation of the International Society for Technology in Education(ISTE), the Quality Matters Organization (QM), the UNESCO, the National Center for Distance Learning Research and Advanced Technologies in the United States of America (DETA) [77], and the Institute of Information Technology in Education of UNESCO IITE. The OECD, in cooperation with Harvard University in public education, prepared the second study [41].

During the two studies, benchmark comparisons were made with more than 193 countries around the world, and the two studies showed the Kingdom's excellence in the diversity of available options, including, for example, the electronic content and satellite channels available for e-learning in public education, provided by the Kingdom [78], and the percentage of countries that succeeded in providing them at the national level, only 38% [79].

The study conducted by the Organization for Economic Cooperation and Development (OECD) and Harvard University included a comparison about the Kingdom's response to education during the Corona pandemic with 37 member states, and the results showed the Kingdom's progress in 13 out of 16 indicators on the average of these countries. The study also showed that there is a great support for overcoming obstacles to activating e-learning indicating that there is a clear strategy for reopening schools in the Kingdom and measuring any losses and treating them with the Ministry of Education [50].

The (OLC) commended the great efforts made by the Ministry of Education in dealing with the crisis in terms of the variety of options available, and the speed of response to changes to ensure the success of the transition to e-learning in an effective manner [59].

It is noteworthy that the two studies recommended 71 proposed development initiatives for public education and 78 proposed developmental initiatives for higher education, and the National Center for E-learning is working in coordination with the Ministry of Education to present the initiatives and start implementation, and the bodies that conducted the two studies will publish their results, given the importance of these studies and the importance of their results,and the second phase of it will complete at the end of the first (current) semester second 2020/2021 [61].

5   Foreseeing the Future of Distance Education and E-learning

The supervision of the future of distance education and e-learning can clarify the problems that accompanied that experience, which can be summarized as follows:

The weakness of the technological infrastructure for the application of distance education and e-learning, it cannot be considered that the ability to connect to the Internet is available and possible for all students, in various countries of the world there are many areas, or some social groups that do not have access to the Internet [80].

The application of emergency education requires the presence of a special personal device dedicated to learning for each student independently, and this has become more difficult in light of the work of parents from home, in addition to the presence of more than one student in the same home and the need for everyone to use computers or mobile devices for the purpose of learning or work [81].

The presence of a large number of distractions during online education and learning, such as constantly browsing of social media sites, or watching videos that are not related to educational content, and other applications that waste time without the desired benefit [82].

Emergency remote education requires the participation of parents for their children in pursuing academic achievement, especially in the case of students in the early stages of education, and this matter concerns two important aspects: one or both parents work from home and the lack of time to follow their children, and the second thing is the extent of their ability to use technological applications and use the Internet to monitor their children's education [83].

Lack of prior preparation for lecturers to use the Internet, technological tools and applications in the field of education, and the need for lecturers to be trained professionally to complete all educational tasks via the Internet [84].

The widespread belief among students and parents that emergency education came as a formality to complete the educational process, and the lack of seriousness in pursuing distance education [85].

In an attempt to overcome these problems and determine the nature of the educational process after Corona, many proposals have appeared to develop the reality of distance education, including what was presented by the Dubai Future Foundation within its vision for the future of education as follows:

Education sector regulators may initiate many transformative changes in traditional education, and develop innovative solutions such as teaching and training for parents, to change regulations proactively instead of responding to unforeseen circumstances [36].

As learning outside the classroom becomes more prevalent, so students will participate in designing learning tools and teaching curricula [86], and schools may need less space when cohorts of students use the school's rotating facilities [87].

A global debate will take place to recognize the advantage of distance education and the qualifications that can be acquired and learned by using online education platforms.

In addition to the above, some proposals can be made that contribute the development of the educational process, enabling all parties to deal with any emergency with greater flexibility [88], as follows:

- The introduction of disaster and crisis management in the Ministry of Education.
- Attracting local experts from general and higher education [77].
- Investing the national competencies of young men and women specializing in educational technology, teaching methods, and educational administration [89].
- Benefiting from successful international experiences and the need to adapt and localize them according to the Saudi environment and the needs of the nation's students [90].
- Converting courses to electronic, activating blended learning, and training all education employees to employ it, such as having direct education 4 days and the fifth day online [91].
- The use of flipped learning or flipped classrooms and training lecturers on their pillars such as active learning and technology [92].
- Achieving equal opportunities and educational justice so that education reaches all regions and villages in all the Kingdom areas in case of of crises by using live and recorded television broadcasting with intensive training for lecturers and service providers on communication skills, technology employment and learning management [93]. Providing mobile Internet and 5G vehicles in cooperation with the Ministry of Communications and Information Technology in places that suffer from weak network or are out of coverage [94].
- Creating a specialized department in educational design made up of qualified of educational technology specialists working on formulating and designing curricula, studying reality and educational needs, designing scenarios, directing, producing and evaluating [95].
- Using artificial intelligence techniques and Internet applications such as facial recognition technology and radio-frequency identification (RFID), which is a technique that uses electromagnetic fields for automatic identification [42], a system that can understanding the context of a specific situation and sharing this context with other systems for response, which describe a specific situation such as the student's condition and his whereabouts, whether on the playground or home to set tests [96].
- Training lecturers and students to apply the Flipped Classroom in the educational process, so that both the lecturer and the student can use technological tools and applications in the teaching and learning process, and according to this method the learning process begins from home or outside the classroom using modern communication techniques, leading to the practice of activities [5]. Learning in the classroom, in addition will enhance the student's sense of responsibility for learning and specifying educational tasks that he should complete while he is at home, and allocating class time for more educational learning activities, which

means doubling the learning time, and the most important aspect of applying this method comes from support the use of technological tools and applications by the student while he is at home, with the aim of learning [97].

- Urging lecturers and students to bring their own devices into a virtual classroom environment [98, 99].
- The adoption of blended education alongside traditional education, so that face to face education in the classroom is combined with the extensive use of handheld devices or portable tablets, applications and technological tools for teaching and learning, provided that this is done in a manner commensurate with the nature of students and their cognitive characteristics in different levels [100, 101].
- Allocating sufficient budgets to distribute handheld devices to students in different stages, devoted to teaching and learning, and producing educational content for all levels in digital form [102].
- Working on launching interactive digital platforms, which include displaying educational content in a different way, via video clips or audio clips, as well as evaluation and feedback activities [33], in addition to launching educational satellite channels to help students who cannot access the Internet [103].
- Training parents to use technology in education, and enhancing their distance learning culture, in light of the possibility of students staying at home, parents have a great role in following up the academic achievement and following up on their children's commitment to accomplish all their educational activities via the Internet [104, 105].
- Benefit from students'passion for using the Internet, portable handheld devices and smart phones, and direct their interests to benefit from the contemporary digital revolution in the field of teaching and learning and reduce the waste of time while browsing the Internet and social networking sites, and other distractions [106–108].

6   Conclusion

It can be said that the experience of introducing the course on Internet search strategies by Saudi schools and universities during the Corona pandemic is considered a successful experience, as all the requirements of each course were presented in an organized electronic way, in a manner that ensures the achievement of the objectives of these courses. Students benefited from these courses on how to build appropriate search strategies to search for information from its various sources, whether from library indexes, databases, search engines, or any other electronic sources. The students also showed their response and enjoyment of the experience.

After the spread of the Corona virus in all countries of the world and the comprehensive closure that most countries followed to limit the spread of the virus, the Kingdom of Saudi Arabia was one of the countries that dealt with the crisis most professionally by imposing a comprehensive closure and then gradually returning to normal life. It followed the distance education system in order to avoid the infection of students and lecturer s with the Coronavirus by following the virtual classroom technology and by launching many electronic platforms of the ministry, which facilitated the process of communication and communication between lecturers and students, such as the Madrasati platform.

The Ministry of Education has made remarkable success in distance education by providing a number of different alternatives. Ain channels are shown on a daily basis to all classes on television, and they continued their work with Ain channels on YouTube.

The study showed that the Ministry of Education in the Kingdom has succeeded in supporting distance education and e-learning programs, as distance education has become a strategic option for the future. What requires continuing work on its development and adopting a culture of change within society to deal with the electronic educational environment without linking it to events or crises, as the current stage is an opportunity for change and development, and to face challenges and overcome many of them, this requires an evaluation of the educational process using e-learning to ensure continuous feedback and continuous development to enhance performance.

The study also showed that the strength of the technical infrastructure in the Kingdom helped a lot to implement this experiment with very simple obstacles. It also gave the Ministry of Education an opportunity to address some of the deficiencies in the e-learning structure, and other positive aspects are that teachers have had to deal with this experience, which made them more willing and accepting to use E-learning techniques in their future educational experience, male and female students may better accept this experience, so they are able to deal with explanations and assignments in a different way, which confirms that this experience can continue later with new ideas and methods to be a strong supporter of traditional education and not a substitute for it.

The study indicated that the educational life after Corona will be completely different during the coming period, given that virtual education will impose itself strongly in light of the spread and renewal of technology day after day, especially in light of the attraction of students in their different stages to this new educational life that encourages them to receive their lessons and lectures through electronic platforms or television programs dedicated to the educational field during time that suits them without restrictions.

Recommendations

Take advantage of the theories and research conducted in the field of designing e-courses, to know the best design methods and learning strategies according to the nature of the learning material and the characteristics of the trainees targeted by it in order to make the most of this technology.

Activating the role of e-courses and e-learning environments in university education stages and making use of Internet applications in displaying and disseminating educational courses on the network.

The necessity of doubling efforts to keep pace with the plans of higher education institutions for e-learning to achieve the goals of development plans and focus on the educational aspect and support self-education, continuous education and lifelong learning.

Work to increase coordination of efforts and projects in the field of e-learning implemented by the ministry or by higher education institutions in the Kingdom to serve the educational system in accordance with international standards and methodological principles.

Including e-learning curricula in higher education institutions in the Kingdom to help develop and change the university learning environment; Thus, creating an educational environment that is more appropriate for the development of the student and increasing the response to his needs.

Holding more training courses for university professors to develop their skills in designing electronic courses, as it enables the faculty member to use various and integrated strategies and employ these tools in the educational process.

References

1. Debattista, M. (2018). A comprehensive rubric for instructional design in e-learning. *International Journal of Information and Learning Technology, 35*(2), 93–104.
2. Haghshenas, M. (2019). A model for utilizing social Softwares in learning management system of E-learning. *Quarterly Journal of Iranian Distance Education, 1*(4), 25–38.
3. Favale, T., Soro, F., Trevisan, M., Drago, I., & Mellia, M. (2020). Campus traffic and eLearning during COVID-19 pandemic. *Computer Networks, 176*, 107290.
4. Alhazzani, N. (2020). MOOC's impact on higher education. *Social Sciences & Humanities Open, 2*, 100030.
5. Zhang, K. A., Basham, L., & Lowrey, J. D. (2020). Foundations for reinventing the global education system: Personalized learning supported through universal design for learning.
6. Allo, M. D. G. (2020). Is the online learning good in the midst of Covid-19 pandemic? The case of EFL learners. *Jurnal Sinestesia, 10*(1), 1–10.
7. Kamarianos, I., Adamopoulou, A., Lambropoulos, H., & Stamelos, G. (2020). Towards and understanding of university students' response in times of pandemic crisis (COVID-19). *European Journal of Education Studies, 7*, 20–40.
8. Alrefaie, Z., Hassanien, M., & Al-Hayani, A. (2020). Monitoring online learning during COVID-19 pandemic; Suggested online learning portfolio (COVID-19 OLP). MedEdPublish, 9(1).
9. Cervi, L., Pérez Tornero, J. M., & Tejedor, S. (2020). The challenge of teaching mobile journalism through MOOCs: A case study. *Sustainability, 12*, 5307.
10. Ching-Ter, C., Hajiyev, J., & Su, C. R. (2017). Examining the students' behavioral intention to use e-learning in Azerbaijan? The general extended technology acceptance model for e-learning approach. *Computers and Education, 111*(1), 128–143.
11. Guerrero-Roldán, A.-E., & Noguera, I. (2018). A model for aligning assessment with competences and learning activities in online courses. *The Internet and Higher Education, 38*, 36–46.
12. Mlekus, L., Bentler, D., Paruzel, A., Kato-Beiderwieden, A. L., & Maier, G. W. (2020). How to raise technology acceptance: User experience characteristics as technology-inherent determinants. *Gr. Interaktion. Organ. Z. Angew. Organ., 51*, 273–283.
13. Martin, F., & Bolliger, D. U. (2018). Engagement matters: Student perceptions on the importance of engagement strategies in the online learning environment. *Online Learning, 22*, 205–222.

14. Kim, D., Yoon, M., Jo, I.-H., & Branch, R. M. (2018). Learning analytics to support self-regulated learning in asynchronous online courses: A case study at a women's university in South Korea. *Computers & Education, 127*, 233–251.
15. Ahmad, N., Quadri, N. N., Qureshi, M. R. N., & Alam, M. M. (2018). Relationship modeling of critical success factors for enhancing sustainability and performance in E-learning. *Sustainability (Switzerland), 10*(12), 1–16.
16. Henderson, D., Woodcock, H., Mehta, J., Khan, N., Shivji, V., Richardson, C., et al. (2020). Keep calm and carry on learning: using Microsoft teams to deliver a medical education programme during the COVID-19 pandemic. *Future Healthcare Journal, 7*, e67.
17. Chandra, Y. (2020). Online education during COVID-19: Perception of academic stress and emotional intelligence coping strategies among college students. *Asian Education and Development Studies.*
18. Murphy, M. P. A. (2020). COVID-19 and emergency eLearning: Consequences of the securitization of higher education for post-pandemic pedagogy. *Contemporary Security Policy, 41*(3), 492–505.
19. Kohl, C., McIntosh, E. J., Unger, S., Haddaway, N. R., Kecke, S., Schiemann, J., & Wilhelm, R. (2018). Online tools supporting the conduct and reporting of systematic reviews and systematic maps: A case study on CADIMA and review of existing tools. *Environmental Evidence, 7*, 8.
20. Abed, M. G. Educational support for Saudi students with learning disabilities in higher education. *Learning Disabilities Research & Practice, 35*(1), 36–44.
21. Carey, K. (2020). Is everybody ready for the big migration to online college? Actually, no. The New York Times. https://www.nytimes.com.
22. Marutschke, D. M., Kryssanov, V., Chaminda, H. T., & Brockmann, P. (2019). Smart education in an interconnected world: Virtual, collaborative, project-based courses to teach global software engineering. *Smart Innovation, Systems and Technologies*, 39–49.
23. Ortagus, J. C. (2017). From the periphery to prominence: An examination of the changing profile of online students in American higher education. *The Internet and Higher Education, 32*, 47–57.
24. Pérez-Pérez, M., Serrano-Bedia, A. M., & García-Piqueres, G. (2020). An analysis of factors affecting students´ perceptions of learning outcomes with Moodle. *Journal of Further and Higher Education, 44*, 1114–1129.
25. Shahmoradi, L., Changizi, V., Mehraeen, E., Bashiri, A., Jannat, B., & Hosseini, M. (2018). The challenges of Elearning system: Higher educational institutions perspective. *Journal of Education and Health Promotion, 7*(1), 1–6.
26. Bralić, A., & Divjak, B. (2018). Integrating MOOCs in traditionally taught courses: achieving learning outcomes with blended learning. *International Journal of Educational Technology in Higher Education, 15*, 2.
27. Rasi, P., & Vuojärvi, H. (2018). Toward personal and emotional connectivity in mobile higher education through asynchronous formative audio feedback. *British Journal of Educational Technology, 49*, 292–304.
28. Al Kurdi, B., Alshurideh, M., & Salloum, S. A. (2020). Investigating a theoretical framework for e-learning technology acceptance. *International Journal of Electrical and Computer Engineering, 10*, 6484–6496.
29. Bozkurt, A., Jung, I., Xiao, J., Vladimirschi, V., Schuwer, R., Egorov, G., et al. (2020). A global outlook to the interruption of education due to COVID-19 Pandemic: Navigating in a time of uncertainty and crisis. *Asian Journal of Distance Education, 15*(1), 1–126.
30. Kovanović, V., Joksimović, S., Poquet, O., Hennis, T., Cukić, I., de Vries, P., et al. (2018). Exploring communities of inquiry in massive open online courses. *Computers & Education, 119*, 44–58.
31. Al-Sharhan, S., Al-Hunaiyyan, A., Alhajri, R., & Al-Huwail, N. (2020). Utilization of learning management system (LMS) among instructors and students. *Lecture Notes in Electrical Engineering, 619*, 15–23.
32. Garrett, R. (2019). Whatever happened to the promise of online learning? *International Journal of Educational, 97*, 2–4.

33. Watted, A., & Barak, M. (2018). Motivating factors of MOOC completers: Comparing between university-affiliated students and general participants. *The Internet and Higher Education, 37*, 11–20.
34. Al Lily, A. E., Ismail, A. F., Abunasser, F. M., & Alqahtani, R. H. A. (2020). Distance education as a response to pandemics: Coronavirus and Arab culture. *Technology in Society, 63*.
35. Vershitskaya, E. R., Mikhaylova, A. V., Gilmanshina, S. I., Dorozhkin, E. M., & Epaneshnikov, V. V. (2020). Present-day management of universities in Russia: Prospects and challenges of e-learning. *Education and Information Technologies., 25*(1), 611–621.
36. Alanazi, A. A., & Alshaalan, Z. M. (2020). Views of faculty members on the use of e-learning in Saudi medical and health colleges during COVID-19 pandemic. *Journal of Natural Science, Biology and Medicine, 3*, 308–317.
37. Baytiyeh, H. (2018). Online learning during post-earthquake school closures, Disaster prevention and management. *An International Journal, 27*(2), 215–227. https://doi.org/10.1108/DPM-07-2017-0173.
38. Chivu, R. G., Turlacu, L. M., Stoica, I., & Radu, A. V. (2018). Identifying the effectiveness of e-learning platforms among students using eye-tracking technology. In *4th International Conference on Higher Education Advances (HEAd'18)* (pp. 621–628).
39. ALHarthi, S. S., Shalabi, M., Tabbasum, A., AlTamimi, M., & Binshabaib, M. (2020). Perception and Perspectives of female undergraduate dental students at the Princess Nourah Bint Abdulrahman University, Saudi Arabia toward problem-based learning methodology: A questionnaire-based study, *Journal of Negro Education, 89* (1), 58–66.
40. Elsheikh, A. H., Saba, A. I., Abd Elaziz, M., Lu, S., Shanmugan, S., Muthuramalingam, T., Kumar, R. et al. (2021). Deep learning-based forecasting model for COVID-19 outbreak in Saudi Arabia. *Process Safety and Environmental Protection, 149*, 223–233.
41. Khan, M. A., Vivek Nabi, M. K., Khojah, M., Tahir, M. (2021). Students' perception towards E-Learning during COVID-19 Pandemic in India: An Empirical Study. *Sustainability, 13*, 57.
42. Vázquez-Cano, E., León Urrutia, M., Parra-González, M. E., & López Meneses, E. (2020). Analysis of Interpersonal Competences in the Use of ICT in the Spanish University Context. *Sustainability, 12*, 476.
43. Layali, K., & Al-Shlowiy, A. (2020). Students' perceptions of e-learning for ESL/EFL in Saudi universities at time of coronavirus: A literature review. *Indonesian EFL Journal, 6*(2), 97–108.
44. Hathaway, D., & Norton, P. (2018). Understanding problems of practice. A Case Study in Design Research; SpringerBriefs in Educational Communications and Technology; Springer International Publishing: Cham, Switzerland. ISBN 978-3-319-77558-6.
45. Alenezi, A. (2020). The role of e-learning materials in enhancing teaching and learning behaviors. *International Journal of Information and Education Technology, 10*(1), 48–56.
46. Hoq, M. Z. (2020). E-Learning during the Period of Pandemic (COVID-19) in the Kingdom of Saudi Arabia: An Empirical Study. *American Journal of Educational Research, 8*(7), 457–464. https://doi.org/10.12691/education-8-7-2
47. Basilaia, G., Dgebuadze, M., Kantaria, M., & Chokhonelidze, G. (2020). Replacing the classic learning form at universities as an immediate response to the COVID-19 virus infection in Georgia. *International Journal for Research in Applied Science & Engineering Technology, 8*(III).
48. Naveed, Q. N., Qureshi, M. R. N., Tairan, N., Mohammad, A., Shaikh, A., Alsayed, A. O., et al. (2020). Evaluating critical success factors in implementing E-learning system using multi-criteria decision making. *PLoS ONE, 15*(5), e0231465, 1–25.
49. Ahn, B., & Bir, D. D. (2018). Student interactions with online videos in a large hybrid mechanics of materials course. *Advances in Engineering Education., 6*(3), 1–24.
50. Rajab, M. H., Gazal, A. M., & Alkattan, K/ (July 02, 2020). Challenges to online medical education during the COVID-19 pandemic. *Cureus, 12*(7), e8966.
51. Shahzad, A., Hassan, R., Aremu, A. Y., Hussain, A., & Lodhi, R. N. (2020). Efects of COVID-19 in E-learning on higher education institution students: The group comparison between male and female, Springer Nature B.V. 2020 (pp. 1–22).

52. Rajab, K. D. (2018). The effectiveness and potential of E-learning in war zones: An empirical comparison of face-to-face and online education in Saudi Arabia. *IEEE Access, 6*, 6783–6794.
53. Affouneh, S., Salha, S. N., & Khlaif, Z. (2020). Designing quality e-learning environments for emergency remote teaching in coronavirus crisis. *Interdisciplinary Journal of Virtual Learning in Medical Sciences, 11*(2), 1–3.
54. Alenezi, F. Y. (2019). The Role of Cloud Computing for the Enhancement of Teaching and Learning in Saudi Arabian Universities in Accordance with the Social Constructivism Theory: A Specialist's Point of View. *International Journal of Emerging Technologies in Learning, 14*(13), 70–87.
55. Badwelan, A., & Bahaddad, A. A. (2017). Cultural Factors that Influence M-Learning for Female University Students: A Saudi Arabian Case Study. *International Journal of Computers and Applications, 166*, 21–32.
56. Bezerra, I. M. P. (2020). State of the art of nursing education and the challenges to use remote technologies in the time of corona virus pandemic. *Journal of Human Growth and Development, 30*, 1–7.
57. Badwelan, A., Drew, S., & Bahaddad, A. A. (2016). Towards acceptance m-learning approach in higher education in Saudi Arabia. *International Journal of Business and Management, 11*, 12.
58. Aldiab, A., Chowdhury, H., Kootsookos, A., Alam, F., & Allhibi, H. (2016). *Prospect of eLearning in higher education sectors of Saudi Arabia: A review, 1st International Conference on Energy and Power, ICEP2016, 14–16 December 2016, RMIT University, Melbourne, Australia*, (pp. 574–580).
59. Khalil, R., Mansour, A. E., Fadda, W. A., Almisnid, K., Aldamegh, M., Al-Nafeesah, A., & Al-Wutayd, O. (2020). The sudden transition to synchronized online learning during the COVID-19 pandemic in Saudi Arabia: a qualitative study exploring medical students' perspectives, l. *BMC Medical Education, 20*, 285, 1–20.
60. Muhammad A., Shaikh A., Naveed Q. N., & Qureshi M. R. N. (2020). Factors affecting academic integrity in e-learning of saudi arabian universities. An investigation Using Delphi and AHP. *IEEE Access, 8*, 16259–16268.
61. Tanveer, M., Bhaumik, A., Hassan, S., Ul Haq, I. (2020). Covid-19 pandemic, outbreak educational sector and students online learning in Saudi Arabia. *Journal of Entrepreneurship Education, 23*(3).
62. Walabe & Luppicini, R. (2020). Exploring e-learning delivery in Saudi Arabian universities. *International Journal of E learning & distance Education, 35*(2), 1–40.
63. Layali, K., & Al-Shlowiy, A. (2020). Students' Perceptions of eLearning for ESL/EFL in Saudi Universities and their Implications during Coronavirus Pandemic: A Review of Literature. *International Journal of English Language & Translation Studies, 8*(1), 64–72.
64. Hasan, N., & Bao, Y. (2020). Impact of "e-Learning crack-up" perception on psychological distress among college students during COVID-19 pandemic: A mediating role of "fear of academic year loss". *Children and Youth Services Review, 118*, 105355.
65. Bovill, C. (2020). Co-creation in learning and teaching: The case for a whole-class approach in higher education. *Higher Education, 79*(1), 1023–1037.
66. Kumar, P., & Kumar, N. (2020). A study of learner's satisfaction from MOOCs through a mediation model. *Procedia Computer Science, 173*, 354–363.
67. Niu, L. (2020). A review of the application of logistic regression in educational research: Common issues, implications, and suggestions. *Educational Review, 72*, 41–67.
68. Sun, A., & Chen, X. (2016). Online education and its effective practice: A research review. *Journal of Information Technology Education: Research*, 157–190.
69. Stenbom, S. (2018). A systematic review of the Community of Inquiry survey. *The Internet and Higher Education, 39*, 22–32.
70. Wise, A. F., & Cui, Y. (2018). Learning communities in the crowd: Characteristics of content related interactions and social relationships in MOOC discussion forums. *Computers & Education, 122*, 221–242.

71. Jung, Y., & Lee, J. (2018). Learning Engagement and Persistence in Massive Open Online Courses (MOOCS). *Computers & Education, 122*, 9–22.
72. Valverde-Berrocoso, J., Garrido-Arroyo, M. d. C., Burgos-Videla, C., & Morales-Cevallos, M. B.(2020). Trends in educational research about e-learning: A systematic literature review (2009–2018). *Sustainability, 12*, 5153.
73. Kimathi, F. A., & Zhang, Y. (2019). Exploring the general extended technology acceptance model for e-learning approach on student's usage intention on e-learning system in University of Dar es Salaam. *Creative Education, 10*(1), 208–223.
74. Thomas, R. A., West, R. E., & Borup, J. (2017). An analysis of instructor social presence in online text and asynchronous video feedback comments. *Internet and Higher Education, 33*, 61–73.
75. Landrum, B., Bannister, J., Garza, G., & Rhame, S. (2020). A class of one: Students' satisfaction with online learning. *The Journal of Education for Business.*
76. Yu, J., Huang, C., Han, Z., He, T., & Li, M. (2020). Investigating the influence of interaction on learning persistence in online settings: Moderation or mediation of academic emotions? *International Journal of Environmental Research and Public Health, 17*(1), 1–21.
77. Gupta, S. B., & Gupta, M. (2020). Technology and E-learning in higher education. *Technology, 29*(4), 1320–1325.
78. Almaiah, M. A., & Al-Khasawneh, A. (2020). Investigating the main determinants of mobile cloud computing adoption in university campus. *Education and Information Technologies*, 1–21.
79. Alashwal, M. (2020). The experience of Saudi students with online learning in U.S. Universities. *Higher Education Research, 5*(1), 31–36.
80. Al-Rahmi, W. M., Yahaya, N., Aldraiweesh, A. A., Alamri, M. M., Aljarboa, N. A., Alturki, U., & Aljeraiwi, A. A. (2019). Integrating technology acceptance model with innovation diffusion theory: An empirical investigation on students' intention to use e-learning systems. *IEEE Access, 1*(1), 99.
81. Briggs, B. (2018). Education under attack and battered by natural disasters in 2018. TheirWorld. https://theirworld.org/.
82. Huang, R., Spector, J. M., & Yang, J. (2019). Design-based research. In *Educational Technology* (pp. 179–188). Singapore: Springer. ISBN 9789811366420.
83. Mahajan, M. V. (2018). A study of students' perception about e-learning. *:Indian Journal of Clinical Anatomy and Physiology, 5.*
84. Owusu-Fordjour, C., Koomson, C. K., & Hanson, D. (2020). The impact of COVID-19 on learning—The perspective of the Ghanaian student. *European Journal of Education Studies, 7*, 1–14.
85. Binyamin, S. S., Rutter, M. J., & Smith, S. (2019). Extending the technology acceptance model to understand students' use of learning management systems in Saudi higher education. *International Journal of Emerging Technologies in Learning, 14*(3), 4–21.
86. Khan, I. A. (2020). Electronic learning management system: Relevance, challenges and preparedness. *Journal of Emerging Technologies and Innovative Research, 7*, 471–480.
87. Brianna, D., Derrian, R., Hunter, H., Kerra, B., & Nancy, C. (2019). Using EdTech to enhance learning. *International Journal of the Whole Child, 4*(2), 57–63.
88. Allam, S. N. S., Hassan, M. S., Mohideen, R. S., Ramlan, A. F., & Kamal, R. M. (2020). Online distance learning readiness during Covid-19 outbreak among undergraduate students. *International Journal of Business and Social Science, 10*, 642–657. https://doi.org/10.6007/ijarbss/v10-i5/7236
89. Jung, I., Choi, S., Lim, C., & Leem, J. (2002). Effects of different types of interaction on learning achievement, satisfaction and participation in web-based instruction. *Innovations in Education and Teaching International, 39*(2), 153–162.
90. Palvia, S., Aeron, P., Gupta, P., Mahapatra, D., Parida, R., Rosner, R., & Sindhi, S. (2018). Online education: Worldwide status, challenges, trends, and implications. *Journal of Global Information Technology Management, 21*, 233–241.

91. Swaggerty, E. A., & Broemmel, A. D. (2017). Authenticity, relevance, and connectedness: Graduate students' learning preferences and experiences in an online reading education course. *The Internet and Higher Education, 32*, 80–86.
92. Juárez Santiago, B., Olivares Ramírez, J. M., Rodríguez-Reséndiz, J., Dector, A., García García, R., González-Durán, J. E. E., & Ferriol Sánchez, F. (2020). Learning management system-based evaluation to determine academic efficiency performance. *Sustainability, 12*, 4256.
93. Liao, C. W., Chen, C. H., & Shih, S. J. (2019). The interactivity of video and collaboration for learning achievement, intrinsic motivation, cognitive load, and behavior patterns in a digital game-based learning environment. *Computers and Education, 133*(1), 43–55.
94. Oguguo, B. C., Nannim, F. A., Agah, J. J., Ugwuanyi, C. S., Ene, C. U., & Nzeadibe, A. C. (2020). Effect of learning management system on Student's performance in educational measurement and evaluation. *Education and Information Technologies.*
95. Salmon, G., Pechenkina, E., Chase, A.-M., & Ross, B. (2017). Designing Massive Open Online Courses to take account of participant motivations and expectations. *British Journal of Educational Technology, 48*, 1284–1294.
96. Eze, S. C., Chinedu-Eze, V. C., Okike, C. K., & Bello, A. O. (2020). Factors influencing the use of e-learning facilities by students in a private Higher Education Institution (HEI) in a developing economy. *Humanities and Social Sciences Communications, 7*, 1–15.
97. Abbasi, S., Ayoob, T., Malik, A., & Memon, S. I. (2020). Perceptions of students regarding E-learning during Covid-19 at a private medical college. *Pakistan Journal of Medical Sciences, 36*(COVID19-S4), 57–61.
98. Moreno, V., Cavazotte, F., & Alves, I. (2017). Explaining university students' effective use of e-learning platforms. *British Journal of Educational Technology, 48*, 995–1009.
99. Al Shehab, N., & Hamdan, A. (2021). Artificial intelligence and women empowerment in Bahrain. *Studies in Computational Intelligence, 2021*(954), 101–121.
100. Salloum, S. A., Al-Emran, M., Shaalan, K., & Tarhini, A. (2019). Factors affecting the E-learning acceptance: A case study from UAE. *Education and Information Technologies, 24*(1), 509–530.
101. Zainal, M. M., & Hamdan, A. (2022). Artificial intelligence in healthcare and medical imaging: Role in fighting the spread of COVID-19. https://doi.org/10.1007/978-3-030-77302-1_10.
102. Muhammad, A. H., Shaikh, A., Naveed, Q. N., & Qureshi, M. R. N. (2020). Factors affecting academic Integrity in E-Learning of Saudi Arabian Universities. An investigation using Delphi and aHP. *IEEE Access, 8*, 16259–16268
103. Almaiah, M. A., & Almulhem, A. (2018). A conceptual framework for determining the success factors of e-learning system implementation using Delphi technique. *Journal of Theoretical and Applied Information Technology, 96*(17).
104. Binaymin, S., Rutter, M. J., & Smith, S. (2018). The moderating effect of education and experience on the use of learning management systems. In *8th International Conference on Educational and Information Technology* (p. 8).
105. Ali Saad, A. Z., Mohd Noor, A. B., & Sharofiddin, A. (2020). Effect of applying total quality management in improving the performance of Al-Waqf of Albr societies in Saudi Arabia: A theoretical framework for "Deming's Model." *International Journal of Business Ethics and Governance, 3*(2), 12–32. https://doi.org/10.51325/ijbeg.v3i2.24
106. Al Fadda, H. (2019). The relationship between self-regulations and online learning in an ESL blended learning context. *English Language for Teaching, 12*(6), 87.
107. Alshurafat, H., Al Shbail, M. O., & Mansour, E. (2021). Strengths and weaknesses of forensic accounting: An implication on the socio-economic development. *Journal of Business and Socio-economic Development, 1*(2), 85–105. https://doi.org/10.1108/JBSED-03-2021-0026
108. Kassim, E., & El Ukosh, A. (2020). Entrepreneurship in technical education colleges: Applied research on university college of applied sciences graduates - Gaza. *International Journal of Business Ethics and Governance, 3*(3), 52–84. https://doi.org/10.51325/ijbeg.v3i3.49

The Impact of E-learning on Education Outcomes in Saudi Universities Considering the COVID-19 Epidemic

Abdelkader Derbali⊙ **and Monia Ben Ltaifa**⊙

1 Introduction

The world has witnessed a remarkable improvement in the ground of communication technology, and the most prominent of these developments is what is recognized as the field of communication and the communication transformation. Perhaps the developments that the world has witnessed today in the field of e-learning have imposed a new reality on the majority of educational institutions, and these institutions have become responsible to everyone for qualifying individuals, raising their efficiency, and graduating individuals. Able to take responsibility and deal with the latest developments in technology and contribute to the advancement and growth of society.

The e-learning is a modern technique of teaching that faces numerous challenges and obstacles, and these challenges have two aspects: the technological readiness side, which is related to information and communication, and the executive readiness side, which pertains to the user, i.e., the extent of the preparations of universities, colleges, companies, government institutions and organizations to use distance education, and there is also a psychological aspect. It is related to university professors, trainers, trainees, and students, as is the current educational system that has been in operation for hundreds of years. It is not surprising that the nature of the human mind is opposed to change.

Higher education faces many transformations and challenges as a result of the social, economic, scientific, and technological transformations and changes that have

A. Derbali (✉)
Finance and Banking at the Department of Administrative and Financial Sciences and Techniques, Community College, Taibah University, Box 2898, Medinah 41461, Saudi Arabia

M. B. Ltaifa
Finance at the Department of Administrative and Financial Program, College of Community in Abqaiq, King Faisal University, P.O Box: 380, AlAhsa 31982, Saudi Arabia

taken place at the international level in general and the Arab level in particular, which makes it need to keep pace with these transformations and changes that have occurred in contemporary societies in order to respond to and confront them [2].

Despite the significant growth in education and training and the remarkable increase in the numbers of higher education, higher education clearly suffers from the lack of educational and training opportunities for multiple groups of people who aspire to achieve their hopes and aspirations without the need to directly enroll in traditional educational institutions, because the conditions of their practical life Economic or social does not help to study full time and enroll in educational institutes far from their places of residence [9].

Education experts have emphasized as well as researchers stressed the need to keep pace with the educational shift, which is considered a revolution in the philosophy and policy of education in this era characterized by high culture, as this type of education allows openness to the world by directly dealing with information sources in an era in which information has become in all its forms. Its forms are available to the whole world through communication channels that penetrated the centralization of information and broke the barrier of secrecy and made information available to all, provided that they benefit from distance education technologies [2].

The world began facing a pandemic, which invaded most countries of the world, it started in Wuhan Province in the Republic of China, then moved to the rest of the countries of the world in varying proportions, and that pandemic began in the Saudi lands from the date (03/14/2020), as a result of which all were suspended Saudi schools and universities in all Saudi regions, and these measures prompted all universities to hasten to lay down plans in order to limit the continuation of providing their services to their students during the home quarantine period imposed by the government to besiege this pandemic.

The world's education systems have witnessed unprecedented turmoil due to the Corona pandemic, so greatest of the world's schools and universities bolted their entrances to supplementary than 1.5 billion schoolchildren, or further than 90% of the total numeral of schoolchildren, giving to current statistics distributed by the UNESCO Institute for Statistics. The education specialists have approved that post-corona teaching will not be the identical as earlier it, especially with the emergence of a highly automated infrastructure using the dataset of the 4th Industrial Revolution and artificial intelligence schemes, and that there are expected variations that will be large and structural in education patterns, methods, trends, and policies, And he organized it, whether at the level of public or university education, and the signs of these transformations are already emerging [5, 7, 10, 12].

As all universities in all parts of the world in general, as well as Saudi universities, have begun to continue and broadcast lectures electronically, which makes it imperative for students to move from traditional education to e-learning, and the fact that this crisis came suddenly and without prior preparation by some universities and students, and that we are working in the field of university education, and by giving us a number of electronic lectures, we have noticed many positives and negatives that accompanied e-learning, so we decided to conduct this study to find out the most important positives in order to strengthen them, and to note the negatives in order to reduce them, as well as to know the outputs of e-learning from the viewpoint of the recipient of the service, and the research problematic can be limited throughout the succeeding key question:

The main question: What is the reality of e-learning in the case Saudi universities and its influence on education outcomes during the COVID-19 epidemic from the students' point of view? and the following sub-questions emerge from it:

(Q1)   *What are the advantages of e-learning in the case of Saudi universities in light of the COVID-19 epidemic?*

(Q2)   *What are the disadvantages of e-learning in the case of Saudi universities in light of the COVID-19 epidemic?*

(Q3)   *What are the outputs of e-learning in the case of Saudi universities in light of the COVID-19 epidemic?*

(Q4)   *Is there a statistically significant effect at the level of significance (α 0.05 $\geq$) on the reality of the e-learning (pros and cons) in the case of Saudi universities through the COVID-19 epidemic on e-learning outcomes?*

(Q5)   *Are there a statistically considerable differences at the threshold of significance in the reality of e-learning in the case of Saudi universities and the impact on education outcomes through the COVID-19 epidemic with different variables: gender, school year, university, location of residence?*

Based on these questions, we can develop the following hypothesis:

H1:   *There are a statistically considerable differences at the threshold of significance in the reality of e-learning in the case of Saudi universities and its impact on educational outcomes during the COVID-19 epidemic, according to different variables: (gender, academic year, university, location of residence)*

H2:   *There is a statistically significant impact at the significance level (α 0.05α) of the positives of e-learning in Saudi universities during the COVID-19 epidemic on e-learning outcomes.*

H:3   *There is a statistically significant impact at the level of significance (α 0.05 $\geq$) of the negatives of e-learning in Saudi universities during the COVID-19 epidemic on e-learning outcomes.*

To respond to these questions: our paper is structured as follow: In Sect. 2, we expose the literature review. Section 3 presents a definition of E-learning. Section 4 exposes the method used in this paper. Section 5 presents the data used in our study. Section 6 presents the empirical findings of this study. Section 7 presents the hypothesis tests. Section 8 concludes. Finally, Sect. 9 presents the policy recommendations of our study.

2   Literature Review

The first study presented in this part is conducted to be designed at studying students' arrogances concerning e-learning at the University of Punjab in India [11]. The data was collected through a survey of (400) graduate students, and the results showed that (76%) of students have Clear positive trends towards e-learning, while the purest (24%) negative trends are towards e-learning, (82%) of the students issue the benefits of the e-learning, and (57%) of the students intend to adopt the e-learning in their work.

Additional paper is conducted to be pointed at reaching the existence of education through the Internet, and its ability to achieve the desired educational goals [13]. The study was an experimental semester that was divided into education in the traditional way, the e-learning method, as well as cooperative education, teamwork, and question-solving, as well as solving problems facing students. The researcher made a questionnaire by which he measured the students' performance and responses after applying e-learning. The number of sample individuals reached (34) students, and they were divided into two parts. The results showed that some students were affected by the absence of a face-to-face teacher, and there was a weakness in the study personnel due to the different personalities.

In the same context, a paper is conducted which designed at revealing the gradation of obtainability of website suggestion ethics on the Yarmouk University website since the perspective of info and announcement knowledge authorities, and the study sample consisted of (32) faculty memberships and (22) computer specialists who work at Yarmouk university in Jordan [1]. To assemble the dataset, the investigators established a survey comprising of (42) things divided into six sections: website operating standards, browsing, language, design, screen appearance, accuracy, and timeliness.

To answer the study questions, they employ the arithmetic means, the standard deviations, and a (T) test. Then, the empirical findings of their study suggest that the degree of availability of design standards came to a large extent. The findings of the study also suggest that there were statistically important variations in the job variable in favor of technicians. They also reveal that there were no statistically important disagreements on the years of experience variable.

Another study is conducted to be aimed at identifying the realism of e-learning in the Palestinian universities during the information organization from the point of interpretation of the faculty memberships in it, and the study sample consisted of (329) faculty members in the universities (Birzeit, Al-Quds and Al-Najah) [6]. The study uses the descriptive methodology, and the survey was used as an instrument to save dataset.

The results indicate that the obtainability of the e-learning situation was in the first residence with an arithmetic mean of (3.91), the field of awareness of the concept of e-learning approached in second place with an arithmetic mean (3.81), and the field of e-learning outcomes came in last place with an arithmetic mean (3.76). The findings find the absence of a significant changes rendering to the gender variable, years of experience, the university level, and the academic qualification, excluding for differences in the scientific qualification variable for An-Najah University, where the differences were substantial and in favor of the doctorate on the master's degree.

A recent paper is prepared to be designed at recognizing the difficulties facing Palestinian university students in the Hebron governorate in the distance learning system (e-learning) during the COVID-19 crisis, and showing the impact of gender, academic year, specialization, and the university on the difficulties facing students. An electronic questionnaire was employed as a tool to collect dataset [4]. The size of the study sample was 102 male and female students, who were randomly chosen, and a descriptive analytical approach was used to analyze the data and come up with the results.

The results of the study indicated that university students in the Hebron governorate suffer from high difficulties in distance learning (e-learning), and these difficulties came in order of importance: (difficulties related to lectures, difficulties with psychological stress, problems related to the curriculum, problems related to the infrastructure, problems related to knowledge in the field of e-learning). Therefore, the findings the absence of a considerable differences at the threshold level of $(0.05 \geq \alpha)$ in the difficulties faced by Palestinian university students in the distance learning system (e-learning) through the COVID-19 crisis, corresponding to the gender and the specialization variables. While the findings show that there are differences corresponding to the academic year variable in preference of first-year students, the university variable, and in benefit of the Al-Quds Open University, the University of Hebron, and the universities outside the Hebron governorate.

In the same study, the findings show the difficulties facing Palestinian university students in the Hebron Governorate in the e-learning system through the COVID-19 crisis [4]. The researcher profited from these researches in detecting the topics of analysis, their variables, and statistical methods in analyzing their results, as well as building the items of the questionnaire, its fields, and paragraphs. The results also benefits from the findings, recommendations, and proposals that these studies came out with [4].

After the previous review of previous studies, the researcher concluded that the topic of e-learning and its importance in communicating with the learner has received the attention of researchers, as the studies dealt with learners'perspectives towards e-learning such as the study of [11]. Other studies also dealt with access to e-learning through The Internet, such as the paper of [13] dealt with the availability of standards for web design, and the study of [6] dealt with the realism of e-learning in Palestinian universities during the knowledge organization from the point of view of faculty memberships.

Perhaps the most important characteristic of this study is that it is concerned with knowing the reality of the e-learning in the case of the Saudi universities and its influence on educational outcomes during the COVID-19 health crisis that most countries of the world are going through, as this topic has not been addressed in this way in previous studies in addition to its uniqueness in examining the reality of e-learning in Saudi universities and its effect on education outputs through the COVID-19 health crisis, which explains to officials in Saudi universities the most important advantages of e-learning to work on strengthening them, and clarifying the negatives of e-learning to work to avoid them from the viewpoint of the students receiving the service, as well as identifying the outputs of e-learning because of its impact on Knowing the extent to which the objectives of the teaching and learning process are achieved.

3 E-learning

The E-learning can be defined by several definitions, including [3]:

- An instructive scheme to deliver educational or training plans to beginners or apprentices at any period and in any residence utilizing communicating info and announcement knowledges such as: Internet, radio, local or satellite canals for television, CDs, telephone, e-mail, computers and teleconferencing to deliver a multi-resource interactive education/learning situation in an instantaneous or asynchronous manner in the schoolroom exclusive of obligating to a precise position contingent on self-learning and communication amongst the student and the educator".
- It is a method of education, and a method for developing a set of different methods of learning using digital technology that opens the way for the spread of learning and provides an opportunity to enhance learning [8].

The e-learning also brings many advantages [3, 14]:

- Increasing prospects for interaction amongst students and collaboration between them, and therefore motivating them to join in the learning subjects presented with alleviate and bravado.
- Opening the possibility for the delegations to the learning procedure in the conversation and discussion of viewpoints on the topics mentioned and getting into understanding the appropriate judgments.
- To consolidate the foundations of fairness and equivalence between students, and to get free of concern of membership and error.
- The prospect of getting a full description of the learning information at any moment and collaborating with the teacher quicker beyond the executive running times. The e-mail can be employed for investigations, and students believe it one of the greatest improvements of e-learning.
- Modifying the study material and introducing it in a technical and discussion manner at the same time.
- Opening the door for the teacher to decide the highly suitable method to provide the learning communication and accomplish its objectives to his students.
- Correspondence to various concentrations of student's information and motivation.
- The abundance of research material over time.

- The capability to get into the curriculum at different periods not including any problem at all.

Despite the advantages of e-learning achieved, there are several disadvantages that were judged as deficiencies connected to it, the best renowned of which are [3]:

- The lack of a strong infrastructure equipped with e-learning means.
- The absence of understanding in e-learning procedures.
- The absence of informative systems favored in the Arabic language.
- The educators have an inadequate knowledge in understanding and employing numerical technology.
- The prospect of the non-existence of the charitable component in the learning procedure.
- Elevated material expense.
- Problem in improving requirements.
- The non-existence of response from the teacher and the postponement in getting it for a long time makes the process lose the advantages of e-learning.

The E-learning is also an educational system, as it is a grouping of several elements that interact in an organized manner in order to achieve goals, and each system can classify its components into Inputs, Outputs and Processes, linked by feedback:

- The inputs of the e-learning system: It is represented in establishing the infrastructure for e-learning, as this requires the provision of devices, lines of communication, the establishment of educational sites, the use of technicians and specialists, the design of electronic courses and their delivery around the clock, the setting of educational goals in a good way, and the qualification of specialists in designing Programs and decisions, equipping classrooms and laboratories, preparing teachers and administrators through training courses, qualifying learners to switch to the electronic system, and preparing parents to accept the new system.
- The processes of the e-learning system: namely, the registration processes and the selection of electronic courses, the implementation of the electronic study, the learners' follow-up of the lessons, either synchronously or asynchronously, and their use of e-learning technologies such as e-mail, video conferences, chat rooms, etc., and the learner's passage of the constructive and formative evaluation.
- The outputs of the e-learning scheme: It is represented in achieving the objectives and for learners reaching the required level of learning, developing the courses and websites of the educational institution, enhancing the role of teachers and administrators, and holding training courses for them.

4 Method

The importance of this study can be summarized by helping Saudi universities to identify the pros and cons of e-learning, and to assist in developing programs and plans to reduce the negatives of e-learning in order to increase its effectiveness.

Where the descriptive analytical method was used for its suitability for such type of studies. As for the study variables, they are:

First: demographic variables:

- Sex, and it has two levels: (male, female).
- The academic year, which has 8 levels: (first, second, third, fourth and more).
- The university, and it has four levels: (Taibah University, Arab Open University, Islamic University in Madinah, and Rayyan Colleges).
- The place of residence, and it has two levels: (city, village).

Second: The independent variable: "The reality of e-learning (pros and cons) during the COVID-19 pandemic."

Third: The dependent variable: "The reality of e-learning outcomes in light of the COVID-19 pandemic."

Also, the most important objectives of this study are to identify:

- The positives of the e-learning in the case of Saudi universities.
- The negatives of e-learning in the case of Saudi universities.
- E-learning outcomes in Saudi universities during the COVID-19 health crisis.
- The extent of the impact of the advantages of the e-learning on e-learning outcomes.
- The extent of the influence of the negative aspects of e-learning on the e-learning outcomes.
- Knowing the extent of the existence of statistically significant differences in the replies of the used sample individuals in the reality of the e-learning in Saudi universities and its relationship to education outcomes through the COVID-19 epidemic according to different variables: (gender, school year, university, location of residence)
- Coming up with suggestions and recommendations to increase the usefulness of the e-learning in the case of Saudi universities. Subsequent paragraphs, however, are indented.

5 Data

The study community consists of all students in Saudi universities in the Medina region, namely: (Taibah University, Arab Open University, Islamic University in Madinah, and Rayyan Colleges). Where we distributed the study tool electronically to a random data sample of students in Saudi universities in Al-Madinah region,

Table 1 The distribution of the sample individuals

Variables		Number	Percentage (%)
Sex	Male	150	75
	Female	50	25
Academic year	First	20	10
	Second	80	40
	Third	20	10
	Fourth and more	80	40
University	Taibah University	100	50
	Arab Open University	20	10
	Islamic University in Madinah	60	30
	Rayyan Colleges	20	10
Location	City	180	90
	Village	20	10

where (200) questionnaires were retrieved, and Table 1 shows the distribution of the sample members.

A questionnaire was prepared to measure "the reality of e-learning in Saudi universities and its impact on education outcomes through the COVID-19 pandemic," based on educational literature and previous studies, and it consisted of two parts:

The first section: this part contains the primary data about the student who fills out the questionnaire, which are: (gender, academic year, university, place of residence).

The second section: it measures the reality of the e-learning in the Saudi universities and its influence on education outcomes during the COVID-19 epidemic, and it consists of (3) areas and (30) paragraphs that dealt with research questions and hypotheses, and the answer to these paragraphs was (strongly agree, agree, neutral, disagree, Strongly Disagree).

Based on the validity and the legality of the instrument used to achieve the goal of our paper, we have presented the questionnaire to a number of specialists and experts in a number of Saudi universities who hold doctoral and master's degrees. The paragraphs of the questionnaire have been amended according to the notes and proposed amendments, and the questionnaire has been reformulated in its final form accordingly to become a number of the paragraphs of the questionnaire in its final form (30) paragraphs.

To verify the stability of the measuring instrument, the internal consistency and stability of the resolution items were examined by calculating the Cronbach 'alpha coefficient, according to Table 2.

Table 2 Cronbach Alpha coefficients

Fields of study	Number of paragraphs	Alpha value
The positives of e-learning	10	0.964
The negatives of e-learning	10	0.905
E-learning outputs	10	0.986
The total degree of the reality of e-learning	30	0.917

Table 2 presents the stability coefficients for the study dimensions of the reality of the e-learning in Saudi universities and its influence on education outcomes during the COVID-19 epidemic, according to Cronbach Alpha coefficients. By looking at Table 2, it becomes evident that the alpha value of the reality of the e-learning in the Saudi universities and its influence on education outcomes through the COVID-19 epidemic ranged among (0.986) and (0.905), and the field of e-learning outputs obtained the highest stability coefficient, while the field of negatives of e-learning on the lowest coefficient of reliability, and the value of alpha on the overall score (0.917), which indicates the accuracy of the measuring tool.

After collection of the data used in this study, we review it in preparation for incoming it into the processor, and it was inserted into the processer by generous it exact statistics, that is, by changing the spoken responses into arithmetical unities, where the answer was given, I agree strongly with five degrees, the answer is four degrees, the answer is three degrees neutral, and the answer is a degree that does not agree with two degrees, The answer strongly disagree with one degree.

Thus, the questionnaire measures the reality of the e-learning in Saudi universities and its influence on the education outcomes through the COVID-19 pandemic from the students' viewpoint in the positive direction. The statistical treatment of the data was done by extracting numbers, arithmetic means, standard deviations, the (T) test, the One Way ANOVA method, Cronbach Alpha stability equation, regression analysis, and (LSD) test for dimensional comparison, using the SPSS Statistical Package for Social Sciences (SPSS) program.

6 Empirical Results

This part deals with a presentation of the results that we have reached through the reaction of the research sample individuals about the existence of e-learning and its impact on education outcomes during the COVID-19 pandemic, according to the study's questions and hypotheses. Then, the value of the arithmetic mean of the phrases in the study tool can be interpreted as follows (Table 3):

In light of processing the data statistically, the study reached the following results:

Table 3 The significance of the arithmetic mean

The arithmetic mean	The significance
1.00–1.79	Very low
1.80–2.59	Low
2.60–3.39	Average
3.40–4.19	High
4.20–5.00	Very high

(Q1) What are the benefits of the e-learning in Saudi universities through the COVID-19 pandemic?

It is evident from Table 4 that the positives of e-learning were average in general with a mean (4.79) and a value of standard deviation (1.795), and the highest paragraphs of the positives of e-learning came the paragraph that states (enable me to retrieve information when it is needed) with a mean (4.45) and a deviation Standard (161) and to a high degree. While the lowest paragraphs of the advantages of e-learning came with a low grade, the paragraph that states (Increased effective communication between me and the faculty member) with a mean (3.03) and a value of standard deviation (1.54), followed by the paragraph that states (E-learning contributed to facilitating my understanding of the educational material) with a mathematics mean (3.05) and a value of standard deviation (1.58), followed by the paragraph that states (Provide equal opportunities for all students) with an a mathematics (3.13) and a value of standard deviation (1.51), and finally the paragraph that states (The educational materials were presented in a better way than traditional education) with a mean (3.16) and a standard deviation (1.56).

(Q2) What are the negatives of e-learning at Saudi universities under COVID-19 pandemic?

It is clear from Table 5 that e-learning negatives are generally very high at an average of (4.82) and a value of standard deviation (1.97), and the highest paragraphs of e-learning conflicts, which stipulates (reduced the interviews between me and two students) with an average account (4.57) Standard (1.14) and a very high degree, followed by paragraph, which provides for (e-learning does not suit the decisions of practical nature) at an average account (4.36) and standard deviation (1.25) and at a high degree, followed by the paragraph that states (increased psychological pressure on me) (4.21) and a standard deviation (1.50) and at a high degree, followed by paragraph, which states (increased the required burden) at an average of my account (4.13) and a standard deviation (1.48) and at a high degree, followed by the item that states (made me feel isolated by non-compliance with me) with an average of our account (4.10) and a standard deviation (1.68) and at a high degree, followed by item, which states (reduced the areas of creativity in answering exams) at an average account (4.05) and a standard deviation (1.61) and at a high degree, followed by the paragraph that states (led education Electronic to the advent of social relations among students) with a mean (3.81) and standard deviation (1.46) and at a high degree, and

Table 4 The statistics of the positives of e-learning during the COVID-19 epidemic (organized in sort of importance)

Paragraph number	Advantages of e-learning	Arithmetic means	Standard deviation	Degree	Rank
5	It enabled me to retrieve information whenever I needed it	4.45	1.61	Very high	1
6	He increased my participation in educational topics with all boldness	3.96	1.71	High	2
8	Save time and effort in learning processes	3.79	1.86	High	3
1	He encouraged me to communicate and exchange experiences in various fields of education	3.66	1.66	High	4
4	E-learning increased my motivation to learn	3.35	1.91	Average	5
9	He worked on providing proper feedback that helped raise my efficiency	3.30	1.38	Average	6
2	The educational materials were presented in a better way than traditional teaching	3.16	1.56	Average	7
10	Provide equal opportunities for all students	3.13	1.51	Average	8
3	E-learning contributed to facilitating my understanding of the educational material	3.05	1.58	Average	9
7	Increase effective communication between me and the faculty member	3.03	1.54	Average	10
The total score for the advantages of e-learning		4.79	1.79	Very high	

Table 5 The statistics for e-learning negatives under the COVID-19 pandemic

Paragraph number	Advantages of e-learning	Arithmetic means	Standard deviation	Degree	Rank
8	Limit the encounters between me and my fellow students	4.57	1.14	Very high	1
3	E-learning is not suitable for practical courses	4.36	1.25	Very high	2
1	It increased the psychological pressure on me	4.21	1.50	Very high	3
7	Increase the burden required of me	4.13	1.48	High	4
6	It made me feel isolated by not meeting my colleagues	4.10	1.68	High	5
10	Reduce my creativity in answering exams	4.05	1.61	High	6
9	E-learning has chilled social relations among students	3.81	1.46	High	7
5	E-learning is not suitable for the offered courses	3.79	1.33	High	8
2	Weaker interaction between me and my fellow students	3.48	1.56	High	9
4	E-learning increased my behavioral crises	3.34	1.43	Average	10
The total score for the negatives of e-learning		4.82	1.97	Very high	

finally the paragraph that provides for (e-learning does not suit the courses raised) at an average account (3.79) and standard deviation (1.33) and at a high degree. While the lowest negative items of e-learning came to a moderate degree, the item that states (e-learning increased my behavioral crises) with a mathematics mean (3.34) and a value of standard deviation (1.43), and finally the item that indicates (weaker than the interaction between me and my fellow students) with an average Arithmetic (3.48) and standard deviation (1.56).

(Q3) What are the outputs of the e-learning in Saudi universities during the COVID-19 epidemic?

It is evident from Table 6 that the e-learning outcomes were high in general with an arithmetic mean (4.15) and a value of standard deviation (1.27), and the highest paragraphs of e-learning outcomes came the paragraph that states (refining my performance skills through the use of educational websites) through an arithmetic average (4.06) and a value of standard deviation (1.23) with a high degree. While the lowest paragraphs of the e-learning outputs came to a moderate degree, the paragraph that states (Increased knowledge exchange between me and my colleagues) by an arithmetic mean (3.18) and a value of standard deviation (1.54), and the paragraph that indicates (I have developed creative thinking skills) with an arithmetic mean (3.24) And a standard deviation (1.36).

7 Hypothesis Tests

Based on the empirical findings of our study, we find a significant difference at the threshold of significance in the reality of the e-learning in Saudi universities and its impact on educational outcomes during the COVID-19 pandemic, according to different variables: (gender, academic year, university, location of residence).

Looking at Table 7, it becomes clear that the results of the study show the absence of significant differences at the threshold of significance of $(0.05 \geq \alpha)$ in the reality of e-learning in Saudi universities and its impact on education outcomes in light of the COVID-19 pandemic depending on the gender variable, as the statistical significance was < 0.05 It is not statistically significant. This is owing to the reality of the e-learning and the results obtained are similar for males and females, and this confirms that the professionals and disadvantages of e-learning are really and experienced by all students, males, and females.

Looking at Table 8, it becomes clear that the results of the study show the absence of a statistically important changes at the threshold of significance of $(0.05 \geq \alpha)$ in the authenticity of the e-learning in Saudi universities and its impact on the education outcomes during the COVID-19 pandemic depending on the school year variable, where the statistical significance was > 0.05, which is not statistically significant. This is due to the fact that e-learning and the results obtained are similar for all students, regardless of their academic level, and this confirms that the pros and cons of e-learning are real and felt by all students.

Looking at Table 9, it bescomes clear that the results of the study show a statistically important changes at the threshold of significance of $(0.05 \geq \alpha)$ in the authenticity of the e-learning in Saudi universities and its impact on the education outcomes during the COVID-19 pandemic depending on the university variable, where the statistical significance was < 0.05, which is Statistical function. To get out the source of the discrepancies, the LSD test is utilized to indicate the differences, according to the Table 10.

Table 6 The statistics of the reality of e-learning outcomes through the COVID-19 epidemic, organized in sort of importance

Paragraph number	Advantages of e-learning	Arithmetic means	Standard deviation	Degree	Rank
5	Hone my performing skills through the use of educational websites	4.06	1.23	High	1
6	Enhance my ability to collect and formulate information	3.76	1.45	High	2
10	Save time and effort	3.54	1.64	High	3
8	I gain academic knowledge through the links he provides	3.53	1.52	High	4
4	It gave me problem-solving skills	3.48	1.47	High	5
7	Enhance my ability to analyze information in a logical way	3.45	1.37	High	6
1	Enhance student control over the educational process	3.34	1.37	Average	7
9	It increased my ability to put what I learn into the correct fields	3.25	1.44	Average	8
3	I developed my creative thinking skills	3.24	1.36	Average	9
2	It increased knowledge exchange between me and my colleagues	3.18	1.54	Average	10
The total score of the e-learning outcomes		4.15	1.27	High	

Table 7 The "T" test for the significance of discrepancies in the reality of the e-learning and its impact on educational outcomes through the COVID-19 epidemic, depending on the gender variable

Sex	Number	Arithmetic means	Standard deviation	Degrees of freedom	T value	Statistical significance
Male	150	3.18	0.861	149	−1.341	0.241
Female	50	3.51	0.702	49		

Looking at Table 10, it becomes clear that the differences were between Islamic University in Madinah and Taibah University and in favor of Islamic University in Madinah with a difference of (0.76900), and between Islamic University in Madinah and Arab Open University and in favor of Islamic University in Madinah with a difference of (0.59200), and between Islamic University in Madinah and Rayyan Colleges, and in favor of the Islamic University in Madinah, with a difference of (0.65200). This is due to the fact that Islamic University in Madinah is one of the first universities to practice e-learning, and therefore its students have more knowledge than others about the pros and cons of e-learning.

Looking at Table 11, it becomes clear that the results of the study show the absence of a statistically important changes at the threshold of significance ($0.05 \geq \alpha$) in the authenticity of the e-learning in Saudi universities and its effect on education outcomes during the COVID-19 pandemic depending on the place of residence, where the statistical significance was > 0.05, which is not statistically significant. This is owing to the reality that the e-learning and the results obtained are similar for all students, regardless of where they live, and this confirms that the pros and cons of e-learning are real and felt by all students.

There is a statistically considerable influence at the significance threshold ($0.05 \geq \alpha$) of the positives of e-learning in Saudi universities based on the impact of the COVID-19 pandemic on e-learning outcomes.

It is evident from Tables 12 and 13 that the correlation coefficient of e-learning positives and its effect on e-learning outcomes is (0.956), which is a strong correlation coefficient, while the value of R square indicates the ratio of what the independent factor explains from the dependent factor, or the ratio of what can be explained by the dependent factor by the independent operator.

The data in Table 14 indicates the significance of the regression, that is, there is a considerable link amongst the positives of the e-learning in Saudi universities and its effect on e-learning outcomes, as the statistical significance was < 0.05.

It is evident from Table 15 that the outputs of e-learning change according to the advantages of the e-learning, so that whenever the reality of the e-learning changes by one degree, the learning outcomes change by (0.968) degrees. That is, the more e-learning advantages increased by one degree, the e-learning outcomes increased by (0.968) degrees, and vice versa.

In this case, there is a statistically considerable influence at the degree of a significance ($\alpha\ 0.05 \geq$) of the negatives of e-learning in Saudi universities in through the impact of the COVID-19 pandemic on e-learning outcomes.

Table 8 The statistical test indicating the differences in the reality of e-learning in Saudi universities and its influence on education outcomes through the COVID-19 pandemic, depending on the school year variable

Academic year	Number	Arithmetic means	Standard deviation	The source of the contrast	Sum of squares	Degrees of freedom	Average of squares	The computed q value	Statistical significance
First	20	3.53	0.506	Between groups	1.408	8	0.493	1.627	0.186
Second	80	3.59	0.521						
Third	20	3.61	0.583	Inside groups	36.254	191	0.384		
Fourth and more	80	3.81	0.519						
Total	200	3.64	0.648	**Total**	38.419	199			

Table 9 The statistical test for the significance of the differences in the reality of e-learning in Saudi universities and its impact on education outcomes during the COVID-19 pandemic, depending on the university variable

Academic year	Number	Arithmetic means	Standard deviation	The basis of the contrast	Sum of squares	Degrees of freedom	Average of squares	The computed q value	Statistical significance
Taibah University	100	3.22	0.738	Between groups	1.649	12	0.257	1.738	0.003
Arab Open University	20	3.45	0.694						
Islamic University in Madinah	60	3.50	0.638	Inside groups	27.458	187	0.374		
Rayyan Colleges	20	3.61	0.570						
Total	200	3.71	0.607	**Total**	28.362	199			

Table 10 LSD test for discrepancies in the reality of e-learning corresponding to the university indicator

University	Taibah University	Arab Open University	Islamic University in Madinah	Rayyan Colleges
Taibah University			−0.76900	
Arab Open University			−0.59200	
Islamic University in Madinah	0.76900	0.59200		0.65200
Rayyan Colleges			−0.65200	

It is evident from Tables 16 and 17 that the negative correlation coefficient of e-learning and its effect on e-learning outcomes is (0.519), which is a weak correlation coefficient, while the value of the R square indicates the ratio of what the independent factor explains from the dependent factor, or the ratio of what can be explained by the dependent factor by The Independent Worker.

The data in Table 18 indicates the significance of the regression, that is, there is a considerable association amongst the negatives of the e-learning in Saudi universities and its effect on e-learning outcomes, as the statistical significance was < 0.05.

It is evident from Table 19 that the outputs of e-learning change according to the disadvantages of the e-learning, so that whenever the disadvantages of the e-learning change by one degree, the learning outcomes change by (0.597) degrees. Any words that increased e-learning negatives by one degree decreased e-learning outcomes by (0.597) degrees, and vice versa.

8 Conclusion

Based on the empirical results obtained in our paper, we find the following results:

- The advantages of e-learning in Saudi universities are generally high.
- E-learning enables the student to retrieve information when needed.
- The negatives of e-learning were generally high.
- E-learning reduced the encounters between students and colleagues.
- E-learning is not suitable for practical courses.
- E-learning increased the psychological pressure on the student, and also increased the burden required of him.
- E-learning made the student feel isolated due to not meeting his colleagues.
- E-learning reduced the students' areas of creativity, especially in the field of exams.

Table 11 The statistical test of the significance of the differences in the reality of e-learning in Saudi universities and its impact on education outcomes through the COVID-19 pandemic, depending on the place of residence

Academic year	Number	Arithmetic means	Standard deviation	The basis of the contrast	Sum of squares	Degrees of freedom	Average of squares	The computed q value	Statistical significance
City	180	3.57	0.473	Between groups	1.732	5	0.325	0.835	0.281
Village	20	3.91	0.462	Inside groups	25.908	194	0.405		
Total	200	3.28	0.638	Total	27.614	199			

Table 12 The arithmetic means and standard deviation for the advantages of e-learning and e-learning outcomes

Variables	Arithmetic means	Standard deviation	Sample size
Advantages of e-learning	3.84	0.851	200
E-learning outputs	3.57	0.839	200

Table 13 (R) values to find the correlation coefficient of the advantages of e-learning and e-learning outcomes

Sample	R value	R square	A modified R box	Estimated standard error
1	0.956	0.938	0.936	0.948

Table 14 Analysis of variance to find out the relationship between the advantages of e-learning and the achievement of e-learning outcomes

Sample	The source of the contrast	Sum of squares	Degrees of freedom	Square rate	F	Statistical significance
A	Decline	98.807	1	98.807	575.001	0.000
	Residual	20.566	198	0.234		
	Total	119.373	199			

Table 15 Transactions that show the impact of the advantages of e-learning on its outcomes

Sample	Unstandardized transactions		Standardized transactions	T-test value	The moral significance
	B	Standard error	Beta		
Constant	0.517	0.140	0.965	3.930	0.000
X	0.968	0.047		21.796	0.000

Table 16 The arithmetic means and standard deviation for the negatives of e-learning and e-learning outcomes

Variables	Arithmetic means	Standard deviation	Sample size
Advantages of e-learning	4.070	0.902	200
E-learning outputs	3.784	0.889	200

Table 17 (R) values to find the correlation coefficient of the negatives of e-learning and e-learning outcomes

Sample	R value	R square	A modified R box	Estimated standard error
1	0.519	0.506	0.503	1.059

Table 18 Analysis of variance to find out the relationship between the negatives of the e-learning in Saudi universities and the achievement of e-learning outcomes

Sample	The basis of the contrast	Sum of squares	Degrees of freedom	Square rate	F	Statistical significance
A	Decline	18.603	1	94.855	69.370	0.000
	Residual	65.660	198	0.225		
	Total	84.263	199			

Table 19 Transactions that demonstrate the impact of e-learning negatives on e-learning outcomes

Sample	Unstandardized transactions		Standardized transactions	T-test value	The moral significance
	B	Standard error	Beta		
Constant	4.487	0.364	−0.522	11.345	0.000
X	−0.597	0.128		−4.793	0.000

- E-learning led to a cooling of social relations between students.
- E-learning is not suitable for the offered courses.
- E-learning outcomes were generally average.
- E-learning led to the refinement of students' performance skills through the use of educational websites.

The empirical findings suggest that there are no statistically important disagreements at the significance level (0.05 ≥ α) in the authenticity of e-learning in Saudi universities and its impact on the education outcomes in light of the COVID-19 pandemic, depending on the variables of sex, school year, and place of residence. While it was observed that there are statistically considerable disagreements corresponding to the university indicator, and in favor of Islamic University in Madinah students.

The results indicated that e-learning outcomes change according to the advantages of e-learning, so that the more e-learning advantages increase by one degree, the e-learning outcomes increase by (0.956) degrees and vice versa. While it was found that there is an adverse effect between the outputs of e-learning and its negatives, so that whenever the negatives of e-learning increased by one degree, the outputs of e-learning decreased by (0.597) degree, and contrariwise.

9 Policy Recommendations

Based on the results showed in our paper, we have the following recommendations:

- Saudi universities should work on designing educational materials that are appropriate for e-learning and keep pace with its requirements.

- Training the lecturers well on the mechanisms and techniques of using e-learning.
- The lecturers should work as much as possible to give equal opportunities to all students during the lectures.
- Lecturers should increase effective communication between them and students.
- Work on designing practical courses that are compatible with e-learning.
- Conducting introductory courses for students on the e-learning mechanism, and reducing the burdens required of the student.

Acknowledgements The Authors are thankful to the anonymous reviewers of the journal for their extremely helpful recommendations to increase the quality of the study.

Conflicts of Interest The authors declare no conflict of interest.

Funding This research received no external funding.

References

1. Al-Ayyad, Y., Al-Omari, M. (2015). The degree of availability of web design standards on Yarmouk University's website from the viewpoint of information and communication technology professionals, (published research). *Al-Manara Magazine, 21*(2), 1–28.
2. Al-Dabbasi, S. I. M. (2002). The effect of using distance learning on female students 'achievement. *Journal of Educational Sciences, 15*(2), 773–795
3. Al-Hiyari, A. (2019). Factors that influence the use of computer assisted audit techniques (CAATs) by internal auditors in Jordan. *Academy of Accounting and Financial Studies Journal, 23*(3), 1–15.
4. Al-Jamal, S. S. A. (2020). The difficulties facing Palestinian university students in the distance learning system (e-learning) in light of the Corona crisis. *International Journal of Research and Studies (IJS), Pioneers of Excellence Academy for Training, Consulting and Human Development, 2*(6), 201–233.
5. Almaiah, M. A., & Al-Khasawneh, A., Althunibat, A. (2020). Exploring the critical challenges and factors influencing the E-learning system usage during COVID-19 pandemic. *Education and Information Technologies, 25*(3), 5261–5280.
6. Al-Titi, M. A. A., & Hamayel, H. J. (2016). The reality of e-learning in Palestinian universities in light of knowledge management from the point of view of its faculty members. *Journal of Al-Quds Open University for Educational and Psychological Research and Studies, 5*(18), 195–210.
14. Deghash, A. (2019). Balanced scorecard application and its challenges. *International Journal of Business Ethics and Governance, 2*(1), 30–56. https://doi.org/10.51325/ijbeg.v2i1.41
7. Favale, T., Soro, F., Trevisan, M., Drago, I., & Mellia, M. Campus traffic and e-Learning during COVID-19 pandemic. *Computer Networks, 176*(7), 107290.
8. Fee, K. (2009). *Delivering eLearning: A complete strategy for design, application, and assessment*. Kogan Page.
9. Kilani, T. (1998). Distance education: Its philosophy, capabilities, foundations and educational media. *Journal of the Association of Arab Universities, 34*, 79–93.
10. Mailizar, A., Almanthari, A., & Maulina, S., Bruce, S. (2020). Secondary school mathematics teachers' views on e-learning implementation barriers during the COVID-19 pandemic: The case of Indonesia. *EURASIA Journal of Mathematics, Science and Technology Education, 16*(7), 1–9.

11. Mehra, V., & Omidian, F. (2011). Examining students attitudes toward e-learning: A case from India. *Malaysian Journal of Education Technology, 11*(2), 13–18.
12. Radha, R., Mahalakshmi, K., Kumar, V. S., & Saravanakumar, A. R. (2020). E-Learning during lockdown of Covid-19 pandemic: A global perspective. *International Journal of Control and Automation, 13*(4), 1088–1099.
13. Sorokina, H (2012). The collaborative study in the virtual classroom: Some practices in distant learning carried out in a Mexican public university: Universidad Autonomy Metropolitan-Azcapotzallco (UAM- A), Mexico City. World Conference on E-Learning in Corp., Govt., Health., & Higher Ed. 1, 1541-1543.

Robots in the Neighborhood: Application and Criminalization of the Artificial Intelligence in Education

Farhana Helal Mehtab and Arif Mahmud

1 Introduction to Artificial Intelligence

When humans and machines work together, the result in many cases had made history. This bonding is ancient and in recent days, humans are trying to make advanced machines in such a model so that they could perform human tasks with intelligence. These particular machines are designated as Artificial Intelligence or AI [1]. There is a misconception of comparing AI with a conscious machine [2–4]. AI is not any advanced machine capable of performing intelligence tasks but it's the system of creating such machines [5]. From a wide viewpoint, McCarthy has recognized Ai as "the science and engineering of making intelligent machines, especially intelligent computer programs. It is related to the similar task of using computers to understand human intelligence, but AI does not have to confine itself to methods that are highly biologically observable" [6]. While taking any decision humans use a cognitive mechanism, on the other hand, the output provided by any AI is calculated and analyzed data, which is not similar to cognition mechanism. The machines are only capable of analyzing human enciphered data within the prescribed protocols. The result is provided by the AI after pointing the data patterns [2–4]. The experience of AI is assembled by analyzing past similar incidents from a huge number of data. Nowadays, "Big Data" plays a crucial role to address new challenges utilizing AI, and thus "Data could be considered as the new oil" [7]. The recent advancement in the field of AI required a more harmonized definition like Kaplan and Haenlein. They have described AI as "a system's ability to correctly interpret external data, to learn from such data, and to use those learning's to achieve specific goals and tasks

F. H. Mehtab · A. Mahmud (✉)
Daffodil Smart City, Daffodil International University, Ashulia, Dhaka, Bangladesh
e-mail: arifmahmud.law@diu.edu.bd

F. H. Mehtab
e-mail: farhana@daffodilvarsity.edu.bd

through flexible adaptation" [8]. The mythical figure of AI engaging in thoughtful conversation or solving critical problems using cognition is created by science fiction writers and the entertainment industry [9]. Only lower-order human activities could be conducted by the AI because of the design and limitations. The higher-order tasks are still beyond the capacity of the AI like decision making or problem-solving etc. [10]. If we get rid of fiction, the capacity of AI is now limited to particular low-setting activities similar to determining right and wrong responses [11].

2 Development of Artificial Intelligence

Ancient philosophers had recognized the process of thought similar to "Automatic manipulation of symbols", and Alan Turing pointed it as the first AI concept [12]. The concept of human-type machines can be traced back to ancient Greek mythologies. Those artificial intelligent machines were claimed to be created by Egyptian engineers [13]. The Research and Development (RAND) Corporation funded a project name "The Logic Theorist" which was capable of solving small problems and considered as the first initiative to create an AI program [14]. It was presented in the "Dartmouth Summer Research Project on Artificial Intelligence" in 1956 by Allen Newell, Cliff Shaw, and Herbert Simon [15]. The term "Artificial Intelligence" was first used by American computer scientist John McCarthy [16] and thus many people considered him as the founding father of AI [17]. A mathematical problem-solving program was invented by him in 1950 which was operated by the computer [18]. Although he is considered the founding father of AI, the modern development was done by a British cryptanalyst named Alan Turing. He was the person who broke the German Enigma codes while the Second World War was ongoing [19]. Both the conception of "Machine learning" and "Computer vision" demanded a mechanism to analyze a huge quantity of information, and this created an obstacle for AI researchers till 1990. This crisis period is known as the "AI Winter" [20]. "IBM's Deep Blue", an AI-based program in May 1997 defeated Garry Kasparov, who was a champion of world cheese [21]. After the beginning of government funding for the development of machine learning and computer hardware around the world, significant changes were observed and the USA was top in the list. The "Big Data" conception evolved when both public and private institutions began storing a large amount of data for their convenience [22]. Tech giant companies like Amazon or Google started using machine learning to store, analyze and process a large quantity of data. In 2014, world-renowned scientist Stephen Hawkings described the creation of AI as one of the crucial and game-changing inventions of all time human history. A chatbot in 2014 won an AI competition in the "Turing test. The test was conducted in the Royal Society in London and the bot was named "Eugene Goostman" [23]. In 2018 an AI program named "Project Debater" was launched by IBM which competed in an abstract topic-based debate competition against two champion debaters. AI-based cybersecurity firms raised on a global scale from 2015 to 2018 and the estimated value was USD3.6 billion [24]. The AI-based programs have already entered into the main tier by engraving a significant mark in our social life.

3 Taxonomy of Artificial Intelligence

AI is capable of performing various acts with different capacities and adaptability and the classification could be made from various aspects [25]. The primary classification or taxonomy depends on the replicating capability of an AI of functions that are by nature exclusively performed by the human. This because the principal goal of AI is to change virtual programs into some beings which are capable of holding intellect [26]. From the point of adaptability and capability, generally, we can divide the AI into 7 types. The first and earliest type is the "Reactive machines" with limited functions and can't provide results analyzing previous analogical data. Another limitation is that it can't store data because of the limited memory [17, 27]. In 1997, Grandmaster Garry Kasparov was defeated by a reactive machine named "IBM ® Deep Blue ®". The second type of AI is the "Limited Memory Machines" which developed experience and knowledge from previous analogical data or pre-provided information [28]. One of the best examples of limited memory machines is automated vehicles. This type of machine is performed on the observational patterns gathered from data fed from the earlier [29]. The third category is the AI with a theory of mind which is capable of communicating with various authorities after calculating their requirement, perception, and thought. The theory of mind is like a secret ingredient or seasoning which assisted the AI program to empathize with the requirements of other entities like humans, machines, or programs [30]. Sophia and Kismet are examples of AI with a theory of mind. The former is developed by Professor Cynthia Breazeal and the latter by Hanson Robotics' (Hanson Robotics, Robots). Both robots are human-like and programmed to give facial expressions on the different emotional peripheral. The fourth kind is known as the "Self-aware AI" which only exists in hypothesis till now because AI holding human consciousness only possible in fiction [31]. Self-preservation is a vital pre-requirement to create self-aware machines and some experts believed that it could extinct the human race [32]. An artificial machine or program will be self-aware when its performance will go beyond the ability to replicate human actions. The idea suggests that self-aware AI would be able to express emotion, or capable of having desires like a living being. "Artificial Narrow Intelligence or ANI" is the fourth kind that covered a large area of machines with limited performance to complicated machines with multi capabilities. The principal characteristic of ANI is that it could independently perform a single task at the same time because of the definite capacity based on previous commands or protocols [33]. The fifth kind is "Artificial General Intelligence or AGI" which is conceptualized as a human-like machine or program which is capable of creating and innovate [34]. The advantage of AI is its speed, it could analyze data way quicker than humans within a very short time but humans are superior for their creativity, reasoning, decision-making power, and innovation [35]. The idea of ANI denotes a program competent to apply to the reason for taking decisions and to solve problems [34]. The last type is called "Artificial Superintelligence or ASI" which would be superior to its human creator. The ability of AI to create or to solve unknown problems is the characteristic of ASI. Nick Bostrom, an Oxford philosopher described ASI as "any intellect that greatly exceeds the cognitive performance of humans in virtually all domains of interest" [33].

4 General Applications of Artificial Intelligence

The AI is comforting our daily life by addressing rigid problems with great pace and efficiency. In our social life, AI is becoming our new neighbor, from household activities to a complex set of tasks, everything is getting new shape by the touch of AI. The application of "Artificial Narrow Intelligence or ANI" is gradually increasing in different industrial sectors like cybersecurity, education, healthcare, banking, and entertainment. "Voice and Speech Recognition or VSR" is capable of identifying the distinctive biological feature of every human's voice and used that for recognizing a specific person. The process of linguistics, acoustics, and natural language is blended to create the VSR mechanism [36]. The development of "Deep Learning" and "Big Data" has created a great influence on the VSR industry and it could be subcategorized into the following domain: Speech recognition systems, Speaker identification, Speaker verification, and Text speech programs. Operating system-based virtual assistances like Siri, Cortana or Bixby use Speech recognition systems to identify the device owner or the commands provided. Immigration and crime investigative agencies use Speaker identification to identify peoples. Nowadays smart home appliances or smart speakers like Amazon Echo or Google Home use Speaker verification systems and programs like Amazon's Ivona or Google text to speech converts texts into voice and vice versa. Machines are capable to differentiate various entities and identifying them, the process is done by image analysis from several sources. That means it is possible for the AI to respond according to the image feed to it and this category is called "Computer Vision", capable of storing data examining images and various associated information. Computer vision is getting popular and widely used in the financial, health care, automotive, agriculture, or manufacturing sectors [37]. In the case of motion detection, recognizing a specific object from several other objects, and crime scene reconstruction; computer vision is used globally [38]. Some other tasks require the application of computer vision like creating 3D models from 2D images, automated inspection, and animal species identification [39]. Another recent application of AI is in the form of "Extended Reality or XR" which is a fusion of VR or Virtual Reality, AR or Augmented reality, and MR or Mixed Reality. This fusion is applied to create a live production environment with an absolute enticing environment by the intermixture of virtual or real worlds. The application of XR provides the viewer a realization to be a part of the virtual ecosystem apart from their physical world while staying in the real-life [40]. It is expected that shortly, the application of XR will be larger in the health care and engineering sector but now it is limited to the entertainment industry like virtual gaming or virtual theater [41]. "Game Playing or General Game Playing or GGP" is another field where AI is used for a long time. The previous programs were created to compete human players in different intellectual games like chess or puzzle solving. The framework's predominant computational speed is abused by the GGP to initiate a brute force algorithm [42]. The algorithm constructs a diverse choice, steps, and outputs as conceivable. The GGP was introduced with a multi-interface mechanism where it was intended to run the gaming programs in more than one interface [43]. Game Definition Language (GDL) is required to guide the AI program. GDL is like a game manual containing the basic description to run the game [44]. E-commerce is becoming very popular in this global village where "Virtual Agents or VA" provides the customer's different types of automated services. While providing such services as assisting peoples to the suitable objects or answering queries, the AI program used digital characters and protocols which are pre-determined. These digital characters are constructed by computer generation, animation, and AI generation [45]. A set of live tool tools for mimicking the behavior of the agent and centralized cloud storage constituted the brain of the VR. "Natural Language Processing or NLP" is a nexus between human language and computer programs that is competent to analyze and process large data sets of natural language. The application of NLP enables machines to collect data after processing human language, in other sense it can be said that it allows AI to read. Like machine learning, NLP is also being very popular as it got a close connection with language, and most of the human functions are somehow connected with language [46]. Some very important applications of NLP mechanisms are document translation, document summarization, and question answering [47].

5   The Impact of Artificial Intelligence in Education

The power of Artificial intelligence (AI) can transform the operation of the educational system by thriving the institutional competitiveness as well as empowering the learners and instructors. Artificial intelligence is capable of creating differences in educational institutions by acting as an equalizer in education. The main requirements to gain a foothold with AI by the institutions are technologies, tools, and expertise. According to Research and Markets, "The analysts forecast the Artificial Intelligence Market in the US Education Sector to grow at a CAGR of 47.77% during the period 2018–2022 [48]. One of the far-reaching impacts of Artificial Intelligence in education is that it could change the teaching and learning strategy by maximizing the success of the students. The teacher or instructor could manage more times for the students with the assistance of the collective intelligence tools as these are capable of performing specific tasks like grading papers, which saves a lot of time [49]. Artificial Intelligence can also help to create customized curriculums for students, translate teachers' presentations into different languages, and granting classroom access to students from remote locations. Artificial Intelligence can play an important role to assist struggling students and direct them to the right path. The Syracuse University's School of Information Studies has developed an Artificial Intelligence-based website name "Our Ability" to conduct experiential learning and for job placement of the students. Artificial Intelligence also creates opportunities for students with hearing or visual impairments or in cases where the student is suffering from illness. Artificial Intelligence has changed the modes of online education in different fields of study [50]. In the past web-based education simply denoted downloading contents, content analysis, and passing by doing assignments [51]. Now Artificial Intelligence-driven online learning includes intelligent and adaptive web-based systems which are capable of adjusting the curriculum by identifying the behavior of the teacher and student [52]. According to Chassignol et al., the role of Artificial intelligence in education could be incorporated into three headings namely: administration, instruction or teaching, and as a medium of learning [53].

5.1 Artificial Intelligence as Education Administrator

In the educational sector, Artificial Intelligence is performing various functions like administrative functions and management activities. Different administrative functions of the teachers like providing feedback after assessing student submissions could be done effectively by Artificial Intelligence. Adaptive and Intelligent Web-based Educational Systems (AIWBEs) is an AI program that is capable of providing grading guides to the teacher for assessment, grading, and delivering feedback [54]. Knewton, an AI program with built functions is capable of providing feedback to students for improvement after evaluating their performance [55]. Several functionalities are provided by AI tutoring systems to assist the teachers in performing administrative tasks like providing grades and feedback for improvement [56]. Besides, plagiarism checking, grading, and rating or improvement feedback facilities are now largely AI-based and several programs are available like TurnItIn, Grammarly, TEXT FIXER, PaperRater, etc.

5.2 Artificial Intelligence as Education Instructor

Using Artificial Intelligence as a tool of pedagogy or instruction has created a major impact on education by improving the effectiveness and quality of the teacher. When relevant content could b delivered in line with the curriculum and according to the needs and capabilities of the learners, the system measurement result is efficient and quality full. On the other hand assessment data on the achievement of the learners show the effectiveness of the system. According to Pokrivcakova as AI is an adaptive system that is based on technology, it is possible to maintain instructional quality and effectiveness by ensuring an optimized learning experience. That means the system provides individualized learning materials and contents base on the needs of the learner [54]. In the web learning environment, Artificial Intelligence is capable of ensuring course material dissemination more effectively both in developing curriculum and providing instruction or delivery of content. Roll and Wylie argued that the tutoring AI program is designed to mitigate various challenges of one-on-one teacher-student tutoring and as a result, it focuses on upgrading teacher or instructor quality [57]. Undoubtedly it can be said that Artificial Intelligence is protecting academic integrity through providing plagiarism detection services as well as proctoring students' activities on different platforms like TurnItIn or Grammarly [58]. Gamification AI programs, team viewer applications, or VR programs are playing a significant role in pursuing instructional effectiveness and efficiency [59]. The human-like appearances and the conversational capabilities of the humanoid robots are also playing a crucial role in fostering instructional quality by promoting successful engagement with learners [60].

5.3 Artificial Intelligence as a Medium of Learning

The students learning experience has been impacted greatly by the use of Artificial Intelligence. According to Rus et al., AI-based learning programs foster deep learning. As an integral part of the system is consisted of conversational agents and AI works with it. For that reason, students are encouraged to provide valid reasoning behind their stands until the explanation is adequate and satisfactory, as a result, receptivity of information has increased [61]. AI easily tracks the progress of learning from and capable of using the outputs for students' capability enhancement. That greatly motivates the students to improve their skills and capability [53]. Pokrivcakova argued that Artificial Intelligence has created a positive impact on learning by developing an adaptive learning system capable of customizing learning contents according to individual students learning needs and the same pattern of teaching is also followed by the program [54]. Teaching students with AI-based learning simulations and similar technologies, they experienced a practical learning experience and that improves the learning quality. Different studies indicate that the learning becomes usable, enjoyable, motivational, and interesting when it has been done using virtual reality-based learning simulation and 3-D technology [62]. Some studies reveal that students' honesty and integrity are encouraged and fostered by the use of Artificial Intelligence-based learning programs through using different writing revision or writing assistance tools such as Pearson's Write-to-Learn [63]. Some adverse effect has been also suggested by the researchers such as there is scope for the students to use paper churning sites or platforms and that could encourage dishonesty and jeopardize academic integrity [64]. The number of students in educational institutions is increasing day by day and Artificial Intelligence can play a significant role in lessening the burden of the teachers by providing customized course materials after scrutinizing the syllabus [53]. After proper evaluation, the program is capable of generating as well as grading assessments and that helps the teacher allocating more time to student performance and development. Individual learning plans could be created by the teacher with the help of the AI program and it could also evaluate examination scripts following rubrics and benchmarks to break human-related bias issues [65]. AI-based computer vision can detect and analyze handwritten scripts in image format and the system could prevent plagiarism and malpractices among students. In the orthodoxical education system, all students are treated equally but the AI can provide specific feedback to every student early in the education after detecting shortcomings [66]. All students cannot be benefited from the same method of teaching but the AI could help to determine the separate method required for an individual student based on their skill, capability, and personality. In this way, suitable career paths for every individual student can be suggested by the AI after collecting personal academic data.

6 Criminalization of Artificial Intelligence

The recent technological advancement has made AI a significant part of our social life to mitigate different day-to-day challenges [67]. Like the other technologies, perpetrators are using AI-based programs to commit crimes and the peace and stability of the society, as well as the individuals, has been threatened. The capacity of AI to perform autonomous action is facilitating the commission of different cybercrimes. Unlike human criminals, in many cases, it is impossible to predict the actions of AI-based crimes [68]. As the AI with machine learning capacity is a human-made program, in many cases we can trace independent development as it is responsive to information and in case of no exact programming [69]. In some cases, this autonomous nature of an AI has resulted in malafide activities although the creator had no idea about it neither is was foreseeable. After close observation of the virtual pattern of behavior, big datasets create preferable victim profiles using "Machine Learning". Chatbots are often used by criminals to steal sensitive information from peoples after long conversations. During these conversations, the bots deceived peoples by providing false identification information and in many cases, the obtained personal and sensitive information's are used for blackmailing [70].

The unpredictable nature of AI has imposed a great challenge to explain some functions, especially autonomous actions. Tracing the online activities of the AI is challenging and the hardest part is to determine the intentions or motives behind any particular activities. An AI creating such deadlock is termed a "Black box" system. We can't blame any person for the autonomous action of an AI, which resulted in a criminal offense and was conducted without the creator's direct control or direction. The complex design and the safety protocols inside an AI, sometimes deny the creators access or command for regaining access, once the AI is in motion [71]. Between various modes of AI-based crimes, the most commons are discussed here. AI has been used to impersonated video and audio. The admissibly of digital evidence before the court has made the digital evidence very credible before the court of law as the conclusive evidence. The admissibility of digital evidence has been threatened because of the fake content created by "Deep Learning" and "Generative Adversarial

Networks". Algorithmic detection could detect audio and video impersonation, but the rate is low and the challenges are huge [72]. Criminals are using automated vehicles as a weapon because it is easily accessible around the world in comparison to the other equipment's necessary for general crimes like chemicals, firearms or explosives. As a result, low organizational overheads requirement of these automated vehicles are preferable by the "lone actor" or "Individual actors" terrorists. Some terrific examples of such attacks are 2017's Manhattan terror attack, 2017's Westminster attack, and 2016's Berlin lorry attack. Vehicle terrorism is alarming because the terrorist could be performed remotely and within a very short time, a single perpetrator could attack several places as the automated vehicles don't require any driver. Another type of AI criminalization is "Highly-tailored phishing". Phishing is a method by which perpetrators collect personal information through fraudulent platforms, websites, emails, or malware installment in the operating system of a computer that is sent from a deceptive source, and the source is purporting to be popular or dependable. People are deceived by the fraudulent looks of the interfaces and shared sensitive information with the perpetrators [73]. The perpetrators use the most popular or trendy brands, vibes, or practices to send generic messages for AI-based phishing attacks [74]. To process the response of the victims, the perpetrators use machine learning and those collected responses are used to response maximization [75]. "Learning-based cyber-attacks" are different from ordinary cyber-attacks. Using sophisticated mechanisms, the general attacks are targeted towards definite people like the DDoS attack [76]. Learning-based cyber-attacks are a kind of hybrid cyber-attack as it is target-oriented and extensive at the same time. The hybrid AI program is capable of bombarding attacks and also identifying the weakness of the system firewall simultaneously. Automated drones are now used for committing several criminal activities as they could be controlled by the pilot from a remote place and through applying radiofrequency. Smugglers use drones for moving arms, drugs, illegal substances, and explosives into prisons or over the borders. For transport disruption, the use of radio-controlled non-autonomous drones are also used in many cases [77]. Now the criminals don't stay inside a confined place to commit crimes, instead, they roam around places by using autonomous or driverless drones, that don't require the pilot to reside inside the transmission range of the drone [78]. Face recognition is widely used now in the mobile or computer operating system and as a security method to prove the identity of the device owner. To track suspected persons or to identify passengers, a face recognition AI system is used by immigration. In the past, we could find the record of the "Morphing" attack, which uses a single photographic ID to bypass the facial recognition security systems by presenting the single ID for passing multiple people [79]. Besides the discussed modes of criminal application of AI, there are some other emerging media or low types of are: Data poisoning, online eviction, burglar bots, AI-authored fake news and false reviews, AI assisted stalking and forgery, terrorism using military robots, snake oil cryptographic method, adversarial perturbations or other methods for evading AI detection.

7   Criminalization of AI Targeting Online Education

During the COVID-19 pandemic Crisis, most of the educational institutions were forced to shift their academic and administrative works online and online learning based on the distant learning management platforms is the new normal. According to the World Economic Forum, by the end of April 2020, almost 1.2 billion children across 186 countries around the globe were impacted by school closures [80]. While distance learning is playing a vital role globally in the response to the pandemic, cybercriminals are using Artificial Intelligence based programs to exploit the victims. The new normal was threatened by AI-based cybersecurity incidents like hacks of distance learning management systems, video conferencing platforms security breaches, ransom attacks on school districts, and phishing scams of learners. A report of Microsoft Security Intelligence shows that in June 2019, the education industry alone encountered 61% of the total number of 7.7 million malware attacks around the globe [81]. In 2019, during the global lockdown, COVID-related email spam distribution increased 6,000% and the target was educators and learners. A report of the MS-ISAC Center for Internet Security revealed that educational institutions were the most impacted sector for demanding payment from educators and learners to give back their stolen data using ransomware [82]. Research shows that in the last six years more than 35 higher educational institutions were attacked by ransomware in the UK although no one paid to the offender [83]. The University of California San Francisco for recovering their academic data in June 2020 paid $1.14 million in Bitcoin [84]. In August 2020, almost half a million dollars was paid to the offenders by New York City-based Monroe College after an AI-based cyberattack targeting the college's website, email, and learning management system [85]. In November 2020, in Baltimore County, Maryland; a ransomware attack shut down public schools and almost 100,000 students were forced to stop online classes [86].

DDoS (distributed denial of service) attacks affected online educational resources and platforms and most of the attack commands were received by Artificial Intelligence bots from the perpetrators. According to a report made by the Kaspersky DDoS Intelligence System, The rate of DDoS attacks affecting online educational institutions or resources increased by 550% in January 2020 when compared to January 2019 [87]. In 2020, from February to June, the number of DDoS attacks targeting online educational resources or institutions grows 350–500% greater in comparison to the same period of 2019. During the same period, various AI-based malware or threats were distributed through the online learning platforms or applications provided for video conferencing. The rate increased almost 20,455% in comparison to the same period of 2019. The perpetrators used Zoom and Moodle for luring victims, spreading threats, or gaining administrative controls. If the attacker could take administrative controls on Moodle, by uploading malicious plug-ins, they could install a PHP backdoor. Sensitive information of the students or teachers could be stolen using backdoors easily [88]. Phishing interfaces stole information under the guise of online platforms which are popular for conducting classes. In Russia, the infection rate was highest, 59 attempts per 1000 users, and second were Germany with the rate of 39 infection attempts per 1000 users in 2020. Phishing is one of the oldest forms of cybercrime and when educational institutions are attacked by it, it is quite unexpected. Following the switch of the popular distant learning management system platform like Zoom or Google Meet, a host of phishing websites disguise theirs programs [89]. Check Point Research discovered almost 2550 domains connected to "Zoom" among those 320 were suspicious and 32 were malicious [90]. Similar suspicious domains were registered for other popular video conferencing platforms like Google Meet and Microsoft Teams. These pages tricked the users to click links that install malicious AI-based software which deceived a person for providing their sensitive information to the cybercriminals. This information could be used by cybercriminals for several malafide purposes like gaining access to victim's other accounts as many of them use the same password for several accounts, stealing personal funds, or launching spam attacks. Universities provide online platforms accessing which faculty and students can seek several academic services and stored resources. In 2020, cybercriminals created phishing pages mimicking university login pages to steal the information of the staff, faculty, and students [91]. Different modes of phishing emails are used to lure the students or faculties that are notified that a meeting has been missed, a particular class has been canceled, or to click the following link for activating your account, etc. Anyone clicking the links run the risk to provide personal information in the hand of the criminals.

8 Challenges for Regulating Artificial Intelligence

Technological development has made the use of AI ubiquitous in our social life to tackle day-to-day challenges [106]. Technological development is not a blessing every time as many technologies are used to the occurrence and facilitation of criminal activities and the independent and autonomous nature of AI has made it more vulnerable [92]. Keeping in mind the abuse of technologies, obstructing the development is not a suitable alternative but for mitigating these challenges, we need to formulate strong policies and strong regulations. The criminalization of AI is generally done by either direct involvement of AI by the criminals for the commission of a crime committed to forming unforeseeable actions or mistakes of the AI. Cybercrimes are different from ordinary crimes and the AI facilitated cyber-attacks are unique from the former types. As a result, common regulation mechanisms are not effective or adequate to tackling the challenges inflicted by AI-based crimes, that's why we need to conduct distinct research and development. The exclusive nature of AI has imposed difficulties and challenges to enact effective regulation. Without physical infrastructure AI research can't be conducted and to protect intellectual property interests confidentiality maintenance is a challenging issue as several contributors from a different location will be engaged [93, 94]. Some of the major challenges are liability addressing, balancing the innovative and regulatory interest, and jurisdiction determination. For mitigating the regulatory challenges, the primary requirement is to determine the technological liability principles [95]. Under modern criminal law, a person should be penalized if he failed to comply with the vicarious liability or has committed an act that is recognized as an offense under the penal law [96]. For establishing criminal liability, the existence of legal personality is a prerequisite but under national or International law, no legal personality has been granted to the AI and it is transferred to the creator or user [97]. The task of imposing liability becomes more difficult when an unforeseeable offense has been committed by an autonomous AI and the creator was completely unaware of it. To establish criminal liability, four criminal liability models for AI have been suggested by Gabriel Hallevy, these are "Direct liability, Perpetration-by-another, Command responsibility, and Natural probable consequence". The direct liability model suggests that "if the design is poor and the outcome faulty, then all the [human] agents involved are deemed responsible" [98]. Any person responsible for making malafide changes to the AI system will be liable, and it could either any person or any legal personality. According to Ronald Arkin, the perpetration-by-another model suggests that while creating and designing an AI, programs should be included so that any criminal command will be rejected by it and the overriding capacity will only be confined to the developer [99]. It would possibly abolish the ambiguity for determining criminal intention as well as liability. The command responsibility model talks about the responsibility of the superiors for the action of this inferior. When the inferior performed any act with malafide intention and the superior after having adequate knowledge about that act, failed to refrain him from performing will be liable and AI would be included in this list in the future [100]. The natural probable consequence model measures the probable consequence of any act, which has resulted in harm, and the user or developer will be liable if the harm is probable and natural. The negligence or recklessness of the user or developer should be proved for establishing liability [101]. Utilitarian innovation could be resulted due to the prolonged development of AI and it's crucial to mitigate the conflict between strict regulations enactment and innovation. In some cases, the software development process is no free from initiating glitches or bugs, and these are considered as the inherent part. Innovation could be obstructed if programmers are made liable for criminal intent too easily and that is a very challenging part for the legislative authorities while enacting strict policy or regulation. A clear and unambiguous definition of the words "Liability", "Foreseeability" and "Causation" [93, 94]. To make the process of research and development more secure and safe, the publication of code might be made mandatory under the law. The nature of cyber-crime has already imposed significant challenges for both law enforcement agencies and legislators. It is possible for a person staying in the northern part of the world to launch a DDoS attack on any server, situated in the southern part of the world. Making a false online identity is very easy and on the opposite locating the criminal is very hard, as a result in AI-based crimes determination of jurisdiction is one of the big challenges. All the most common types are categorized under seven heading by the "ITU toolkit", which is considered as the first AI jurisdiction framework [102]. The ITU toolkit has suggested some legal doctrines to enacting an effective legal framework globally and to create harmony between the states to settle jurisdictional disputes in AI crimes. The first doctrine is the "Territoriality principle" which enables the state to conduct a trial on the national court where the offense is committed within its territory. When the crime has primarily been committed within a state territory but was completed outside the state territory, in that particular case the "Ubiquity doctrine" permits a state to claim jurisdiction to try such offenses. When an offense has been committed inside the state territory (including aid and abet), based on that effect, a state can claim jurisdiction under the "Effects doctrine". "The flag principle", extends the territorial principle to ships and aircrafts flying under the flag of the country.

9 Conclusion

The use of Artificial Intelligence in different segments of our social life is increasing at a great speed. By providing support to the instructors and learners, Artificial Intelligence has started to make its marks as an excellent assistant. We are entering into a technological regime where an ever-increasing variety of tasks are performed by autonomous machines with a self-learning capability and the education industry is not an exception. From the modes of computer-based program, Artificial Intelligence in education is taking the form of remote learning, distant learning management platform, or web-based educational platforms. Now a day's chatbots are performing the task of an instructor or teacher. With the application of the embedded system, robots in the modes of humanoid robots or cobots are now playing the role of an instructor colleagues, or instructor. The instruction quality of the teacher has improved because the application of these tools and programs has strengthened the effectiveness as well as efficiency. As Artificial Intelligence is capable of providing customized syllabus and course materials to the individual needs of students, students are experiencing an improved learning environment. Despite the significant development and advancement of AI, the AI consciousness is still fiction. Like the dark side of every technology, AI is now used by criminals to fuel cybercrimes and to commit cyber-attacks on a larger scale. The secure society and the people, we need new regulations which are adaptive to the change. As the AI is incapable of holding consciousness, in case of any crime committed, the creator or developer could not escape the liability. While distance learning is playing a vital role globally in the response to the pandemic, cybercriminals are using Artificial Intelligence based programs to exploit the victims. The new normal was threatened by AI-based cybersecurity incidents like hacks of distance learning management systems, video conferencing platforms security breaches, ransom attacks on school districts, and phishing scams of learners. Cybercriminals are using Artificial Intelligence to targeting a huge amount of victims in shorter times and machine learning has been used to create target profiles. Phishing, malware distribution, spyware spreading, and many other emerging forms of cybercrimes are taking place with the aid of Artificial Intelligence. As online life and education is the new normal, it is high time to formulate effective regulation to punish AI crimes and in the long run, to prevent such technological malfeasance. Sooner rather than later, we have to scrutinize the AI regulation's advantages and loopholes. To determine the criminal liability and for passing strong regulations, the recommended doctrines could play an important role. We believe that our findings will allow other peoples to rethink their positions and precept on AI and while enacting a regulation for preventing AI-based crimes.

References

1. Russell, S., & Norvig, P. (2010). *Artificial intelligence: A modern approach* (3rd ed., p. 1). Pearson.
2. Surden, H. (2014). Machine learning and law. *Washington Law Review, 81,* 89. https://scholar.law.colorado.edu/articles/81/.
3. Surden, H. (2014). Machine learning and law. *Washington Law Review, 81,* 87, 89. https://scholar.law.colorado.edu/articles/81/.
4. Surden, H. (2014). Machine learning and law. *Washington Law Review, 81,* 89–90. https://scholar.law.colorado.edu/articles/81/.
5. Kurzweil, R. (1990). *The age of intelligent machines.* MIT Press.
6. McCarthy, J. (2007, March 5). What is artificial intelligence? Stanford University. Retrieved from http://www-formal.stanford.edu/jmc/whatisai/.
7. Arthur, C. (2013, August 24). Tech giants may be huge, but nothing matches big data. The Guardian. Retrieved from https://www.theguardian.com.
8. Kalpan, A., & Haenlein, M. (2019). Siri, Siri, in my hand: Who's the fairest in the land? On the interpretations, illustrations, and implications of artificial intelligence. *Business Horizon, 62*(1), 15–25.
9. Ebiri, B. (Mar. 6, 2015). The 15 best robot movies of all time. Vulture. Retrieved from https://www.vulture.com/2015/03/15-best-robot-movies-of-all-time.html.
10. Krupansky, J. (June 13, 2017). Untangling the definitions of artificial intelligence, machine intelligence, and machine learning. Medium. Retrieved from https://medium.com/@jackkrupansky/untangling-thedefinitions-of-artificial-intelligence-machine-intelligence-and-machine-learning-7244882f04c7.
11. Deshai, R. (2017, March 23). Artificial Intelligence (AI). Retrieved from http://drrajivdesaimd.com/2017/03/23/artificial-intelligence-ai/.
12. Turing, A. M. (1950). Computing machinery and intelligence. *Mind, LIX, 236,* 433–460. https://doi.org/10.1093/mind/lix.236.433
13. Lewis, T. (2014, December 1). A brief history of artificial intelligence. *Live Science.* Retrieved from https://www.livescience.com/49007-history-of-artificial-intelligence.html#:~:text=The%20beginnings%20of%20modern%20AI,%22artificial%20intelligence%22%20was%20coined.
14. Anyoha, R. (2017, August 28). The history of artificial intelligence. Harvard University Blog, special edition on artificial intelligence
15. McCarthy, J., Minsky, M., Rochester, N., & Shannon, C. E. (1955). A proposal for the Dartmouth summer research project on artificial intelligence. Retrieved from http://raysolomonoff.com/dartmouth/boxa/dart564props.pdf.
16. Moor, J. (2006). The Dartmouth College artificial intelligence conference: the next fifty years. *AI Magazine, 27*(4), 87–91.
17. Ray, S. (2018, August 11). History of AI? Towards data science. Retrieved from https://towardsdatascience.com.
18. McCarthy, J. (1978). History of LISP. History of programming languages, 173–185. doi:https://doi.org/10.1145/800025.1198360
19. Copeland, B., & Proudfoot, D. (1999). Alan turing's forgotten ideas in computer science. *Scientific American, 280*(4), 98–103. https://doi.org/10.1038/scientificamerican0499-98
20. Buchanan, B. G. (2006). A (very) brief history of artificial intelligence. *AI Magazine, 26*(4), 53–60.

21. Schultebraucks, L. (2018, December 5). A short history of artificial intelligence. Dev. Retrieved from https://dev.to.
22. Manyika, J. (2011). Big data: The next frontier for innovation, competition, and productivity. Retrieved from McKinsey Global Institute website: https://www.mckinsey.com/~/media/McKinsey/Business%20Functions/McKinsey%20Digital/Our%20Insights/Big%20data%20The%20next%20frontier%20for%20innovation/MGI_big_data_exec_summary.ashx
23. Sample, I., & Hern, A. (2014, June 9). Scientists dispute whether computer 'Eugene Goostman' passed Turing test. Retrieved from https://www.theguardian.com/technology/2014/jun/09/scientists-disagree-over-whether-turing-test-has-been-passed.
24. Chaturvedi, A. (2018, October 7). 13 major Artificial Intelligence trends to watch for in 2018. Geospatial World. Retrieved from https://www.geospatialworld.net
25. Avitz, D. (2020, February 14). What is classification in artificial intelligence?. Aglopix. Retrieved from https://algopix.com/questions/what-is-classification-in-artificial-intelligence.
26. Joshi, N. (2019, Jun 19). 7 types of artificial intelligence. Forbes. Retrieved from https://www.forbes.com/sites/cognitiveworld/2019/06/19/7-types-of-artificial-intelligence/?sh=3e3fa71d233e.
27. Ray, S. (2018, August 18). Four Types of AI. Codeburst.io. Retrieved from https://codeburst.io/four-types-of-ai-6aab2ce57c19.
28. Reynoso, R. (2019. March 27). 4 main types of artificial intelligence. Learning Hub. Retrieved from https://learn.g2.com/types-of-artificial-intelligence.
29. University of Michigan, Center for Sustainable Systems. (2020). Autonomous Vehicles Factsheet. Retrieved from http://css.umich.edu/factsheets/autonomous-vehicles-factsheet.
30. Fan, S. (2018, September 19). Thinking like a human: what it means to give AI a theory of mind. SingularityHub. Retrieved from https://singularityhub.com/2018/09/19/thinking-like-a-human-what-it-means-to-give-ai-a-theory-of-mind/.
31. Chatila, R., Renaudo, E., Andries, M., Gottstein, R., Alami, R., Clodic, A., et al. (2018). Toward self-aware robots. Frontiers. Retrieved from https://doi.org/10.3389/frobt.2018.00088.
32. Musser, J. (2017, June 19). What the rise of sentient robots will mean for human beings. Retrieved from https://www.nbcnews.com/mach/tech/what-rise-sentient-robots-will-mean-human-beings-ncna773146.
33. Jajal, T. (2018, May 21). Distinguishing between Narrow AI, General AI and Super AI. Mapping out 2050. Retrieved from https://medium.com/mapping-out-2050/distinguishing-between-narrow-ai-general-ai-and-super-ai-a4bc44172e22.
34. Berruti, F., Nel, P., & Whiteman, R. (2020, April 29). An executive primer on artificial general intelligence. McKinsey&Company. Retrieved from https://www.mckinsey.com/business fun ctions/operations/our-insights/an-executive-primer-on-artificial-general-intelligence.
35. Goertzel, B. (2015). Artificial general intelligence. *Scholarpedia*. https://doi.org/10.4249/scholarpedia.31847
36. Russell, S. J., & Norvig, P. (2003). *Artificial intelligence: A modern approach* (2nd ed.). Prentice Hall.
37. Peregud, I., & Zharovskikh, A. (2020, August 19). Computer vision applications examples across different industries. InData Labs. Retrieved from https://indatalabs.com/blog/applications-computer-vision-across-industries?cli_action=1606670210.172.
38. Morris, T. (2004). *computer vision and image processing*. Palgrave Macmillan.
39. Chen, C. H. ed. (2015). Handbook of pattern recognition and computer vision. World Scientific.
40. Culp, S. (2018, April 30). Virtual reality plus artificial intelligence can transform risk management. Forbes. Retrieved from https://www.forbes.com.
41. Riva, G. (2004). Distributed virtual reality in health care. *CyberPsychology & Behavior, 3*(6). https://doi.org/10.1089/109493100452255.
42. University of Toronto. (1999). Artificial Intelligence I Game Playing. Retrieved from http://psych.utoronto.ca/users/reingold/courses/ai/games.html.

43. Genesereth, M., Love, N., & Pell, B. (2005). General game playing: Overview of the AAAI competition. *AI Magazine, 26*(2), 62–72.
44. Love, N., Hinrichs, T., & Genesereth, M. (2006). General game playing: Game description language specification. Stanford Logic Group LG-2006-01. Computer Science Department, Stanford University, Stanford, Calif, USA.
45. Adam, C., & Gaudou, B. (2016). BDI agents in social simulations: A survey. *The Knowledge Engineering Review, 31*(3), 207–238.
46. Expert System. (2017, October 5). Natural language processing applications. Retrieved fromhttps://www.expertsystem.com/natural-language-processing-applications/.
47. Davydova, O. (2017, August 17). 10 Applications of artificial neural networks in natural language processing. Medium. Retrieved from https://medium.com.
48. Plitnichenko, L. (2020, May 30). 5 main roles of artificial intelligence in education. *eLearning Industry.* Retrieved from https://elearningindustry.com/5-main-roles-artificial-intelligence-in-education.
49. Ayoub, D. (2020, March 4). Unleashing the power of AI for education. MIT Technology Review. Retrieved from https://www.technologyreview.com/2020/03/04/905535/unleashing-the-power-of-ai-for-education/.
50. Devedžic, V. (2004). Web intelligence and artificial intelligence in education. *Educational Technology & Society, 7*(4), 29–39.
51. Kahraman, H., Sagiroglu, S., & Colak, I. (2010). Development of adaptive and intelligent Web-based educational systems. In *Proceedings of 4th International Conference on Application of Information and Communication Technologies* (pp. 1–5).
52. Peredo, R., Canales, A., Menchaca, A., & Peredo, I. (2011). Intelligent Webbased education system for adaptive learning. *Expert Systems with Applications, 38*(12), 14690–14702.
53. Chassignol, M., Khoroshavin, A., Klimova, A., & Bilyatdinova, A. (2018). Artificial intelligence trends in education: A narrative overview. *Procedia Computer Sciencess, 136,* 16–24.
54. Pokrivcakova, S. (2019). Preparing teachers for the application of AI-powered technologies in foreign language education. *Journal of Language and Cultural Education, 7*(3), 135–153.
55. Sharma, R., Kawachi, P., & Bozkurt, A. (2019). The landscape of artificial intelligence in open, online and distance education: Promises and concerns. *Asian Journal of Distance Education, 14*(2), 1–2.
56. Rus, V., D'Mello, S., Hu, X., & Graesser, A. (2013). Recent advances in conversational intelligent tutoring systems. *AI Magazine, 34*(3), 42–54.
57. Roll, I., & Wylie, R. (2016). Evolution and revolution in artificial intelligence in education. *International Journal of Artificial Intelligence in Education, 26*(2), 582–599.
58. Sutton, H. (2019). Minimize online cheating through proctoring, consequences. *Recruiting Retaining Adult Learners, 21*(5), 1–5.
59. Kiesler, S., Kraut, R., Koedinger, k., Aleven, V., & Mclaren, B. (2011). Gamification in education: What, how, why bother. *Academic Exchange Quarterly, 15*(2), 1–5..
60. Weiguo, W. (2015). Research progress of humanoid robots for mobile operation and artificial intelligence. *Journal of Harbin Institute of Technology, 47*(7), 1–19.
61. Kim, Y., Soyata, T., & Behnagh, F. (2018). Towards emotionally aware AI smart classroom: Current issues and directions for engineering and education. *IEEE Access, 6,* 5308–5331.
62. Mikropoulos, T., & Natsis, A. (2011). Educational virtual environments: A ten-year review of empirical research (1999–2009). *Computer Education, 56*(3), 769–780.
63. Murphy, R. (2019). Artificial intelligence applications to support K–1 2 teachers and teaching. RAND Corporation, Santa Monica, CA, USA, Tech. Rep. PE135. https://doi.org/10.7249/PE315.
64. Crowe, D., LaPierre, M., & Kebritchi, M. (2017). Knowledge based artificial augmentation intelligence technology: Next step in academic instructional tools for distance learning. *TechTrends, 61*(5), 494–506.
65. Estevez, J., Garate, G., & Graña, M. (2019). Gentle introduction to artificial intelligence for high-school students using scratch. *IEEE Access, 7,* 179027–179036.

66. Global Development of AI-Based Education. (2019). Deloitte Res., Deloitte China, Deloitte Company. Retrieved from https://www2.deloitte.com/cn/en/pages/technology-media-and-tel ecommunications/articles/development-of-ai-based-education-in-china.html.
67. West, D. M. (2018). *The Future of Work: Robots, AI, and Automation.* Brookings Institution Press.
68. Yasseri, T. (2017, August 25). Never mind killer robots — Even the good ones are scarily unpredictable. PHYS.ORG. Retrieved from https://phys.org/news/2017-08-mind-killer-rob ots-good-scarily.html.
69. Castelvecchi, D. (2016, October 5). Can we open the black box of AI?. Nature. Retrieved from https://www.nature.com/news/can-we-open-the-black-box-of-ai-1.20731.
70. Miles, B., & Avin, S. (2018). The malicious use of artificial intelligence: Forecasting, prevention, and mitigation. ReseachGate. Retrieved from https://www.researchgate.net/pub lication/323302750_The_Malicious_Use_of_Artificial_Intelligence_Forecasting_Preven tion_and_Mitigation.
71. Falkon, S. (2017, December 24). The story of the DAO — Its history and consequences, the startup. Retrieved from https://medium.com/swlh/the-story-of-the-dao-itshistory-and-conseq uences-71e6a8a551ee.
72. Güera, D., & Delp, E. J. (2018). Deepfake video detection using recurrent neural networks. In *Presented at the 2018 15th IEEE International Conference on Advanced Video and Signal Based Surveillance.*
73. Boddy, M. (2018). Phishing 2.0: the new evolution in cybercrime. *Computer Fraud & Security Bulletin, 2018*(11), 8–10.
74. Vergelis, M., Shcherbakova, T., & Sidorina, T. (2019, March 12). Spam and Phishing in 2018. Securelist.com. Retrieved from https://securelist.com/spam-and-phishing-in-2018/89701/.
75. Bahnsen, A. C., Torroledo, I., Camacho, L. D., & Villegas, S. (2018). DeepPhish: Simulating malicious AI. In *Presented at the APWG symposium on electronic crime research* (pp. 1–9).
76. Kushner, D. (2013, February 26). The real story of Stuxnet. IEEE Spectrum.
77. Weaver, M., Gayle, D., Greenfield, P., & Perraudin, F. (2018). Military called into help with Gatwick drone crisis. The Guardian. Retrieved from https://www.theguardian.com/uk-news/ 2018/dec/19/gatwick-flights-halted-after-drone-sighting.
78. Peters, K. M. (2019). *21st century crime: How malicious artificial intelligence will impact homeland security.* California: Monterey.
79. Andrews, J. T. A., Tanay, T., & Griffin, L. D. (2019). Multiple-identity image attacks against face-based identity verification. arXiv.org, vol. cs.CV (pp. 1–13).
80. Cathay, L., & Farah, L. (2020, April 29). The COVID-19 pandemic has changed education forever. *World Economic Forum.* Retrieved from https://www.weforum.org/agenda/2020/04/ coronavirus-education-global-covid19-online digital learning/.
81. Castelo, M. (2020, June 17). Cyberattacks increasingly threaten schools — here's what to know. EdTech Magazine Retrieved from https://edtechmagazine.com/k12/article/2020/06/ cyberattacks-increasingly-threaten-schools-heres-what-know-perfcon.
82. Krueger, K. (2019). cyber criminals are targeting students & teachers. education and career news. Retrieved from https://www.educationandcareernews.com/education-techno logy/cyber-criminals-are-targeting-students-teachers/#.
83. Insights. (2020, August 3). UK university ransomware FoI results. TopLine. Retrieved from https://toplinecomms.com/insights/uk-university-ransomware-foi-results
84. Mellor, C. (2020, August 18). UCSF ransomware attack: University had data protection but it wasn't used on affected systems. Blocks & Files. Retrieved from https://blocksandfiles.com/ 2020/08/18/ucsf-ransomware-attack-data-protection/.
85. Pierce, S. (2020, August 21). University of Utah pays more than $450,000 in ransomware attack on its computers. The Salt Lake Tribune. Retrieved from https://www.sltrib.com/news/ 2020/08/21/university-utah-pays-more/.
86. Oxenden, M. (2020, November 30). Baltimore County schools suffered a ransomware attack. Here's what you need to know. The Baltimore Sun. Retrieved from https://www.baltimore sun.com/maryland/baltimore-county/bs-md-co-what-to-know-schools-ransomware-attack- 20201130-2j3ws6yffzcrrkfzzf3m43zxma-story.html.

87. KASPERSKY. (2020, September 4). Digital education: The cyberrisks of the online classroom. Securelist. Retrieved from https://securelist.com/digital-education-the-cyberrisks-of-the-online-classroom/98380/.
88. Constantin, L. (2017, March 21). Flaws in Moodle CMS put thousands of e-learning websites at risk. CSO. Retrieved from https://www.csoonline.com/article/3183533/flaws-in-moodle-cms-put-thousands-of-e-learning-websites-at-risk.html.
89. Gatlan, S. (2020, March 30). Hackers take advantage of zoom's popularity to push malware. Bleepingcomputer. Retrieved from https://www.bleepingcomputer.com/news/security/hackers-take-advantage-of-zooms-popularity-to-push-malware/.
90. Peters, J. (2020, May 12). Hackers are impersonating Zoom, Microsoft Teams, and Google Meet for phishing scams. THE VERGE. Retrieved from https://www.theverge.com/2020/5/12/21254921/hacker-domains-impersonating-zoom-microsoft-teams-google-meet-phishing-covid-19.
91. Foley. (2019, March 4). Cornell university is hit by a phishing scam that leads to stolen passwords. CYCLONIS. Retrieved from https://www.cyclonis.com/cornell-university-hit-phishing-scam-leads-to-stolen-passwords/.
92. King, T. C., Aggarwal, N., Taddeo, M., & Floridi, L. (2019). Artificial intelligence crime: An interdisciplinary analysis of foreseeable threats and solutions. *Science and Engineering Ethics*. https://doi.org/10.1007/s11948-018-00081-0
93. Scherer, M. U. (2016). Regulating artificial intelligence systems: Risks, challenges, competencies, and strategies. *Harvard Journal of Law and Technology, 29*, 369.
94. Scherer, M. U. (2016). Regulating artificial intelligence systems: Risks, challenges, competencies, and strategies. *Harvard Journal of Law and Technology, 29*, 353.
95. Holder, C., Khurana, V., Harrison, F., & Jacobs, L. (2016). Robotics and law: Key legal and regulatory implications of the robotics age (Part I of II). *Computer Law & Security Review, 32*, 383–402. https://doi.org/10.1016/j.clsr.2016.03.001
96. Čerka, P., Grigienė, J., & Sirbikytė, G. (2015). Liability for damages caused by artificial intelligence. *Computer Law & Security Review, 31*, 376–389. https://doi.org/10.1016/j.clsr.2015.03.008
97. Hallevy, G. (2010). The criminal liability of artificial intelligence entities – from science fiction to legal social control. *Akron Intellectual Property Journal, 4*(2), 171–201.
98. Floridi, L. (2016). Faultless responsibility: On the nature and allocation of moral responsibility for distributed moral actions. *Royal Society's Philosophical Transactions A: Mathematical, Physical and Engineering Sciences, 374*(2083), 1–22. https://doi.org/10.1098/rsta.2016.0112.
99. Arkin, R. C., & Ulam, P. (2012). Overriding ethical constraints in lethal autonomous systems. Technical report GIT-MRL-12–01 (pp. 1–8). https ://pdfs.seman ticsc holar.org/d232/4a80d 870e0 1db4a c02ed 32cd3 3a8ed f2bbb 7.pdf.
100. McAllister, A. (2017). Stranger than science fiction: The rise of AI interrogation in the dawn of autonomous robots and the need for an additional protocol to the UN convention against torture. *Minnesota Law Review, 101*, 2527–2573. https://doi.org/10.3366/ajicl.2011.0005.
101. Wellman, M. P., & Rajan, U. (2017). Ethical issues for autonomous trading agents. *Minds and Machines, 27*(4), 609–624.
102. Tafazzoli, T. (2018, May). Cyber crime legislation. Retrieved from https://www.itu.int/en/ITU-D/Regional-Presence/AsiaPacific/SiteAssets/Pages/Events/2018/CybersecurityASPCOE/cybersecurity/Tafazzoli-cybercrime%20legislations.pdf.

Teaching Using Online Classroom Applications: A Case Study on Educational Institutes in Bangalore City

Kartikey Koti and Rupasi M. Krishnamurthy

1 Introduction

Regular teaching has been seen globally for quite a while. Review corridor instruction expects a predominant part, which has been reached out for a troublesome day directly from primary coaching, school, and specialists. Standard teacher-focused tech-niques focused on redundancy learning and maintenance should be abandoned, including understudy-centred and task-based approaches to managing to learn. Ordinary schooling is connected with significantly more grounded segments of terrorizing than has all the earmarks of being commendable now in numerous social orders. In web-based learning, understudies go to classes on the web and include genuine interactions with instructors and understudies at the opposite end. Web-based learning is a reality and gratuitously turning out to be important for formal instruction. Give an open trade of data among educators and understudies. Understudies can see learning material at their relaxation or even go to booked meetings or talks. The coming of online schooling has made it workable for understudies with occupied lives and restricted adaptability to acquire quality instruction. An examination was held in 2009; the examination informs us concerning the contrast among on the web and customary instructing. Universally electronic guidance has made it conceivable to bring to the table classes worldwide through a solitary Internet association. Up close and personal guidance doesn't depend upon organized frameworks. In web-based learning, the understudy is subject to admittance to an unhampered Internet association. Understudies can depend upon overseers to help in course choice and give scholarly suggestions.

K. Koti (✉)
Faculty, Department of Commerce, PES University, Bangalore, India

R. M. Krishnamurthy
Department of Commerce, HOD, PES University, Bangalore, India
e-mail: rupasimkrishnamurthy@pes.edu

© The Author(s), under exclusive license to Springer Nature Switzerland AG 2022　　411
A. Hamdan et al. (eds.), *Technologies, Artificial Intelligence and the Future of Learning Post-COVID-19*, Studies in Computational Intelligence 1019,
https://doi.org/10.1007/978-3-030-93921-2_23

Regular guidance is moreover called standard tutoring or customary preparing. The standard reasoning of ordinary tutoring is to pass on the characteristics, propensities, capacities and social practice to the future, which is significant for their perseverance. In traditional tutoring, the understudy gets some answers concerning the practices and custom of the overall population wherein he lives. This kind of guidance is generally conceded to the understudies using oral recitation.

There is no created work or realistic work [1]. The understudies fundamentally plunk down together and check out the teacher or another who will relate the activity. The ordinary prohibits made tests. Anyway, it consolidates some oral tests, which are not very formal. Ordinary preparation is far from the use of science and advancement. The guidance about the sciences we move today in unbelievable detail is given in the traditional tutoring system. The standard guidance system included data about customs, customs, and religions. That is the explanation. It is called customary tutoring.

1.1   Online Education System in India

There is a worldwide acknowledgement of the requirement for comprehensive instruction approaches during the pandemic. To make online schooling more powerful, open and more secure, different online assets, preparing projects and plans have been created by the Government of India for understudies, instructors and instructive organizations. The local instructing area has met up to shape a cross country casual and deliberate organization of educators, called the Discussion Forum of Online Teaching (DFOT), to examine different parts of internet educating and make vaults of fundamental assets. Cut-chime edge advancements like man-made reasoning (AI) has opened additional opportunities for inventive and customized approaches obliging distinctive learning capacities.

1.2   COVID Impact in Terms of Teaching Aspect

The Coronavirus has affected across the world, prompting the shutdown of schools and universities. Over 1.2 billion understudies are out of the homeroom in light of an emotional change across the world. Some learning applications are sans giving assistant to their administration. Hence, guidance has changed essentially, with the undeniable rising of e-learning, whereby teaching is endeavoured remotely and on cutting edge stages. With this sudden shift away from the homeroom in various bits of the globe, some riddle about whether the gathering of web learning will keep on persevering post-pandemic and what such a shift would mean for the general instruction market. The practicality of electronic learning shifts among age social occasions; a couple of understudies without reliable web access and advancement fight to look into modernized learning; it has been used extensively since mid-February after the

Chinese government instructed a fourth of a billion full-time understudies to proceed with their examinations through online stages.

2 Literature Review

Shivangi Dhawan 2020 Educational foundations (colleges) in India are presently dependent upon conventional methodology for comprehension; they keep the standard set up of extremely close talks in an assessment passageway. Even though different scholarly units have likewise begun mixed learning, many are still left with old strategies. The sudden scene of a dangerous disease called Covid-19, accomplished by a Corona Virus (SARS-CoV-2), shook the whole world. The World Health Organization explained it as a pandemic. The current circumstance attempted the planning framework across the world and constrained educators to move to an online technique for training until additional notification. Different scholarly affiliations that were hesitant to change their standard edifying methodology had no alternative to web educating learning. The article intertwines the significance of internet learning and Strengths, Weaknesses, Opportunities, and Challenges (SWOC) evaluation of e-learning modes in the hour of emergency. This article likewise reveals insight into EdTech new businesses' improvement during the hour of pandemic and disastrous events and joins considerations for scholarly relationship on the most proficient method to direct difficulties identified with web-based learning. As per the World Economic Forum, the Covid-19 pandemic has changed several get-togethers get and give arranging. To find new solutions for our issues, we may get some truly critical developments and change. Educators have gotten steady to standard systems for instructing as versus addresses, and likewise, they delay in enduring through any change. By and by, we have no other elective left amidst this crisis other than acclimating to the amazing condition and continuing through the change. It will be worthwhile for the orchestrate mentioned and could bring a titanic stack of amazing developments. We can't disregard and negligence to survey the understudies who don't progress toward all online progress. These understudies are less prosperous and have a spot with less-showed families with cash related resources necessities; subsequently, they may leave behind a remarkable possibility when classes occur on the web. They may leave behind a record of the enormous expenses related to automated contraptions and web data plans. This verifiable level package may grow the openings of trim sidedness.

Curtis Bonk 2006 Foundations of significant level preparing have progressively recognized online direction. The number of understudies picked distance programs is quickly moving in schools and colleges all through the United States. Considering these progressions in enlistment requests, different states, establishments, and affiliations have been chipping away at major intends to do web mentoring. Simultaneously, misinterpretations and legends identified with the trouble of training and learning on the web, advances accessible to help online bearing, the help and pay required for extraordinary teachers, and the necessities of online understudies make difficulties for such vision clarifications and planning documents. This assessment didn't explore authentic web-based preparing and learning rehearses. A few reactions were identified with late furies that might be reasonable.

What's more, we didn't study understudies to view internet learning patterns and potential outcomes. An investigation of understudies may show that they consider various innovations significant and on the cusp of critical development. In a student-focused world, who can more readily anticipate innovation drifts today—teachers or understudies? This examination also showed that mixed learning would be a more critical development territory than completely internet learning. Follow-up investigations may zero in on parts of mixed discovering that foundations need to address, like sorts of mixed learning, exercises that lead to mixed learning achievement, and educator preparing for mixed learning circumstances.

Elsie Sophia Janse van Rensburg [2] internet encouraging interest has expanded to guarantee openness and moderateness of advanced education. Understudy need backing to help with acclimation to the online setting. Instructors in the web-based training setting are confronting expanded understudy numbers, bringing about a higher responsibility. Successful internet encouraging practices can improve understudy and instructor exhibitions in wellbeing sciences. The point of this incorporated audit was to recognize compelling internet instructing and learning rehearses for undergrad wellbeing sciences understudies and educators. It is prescribed that exploration is done to explain the distinctions in online talks (content) and online modules that support dynamic cooperation. Explaining these online intercessions can help with the materialness thereof in a determined undergrad wellbeing science setting. As Tsang and Law (2018) [3] recommended, future exploration is expected to research explicit online plans that can explicitly improve learning results for undergrad wellbeing sciences understudies. Backing for understudies and instructors is required in the web-based education and learning setting to address difficulties and guarantee positive internet instructing and learning rehearses. These positive practices make a protected, online stage where teachers can encourage communitarian learning through dynamic interest to improve information and abilities advancement of undergrad wellbeing sciences understudies.

Mishra, Gupta and Shree, [4] The entire educational framework from easy to tertiary level was imploded during the novel Covid illness lockdown time 2019 (COVID-19) India similarly as across the globe. This assessment depicts internet encouraging learning modes embraced by Mizoram University to teach learning measure and coming about semester examinations. It anticipates a mentally advanced possibility for additional future scholarly unique during any difficulty. The organized motivation driving this paper endeavours to address the key essentialities of web instructing learning in planning amidst the COVID-19 pandemic and how could it be conceivable that current would assets of enlightening establishments sensibly change formal coaching into online direction with the assistance of virtual classes and other pushing on the web gadgets in this continually moving instructive scene. The paper utilizes both quantitative and conceptual ways to deal with the impression of educators and understudies on electronic teaching–learning modes and featured the execution example of web preparing learning modes. Progression, Privatization and

Globalization of getting ready has been weakened astoundingly considering bound adaptability and limitedly kept trade adventures of instructive exercises among the nations during the COVID-19 lockdown. The youthful nations are confronting procedure loss of development in managing the unexpected moving situation of informational arranging, the bosses and relationship during this pandemic with their broke explicit design, scholarly lack and nonattendance of assets; particularly among them, low and center compensation nations would drive forward through the difficulties most as they were by then running out of money (Tho-mas, 2020). Be that as it may, unmistakably, everybody should figure out some lifestyle choice and make do with the current emergency as it is the start just; no one can manage the cost of the neglectful Ness towards bleeding edge over the long haul change in HEIs. Making multimodal ways to deal with oversee accomplish course content concentrates for better learning result can be a preferred game plan over manage the diverse idea of web getting ready.

Sun and Chen [5] Using a hypothetical substance evaluation approach, this appraisal overviewed 47 spread appraisals and assessment on the web getting ready and learning since 2008, essentially focusing on how hypotheses, practices, and evaluations apply the electronic learning environment. The inspiration driving this paper is to offer sensible plans to people who are signed to choose online courses to make taught decisions in the execution affiliation. Considering the disclosures, the makers battled that convincing on the web heading is poor upon (1) particularly coordinated course content, prompted connection between the teacher and understudies, strong and solid and cared for educators; (2) arrangement of an impression of web learning territory; (3) quick advancement of progress. In doing this, it is acknowledged that this will stimulate a nonstop discussion of suitable approaches that can revive schools and labour force achievement in changing to teach on the web. Under current conversations on the cost and nature of huge level setting up, this evaluation could help improve unquestionable level getting ready and understudy enrolment and support. Internet arranging is here and is sure to remain and make. The framework of its game-plan of encounters irrefutably shows online course has developed immediately, filled by Internet accessibility, plan to set favourable to aggression and monster market. It has progressed from nineteenth-century correspondence tasks to the twenty-first century's dynamic and all-around coordinated institutional online submitments. We can expect that web arranging will continue to foster its quality and impact critical level getting ready through a substitute relationship of reshaping, refining, and evolving. It is unthinkable, regardless, to abrogate standard unquestionable level planning yet fundamentally to be another choice. Regardless, in light of its versatility, transparency, and moderateness, electronic coaching is getting in standing, especially for people who are throughout unacceptable to prepare to contemplate the certified distance, plan conflicts, and irrationally extreme costs.

Rapanta et al. [6] The Covid-19 pandemic has raised enormous difficulties for the general preparing neighbourhood. A specific test has been the urgent and startling mentioning for al-arranged inverse school courses to be told on the web. Electronic tutoring and learning propose a specific instructive substance (PCK), for the most part, identified with ar-going and gathering for better learning encounters and developing unmistakable learning conditions, with the assistance of modernized turns of events. This article gives some master experiences into this electronic learning-related PCK to help non-pro teachers (for example, individuals who have little commitment in web figuring out how) to research these problematic occasions. Our exposures feature learning practices with unequivocal characteristics, the blend of three sorts of exemplification (social, mental and facilitatory) and the essential for changing assessment to the new learning necessities. We end with contemplating how reacting to an emergency (amazingly well) may accelerate improved preparing and learning rehearses in the post-automated period.

From one perspective, the game plan of productive learning conditions and embeddings on the web advances can fill in as impetuses for educators to test new things, look at innovative changed other options and think about their practices. On the other hand, the request for quality guidance at undeniable level preparing affiliations, trailed by satisfactory appearance evaluation systems, is these days more pressing than as of now, as the post-pandemic measures of new and abroad understudies' cooperations in numerous colleges have essentially dropped For significant level preparing establishments all through the planet to be dead serious confirmation of workforce accessibility to the degree demonstrable ability is central. Web instructing is a focal piece of such expert status, in any case, not alone. Before long, like never before, schools should put resources into educator fit improvement of their staff to be strengthened on productive, instructive frameworks with or without online turns of events.

Scherer et al., [7] The COVID-19 pandemic has obliged progress to electronic preparing and learning (OTL) schools and colleges across the globe, expecting that instructors should change their educating in an exceptionally brief timeframe—self-administering of whether they were arranged. Drawing from a general delineation of $N = 739$ significant level preparing educators in 58 nations, the current assessment uncovers information into teachers' basis for OTL at the hour of the pandemic by (a) perceiving teacher profiles dependent upon a ton of key fragments of openness; (b) clarifying profile collaboration by specific teacher attributes, setting centered bits of the progress to OTL, and nation level markers watching out for enlightening new development and social heading. We drove inactive profile evaluation and perceived three instructor profiles with dependably high or low arranging or a conflicting availability profile—henceforth, teachers insignificant level preparing is not a homogeneous get-together. Key individual and setting centered segments, like instructors' sex and earlier OTL experience, the setting of the OTL move, the movement potential in planning, and social bearing, clarified profile enrolment. We talk about these exposures concerning the profiles, how they can be found concerning key determinants and their thoughts for OTL in high-level training. As a rule, the deferred outcomes of our exams-country recommend that educators insignificant level preparing are not a homogeneous gathering concerning their detailed preparation for OTL—yet, various subgroups of instructors exist which may require various techniques for help. Perceiving such favourable documents is essential to clarifying the heterogeneity among instructors and, at last, engage altered assistance for executing OTL. The profiles we saw in our assessment didn't just show constantly high or low degrees of openness yet, furthermore, showed that individual and setting centred status presumably would not go together. Like this, we fight that the preparation improvement is not diverse and requires taking an individual and setting centred viewpoint.

Similarly, the determinants clarifying profile backing may not affect all educators equivalently. The various tremendous affiliations influencing one party of talks might be intriguing for another, given various foundations and consideration in OTL. Additionally, as one of the profiles in our appraisal appeared, we saw backing and self-common sense most likely will not go vaguely. Unequivocally, phenomenal institutional help may not make up for little trust in training on the web. This wisdom revolves around the necessity for the two focuses to be tended to in the system to help instructors amidst OTL. To get a handle on these profiles, a broad degree on the potential determinants should be taken, going past specific instructor attributes and including factors that depict the setting of OTL, culture, and progress. In particular, we battle that the real level is fundamental to engage the limit working at establishments to help OTL. Irrefutably, unmatched data into instructors' plan profile is an enormous improvement towards seeing how to help them best progression to electronic learning. OTL educators' status goes past their self-possibility and showing presence and relies on the institutional, social, and improvement setting.

3 Objective of the Study

1. To study the behaviour of the faculties (Course facilitator) using online classroom application
2. To examine the perception of the faculties (Course facilitator) using online classroom application.

4 Research Methodology

The researchers have collected primary data for the study conducted in Bangalore city. The sample was collected based on the convenience of the researchers. The data were collected from various universities of Bangalore city, especially from private universities and institutes. For analyzing data, the researchers have used fundamentals statistical tools like frequency, percentages. And SPSS to find mean scores and another analytical part of the study.

4.1 Need for the Study

The need to examine this subject is to know how resources felt while instructing web-based during the pandemic and the assessments of different universities about internet educating through OSAs. Correspondence of information has been collected given investigation and examination of data and issues. A segment of the necessities to scrutinize this article is to decide definitional ambiguities and design the degree of the subject, give an organized, synthesized blueprint of the current status of data, survey existing methodological approaches and exceptional pieces of information, make determined constructions to oblige and loosen up past research encounters, existing openings, and future.

4.2 Scope for the Study

The study covers the comfortless of faculties in teaching online classes during covid-19. Samples were collected from various Universities. It focuses on how many problems the faculties faced and tells about the online classroom apps they have used so far. The location covered for taking the sample is only Bangalore. The primary motivation behind schooling is to accomplish up portability. Online courses affirmation programs have given cheap training to the majority and save time, energy, and cash. Electronic learning through affirmed online courses gives a wide scope of courses that consider the understudy's centre advantages, accordingly making a rich field for future progression. There is a lost thought that businesses favour understudies with conventional physical advanced educations. Despite what is generally expected, corporate associations in India perceive the high ability levels of understudies who have gone through online courses confirmation programs from profoundly acclaimed instructive foundations.

4.3 Limitation of the Study

The restriction on location was we could not exceed the place beyond Bangalore, where we chose some of the top colleges in Bangalore. Lack of cooperation while collecting the sample. The faculties gave limited responses. Respondents did not felt encouraging to provide accurate answers. Lack of understanding by the respondents. Lack of patience to fill the questionnaire. Time consumption is more as faculties show a lack of interest. Colleges did not provide permission to carry out the research. Lecturers were unavailable during the research. There were too many incomplete entries found after reviewing the samples. Difficulties in collecting the sample from the lecturers.

5 Result and Discussion

a. It was observed that 80% of the faculties conduct the online course, and only 20% of the respondents didn't conduct courses online before COVID 19

b. The applications used to conduct online classes were Zoom session which was 45%, G-Meet were 25%, WhatsApp were 20%, and MS Team were 4%, and Jio meets 6%.

c. The majority of the educational institute subscribed to Zoom, G Meet, 45% and 25%, respectively.

5.1 Usage of Online Classroom Apps—Part A

a. During the study, it was observed that 54% of respondents agreed with the fact that they think teaching via OCAs during the COVID-19 pandemic is secure from COVID.

b. Based on research, we can conclude that 76% of respondents think teaching via OCAs during the COVID-19 pandemic is reliable.

c. From the above sample, 56% of the applicants believe teaching via OCAs, even using computer devices shared by colleagues, during the COVID-19 pandemic is safe.

d. During the survey, 58% of respondents trust OCAs for teaching during the COVID-19 pandemic.

e. Overall, 72% of applicants think teaching online using OCAs during Covid19 will not harm them during the study.

f. In the research conducted, it was observed that 62% of faculties find that the level of student participation in class discussions to be low on OCAs.

g. From the question, difficulty keeping the students involved throughout the session on OCAs was only 38%.

h. From the above sample, it can be concluded that 54% of the faculties find it difficult to motivate the students in the class on OCA.

i. In the research conducted, we can notice that 54% of the respondents think students attending classes via OCAs are comparatively passive in contacting them regarding course-related matters.

j. 64% of the respondents find their workload to be higher when teaching on OCAs than traditional teaching.

k. During the study, it was noticed that 82% of the faculties need to be more creative in terms of the resources used for the courses on OCAs.

l. Based on the study, 84% of the applicants have to alter their teaching methods.

m. 74% of the faculties have to prepare more for OCA's as compared to traditional teaching.

n. During the survey, 70% of applicants find it difficult to use some of the technical methods.

o. 56% of the faculties do not find it stressful to teach using OCA.

p. 78% of the faculties are supported and encouraged by their institution to sue the OCA.
q. From the above sample, 70% of the respondents believe the institution's workshops and training helped them get familiar with OCA's.
r. 60% of the faculties think that their institution is keen to see that they are comfortable using the OCA.
s. We can notice that 62% of the respondents can rely on the technical support provided by their institution.
t. 32% of the respondents are neutral on whether to continue using OCA's for teaching even after COVID-19.
u. 40% of the respondents are not willing to continue teaching using OCA after COVID-19.
v. 38% of the applicants are not excited about using OCA's for teaching after COVID-19.
w. 50% of the people on whom the survey was conducted feel happy by teaching using OCA despite the COVID-19 fear.
x. During the survey, 51% of the respondents were not frustrated, bored and angry while teaching using OCA during the COVID-19 outbreak.

5.2 Perception of Using OCA'S—Part B

a. 85% of the faculties have the required technical skills to work with different OCA for teaching.
b. 66% of the respondents know the different features and different OCA's available for teaching in the market.
c. During the survey, 55% of the faculties think teaching using OCA's is relevant, convenient, and meaningful.
d. 56% of the faculties think that a part of class learning is ignored in the OCA class.
e. 70% of the respondents think teaching using OCA's is not effective and comprehensive in delivering student learning.
f. Based on research, 66% of the applicants think their teaching materials on OCA's can be shared and easily saved.
g. From the above sample, 36% of the faculties think that the negative effects and infectious nature of the COVID-19 are severe in educational institutions.
h. 42% of the staff desire to continue teaching using OCA's, whereas 25% avoid teaching using OCA.
i. Most applicants disagree that they are not teaching even using OCA due to avoiding interacting with others during a pandemic.
j. During the survey, 76% of the applicants expected the COVID-19 problem to be normal in Indian educational institutions very soon.
k. During the study, 80% of the respondents were female, and 20% were male.

l.   44% of the respondents' age were between 40 and 50 years, 32% were between 30 and 40 years, 22% were between 23 and 30 years.

m.   Faculties teaching Experience 42% of the respondents teaching experiences were between 5 and 10 years, 38% of the respondents teaching experiences were between 10 and 20 years.

n.   All the faculties were on a full-time basis in the private Universities and institute working.

o.   The respondents were working in 50% in private universities, 22% state University, Central Government University which was 6% and other 6% in private institutes. Other education organization and institutes were 16%.

6 Recommendations

a.   Most of the faculties lacked technical knowledge. The institution should provide ample knowledge on the OCA's to teachers to be updated with the required amount of knowledge.

b.   Motivating and grasping the attention of students was an arduous task. Therefore, the faculties should find out creative and motivating content to keep the attention of the students.

c.   Traditional methods did not apply to online classes. Hence, there was a need for alteration in the teaching methods by the faculties.

d.   56% of the respondents find it stressful to teach using OCA. The institution should provide stress-free methods or exercises to motivate and boost the faculties' energy levels. The common things in the pandemic that motivated people were Yoga, healthy gym competition, walking challenges, meditation etc.

e.   Online bullying and abuse of faculties took place during the tough times through OCA. Hence, the institution should take necessary and secure steps to maintain the dignity and respect of the faculties.

f.   Interaction and coordination with the students was a major drawback faced by the faculties through OCA. Student-friendly topics should be cultivated, and other necessary information should be provided to keep the students' interest.

g.   52% of the respondents feel uncomfortable with the idea of teaching using OCA due to the pandemic. As the COVID-19 was at its peak, there was no other option suitable other than this.

h.   Information security training should be customized to the audience.

7 Conclusion

As indicated by the World Economic Forum, the Covid-19 pandemic has in like manner changed how a couple of gathering get and give tutoring. We may get some really vital turns of events and change to find new solutions for our issues. Teachers have gotten normal to standard appearance techniques as eye-to-eye addresses, and in this manner, they puzzle over whether to recognize any change. However, we have no other elective left amidst this crisis other than acclimating to the remarkable condition and enduring the change. It will be profitable for the tutoring region and could bring a lot of surprising turns of events. This assessment didn't explore genuine online instructing and learning practices. A couple of responses were likely related to continuous patterns that may potentially be.

References

1. Joe (2020). References online learning: A panacea in the time of COVID-19 crisis. *Jounal of Educational Technology Systems*, *49*(1), 5–22; Article first published online: June 20, 2020; Issue published: September 1, 2020.
2. van Rensburg, E. S. J. (2006). The future of online teaching and learning in higher education. The Survey Says Published in number:4, 2006 at *Educause Quarterly*.
3. Tsang, O. T. Y., & Law, C. K. (2018). "Effective online teaching and learning practices for undergraduate health sciences students" An integrative review. *International Journal of Africa Nursing Sciences, 9*, 73–80.
4. Mishra, L., Gupta, T., Shree, A. Research-based teacher education from stake holders point: A case study. *Universal Journal of Educational Research, 8*(9), 3895–3901
5. Sun, A. Q., Chen, X. "Online education and its effective practice: A research review. *Journal of Information Technology Education: Research, 12*(3), .345–366.
6. Chrysi Rapanta, Luca Botturi, Peter Goodyear, Lourdes Guardia "Online university teaching during and after the Covid-19 crisis: Refocusing teacher presence and learning activity. *Postdigital Science and Education.* https://doi.org/10.1007/s42438-020-00155-y.
7. Scherer, R., Howard, S. K., Tondeur, J., Siddiq, F. Profiling teachers' readiness for online teaching in secondary education. *Universal Journal of Educational Research, 4*(7), 895–901.

Researchers' Motivation and Its Correlates: An Empirical Study Amid COVID-19 Pandemic in Arab Region

Ahmed H. Ebrahim⬤, Mai Helmy ⬤, Shahenaz Najjar⬤, Omar Alhaj⬤, Khaled Trabelsi⬤, Maha AlRasheed⬤, and Haitham Jahrami⬤

1 Introduction

Research is a fundamental catalyst and central for socio-economic growth, national competitiveness, and the global stock of knowledge advancement [53, 41], hence, public and private investment in research and development (R&D), as part of the sustainable development goals (SDGs), has become a national priority and an integral part of the knowledge economy worldwide. This includes efforts to increase the global share of researchers and stimulate the growth of human capital in areas of research, education, and innovation. Any society's development is measured by

A. H. Ebrahim (✉) · H. Jahrami
Ministry of Health, Manama, Kingdom of Bahrain

A. H. Ebrahim
Ahlia University, Manama, Kingdom of Bahrain

M. Helmy
Faculty of Arts, Psychology Department, Menoufia University, Shibin el Kom, Egypt

S. Najjar
Health Sciences Department, Arab American University, Graduate Studies, Ramallah, Palestine

O. Alhaj
Department of Nutrition, Faculty of Pharmacy and Medical Sciences, University of Petra, Amman, Jordan

K. Trabelsi
High Institute of Sport and Physical Education of Sfax, University of Sfax, 3000 Sfax, Tunisia

M. AlRasheed
Clinical Pharmacy Department, College of Pharmacy, King Saud University, Riyadh, Saudi Arabia

H. Jahrami
College of Medicine and Medical Sciences, Arabian Gulf University, Manama, Kingdom of Bahrain

© The Author(s), under exclusive license to Springer Nature Switzerland AG 2022   423
A. Hamdan et al. (eds.), *Technologies, Artificial Intelligence and the Future of Learning Post-COVID-19*, Studies in Computational Intelligence 1019,
https://doi.org/10.1007/978-3-030-93921-2_24

how far it has advanced in the domain of research and development. As a result, one may claim that society and countries differ based on their research and development (R & D) achievements [1]. According to the UNESCO data, the region of Arab states has a meager share of the world's researchers (1.9%) and low R&D spending as% of their GDP (0.5%), compared to the region of North America and Western Europe, which has 2.4% and 39.7% in terms of these parameters respectively [51]. Despite the underinvestment in research and low combined publication output in the Arab region (estimated at 24 scientific papers per 1,000 academic professors and full-time researchers), the rates of scientific production have witnessed a rapid and dramatic growth in many Arab states over the past decade; majorly because of the increased number of higher education institutions, stimulated funds for research and increased reliance on research international collaboration networks [8, 26, 29]. Such figures represent critical signs for (1) assessing the infrastructure (landscape of higher education and research), intellectual climate, and human competence for research and knowledge production in the Arabian region, and (2) importantly studying current researchers' productivity and motivations towards publication. This study investigates researchers' motivation, and its correlates through an empirical examination amid the COVID-19 pandemic in the Arab region. The paper comprises this introduction, background, methods, findings, discussion, followed by three sections aiming to (1) stress the importance of researching motivations in the context of technology, (2) highlight the limitations and suggestions for future research, and (3) conclude the study.

2   Background

2.1   Motivation for Research

Establishing a capital of competent, talented, and motivated researchers is one of the principal keys to a productive and high-performing national research system [45, 58]. Researchers' motivation plays a central role in augmenting the excellence and quality of knowledge production and publications; and subsequent positive implications on the scientific reputation and competitiveness of the institution [58]. Recent research implies that lack of extrinsic motivational factors—like research rewards/remuneration, professional development opportunities or encouraging culture/climate—has been linked with the brain drain phenomena in Arab countries [2]. Further, failure to warrant an effective research environment could demotivate scholars and scientists from continuing research at their home institutions [36, 3]. And despite the role of intrinsic motivational factors—like the ones related to personal investment, sense of self-determination, and engagement on attaining innovative outcomes [21, 19], there is a severe lack of research exploring how intrinsically Arab researchers are driven towards research conduct. Therefore, investigating motivations amongst the Arab research community for taking up and continuing research activities should be a priority for future higher education and research improvement strategies and policy-making.

2.2 Preferences and Attitudes Towards Research

Scholars' attitudes, preferences, and inclinations towards research are essential to determine the quality of produced knowledge and continuity in researching activities [40, 30]. Pragmatically, research attitudes are influenced by different factors, including but not limited to job satisfaction, motivation for research, and accessibility to resources and support (Ibid). Therefore, it is always necessary to dimensionally visualize the entire picture when investigating the subject of research attitudes, which could range from research topic selection, team selection, to journal selection. In recognition of the role of research attitudes in knowledge acquisition and dissemination, trends in research conduct have been studied from various dimensions. From the dimension of journal selection, a recent cross-sectional study in the Middle East and Africa found that authors' decision in selecting a target journal was significantly influenced by the journal's impact factor, indexation status, exclusion of article-processing charges, and the journal's international status [12]. Further, evidence shows that researchers' inclination towards publishing in an open-access journal could be potentially associated with beliefs of how likely those researchers undermine the value to the importance of peer review or intend to boost the accessibility of publication [44]. From the dimension of research collaboration, there has been increasing attention on scholarly work on research collaboration because of its importance in the advancement of a knowledge-based economy and innovation spillovers [35]. Additional, collaborative authorship, especially at the interinstitutional and international levels, has been recognized as a catalyst for improving research quality beyond the individualistic approach of research conduct [38]. Recent research focused on the Arab region, particularly the Arabian Gulf Countries, has largely stressed building a Collaborative Research Network (CRN) to expedite opportunities of knowledge creation and creativity stimulation. This has been a practical avenue to overcome issues of fragment research efforts and R&D-associated costs [5]. Worth noting, current literature has focused on studying graduate/doctoral students' attitudes toward research, while there is a drastic lack of research targeting scholars, academic staff, and senior researchers for understanding their desires and inclinations when conducting research.

2.3 Support for Research

Research support and resources represent a backbone for building scientific capacity. Many modern governments and institutions of research and higher education in developed countries have steered investments towards strengthening research support services, ratifying research incentives schemes, dedicating financial funds and grants, and warranting state-of-the-art technological and libraries infrastructure [31, 46]. Research surveyed Arab researchers about barriers and obstacles hindering their performance or stimulating their intention for emigration is plentiful. The outcomes of such research point towards the presence of complex issues that many researchers may encounter in this region; majorly, these include lack of a scientific milieu, negligence of indigenous knowledge development, inadequate research training, limited funding of research, and low incentive and credits [14, 16, 17, 33, 48, 49, 54]. These factors might ultimately demotivate both early career and senior researchers and mitigate opportunities for retaining competent scholars.

Based on the above context, assessing researchers' motivation, preferences, and attitude towards research in the Arab region have been seized as a primary problem of this research. This also includes exploring the impact of research support and resources on the attitude and motivation dimensions. An inclusive understanding of push and pull factors in research conduct could help in tailoring solutions and proposing plans aimed at enhancing the interest and approaches in scholarly activities among Arab researchers and scientific communities.

3 Methods

3.1 Participants, Sampling, and Ethical Considerations

This cross-sectional study targeted researchers in the Arab region by applying convenience sampling techniques of both snowballing and self-selection. The distribution of an online survey was based on both personal and academic social networks, from April 1, 2021, to May 06, 2021. The sample size comprised 205 researchers. Electronic informed consent was introduced and obtained from each participant with information about the purpose of the study, anonymity, confidentially of collected data, absence of monetary and non-monetary compensations, autonomy for withdrawal, and time required to complete the survey. The inclusion criteria for this study were: (1) participants undertaking research roles; (2) being a researcher in the Arab region (3) agreed to take part in the survey. The exclusion criteria were: (1) researchers or academics in the non-Arab region; (2) incomplete data on the online questionnaires. The study was approved by the Research Ethics Committee of Menoufia University. Additional, research ethical considerations and principles were embraced following the American Psychological Association (APA) and Declaration of Helsinki for human research [47, 9].

3.2 Measures and Procedures

An Arabic language, self-administered questionnaire was used to collect the data, as the native communication language in the Arab region is Arabic. The questionnaire comprised structured, closed-ended questions. There were no open-ended or continuing questions, for reasons of simplicity and response speed. Based on a pilot test, the research team estimated that it would take each participant around 6–8 min to fully answer the survey. The questionnaire was developed by the researchers in English based on reviewing relevant literature, then it has been translated into the Arabic language; the process of translation followed the World Health Organization's [56] method of translation and adaptation of instruments.

The questionnaire included the following six domains: (1) a socio-demographic domain to collect data about participant's country of residence, age, sex, academic specialization, highest qualification, work field, and average income (USD/month); (2) a research background domain encompassing: years of researching experience, experience of reviewing in a journal (dichotomous [yes/no]), number of articles published over the last three years, and average period for accomplishing a publication; (3) a domain of eight questions to understand preferences in research including: role in research conduct (multi select answer options), research methods, type of publishing (articles or/and books), online research information management (RIM) system(s) (e.g. ResearchGate, Academia.edu, Google Scholar, or others), posting a research synopsis on social media platforms [dichotomous—yes/no], publishing mode (Open access "OA" or/and traditional publishing), preferred journal site (domestic, regional or international) [3 items, 4-point rating scale of perceived priority] and preferred journal characteristics (like impact factor, readership, acceptance rate, etc.) [6 items, 6-point rating scale of perceived influence]; (4) a research attitudes domain—[a grid matrix of 10 items, 5-point rating scale of agreement level]—to investigate participant's inclination towards: lead authorship, continuity in research conduct, innovativeness in applying research strategies, conducting research in area of practical experience, studying gaps in professional practice/community, reliance on previous research ideas and recommendations, involving in collaborative research teams, intra-institutional (internal teams) research collaboration, interinstitutional research collaboration, and international research collaboration; (5) a domain of a grid-matrix question [8 items, 5-point excellent-to-poor rating scale] to rate the research resources and support offered by the participant's institution, this included: financial support, technological and database resources, research ethical review, professional collaboration, establishing research alliances, creating research culture, using research findings, and disseminating research findings to the broad scientific community; (6) a domain for motivations towards research, comprising two sub-domains: intrinsic motivation (a grid matrix of 16 items, 5-point rating scale of agreement level) and extrinsic motivation [a grid matrix of 14 items, 5-point rating scale of agreement level]. The items of intrinsic and extrinsic motivation were developed based on adapting and changing Zhang's [59] instruments for measuring researchers' motivation towards conducting research.

3.3 Sample Characteristics

The overall sample characteristics are shown in Table 1.

Table 1 Sample characteristics (n = 205)

Domain		Frequency	Percentage (%)
Country	Bahrain	11	5.4
	Egypt	137	66.8
	Iraq	1	0.5
	Jordan	12	5.9
	Morocco	1	0.5
	Palestine	1	0.5
	Saudi Arabia	17	8.3
	Tunisia	21	10.2
	United Arab of Emirates	4	2.0
Gender	Male	67	32.7
	Female	138	67.3
Age	22–30	46	22.4
	31–40	68	33.2
	41–50	66	32.2
	51–60	18	8.8
	> 60	7	3.4
Highest qualification	Doctoral	141	68.8
	Master	46	22.4
	Bachelor	18	8.8
Academic specialization	Business & management	4	2.0
	Humanity	35	17.1
	Natural & Applied	20	9.8
	Engineering	6	2.9
	Health and medical	82	40.0
	Social and psychology	58	28.3
Work field	Academic	153	74.6
	Professional	21	10.2
	Professional with academic affiliation	15	7.3
	Administration	7	3.4
	Research and development (R&D)	8	3.9
	Autonomous researcher	10	4.9

(continued)

Table 1 (continued)

Domain		Frequency	Percentage (%)
	Postgraduate research student	30	14.6
	Others	7	3.4
	Cumulative*	251	122.3
Average income (USD/month)	$10 k or less	149	72.7
	$10.001–$20 k	20	9.8
	$ 20,001–$30 k	7	3.4
	$30,001–$40 k	6	2.9
	$40,001–$50 k	3	1.5
	$ 50,001–$60 k	6	2.9
	$60,001–$70 k	2	1.0
	$70,001–$80 k	3	1.5
	$ 80,001–$90 k	2	1.0
	$ 90,001–$100 k	0	0
	$ 100,001& above	7	3.4

* Response percentage exceeds 100% as the question allows respondents to select multiple answers

3.4 Data Analysis

The electronically collected data set from the 205 respondents was complete as all the questions were selected with a mandatory answer option. The online data were downloaded into Microsoft Excel 2017 version and then transferred to Statistical Package for Social Science (SPSS) software program (IBM SPSS v.25 Inc., Chicago Il, USA) for conducting descriptive and inferential analysis of the data. Descriptively, the results were presented as percentages and frequency distribution. The arithmetic mean (M) and standard deviation (SD) were reported for continuous variables, and counts and percentages were reported for categorical variables. On the side, applied inferential tests included t-test, Pearson product-moment correlation coefficient r to measure the relationship strength between variables, and simple linear regression to assess the prediction or influence of motivation on research attitude. A p-value of < 0.05 was considered being significant.

3.5 Reliability and Validity Measures

The face validity of the developed questionnaire was ensured through revising the questions, items, and scales with two expert scholars in research sciences. Based on the obtained feedback, improvement has been made where necessary. Further, a pilot study was conducted by recruiting 12 researchers from the personal network of the authors, based on their comments on the overall questionnaire style and content; some amendments had taken place to enhance respondents' understanding and interpretation of each question.

The Kaiser–Meyer–Olkin (KMO) value was 0.834, showing that the variables of the sample are adequate to correlate or to proceed with factor analysis. Bartlett's test showed a highly significant relationship between the variables with p = 0.000 (p < 0.05). Thus, the questionnaire form of the study was valid as well as the factor analysis being meaningful and acceptable [13]. The internal consistency was scrutinized using Cronbach's alpha given coefficients of 0.916 and 0.846 for the constructs of intrinsic motivation, and extrinsic motivation, respectively; hence, the reliability of each motivation variable was highly acceptable [50].

4 Findings

4.1 Descriptive Analysis

From a five-point scale, the study has taken a position to interpret the generated mean scores according to Garrett's [23] numerical boundary limits [23]. The results show that the mean scores of extrinsic motivation and intrinsic motivation are (M: 3.67, SD: 0.55) and (M: 4.05, SD: 0.58), respectively. 72% (mean scores ≥ 3.4) of respondents were extrinsically motivated, and 87% of the respondents were intrinsically motivated. Overall, these descriptive results to show that most respondents exhibited acceptable levels of intrinsic and extrinsic motivation towards research conduct. Out of the 16 factors of extrinsic motivation, the following three motivators received the highest ratings: building status in research and academic community (M: 4.31, SD: 0.78), obtaining promotion (M: 4.25, SD: 0.98), and expanding personal research network (M: 4.23, SD: 0.73). On the other side, the intrinsic motivation ratings were found the highest within the following factors: seeking to develop research skills and knowledge (M: 4.31, SD: 0.69), contributing to society (M: 4.21, SD: 0.82), followed by the factor of a sense of achievement (M: 4.20, SD: 0.75).

4.2 Inferential Analysis

Motivation and Research Attitudes. Table 2 shows the results of predicting research attitudes (dependent variables) from the extrinsic and intrinsic motivation (independent variables) by applying simple linear regression analysis. Only significantly correlated variables (p < 0.05) were included in the regression analysis. Remarkably, six variables of research attitudes were predicted from intrinsic motivation, while only two research attitudes were predicted from extrinsic motivation. Also, worth noting, the intrinsic motivation correlated positively and significantly with extrinsic

Table 2 Motivation and research attitudes (correlational and regression analysis)

Research attitudes as dependent variables	Independent variables	
	Extrinsic motivation	Intrinsic motivation
Being a lead author	(r [203] = 0.07, p = 0.320)	(r = 0.227, F [1, 203] = 11.028, p < 0.01)**
Continuity in conducting research	(r [203] = 0.12, p = 0.085)	(r = 0.319, F [1, 203] = 23.014, p < 0.001)**
Applying novel research strategies	(r [203] = 0.014, p = 0.847)	(r = 0.335, F [1, 203] = 25.706, p < 0.001)**
Researching in areas of practical expertise	(r = 0.210, F [1, 203] = 9.325, p < 0.01)**	(r = 0.266, F [1, 203] = 15.480, p < 0.001)**
Steering research towards practice-related gaps	(r [203] = -0.015, p = 0.831)	(r = 0.239, F [1, 203] = 12.351, p < 0.01)**
Being driven by previous research ideas	(r = 0.142, F [1, 203] = 4.207, p < 0.05)*	(r [203] = 0.067, p = 0.337)
Joining a team for conducting research studies	(r [203] = 0.002, p = 0.979)	(r [203] = -0.087, p = 0.213)
Collaborating with domestic academic institutions	(r [203] = 0.091, p = 0.196)	(r = 0.139, F [1, 203] = 4.009, p < 0.05)*
Joining internal research teams in the employing institution	(r [203] = 0.009, p = 0.902)	(r [203] = -0.080, p = 0.254)
Joining international research networks	(r [203] = 0.081, p = 0.249)	(r [203] = 0.124, p = 0.076)

*, ** denotes significant levels at 1% (2-tailed), 5% (2-tailed), respectively

motivation (r [203] = 0.378, p <0.001). In summary, intrinsic motivation is found a potent enabler for researchers to pursue a leading role in research, maintain continuous research activity, consider innovative research strategies, and leverage research through practical expertise and solving practice-related issues.

Motivation and Characteristics of Respondents' Research Profile. Table 3 shows the results of testing the correlation between variables of respondents' research profile and each of extrinsic and intrinsic motivation. Interestingly, different from extrinsic motivation, intrinsic motivation was positively associated with the volume of produced articles. Also, the results imply that the greater the motivation is (whether intrinsic or extrinsic), the lesser period a researcher may take to accomplish a publication. By comparing the means of motivation between those who pursued the role of a journal reviewer and those who didn't, there was no significant difference for both intrinsic and extrinsic motivation.

Research Support and Motivation. As shown in Table 4, extrinsic motivation is significantly and positively associated with the quality of different research support factors; however, the strength of these associations is almost weak. Similarly, significant and positive correlations are reported between intrinsic motivation and all the research support factors except with the variable of 'creating a culture for research' which might have an indirect association with intrinsic motivation through extrinsic motivation; such finding may imply that organizations can't motivate researchers to certain target levels since motivation stems from the inner-self.

Researcher's Preferences and Motivation. Through applying t-test, it has been found that intrinsic motivation towards research was significantly higher in researchers who undertake the following roles than those who don't: conceptualization and hypotheses development [(M: 4.14, SD:0.54) vs. (M: 3.96, SD:0.60), t [203] = −2.313, p = 0.022], research design [(M: 4.12, SD:0.57) vs. (M: 3.89, SD:0.57), t [203] = −2.746, p = 0.007], data analysis [(M: 4.19, SD:0.51) vs. (M: 3.94, SD:0.60), t [203] = −3.061, p = 0.003], statistical analysis [(M: 4.1796, SD:0.56238) vs. (M: 3.9584, SD:0.56991), t [203] = −2.749, p = 0.007], manuscript review [(M: 4.21, SD:0.53) vs. (M: 3.98, SD:0.58), t [203] = −2.633, p = 0.009], and research guarantor [(M: 4.21, SD:0.59) vs. (M: 3.99, SD:0.56), t [203] = −2.547, p = 0.012]. On the other side, this study results show that there was no significant difference regarding researchers' extrinsic motivation based on embracing different research roles; except for those who undertake the role of statistical analysis, the findings indicate that they demonstrate higher extrinsic motivation (M: 3.77, SD:0.56) than those who don't embrace this role (M: 3.61, SD:0.54), t [203] = −2.136, p = 0.034]. In the overall picture, intrinsic motivation seems of more potential influence than extrinsic motivation on researchers' selection and preferences of research roles.

43.4%, 50.7%, and 26.3% of respondents relied mainly on quantitative methods, mixed-methods, and qualitative methods for their research conduct, respectively. The respondents' preferred scientific production concentrated on writing 'articles' (n = 135, 65.9%), 'both articles and books' (n = 67, 32.7%), and the least was for 'books/chapters' (n = 3, 1.5%). Also, their preferred online research information

Table 3 Motivation and respondents' research profile: correlations and means

Variables		Descriptive Statistics		Inferential Statistics	
		Frequency	Percentage (%)	Extrinsic motivation	Intrinsic motivation
Researching experience	Less than 2 yrs	15	7.3	(r [203] = −0.118, p = 0.091)	(r [203] = 0.036, p = 0.608)
	2–5 yrs	40	19.5		
	6–10 yrs	48	23.4		
	11–15 yrs	39	19.0		
	16–20 yrs	27	13.2		
	21 yrs. & above	36	17.6		
	Cumulative	205	100.0		
No. of articles published over the last three years	1	64	31.2	(r [203] = 0.046, p = 0.516)	(r [203] = 0.189, p = 0.007)**
	2–5	81	39.5		
	6–10	29	14.1		
	11–15	11	5.4		
	16–20	8	3.9		
	21 & above	12	5.9		
	Cumulative	205	100.0		
Average period for accomplishing a publication	1 month or less	5	2.4	(r [203] = −0.195 , p = 0.005)**	(r [203] = −0.144 , p = 0.039)*
	2–3 months	22	10.7		
	4–5 months	32	15.6		
	6–8 months	42	20.5		
	9–10 months	25	12.2		
	11–12 months	19	9.3		
	more than 1 yr	60	29.3		
	Cumulative	205	100.0		

(continued)

Table 3 (continued)

Variables		Descriptive Statistics		Inferential Statistics		
		Frequency	Percentage (%)	Extrinsic motivation		Intrinsic motivation
Experience of reviewing in a journal	Yes	93	45.4	(M = 3.65, SD = 0.57)	t [203] = −0.519, p = 0.605	(M = 4.13, SD = 0.57)	t [203] = 1.828, p = 0.069
	No	112	54.6	(M = 3.69, SD = 0.55)		(M = 3.98, SD = 0.58)
	Cumulative	205	100.0			

*, ** denotes significant levels at 1% (2-tailed), 5% (2-tailed), respectively

Table 4 Correlations between motivation and research support factors

Research support factors (M ± SD)	Ext. motivation		Int. motivation	
	r	P value	r	P value
Research funding (1.85 ± 1.14)	0.234**	0.001	0.188**	0.007
Access to quality resources (technology, access to databases, etc.) for research (2.26 ± 1.21)	0.197**	0.005	0.186**	0.007
Support for the application and process of research ethics review (2.61 ± 1.15)	0.211**	0.002	0.177*	0.011
Cooperation at professional and interpersonal level (2.71 ± 1.16)	0.166*	0.017	0.185**	0.008
Establishing partnerships and collaborations with external institutions (2.28 ± 1.17)	0.200**	0.004	0.152*	0.029
Creating a culture for research (like adopting reward scheme) (2.23 ± 1.21)	0.199**	0.004	0.124	0.075
Applying the results of research to induce change, or inform policymakers. (2.22 ± 1.18)	0.233**	0.001	0.179*	0.010
Grant facilitations to present/communicate results of research to the wider community (like through conferences) (2.33 ± 1.23)	0.262**	0.000	0.205**	0.003

*, ** denotes significant levels at 1% (2-tailed), 5% (2-tailed), respectively

management (RIM) systems were Google Scholar (n = 146, 71.2%), ResearchGate (n = 120, 58.5%), and Academia.edu (n = 31, 15.1%). From another angle, less than a quarter of the respondents posted a synopsis of their published research on social media platforms (n = 43, 21%). To publish mode, 16.1% of respondents relied solely on open access publishing, 39.5% on traditional publishing, and 44.4% on both publishing modes. In terms of preferred journal site and preferred journal characteristics, Table 5 exhibits the main findings concerning motivation constructs. In summary, the greatest priority was given by the respondents to the international journals when seeking to publish their scholarly work. Also, the preference for international publishing was significantly associated with both intrinsic and extrinsic motivations. The journal impact factor received the highest rating as the most perceived influencing factor when selecting a journal and showed a significant correlation with both extrinsic and intrinsic motivation.

5 Discussion

This study is the first of its kind in the Arab region; it has examined researchers' involvement in research projects in a spectrum of Arab countries. The intrinsic and extrinsic motivations towards research were descriptively assessed; alongside the association of these motivations with research attitudes and preferences, were empirically tested. The findings of the study could strengthen efforts of research policy

Table 5 Correlations between motivation and preferred publisher and journal characteristics

Variables (M ± SD)		Correlation with			
		Ext. motivation		Int. motivation	
		r	P value	r	P value
Preferred Publisher Level (4-point rating scale of perceived priority)	Domestic Publisher (2.62 ± 1.18)	0.054	0.440	0.052	0.459
	Arab region publisher (2.48 ± 1.13)	0.008	0.909	0.023	0.747
	International Publishers (3.64 ± 0.76)	0.184**	0.008	0.202**	0.004
Influence of Journal characteristics (6-point rating scale of perceived influence)	Journal impact factor (5.10 ± 1.01)	0.264**	0.000	0.253**	0.000
	Journal readership (4.58 ± 1.14)	0.172*	0.014	0.218**	0.002
	Journal acceptance rate (4.67 ± 1.22)	0.170*	0.015	0.077	0.270
	Journal publication speed (4.80 ± 1.23)	0.258**	0.000	0.046	0.510
	Journal article-processing charges (4.89 ± 1.24)	0.114	0.103	−0.030	0.672
	Journal peer-review quality (4.80 ± 1.04)	0.180**	0.010	0.288**	0.000

*, ** denotes significant levels at 1% (2-tailed), 5% (2-tailed), respectively

devoted to promoting Arab researchers' involvement, attitude, and role in scholarly activities. Overall, we found that most the respondents had acceptable levels of intrinsic and extrinsic motivation towards research conduct. 87% of total respondents were intrinsically motivated, while 72% of them were extrinsically motivated. Also, the potency of intrinsic motivation in defining a researcher's pattern of involvement with research has been verified and endorsed. Being intrinsically motivated could be more influential than extrinsic motivation to lead research, innovate, and solve real-world issues, with a higher level of knowledge productivity in less time.

The highest three extrinsic motivators amongst the participants were building status in the research and academic community, obtaining a promotion, and expanding a personal research network. Hence our work shows the paramount importance and interest of developing and maintaining a personal reputation in research and education which could reinforce the researchers to get rewards, better opportunities, and being promoters for the image of their institutions in society [52]. In line with previous studies, our findings underscore academic promotion as an essential component for a reward system and metrics assessment, ultimately determining researchers' motivation for the desired behaviors of research creativity [34], and their academic engagement with policy and decision-making roles [28]. Hence, perhaps these identified extrinsic motivators need to be well-recognized by research-focused institutions in Arab countries to warrant supportive research environments that enable researchers to flourish, build networks, and get rewarded.

On the other side, the intrinsic motivations that ranked the highest included: seeking to develop research skills and knowledge, contributing to society, followed by the factor of sense of achievement. Our results show and imply that a large segment of surveyed academics and researchers find their work intrinsically satisfying and enjoy being engaged in continuous researching competence development. Such observed quality of intrinsic motivation reflects the need for self-achievement and creating value for the broader community. Consistent with previous research findings, being intrinsically motivated is a critical predictor of performance, creativity, and innovation [18, 25, 32]. The present study found that intrinsic motivation positively associates with the volume of published articles (r [203] = 0.189, p = 0.007). Unlike extrinsic motivation, intrinsic motivation predicts leadership in researching the behavior, continuity in researching activity, and innovativeness. However and despite the greater influential role of intrinsic motivation compared to extrinsic motivation on certain research intentional and behavioral aspects, both motivational components are fundamental because of their mutual spillovers, clear through the observed positive correlation between the intrinsic motivation and extrinsic motivation (r [203] = 0.378, p <0.001) in this study, and also in other previous studies [21, 34]. With keeping in mind that motivation based on tangible rewards could be hard to sustain [10], there is a necessity to give more focus on the intrinsic motivation factors to promote researchers' engagement and productivity in scholarly work.

In this study, institutional, technological, and financial research support was found vital to promote researchers' extrinsic and intrinsic motivations. This finding is consistent with current evidence inferring the well-acknowledged role of research resources and capabilities to motivate University faculty to be involved in research [37]. Worth mentioning, most of the rating responses for the different investigated research support factors in this study were at the mid-point of the scale or lower, showing the sufficiency of the research support services. However, these results should not be generalized to the Arab region as the greatest portion of the respondents was from Egypt, besides our reliance on the convenience sampling approach. Considering the general situation in the Arab region [7, 16, 17, 33, 48, 49, 54], research and academic institutions in the Arab region should prioritize designing and implementing effective strategies to promote their research support services and resources and develop their research infrastructure up to acceptable international standards. This strategic orientation can be of great value in building the Arab "knowledge economies" or creating globally competitive research institutions.

6 Researchers' Motivation Towards Technology: A Call for Further Research

In the twenty-first century, advanced technology has been intensively integrated into academic and scientific activities. The association between technology and learners' motivation in a classroom context is well-researched in current literature [22, 27]. However, limited scholarly attention has been steered towards understanding the impacts of technology on professional and academic researchers despite the growing diversity of research-related technologies. Web-based surveys, ICT–based research collaborations, advance and automated statistical analysis software, computer-assisted interviewing systems, reference management systems (e.g., End Notes), Web 3.0 tools, research collaborative and social networking platforms, plagiarism and editing online software, online databases, and search engines are some examples of modern technology, which have become crucial elements for improving education and research processes and outcomes [39, 15, 57]. Further, there is limited knowledge of how the researchers utilize these technologies during the era of the COVID-19 pandemic and what technological trends would be accentuated post the pandemic. Alongside these gaps in the literature, there are many other several issues potentially affecting the researchers' behaviors, including but not limited to technophobia, digital divide, artificial intelligence (AI) literacy, and internet or technology addiction [55, 42, 11]. These technology-related issues can shape researchers' performance and capacity to communicate research results and influence their motivation levels to undertake specific research activities and projects. Motivations towards leveraging technological capacities can reinforce researchers not only to collect data or organize the content in better ways but enable creativity and innovativeness in making research more efficient, impactful, and useful. Hence, drawing from this background, studying researchers' interactions, attitudes, preferences, and motivation towards the latest research technologies must be placed at the forefront of future research and academic initiatives in the Arab region.

As per the correlation testing results (Table 4), this study underscores the importance of access and availability of research-supporting technologies to promote intrinsic and extrinsic motivation for research conduct. However, the extent to which researchers and academicians use the different options of modern research technology through their institutions' offered resources or self-orientation and learning begs for empirical explorations. Also, with the increasing dependence on learning information systems (LMS) during the pandemic, it has become imperative to monitor and encourage motivation towards using advanced technologies. Signs alarming for such empirical perspective can be visualized through recent research highlighted issues of technophobia, techno-stress, and digital incompetence amongst faculty members during the COVID-19 in countries like Saudi Arabia and Algeria [4, 6, 43, 24]. This inevitably calls for designing tailored technology engagement assessment measures and strategies for creating resilient education environments for change acceptance, coping with the innovative teaching methods, and ultimately fostering the motivation towards technology-centric education and research practices.

7 Limitations and Suggestions for Future Research

This study was confronted with several limitations. First, the research was based on a cross-sectional design using self-measurement as a primary data source. The results may be contaminated by the common method bias on how truthful the provided information was. In this kind of study, participants might show in their responses how they would like their motivation and academic performance to be perceived. Future studies using mixed methods of quantitative and qualitative approach might be useful to understand motivational factors deeply and enrich the database. Second, since our study was cross-sectional, causality could not be inferred. However, our study highlights several factors associated with intrinsic and extrinsic motivation factors towards research and their association with research attitudes and preferences. Third, this research has analyzed intrinsic motivation and extrinsic reward motivators towards research in Arab countries using a convenience sample through a non-randomized snowball and self-selection technique sample; hence, generalizability is limited. Further research is recommended on a large sample to verify the association with intrinsic and extrinsic motivation factors towards research and their association with research attitudes and preferences. The study opens new avenues for future researchers to consider investigating determinants of intrinsic motivation towards research. Also, identifying the mediating and moderating factors relating to both intrinsic and extrinsic motivation would be a valuable area of research, particularly concerning research culture agents that may include institutional research policy, budget allocations for research, working conditions, and collaboration frameworks. In a larger view and beyond the personal level, exploring the motivations of organizations in the Arab region should also be paid attention to understand how scientific research is being adopted for improvement and avoidance of arbitrary decisions that are not acceptable in many circumstances [20].

8 Conclusion

This research, the first of its kind in the Arab region, represents a baseline study to monitor and assess trends of Arab researchers' motivation, likely to be reshaped by the post-pandemic transformations and constantly growing technological advancements. Further, the study provides some evidence in the Arab context to inform policy-makers, academic decision-makers, and research infrastructure developers to foster researchers' motivation and subsequent productivity and affluent involvement in research activities. Despite the devastating impacts of COVID-19 on the educational and economic systems, high levels of intrinsic and extrinsic motivation amongst the surveyed researchers were observed. However, there is a critical need for a parallel institutional motivation for capitalizing on researchers' motivation to strengthen research outcomes and promote their morale. Importantly, research-enhancing technologies need to be concretely scaled-up and integrated into education and research works, with greater attention on formulating strategies for promoting users' digital competence and motivation.

Our research suggests that intrinsic motivation is integral in encouraging researchers to lead research, adopt innovative research strategies, and sustain high research performance and productivity. Extrinsic motivation has the potency to drive intrinsic motivation, but extensive research is needed to understand this relationship further. These results are based on cross-sectional study and convenience sampling that may be biased, thus limiting their interpretation.

References

1. Ahmed, D., & Albuarki, J. (2017). Review of the challenges of scientific research in the Arab world and its influence on inspiration driven economy. *International Journal of Inspiration & Resilience Economy, 1*(1), 28–34.
2. Almansour, S. (2016). The crisis of research and global recognition in Arab universities. *Near and Middle Eastern Journal of Research in Education , 2016*(1), 1.
3. Alrahlah, A. A. (2016). The impact of motivational factors on research productivity of dental faculty members: A qualitative study. *Journal of Taibah University Medical Sciences, 11*(5), 448–455.
4. Alshaikh, K., Maasher, S., Bayazed, A., Saleem, F., Badri, S., & Fakieh, B. (2021). Impact of COVID-19 on the educational process in Saudi Arabia: A technology–organization–environment framework. *Sustainability, 13*(13), 7103.
5. Al-Soufi, A., & Al-Ammary, J. (2011). The role of a collaborative research network (CRN) in improving the arabian gulf countries' performance in research and innovation. *International Journal of Technology Diffusion (IJTD), 2*(3), 24–35.
6. Alvi, A. H., Muhammad Bilal, S., & Abdul Rahim Alvi, A. (2021). Technology, pedagogy & assessment: Challenges of COVID19-imposed E-teaching of ESP to Saudi female PY students. *Arab World English Journal (AWEJ) Special Issue on Covid19 Challenges.*
7. Alzaneen, R., & Mahmoud, A. (2019). The role of management information systems in strengthening the administrative governance in ministry of education and higher education in gaza. *International Journal of Business Ethics and Governance, 2*(3), 1–43. https://doi.org/10.51325/ijbeg.v2i3.44
8. Al-Zoubi, A., & Abu-Orabi, S. (2019). Impact of internationalization on Arab higher education the role of association of Arab universities. *Journal of Education and Human Development, 8*(1), 69–85.
9. APA (American Psychological Association). (2016). Revision of Ethical Standard 3.04 of the" Ethical Principles of Psychologists and Code of Conduct"(2002, as amended 2010). *The American Psychologist, 71*(9), 900.
10. Ben-Hur, S., & Kinley, N. , (2015). Changing employee behavior: Do extrinsic motivators really not work? Tomorrow's Challenges, IMD.
11. Ben-Jacob, M. G., & Liebman, J. T. (2009). Technophobia and the effective use of library resources at the college/university level. *Journal of Educational Technology Systems, 38*(1), 35–38.
12. Beshyah, S. A. (2019). Authors' selection of target journals and their attitudes to emerging journals: A survey from two developing regions. *Sultan Qaboos University Medical Journal, 19*(1), 51.
13. Cleff, T. (2019). *Applied statistics and multivariate data analysis for business and economics.* Springer International Publishing.

14. Costandi, S., Hamdan, A., Alareeni, B., & Hassan, A. (2019). Educational governance and challenges to universities in the Arabian Gulf region. *Educational Philosophy and Theory, 51*(1), 70–86.
15. Cuff, E. (2014). The effect and importance of technology in the research process. *Journal of Educational Technology Systems, 43*(1), 75–97.
16. Currie-Alder, B., Arvanitis, R., & Hanafi, S. (2018). Research in Arabic-speaking countries: Funding competitions, international collaboration, and career incentives. *Science and Public Policy, 45*(1), 74–82.
17. Currie-Alder, B. (2015). Research funding in Arab countries: Insights emerging from a forum of research funders held in Cairo in December 2014.
18. De Jesus, S. N., Rus, C. L., Lens, W., & Imaginário, S. (2013). Intrinsic motivation and creativity related to product: A meta-analysis of the studies published between 1990–2010. *Creativity Research Journal, 25*(1), 80–84.
19. Di Domenico, S. I., & Ryan, R. M. (2017). The emerging neuroscience of intrinsic motivation: A new frontier in self-determination research. *Frontiers in Human Neuroscience, 11*, 145.
20. Elsayed, A. M., & Sabtan, Y. (2019). Reality of scientific research in the Arab world and suggestions for development. *International Journal of Advances in Social Science and Humanities, 7*(1), 31–35.
21. Fischer, C., Malycha, C. P., & Schafmann, E. (2019). The influence of intrinsic motivation and synergistic extrinsic motivators on creativity and innovation. *Frontiers in Psychology, 10*, 137.
22. Francis, J. (2017). The effects of technology on student motivation and engagement in classroom-based learning.
23. Garrett, H. (1966). *Statistics in psychology and education.* Green and Co.
24. Ghounane, N. (2020). Moodle or social networks: What alternative refuge is appropriate to Algerian EFL students to learn during COVID-19 pandemic. *Arab World English Journal, 11*(3), 21–41.
25. Hammond, M. M., Neff, N. L., Farr, J. L., Schwall, A. R., & Zhao, X. (2011). Predictors of individual-level innovation at work: A meta-analysis. *Psychology of Aesthetics, Creativity, and the Arts, 5*, 90–105.
26. Hanafi, S., Arvanitis, R., & Hanafi, O. (2013). The broken cycle: Universities, research and society in the Arab region: Proposal for change. ESCWA, Beyrouth.
27. Harris, J., & Al-Bataineh, A. (2015). One to one technology and its effect on student academic achievement and motivation. In *Global learn, association for the advancement of computing in education (AACE)* (pp. 579–584).
28. Jessani, N. S., Valmeekanathan, A., Babcock, C. M., & Ling, B. (2020). Academic incentives for enhancing faculty engagement with decision-makers—considerations and recommendations from one School of Public Health. *Humanities and Social Sciences Communications, 7*(1), 1–13.
29. Kent, K. J. (2021) Nature Index 2019: A year of Arab science in numbers. Retrieved May 02, 2021, from https://www.natureasia.com/en/nmiddleeast/article/, https://doi.org/10.1038/nmiddleeast.2019.115.
30. Khan, S., Shah, S. M. H., & Khan, T. M. (2018). An investigation of attitudes towards the research activities of university teachers. *Bulletin of Education and Research, 40*(1), 215–230.
31. Larsen, A. V., Dorch, B., Nyman, M., Thomsen, K., & Drachen, T. M. (2010). *Analysis of research support services at international best practice institutions.* Copenhagen University Library and Information Service (CULIS).
32. Liu, D., Jiang, K., Shalley, C. E., Keem, S., & Zhou, J. (2016). Motivational mechanisms of employee creativity: A meta-analytic examination and theoretical extension of the creativity literature. *Organizational Behavior and Human Decision Processes, 137*, 236–263.
33. Maalouf, F. T., Alamiri, B., Atweh, S., Becker, A. E., Cheour, M., Darwish, H., et al. (2019). Mental health research in the Arab region: Challenges and call for action. *The Lancet Psychiatry, 6*(11), 961–966.
34. Malik, M. A. R., Butt, A. N., & Choi, J. N. (2015). Rewards and employee creative performance: Moderating effects of creative self-efficacy, reward importance, and locus of control. *Journal of Organizational Behavior, 36*(1), 59–74.

35. Mensah, M. S. B., & Enu-Kwesi, F. (2018). Research collaboration for a knowledge-based economy: Towards a conceptual framework. *Triple Helix, 5*(1), 1–17.
36. Muthanna, A., & Sang, G. (2018). Brain drain in higher education: Critical voices on teacher education in Yemen. *London Review of Education, 16*(2), 296–307.
37. Narbarte, M. P. (2018). Research Involvement, Motivation, and University Initiatives as Agents for Enhancing Research Culture and Quality. *Human Behavior, Development and Society, 17*, 68–78.
38. Ngussa, B. M. (2015). Trends in Research Collaboration: Experiences in Tanzanian Institutions of Higher Learning. *International Journal of Academic Research in Business and Social Sciences, 5*(1), 187.
39. Ohei, K. N., & Brink, R. (2019). Web 3.0 and web 2.0 technologies in higher educational institute: Methodological concept towards a framework development for adoption. *International journal for Infonomics (IJI), 12*(1), 1841–1853.
40. Okoduwa, S. I., Abe, J. O., Samuel, B. I., Chris, A. O., Oladimeji, R. A., Idowu, O. O., & Okoduwa, U. J. (2018). Attitudes, perceptions, and barriers to research and publishing among research and teaching staff in a Nigerian Research Institute. *Frontiers in Research Metrics and Analytics, 3*, 26.
41. Okokpujie, I. P., Fayomi, O. S. I., & Leramo, R. O. (2018). The role of research in economic development. In *IOP Conference Series: Materials Science and Engineering, 413*(1), 012060.
42. Pedro, F., Subosa, M., Rivas, A., & Valverde, P. (2017). Artificial intelligence in education: Challenges and opportunities for sustainable development.
43. Rajab, M. H., Gazal, A. M., & Alkattan, K. (2020). Challenges to online medical education during the COVID-19 pandemic. *Cureus, 12*(7).
44. Rowlands, I., Nicholas, D., & Huntington, P. (2004). Researchers' attitudes towards new journal publishing models. *Learned Publishing, 17*.
45. Rumman, A. A. A., & Alheet, A. F. (2019). The role of researcher competencies in delivering successful research. *Information and Knowledge Management, 9*(1), 15–19.
46. Schiermeier, Q. (2019). Europe is a top destination for many researchers. *Nature, 569*(7758), 589–592.
47. Schmidt, U., Frewer, A., & Sprumont, D. (2020). *Ethical Research: The Declaration of Helsinki, and the Past, Present, and Future of Human Experimentation*. Oxford University Press.
48. Sheblaq, N., & Al Najjar, A. (2019). The challenges in conducting research studies in Arabic countries. *Open Access Journal of Clinical Trials, 11*, 57–66.
49. Sultana, K., Al Jeraisy, M., Al Ammari, M., Patel, R., & Zaidi, S. T. R. (2016). Attitude, barriers and facilitators to practice-based research: Cross-sectional survey of hospital pharmacists in Saudi Arabia. *Journal of Pharmaceutical Policy and Practice, 9*(1), 1–8.
50. Taber, K. S. (2018). The use of Cronbach's alpha when developing and reporting research instruments in science education. *Research in Science Education, 48*(6), 1273–1296.
51. UNESCO (The United Nations Educational, Scientific and Cultural Organization): How much does your country invest in R&D?. Retrieved May 02, 2021, from http://uis.unesco.org/apps/visualisations/research-and-development-spending/.
52. Urip, S.R., Kurniawati, N.: The Concept of Maintaining Personal Reputation in Educational Institutions. KnE Social Sciences, 522–530 (2020).
53. Verner, T. (2011). National competitiveness and expenditure on education, research and development. *Journal of Competitiveness, 3*(2).
54. Waast, R. (2010). Research in Arab countries (North Africa and West Asia). *Science, Technology and Society, 15*(2), 187–231.
55. Wang, H. Y., Sigerson, L., & Cheng, C. (2019). Digital nativity and information technology addiction: Age cohort versus individual difference approaches. *Computers in Human Behavior, 90*, 1–9.
56. World Health Organization. (2009). Process of translation and adaptation of instruments, at. Retrieved April 15, 2021, from http://www.who.int/substance_abuse/research_tools/translation/en/.

57. Wright, D. A. (2005). Using technology to conduct research in education. *IPSI Transactions on Internet Research, 1*(1), 90–96.
58. Zain, S. M., Ab-Rahman, M. S., Ihsan, A. K. A. M., Zahrim, A., Nor, M. J. M., Zain, M. F. M., Hipni, A., Ramli, N. L., & Ghopa, W. A. W. (2011). Motivation for research and publication: Experience as a researcher and an academic. *Procedia-Social and Behavioral Sciences, 18*, 213–219.
59. Zhang, X. (2014). Factors that motivate academic staff to conduct research and influence research productivity in Chinese project 211 universities. Doctoral dissertation, University of Canberra.

Applying Simulation Tools in Education During the COVID-19 Pandemic

Towards Academic Integrity: Using Bloom's Taxonomy and Technology to Deter Cheating in Online Courses

Kakul Agha⬤, Xia Zhu⬤, and Gladson Chikwa⬤

1 Introduction

There has been a plethora of literature documenting academic dishonesty as a widespread global phenomenon among students in Higher Education Institutions [1]. Scholars from many parts of the world, such as UK, Canada, USA, and Australia have witnessed a significant increase of academically dishonest behaviour over the years [2, 3]. The cheating behaviour is prevalent across many disciplines, such as STEM [4], business and management studies [5], humanities [6], engineering [7] and medicine and health science [8]. During the Covid-19 pandemic, there has been a fast-paced transition of teaching and learning from offline to online in order to minimize education disruption, at the same time, assessments have also been rapidly changed from face-to-face to be taken virtually and remotely [3]. A recent survey across Germany shows that the probability of students' cheating behaviour doubled in online exams [9]. The exponential growth of online cheating behaviour has also been recognised in the UK and Australia [3, 4]. Overall, academic dishonesty has been exacerbated in the remote online teaching and learning setting.

Academically dishonest behaviour may take various forms [10]. For example, students may access forbidden resources, especially in 'open book' exams or students may communicate and share answers with one another which is known as 'collusion' [3]. At the same time, students may use internet websites, such as Chegg, Course Hero, OneClass and Thinkswap, to share assessment briefings and answers upon

K. Agha (✉)
Skyline University College, Sharjah, United Arab Emirates
e-mail: kakul.agha@skylineuniversity.ac.ae

X. Zhu
The Open University, Milton Keynes, UK

G. Chikwa
University of Bradford, Bradford, UK

© The Author(s), under exclusive license to Springer Nature Switzerland AG 2022   447
A. Hamdan et al. (eds.), *Technologies, Artificial Intelligence and the Future of Learning Post-COVID-19*, Studies in Computational Intelligence 1019,
https://doi.org/10.1007/978-3-030-93921-2_25

different financial agreement (with or without a fee), which is known as 'file sharing' [4]. Sometimes, students' work might be partially or wholly conducted by family, relatives or friends, despite the absence of any commercial transaction involved in these situations, it is classified as 'contract cheating' [11], although 'contract cheating' is more often known as students outsourcing or commissioning a third party (e.g. essay mill) to produce work and complete their assignments with agreement on a fee for those 'services' [11]. In addition, students lying or using inaccurate information, such as providing a false excuse in order to take an assessment at a later time, is recognised as 'perjury', and students' reuse of part or all of one's own pervious work is known as 'recycling' [10]. Moreover, there are more passive deceitful behaviour, such as 'slacking' when students receive credits for the little work they have contributed to the group assignment [10]. Furthermore, plagiarism is broadly defined as 'where a student passes off someone else's ideas and/or words, intentionally or unintentionally, as their own, for their own benefits' [20, p. 46]. Students may engage in academically dishonest behaviour unintentionally when there is ambiguity in their understanding of what classifies academic dishonesty [12]. Thus, McClung and Schneider (2015) identified 18 categories of academically dishonest behaviour as a framework and urged a clearer understanding of what constitutes academically dishonest behaviour in order to educate students and provide practical guidelines to reduce academic malpractice [10].

There are multiple reasons and motives behind students' cheating behaviour [2, 13]. For example, students without proper academic skills training may have a poor understanding of what constitutes academic misconduct or lack the awareness of committing academic integrity violation [14]. Especially in the online exams, there are confusions and lack of clarification on what resources are permissible [3]. International students whose first language is not English may have a different cultural interpretation regarding what is plagiarism [12]. Students with poor academic performance may be targeted by essay mills [15]. Moreover, students under pressure, lacking time management skills, or feeling being isolated without support are vulnerable and more tempted to cheating [14]. In addition, perceived lack of supervision in assessments [3], the inadequate number of invigilators during examinations, the lack of modern technology infrastructure or similar assessment questions used year after year may all encourage cheating behaviour [16].

The consequences of academic dishonesty have wide implications [11]. At the individual level, it victimizes students, undermines students' learning, damages a clear diagnosis on students' learning gaps and causes unfairness to those who do not get involved in cheating [7]. At the same time, students' personal data stored by essay mills may be exposed with risks to identity theft and students may suffer from blackmailing for further payment [11]. At the institutional level, it poses threats to academic standards and quality, devalues a university degree, damages a university's reputation, as well as the potential loss of university's accreditation from external bodies and public confidence in higher education [7, 17]. At the societal level, academic dishonesty may cause risks to the wider public especially if courses are professionally accredited, such as accounting, law, medicine and engineering, where the inadequate professional skills in practice may lead to public safety concerns [11]. For instance, cheating in medicine and health sciences may have serious consequences for human life [8]. Studies have shown a close link between academic dishonesty and workplace dishonesty [5]. Especially for those professions that are based on trust, the seriousness of social impact and detrimental long-term implications of academic dishonesty on students' professional career demands a clear communication [5, 14].

Academics and scholars have made an urgent call for a deeper understanding of the complexity of academically dishonesty behaviour [2, 4] in order to reduce academic integrity violation [14]. According to [11], academic integrity encompasses six fundamental values: honesty, trust, fairness, respect, responsibility and courage. Academic practices fully embracing those values would not only combat dishonest behaviour but also promote academic integrity. In order to counter online cheating, multiple stakeholders (e.g. students, university staff, policy makers, educational regulatory bodies) need to be informed and educated with a consistent understanding of academic integrity and to foster the sense of social responsibility [18, 19].

Previous studies have suggested a number of strategies to curb academically dishonest behaviour [13], such as raising awareness and communicating the expectations of academic integrity to both staff and students [3], effectively implementing an honor code and encouraging students to abide by rules of conduct [1], making academic integrity as part of teaching [20], modifying classroom environment and increasing students' engagement to develop a 'cheat-free' classroom [21], constructing appropriate and effective assessments that would reduce opportunities for cheating [13], employing technology to detect plagiarism and authenticate identity [22], developing institutional policies as well as investing resources and staff support to counter academic dishonesty [11], and making changes to the law to make commercial contract cheating services illegal [15]. This chapter focuses on the use of effective assessment design and the creative use of innovative technology to curb online cheating and promote academic integrity in the higher education sector, more specifically by investigating two following questions:

- Does the use of Bloom's Taxonomy in a judicious mix of assessments enable the curbing of academic dishonesty in online assessments?
- How can technology be harnessed to deter academic dishonesty in online assessments?

The remaining sections of the chapter are structured as follows: Sect. 2 presents Bloom's Taxonomy, reviews its historical development and explains its application in assessment design. Section 3 introduces several innovative technologies and examines their usage for proctoring, authentication and verification of student identity, and detection of plagiarism. Section 4 discusses the potential use and challenges of using Bloom's Taxonomy and modern technology to curb cheating in online courses. The chapter ends with concluding comments including a consideration of implications for practice where a call is made for higher education institutions to uphold academic integrity and fulfill their social mission in Sect. 5. Lastly but not least, Sect. 6 focuses on the way forward with readers being invited to reflect on the next steps towards the enhancement of academic integrity in online courses.

2 Using Bloom's Taxonomy to Reduce Academic Dishonesty

As we are aware, curbing academic dishonesty and promoting academic integrity are two faces of the same coin. Each of these need to be monitored well and effective use of Bloom's Taxonomy in framing of assessments can be one of the reliable methods for curbing academic dishonesty.

2.1 Understanding Bloom's Taxonomy

Benjamin S. Bloom along with his colleagues developed the framework for classifying educational objectives which is commonly known as the Bloom's Taxonomy in 1948 in order to make easy student comparisons by standardized testing of their achievement in terms of learning [23]. Bloom's Taxonomy is a meta-cognitive framework with a six-level classification system [23]. The taxonomy moves in a progressive manner from basic level to a more complex one namely knowledge, comprehension, application, analysis, evaluation and creation. This hierarchical model is not a prescriptive but rather an in-depth framework to be understood and used effectively by academicians. It has evolved from merely being used as an assessment framework to a tool for promoting higher order critical thinking among learners where they are able to create meaning in an environment [23–25]. Critical thinking is directly related to deep learning and can be tested mostly through open-ended questions [24]. This section highlights the use of Bloom's Taxonomy in higher education for student assessment instruction evaluation and curbing plagiarism [23]. In order to develop a higher sense of academic integrity among students during the contemporary times, it is pertinent to train students to use the Bloom's Taxonomy in classroom activities as well as formative and summative assessments. It is widely used as a tool for cognitive development among students of all levels, which in itself supports the students' learning process. One cannot deny that several methods can be used to support the reduction of academic dishonesty among students during online classes, but Bloom's Taxonomy can be widely used as

(i)   it is a familiar tool for most academics around the world in the higher education sector
(ii)   it is a generic and widely applicable tool across different domains of study and more importantly
(iii) it is an extremely user-friendly tool to employ and administer [26].

Bloom proposed this tool to study the achievement of educational objectives, student intellectual behaviour and add specific terminology to assess and measure the learning process. The cognitive, affective and psychomotor domains of learning were identified by Benjamin Bloom. Although Bloom worked on cognitive and affective domains, the psychomotor domain was developed during 1972 by Simpson [27].

Bloom classified the cognitive domain into a hierarchical pattern using knowledge, comprehension, application, analysis, synthesis and evaluation [28]. Further, the explanation of each details the kind of actions to be taken at each level while framing assessments questions and the verbs that could be used by academic practitioners [28]. 'Knowledge' pertains to students' ability to recall and remember information that they have learned before in and outside the classroom. 'Comprehension' level demands students to rephrase, explain and exhibit information based on their own understanding, giving examples to demonstrate their knowledge developed during the learning process. 'Application' level expects the students to be able to apply already known algorithms and rules in order to solve problems and find appropriate and accurate solutions. 'Analysis' level is a higher order that demands students to be able to separate parts of the whole and be able to identify the relationship between the parts or sections. Comparing, contrasting and distinguishing elements from each other is carried out at this level. 'Synthesis' being the next level expects students to be able to design, construct, develop and formulate solutions in order to solve problems. The highest level as framed by Bloom is 'Evaluation' where the learner is expected to justify, evaluate and judge the merits of ideas in order to choose better options of solutions from given problems [24, 26, 29]. So, in all three levels—lower, intermediate and higher levels can be differentiated, where knowledge and comprehension fall under lower order cognitive questions (LOCQ), application and analysis will be intermediate order cognitive questions (IOCQ) and synthesis and evaluation will be group as higher order cognitive questions (HOCQ) [24, 26]. The categories of Bloom's Taxonomy are shown in Fig. 1.

Later in the 1900s, Lorin Anderson, a student of Benjamin Bloom was commissioned the task of revamping the original Bloom's Taxonomy in line with the requirements of the twenty-first Century. The team spent six years and in the year 2001 the team proposed minor but significant changes in the taxonomy. The changes were focused on three aspects namely terminology, structure and emphasis. The most obvious difference was in the terminology changing the noun forms to verb forms (see Fig. 2). The new forms are *remember, understand, apply, analyze, evaluate* and *create* [31]. The revised taxonomy focused on building enhanced student-learning experience through student-centered learning so as to build better cognition and critical thinking among students [32].

2.2 Using Bloom's Taxonomy in Designing Assessments

To understand the use of Bloom's Taxonomy, one needs to know how the world has been utilising the six levels of the hierarchical taxonomy. The key question here is how academicians are using the taxonomy in designing assessments that support enhanced student learning and simultaneously help in curbing cheating and other dishonest practices used by students. Mostly during the COVID-19 pandemic, academicians are seen in a state of panic as most students resorted to utilising tools that showcase lack of academic integrity and add on the unethical academic practices.

Blooms Category	Definition	Sample Keywords / Verb
Knowledge	Draws out factual answer, basic concepts, testing recall and recognition of facts	Define, Duplicate, Identify, List, Memorize, Name, Quote, Recite, Recall, State, Tabulate, Tell,
Comprehension	Demonstrates an understanding of the meaning of information	Arrange, Classify, Describe, Demonstrate, Distinguish, Explain, Review, Select, Summarize, Translate,
Application	Ability to apply or use knowledge in actual situations	Apply, Compute, Construct, Choose, Illustrate, Modify, Operate, Prepare, Predict, Schedule, Solve, Use
Analysis	To break down into parts or forms and then draw connections between the parts; Make a relation to the assumptions, and find evidence to support generalisations	Analyse, Compare, Contrast, Distinguish, Examine, Identify, Categorize
Synthesis	Rearrange component ideas into a new whole. Propose or develop patterns and structures from diverse components. Propose alternative solutions.	Arrange, Combine, Compile, Compose, Create, Combine, Develop, Devise, Design, Generate, Prepare, Rewrite, Synthesize, Write
Evaluation	Make judgements based on evidence or value using definite criteria	Argue, Appraise, Assess, Critique, Conclude, Decide, Judge, Justify,

Fig. 1 Summary of categories in bloom's taxonomy; adapted from [30]

In order to understand this phenomenon, an extant of contemporary literature was reviewed. Using the taxonomy in formative and summative assessments is a great way of ensuring students understand the different levels of work required in assessments and this could actually support a higher level of skill development among students while they are in the university.

One of the critical methods for helping students to succeed in their learning process is scaffolding. Scaffolding, in simple words, is a higher order intellectual support to students so that they can function in the best interest of their cognitive development [23]. Another advantage of scaffolding includes letting students build better critical thinking skills in all types of courses that require application and synthesis of knowledge; complex thinking and comprehension [23, 33]. It is also found by the research of [23] that when students are challenged to display higher order comprehension

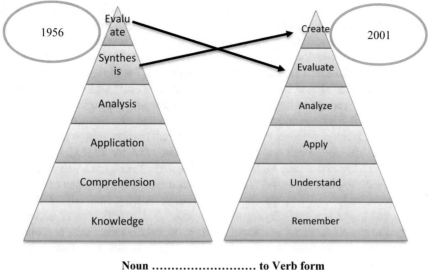

Noun .... **to Verb form**

Fig. 2 Edited bloom's taxonomy; adapted from [31]

and analysis or are expected to create and synthesize content in the academics and workplace, they get stressed but if they have been scaffolded to develop these abilities with the help of Bloom's Taxonomy, then they perform better in reality. So here the Bloom's Taxonomy comes in handy when the teachers utilise it for formative assessments in the class. Formative assessments could be of many types including tests, open ended questions or even case studies where higher order skills need to be utilised. Students can be taught in the class how to access questions of different types and how to utilise keywords to answer those questions. Even explaining the meaning of the keywords could be a great strategy for students in the classroom. Further, it has been added that making students understand and reciprocate to queries and discussions in the classroom setting can enable them to discover course information through enquiry and prevent them from being mere passive absorbers of information during the classroom [29].

An impressive and much researched area is to develop assessments that do not just focus on lower levels of Bloom's Taxonomy but higher order one as well is using Bloom's Taxonomy in examinations [28]. However, students tend to indulge in cheating when only lower levels of Bloom's Taxonomy are assessed [31]. Academic pieces of work that compel students to think well and engage deeply using higher order skills and level of Bloom's Taxonomy support in more academic integrity and less cheating during examinations and assessments [13]. It has been researched and recognised that online examinations with lesser levels of thinking and predictable answers let students use freely available online content to complete the exam, without feeling that they were academically dishonest.

In online examinations, conducting short objective tests is a helpful way to measure the knowledge of the students, however if the questions are framed as a stand-alone method with no building on prior course work, it leads to more instances of cheating and collusion [34, 35]. In order to curb this, [34] suggests designing questions using Bloom's Taxonomy higher order levels. He further advises that rote memory questions should be avoided for measuring mastery of knowledge, rather could be used to recognize whether the students are learning and understanding or not [34]. Further, it has been added that for assessments and examinations that involve calculations, faculty members could provide different number sets so as to upgrade the level of difficulty in carrying out acts of academic dishonesty [13].

It has been found that case studies are valuable to use the higher levels of Bloom's Taxonomy during assessments. Even during online examinations, open ended tasks like a case study will help to check the problem solving and critical thinking approaches in the students. Trying to solve cases studies will compel students to maintain academic integrity and focus on their own skills rather than on collusion and plagiarism through the internet [32]. In many fields of study academicians tend to utilise case studies as a higher order tool.

Another theme which may be relevant for this discussion is setting assessment tasks that reflect the real-life practical solutions [36] so that they enable students to use their critical thinking abilities instead of cheating as a tool to write an appropriate answer. It helps to test the problem solving and higher order thinking which according to Bloom's Taxonomy will trigger student learning [36].

Personalised testing tools, (multiple choice questions MCQs or essay type questions) may be used as a method to curb plagiarism for online courses during the COVID times [37]. Personalised assessments may use three dominant approaches namely (1) Parameterization—substitution of parameters with random values (2) Databank—choice of certain questions from a large databank of questions, and (3) Macro—Replacement of macros in questions [37]. With these tools it is certain that a variety of ways in which we can reduce the kind of cheating students carry out during their online assessments. Most assessments should be conducted as a meaningful and judicious mix of assessment forms so as to create a holistic examination situation providing a fair chance to all students to display their abilities [38].

3   Using Technology to Avoid Cheating in Online Courses

As highlighted above, cheating, plagiarism and academic integrity are some of the key issues in literature focusing on online course design and implementation [39, 40]. Arguably, addressing cheating is of paramount importance if Higher Education Institutions (HEIs) are to maintain the integrity of their educational programmes [5, 41]. In this part of our chapter, we will explore the role of educational technologies in mitigating cheating in online courses, that is, how technology can be harnessed to either catch, prevent or discourage students from cheating [42]. The issues around cheating in online courses have become more prevalent especially following the pivot to online teaching and assessment because of the Covid-19 pandemic [43, 44]. HEIs across the globe have had to implement some strategies aimed at ensuring the secure delivery of online assessments to maintain academic standards. It is generally acknowledged that the problem of academic dishonesty is a major concern in any learning environment, however, it is more widespread in online courses [5, 45]. Some authorities posit that it is difficult to ascertain the identity of students taking assessments online given the fact that they work independently and anonymously in most cases [45]. It is, therefore, imperative for HEIs to find reliable solutions to combat this problem which threatens to affect the standards and quality of educational provision and the integrity of degree programmes. Arguably, there is no easy answer to this problem. As contested by QAA it is difficult to design cheating out of assessment [46]. In the same vein, some authorities have raised concern regarding the lack of studies focusing on approaches that can be employed to promote academic integrity in online environments [47]. It is widely discussed in literature that one of the challenges is that most of the techniques used to prevent cheating in face-to-face environments are not effective in online environments [47]. For instance, a study conducted by LoSchiavo and Shatz cited in [47] established that asking students to agree to an honour code did not impact on the rate of cheating in a particular online task. We, therefore, believe that it is important to identify some of the technologies that can be used and to explore how they can be employed to avoid cheating including some of the technical, logistical and ethical considerations that educators should bear in mind when deciding to use different technologies to avoid cheating in online courses.

3.1 Using Proctoring Technologies

To ensure the reliability of online assessments, most HEIs have resorted to the use of remote proctoring technologies. In general, invigilating assessments in an online environment is achieved by the use of innovative technologies to monitor the activities of students during online assessment activities over the internet through a web camera [48]. It has been shown that these proctoring systems rely on tools that are readily available on students' devices including webcameras and microphones to monitor them and foster academic integrity [44]. This proctoring system was first developed in 2006 by Kryterion and was introduced on the market as a reliable technological solution in 2008 [44]. Several proctoring systems are now available on the market, however, there is limited evidence of widespread use in HEIs. This makes it difficult when it comes to decisions about the most appropriate option for an institution. It is worth considering the cost and the other technical requirements involved when such decisions are made. Another study highlighted, three factors that should be borne in mind when selecting an appropriate proctoring system at institutional level and these include the cost, security as well as the educator and student comfort with the use of the technology [43]. The same views are echoed by [44] who also talk about the importance of ensuring that adequate training should be provided to facilitate the use of these online proctoring systems and the need to ensure that the systems are not overly intrusive as this might raise some ethical concerns around student privacy. When such decisions are made, it is worth consulting all the key stakeholders who include students, academic staff and the technical staff that support the deployment of new technologies across the educational institutions. At the moment it is not fully understood whether the use of online proctoring systems produces more benefits than risks in the students' learning experience [44], however, there is evidence that these systems are beneficial in terms of fostering academic integrity.

As highlighted above, the use of proctoring systems facilitates the delivery of online assessments while upholding the sanctity, security, trustworthiness and reliability of the online assessments. According to [49] the appropriate use of these proctoring systems reduces the instances of student cheating in online courses. In the same vein, [50] assert that students will behave properly and avoid cheating if they are aware that the academic staff will review their recorded assessment sessions. As will be explained in more detail in the subsequent paragraphs, some of the proctoring systems ensure the authentication and the verification of the student identity, thereby fostering academic integrity in an online environment [44]. For the benefit of colleagues interested to make use of these systems, it is necessary to understand their hardware and how they work.

Each proctoring system has got two main components including the web camera and the computer or browser lockdown. The web camera on the student's device should be enabled to record all the activities of the student including their surroundings when undertaking assessments. This enables the proctor or the examiner to review the video recording from the other end. This way, the proctor is able to identify any behaviours including potential cheating activities such as talking to someone in the same environment, consulting from a book or a mobile device for answers. The proctor can identify any suspicious movements and this can play an important role in combating cheating during online assessments [43]. As indicated earlier, the second component of any proctoring system is the computer or browser lockdown. This is important in ensuring that a student cannot use internet applications and other processes including copying, pasting or printing which might lead to cheating under assessment conditions in an online environment [51]. It is important to note that proctoring systems can facilitate the recording of all the behaviours and students' internet activities during the assessment including the websites that the student tries to access. Such a video record can be used by the examiners or proctors either during the assessment period or afterwards to monitor and identify any potential cheating.

There exist three different types of online proctoring systems that can be considered for use by educational institutions and the main categories are discussed below [44].

- Live proctoring: this involves the human proctor monitoring the assessment in real-time but virtually in an online environment. Such systems facilitate the scheduling of assessments at a specific time so students enrolled on a course can all do the assessment on the same day and at the same time. Arguably, the human proctors will need training to be adept in the use of technology so they can be able to monitor and identify any potential cheating.

- Recorded proctoring: as explained earlier, this approach consists of the recording of students' activities and their surroundings while undertaking an online assessment. The human proctor or examiner will be able to review the recording after the assessment to evaluate the integrity of the assessment, that is, to find out if any suspicious behaviour or cheating has been committed during the assessment. The advantage of recording is that students can actually do the assessments at different times and this can facilitate the conduct of different assessments simultaneously. The downside of this is that it can be costly as human intervention for the review of the recordings is necessary.
- Automated proctoring: this is a cost-effective system that does not involve human intervention for the review of recordings. The system is able to identify potential cheating activities. Unlike the other systems, students are more likely to find this system friendlier given that they are not going to be undertaking assessment under the surveillance of a human proctor, neither do they need to make arrangements for live human proctors for their assessments. However, this system presents challenges including the potential for students to find out ways of preventing the detection of cheating. In addition, the system can easily produce false positives, that is, identifying innocent activities as potential cheating.

The major features of the online proctoring systems that we also consider important to flag up have been discussed in the literature [43, 44]. These include

(i)   authentication: this process ensures that the student identity is verified making it possible to ascertain that the student taking the assessment is the registered one,

(ii)   browser lockdown: this feature makes it possible to block students' access to other online resources when undertaking an online assessment on their devices-they cannot search for answers during the assessment,

(iii) remote authorizing and control: this is an important feature that gives the proctor control during an online assessment activity-the proctor can start, pause and end an assessment at a distance, and

(iv)   report generation: this facilitates the production of reports capturing all the activities of a student during an invigilated online assessment.

These different features ensure that the proctoring systems are quite useful in combating cheating in online courses. Several online proctoring systems have been designed and can be accessed on the market. Some examples of the proctoring systems that exist in some HEIs are: ProctorU, HonorLock, ProctorExam, Respondus, ProctorCam, B Virtual, Proctorio, Kryterion, and LearnerVerified [5, 44, 48]. It is beyond the scope of this chapter to evaluate the effectiveness of the different proctoring systems; however, we would like to reiterate that when selecting a system that is fit-for-purpose, you need to bear in mind several factors and to engage all the key stakeholders. In addition, the studies that have been conducted to date to assess the impact of using proctoring systems on student performance have produced mixed results. Some studies show that students who undertake assessments in non-proctored online assessments perform better than those who take proctored online assessments, and on the other hand, some studies reveal that there is no statistically significant difference between the students' performance in proctored and unproctored online assessments [43, 51, 52]. In the same vein, it remains to be verified whether the use of online proctoring systems has more advantages than disadvantages for students [44].

3.2 Using Technology for Identity Verification

As highlighted earlier, the proliferation of online courses poses some challenges with regards to academic integrity, in particular, issues around identity verification. While in physical spaces it is relatively easy for invigilators to ask students to provide an ID to verify their identity, identity verification can present enormous challenges in an online environment. When HEIs started using the Zoom platform for delivering synchronous lectures during the pandemic, one of the problems that most lecturers faced was the interruption of meetings with strangers who were able to access and join meetings online. This was, however, resolved by enabling the authentication function in Zoom, which meant that only students registered on a particular course were able to access the sessions using their university email addresses and log-in details. By doing so, it was possible to address the problem of being invaded in meetings by some strangers.

When offering assessments online which involve certification and accreditation it is important to ensure that the identity of students undertaking the assessments is verified so that only students registered on the courses can access and receive credits [48, 53]. In the same vein, it is also important to ensure that the students adhere to the set academic standards and demonstrate acceptable behaviour during their engagement with the course activities including online assessments. In other words, to maintain the academic integrity, it is necessary to put in place some strategies to make sure that students do not solicit the help of others or cheat to pass the online courses.

Technology can be used to authenticate and verify the identity of students working in an online environment. Some universities are using Zoom or Microsoft Teams to monitor students writing exams online. However, there are some proctoring systems that are designed to provide authentication as well as continuous monitoring of students in an examination situation. As argued by [54] authentication constitutes the first line of defence in the enforcement of academic integrity. Some authorities described a proctoring system that helps with online student authentication based on multimodal biometrics technology [48]. In addition to identity verification, the same system can also provide automatic continuous image and audio processing a feature which allows online courses to gain value that is beneficial to both institutions and students. In a live online proctoring session, students connect their screens with a live human proctor in an online setting. The human proctor can ask the students to show a photo ID to verify their identity and this helps to foster academic integrity.

3.3 Using Technology to Detect Plagiarism

One of the major forms of academic dishonesty is plagiarism. Anecdotal evidence shows that many HEIs had to deal with increased cases of plagiarism during the pandemic when most of the teaching and assessment activities were migrated to online environments. Apparently when students are experiencing stress they tend to commit academic dishonesty. Some students choose to copy and paste entire paragraphs from the internet without acknowledging the sources when writing essays or short answer assessments [45]. We call that academic shoplifting and this constitutes a gross academic misconduct. Many HEIs have developed some policies around plagiarism, however, some students do not bother to familiarize themselves with such policies and choose to continue to use other people's work without giving them proper credit. Before the advent of technology, it was difficult to detect plagiarism. Educators had to depend on experience and their ability to detect copied work. However, with the advent of technology, it is now possible to deploy some similarity detection software. Some of the commonly used software are Turnitin (http://www.turnitin.com), Writecheck (http://writecheck.com), Duplichecker (http://www.duplichecker.com), and others. These tools enable all potential plagiarism to be detected. However, it is important to remember that the educator's professional judgement plays an important role in terms of determining or investigating further if a student has committed plagiarism or not. We feel that students must be supported to avoid plagiarism by developing their academic writing skills including the ability to cite consulted sources in a critical manner.

3.4 Using Test Bank Questions

The use of test bank questions constitutes a viable option to mitigate cheating in online assessments. These test bank questions can be developed using tools available in virtual learning environments such as Blackboard, Moodle or Canvas, among others. This allows the randomization of questions and answers given to students in multiple choice exams, making it difficult for them to collude while undertaking online assessment. However, it is worth noting that the use of test bank questions is fraught with problems of academic dishonesty. The tech-savvy students we have today can search for answers on internet during the online assessment. While this can be combatted by locking down the browser on the student's device, students can still make use of a different device such as a mobile phone to search for answers especially if they are undertaking the assessment at a distance. One way to address this challenge is to paraphrase the test bank questions. It has been observed that paraphrasing the questions in a test bank can reinforce academic integrity as students can find it difficult to search for answers online. In a study conducted by [5] it was established that students tend to perform better when using test bank questions that are not paraphrased, hence, it is necessary to avoid using verbatim questions in online assessments. This is an important result for HEIs that are keen to prevent academic dishonesty and for educators who seek to address the challenges of using test banks. This shows that in addition to the use of technology, it is also important for academics to be supported to design assessments in ways that help to deter cheating. An example of this has been discussed earlier when we explored the use of Bloom's Taxonomy to design questions in assessments.

3.5 Other Technologies

We have highlighted some examples of how technology can be harnessed to avoid cheating in online courses. The examples cited in our chapter are by no means conclusive, there exist other strategies that have been developed to reinforce academic integrity. For instance, [55] developed an online lab examination management system (OLEMS) which is able to generate different questions for the online lab exams to ensure that students are discouraged from cheating through exchange of information. For instance, for multiple choice questions, the system can change the answers for each of the students undertaking the online assessment. Several other technologies have been developed, hence it is worth exploring the different technologies and their affordances as we seek to combat cheating in online courses.

4 Discussion

The pandemic has witnessed the leapfrogging of digital exams, remote assessments, and an increase of academically dishonesty behaviour. The pandemic may have altered the delivery of teaching and learning, but not changed the core values of honesty, trust, fairness, respect, responsibility and courage in academic practices [11] and the societal mission of higher education institutions [56]. This chapter explored the use of effective assessment design and technology to combat academic malpractice and foster academic integrity in online courses.

Previous studies show that the prevalence of cheating is affected by types of assessment [1, 7]. For instance, students are most likely to cheat on heavily weighted assessments and those with pressurized deadlines [57]. In contrast, students are least likely to cheat on four types of assessments, namely, "in-class tasks, personalized and unique tasks, vivas and reflections on practical placements" [57, p. 685]. Reference [3] suggested designing assessments by taking into account that students have access to a wide range of online resources. In addition, online cheating may be effectively deterred when assessment questions are designed in a way that requires students to apply knowledge, this compels students to demonstrate sufficient understanding of concepts as a prerequisite for the provision of answers [3]. In other words, assessments that require higher-order thinking skills rather than lower-order cognition such as recalling facts or memory testing can be effective deterrence strategies for cheating behaviour in the digital age [3]. Hale [21] pointed out that visual and spatial assignments, such as storyboards, poster presentations and mind mapping are effective in thwarting plagiarism. Moreover, instead of having a traditional high-stake summative assessment, a regular low-stake formative assessment may effectively reduce students' stress and related academically dishonesty behaviour and build up students' confidence [14]. Furthermore, both students and staff need to be supported in the transition to effective and ethical assessment online [44]. For example, a clarification needs to be made on what constitutes academically dishonest behaviour for each type of assessment [7] and providing clear definitions and instructions of what resources are permitted in the assessment guidelines [3]. Appropriate assessment design may effectively reduce academic integrity violation [14].

Tackling online cheating is time consuming and difficult [58]. For instance, traditional exams require face-to-face invigilation, while proctored online examination is more challenging to implement and there are equity and privacy concerns around online proctoring software [3]. Technologies can be used for identity authentication [13] and software such as Turnitin (http://www.turnitin.com), Writecheck (http://writecheck.com), Duplichecker (http://www.duplichecker.com) may detect and reveal sources of plagiarism. However, ultimately, technologies are tools which need to be harnessed, embracing the core values of honesty, trust, fairness, respect, responsibility and courage [11, 59] in order to curb online cheating and foster academic integrity. Given the wide range of proctoring technologies on the market, it is necessary to consider the cost, security and the ease of use for both staff and students when deciding which tool to invest in [43]. It is also worth considering the risks associated with the technologies, especially to ensure that the technologies are not overly intrusive as this might create some ethical problems [44]. We also recognise that the use of these technologies should not disadvantage any student who may not have access to the relevant technologies, hence the need for inclusive and equitable policies to be put in place. Although there are mixed results regarding the impact of using proctoring technologies on student performance [51, 52, 60], there is a general consensus that these technologies help to uphold the sanctity of online assessments [49]. It is, therefore, possible to implement reliable online assessments. This makes it possible for educational institutions to expand the designing and implementation of online courses. Technology can be used to create test bank questions that can be randomized in assessments making it difficult for students to cheat online. Arguably, if the same questions are used repeatedly, students who are tech-savvy can easily search for answers online. However, this can be prevented by ensuring that the questions are paraphrased to make it difficult for students to find answers online [5]. This raises an important issue around academic development; academic staff should be supported to design questions in ways that make it difficult for students to search and find answers easily online. Although it is difficult to eliminate cheating completely, use of the Bloom's Taxonomy and the creative use of innovative technology can help to mitigate cheating in online courses.

5 Conclusion

The widespread phenomenon of academic dishonesty among students in Higher Education Institutions is not surprising. However, Covid-19 pandemic has been an incubator for the cheating behaviour to be exacerbated as teaching and learning have been swiftly transitioned online in response to the pandemic. The destructive and harmful consequences of academic dishonesty have severe impacts at individual, institutional and societal levels. Thus, to tackle academic dishonesty is pressing as teaching and learning are moving towards the digital domain. The use of Bloom's Taxonomy for assessment design and use of technology to detect and deter online cheating are effective ways to curb academically dishonest behaviour. Decisions around investing in technologies to reinforce academic integrity should involve all the different stakeholders within the educational institutions. It is important to ensure that the technologies are cost effective and easy to use by the academic staff and students without creating unethical practices and unnecessary inequalities. It is, therefore, important for educational institutions to put in place policy infrastructure that encourages inclusivity and equity. In the same vein, academic staff and students should be supported to make use of the available technologies. Academic development opportunities should be provided strategically to support staff to know how to embrace toolkits such as the Bloom's Taxonomy to facilitate the effective designing of assessments. This way, cheating in online courses can be mitigated and reliable online courses can continue to be delivered. Ultimately, to combat these online cheating practices there is need to embrace values of honesty, trust, fairness, respect, responsibility and courage to promote academic integrity in the higher education sector. These practices are not only relevant in the Covid-19 transitional period, but also relevant in all educational contexts in the post-pandemic era. It is an obligation for Higher Education Institutions to provide students with the moral compass, develop ethical behaviours, cultivate a culture of integrity and prepare students for their citizenship and service to the society.

6 Way Forward

This chapter highlighted the importance of curbing academic dishonesty and fostering academic integrity both during the COVID-19 and post-pandemic era. The authors have made a rigorous effort to highlight some of the reliable approaches that can be judiciously deployed by the academicians to uphold academic integrity. Understandably, contextual factors play an important role in making decisions around the appropriateness of different methods that can be embraced by academicians. It is important to ensure that there is adequate institutional support to help academics to apply Bloom's Taxonomy in designing assessments and to provide the technical help needed to use the available technologies. Lack of technical expertise can undermine the use of the available technologies. Yet, digital poverty can also constitute a major barrier in the use of these innovative tools. One of the key questions for you is: 'So what is the way forward? Are you ready to embrace some of the ideas discussed in this chapter? What more can be done to uphold academic integrity in online courses?'.

References

1. McCabe, D. L., Treviño, L. K., & Butterfield, K. D. (2001). Cheating in academic institutions: A decade of research. *Ethics and Behavior, 11*(3), 219–232. https://doi.org/10.1207/S15327 019EB1103_2
2. Newton, P. M. (2018). How common is commercial contract cheating in higher education and is it increasing? A systematic review. *Frontiers in Education, 3*. https://doi.org/10.3389/feduc. 2018.00067.
3. Reedy, A., Pfitzner, D., Rook, L., & Ellis, L. (2021). Responding to the COVID-19 emergency: Student and academic staff perceptions of academic integrity in the transition to online exams at three Australian universities. *International Journal of Educational Integrity, 17*(1), 1–32. https://doi.org/10.1007/s40979-021-00075-9
4. Lancaster, T., & Cotarlan, C. (2021). Contract cheating by STEM students through a file sharing website: A Covid-19 pandemic perspective. *International Journal of Educational Integrity, 17*(1), 1–16. https://doi.org/10.1007/s40979-021-00070-0
5. Golden, J., & Kohlbeck, M. (2020). Addressing cheating when using test bank questions in online classes. *Journal of Accounting Education, 52*,. https://doi.org/10.1016/j.jaccedu.2020. 100671
6. Harding, T. S., Mayhew, M. J., Finelli, C. J., & Carpenter, D. D. (2007). The theory of planned behavior as a model of academic dishonesty in engineering and humanities undergraduates. *Ethics and Behavior, 17*(3), 255–279. https://doi.org/10.1080/10508420701519239
7. Passow, H. J., Mayhew, M. J., Finelli, C. J., Harding, T. S., & Carpenter, D. D. (2006). Factors influencing engineering students' decisions to cheat by type of assessment. *Research In Higher Education, 47*(6), 643–684. https://doi.org/10.1007/s11162-006-9010-y
8. Desalegn, A. A., & Berhan, A. (2014). Cheating on examinations and its predictors among undergraduate students at Hawassa University College of Medicine and Health Science, Hawassa, Ethiopia. *BMC Medical Education, 14*(1), 1–11. https://doi.org/10.1186/1472-6920- 14-89
9. McKie, A. (2021). *Students 'twice as likely to cheat' in online exams.* https://www.timeshigh ereducation.com/news/students-twice-likely-cheat-online-exams. Accessed 10 Jul 2021.
10. McClung, E. L., & Schneider, J. K. (2015). A concept synthesis of academically dishonest behaviors. *Journal of Academic Ethics, 13*(1), 1–11. https://doi.org/10.1007/s10805-014- 9222-1
11. Quality Assurance Agency. (2020). *Contracting to cheat in higher education: How to address essay mills and contract cheating* (2nd edn, p. 35). https://www.qaa.ac.uk/docs/qaa/guidance/ contracting-to-cheat-in-higher-education-2nd-edition.pdf.
12. Busch, P., & Bilgin, A. (2014). Student and staff understanding and reaction: Academic integrity in an australian university. *Journal of Academic Ethics, 12*(3), 227–243. https://doi.org/10.1007/ s10805-014-9214-2
13. McGee, P. (2013). Supporting academic honesty in online courses. *Journal of Educators Online, 10*(1). https://doi.org/10.9743/JEO.2013.1.6.

14. Roberts, C. (2021). *Fair assessment: tackling the rise in online cheating.* https://www.timeshigh ereducation.com/campus/fair-assessment-tackling-rise-online-cheating. Accessed 10 Jul 2021.
15. Skidmore, C. (2021). *Essay mills (Prohibition)* (vol. 689). Hansard—UK Parliament, UK.
16. Lefoka, P. J. (2020). The prevalence of and factors contributing to assessment malpractice at the National University of Lesotho. *Humanities and Social Science Research, 3*(3). https://doi.org/10.30560/hssr.v3n3p10
17. Quality Assurance Agency. (2020). Assessing with integrity in digital delivery introduction What do we mean by 'academic integrity'?, pp. 1–7.
18. Abdalqhadr, A. (2020). Academic integrity is an extension of your own personal integrity. *Economics, Finance and Management Review, 1*, 93–98. https://doi.org/10.36690/2674-5208-2020-1-93-98
19. Smith, D. (2014). Fostering collective ethical capacity within the teaching profession. *Journal of Academic Ethics, 12*(4), 271–286. https://doi.org/10.1007/s10805-014-9218-y
20. Peters, M., Boies, T., & Morin, S. (2019). Teaching academic integrity in Quebec Universities: Roles professors adopt. *Frontiers in Education, 4*(99), 1–13. https://doi.org/10.3389/feduc.2019.00099
21. Hale, M. (2018). Thwarting plagiarism in the humanities classroom: Storyboards, scaffolding, and a death fair. *Journal of the Scholarshop of Teaching and Learning, 18*(4), 86–110. https://doi.org/10.14434/josotl.v18i4.23174
22. Bedford, W., Gregg, J., & Clinton, S. (2009). Implementing technology to prevent online cheating: a case study at a small southern regional university (SSRU). *Journal of Online Learning and Teaching, 5*(2), 230–238. http://jolt.merlot.org/vol5no2/gregg_0609.pdf.
23. Athanassiou, N., Mcnett, J. M., & Harvey, C. (2003). Critical thinking in the management classroom: Bloom's taxonomy as a learning tool. *Journal of Management Education, 27*(5), 533–555. https://doi.org/10.1177/1052562903252515
24. Swart, A. J. (2010). Evaluation of final examination papers in engineering: A case study using Bloom's taxonomy. *IEEE Transactions on Education, 53*(2), 257–264. https://doi.org/10.1109/TE.2009.2014221
25. Zaidi, N. B., Hwang, C., Scott, S., Stallard, S., Purkiss, J., & Hortsch, M. (2017). Climbing Bloom's taxonomy pyramid: Lessons from a graduate histology course. *Anatomical Sciences Education, 10*(5), 456–464. https://doi.org/10.1002/ase.1685
26. Jones, K. O., Harland, J., Reid, J. M. V., & Bartlett, R. (2009). Relationship between examination questions and Bloom's taxonomy. In *Proceedings—39th ASEE/IEEE Frontiers in Education Conference,* 18–21 October. https://doi.org/10.1109/FIE.2009.5350598.
27. Cullinane, A. (2009). Bloom's taxonomy and its use in classroom assessment. *NCE-MSTL Resource and Research Guides, 1*(13). https://www.researchgate.net/publication/283328372_Blooms_Taxonomy_and_its_Use_in_Classroom_Assessment.
28. Kim, M. K., Patel, R. A., Uchizono, J. A., & Beck, L. (2012). Incorporation of Bloom's taxonomy into multiple-choice examination questions for a pharmacotherapeutics course. *American Journal of Pharmaceutical Education, 76*(6). https://doi.org/10.5688/ajpe766114.
29. Lord, T., & Baviskar, S. (2007). Moving students from information recitation to information understanding: Exploiting Bloom's taxonomy in creating science questions. *Journal of College Science Teaching, 36*(5), 40–44.
30. Omar, N., Haris, S. S., Hassan, R., Arshad, H., Rahmat, M., Zainal, N. F. A., & Zulkifli, R. (2012). Automated analysis of exam questions according to Bloom's taxonomy. *Procedia—Social Behavioral Sciences, 59*, 297–303. https://doi.org/10.1016/j.sbspro.2012.09.278
31. Wilson, L. O. (2016). Anderson and Krathwohl Bloom's taxonomy revised understanding the new version of Bloom's taxonomy. *Second Principles* (pp. 1–8). https://www.quincycollege.edu/content/uploads/Anderson-and-Krathwohl_Revised-Blooms-Taxonomy.pdf%0Ahttps://thesecondprinciple.com/teaching-essentials/beyond-bloom-cognitive-taxonomy-revised/%0Ahttp://thesecondprinciple.com/teaching-essentials/beyond-bloom-cog.
32. Nkhoma, M., Lam, T., Richardson, J., Kam, B., & Lau, K. H. (2016). Developing case-based learning activities based on the revised Bloom's taxonomy. In *Proceedings of Informing Science and IT Education Conference (InSITE) Conference,* pp. 85–93. https://doi.org/10.28945/3496.

33. Zheng, A. Y., Lawhorn, J. K., Lumley, T., & Freeman, S. (2008). Debunks the MCAT myth. *Science, 319*, 2–3.
34. Krsak, A. M. (2007). Curbing academic dishonesty in online courses. In *TCC 2007 Proceedings*, pp. 159–170. http://www.ncta-testing.org/cctc/.
35. Paullet, K. (2020). Student and faculty perceptions of academic dishonesty in online classes. *Issues in Information Systems, 21*(3), 327–333. https://doi.org/10.48009/3_iis_2020_327-333
36. Abdul Rahim, A. F. (2020). Guidelines for online assessment in emergency remote teaching during the COVID-19 pandemic. *Education in Medical Journal, 12*(2), 59–680. https://doi.org/10.21315/eimj2020.12.2.6.
37. Manoharan, S. (1997). Cheat-resistant multiple-choice examinations using personalization. *Computers and Education, 130*, 139–151.
38. Jantos, A. (2021). Motives for cheating in summative e-assessment in higher education - a quantitative analysis. *EDULEARN21 Proceedings, 1*, 8766–8776. https://doi.org/10.21125/edulearn.2021.1764.
39. Cramp, J., Medlin, J. F., Lake, P., & Sharp, C. (2019). Lessons learned from implementing remotely invigilated online exams. *Journal of University Teaching and Learning Practice, 16*(1).
40. D'Souza, K. A., & Siegfeldt, D. V. (2017). A conceptual framework for detecting cheating in online and take-home exams. *Decision Sciences Journal of Innovative Education, 15*(4), 370–391. https://doi.org/10.1111/dsji.12140
41. Hollister, K. K., & Berenson, M. L. (2009). Proctored versus unproctored online exams: Studying the impact of exam environment on student performance. *Decision Sciences Journal of Innovative Education, 7*(1), 271–294. https://doi.org/10.1111/j.1540-4609.2008.00220.x
42. Michael, T., & Williams, M. (2013). Student equity: discouraging cheating in online courses. *Administrative Issues Journal, 3*(2). https://doi.org/10.5929/2013.3.2.8.
43. Hussein, M. J., Yusuf, J., Deb, A. S., Fong, L., & Naidu, S. (2020). An evaluation of online proctoring tools. *Open Praxis, 12*(4), 509. https://doi.org/10.5944/openpraxis.12.4.1113
44. Nigam, A., Pasricha, R., Singh, T., Churi, P. (2021). A systematic review on AI-based proctoring systems: Past, present and future. *Education and Information Technologies*, 0123456789. https://doi.org/10.1007/s10639-021-10597-x.
45. Moten, J., Fitterer, A., Brazier, E., Leonard, J., & Brown, A. (2013). Examining online college cyber cheating methods and prevention measures. *The Electronic Journal of e-Learning, 11*(2), 139–146.
46. Quality Assurance Agency. (2021). Digital assessment security advice on online invigilation and other solutions for ensuring good academic conduct in online assessment (pp. 1–11).
47. Corrigan-Gibbs, H., Gupta, N., Northcutt, C., Cutrell, F., & Thies, W. (2015). Deterring cheating in online environments. *ACM Transactions on Computer-Human Interaction, 22*(6). https://doi.org/10.1145/2810239.
48. Labayen, M., Vea, R., Florez, J., Aginako, N., & Sierra, B. (2021). Online student authentication and proctoring system based on multimodal biometrics technology. *IEEE Access, 9*, 72398–72411. https://doi.org/10.1109/ACCESS.2021.3079375
49. Karim, M. N., Kaminsky, S. E., & Behrend, T. S. (2014). Cheating, reactions, and performance in remotely proctored testing: An exploratory experimental study. *Journal of Business Psychology, 29*(4), 555–572. https://doi.org/10.1007/s10869-014-9343-z
50. Kolski, T., & Weible, J. L. (2019). Do community college students demonstrate different behaviors from four-year university students on virtual proctored exams? *Community College Journal of Research and Practice, 43*(10–11), 690–701. https://doi.org/10.1080/10668926.2019.1600615
51. Alessio, H. M., Malay, N., Maurer, K., Bailer, A. J., & Rubin, B. (2017). Examining the effect of proctoring on online test scores. *Online Learning, 21*(1). https://doi.org/10.24059/olj.v21i1.885.
52. Beck, V. (2014). Testing a model to predict online cheating-Much ado about nothing. *Active Learning in Higher Education, 15*(1), 65–75. https://doi.org/10.1177/1469787413514646

53. Atoum, Y., Chen, L., Liu, A. X., Hsu, S. D. H., & Liu, X. (2017). Automated online exam proctoring. *IEEE Transactions on Multimedia, 19*(7), 1609–1624. https://doi.org/10.1109/TMM. 2017.2656064

54. Norris, M. (2019). University online cheating - how to mitigate the damage. *Research in Higher Education Journal, 37*, 1–20.

55. Kolhar, M., Alameen, A., & Gharsseldien, Z. M. (2018). An online lab examination management system (OLEMS) to avoid malpractice. *Science and Engineering Ethics, 24*(4), 1367–1369. https://doi.org/10.1007/s11948-017-9889-z

56. King, P. M., & Mayhew, M. J. (2002). Moral judgement development in higher education: Insights from the defining issues test. *Journal of Moral Education, 31*(3), 247–270. https://doi. org/10.1080/0305724022000008106

57. Bretag, T., et al. (2019). Contract cheating and assessment design: Exploring the relationship. *Assessment and Evaluation in Higher Education, 44*(5), 676–691. https://doi.org/10.1080/026 02938.2018.1527892

58. Quality Assurance Agency. (2016). *Plagiarism in higher education.*

59. International Center for Academic Integrity [ICAI]. (2021). The fundamental values of academic integrity (3rd edn.). 20019_ICAI-Fundamental-Values_R12.pdf (academicintegrity.org). Accessed 10 Jul 2021.

60. Alzaneen, R., & Mahmoud, A. (2019). The role of management information systems in strengthening the administrative governance in ministry of education and higher education in Gaza. *International Journal of Business Ethics and Governance, 2*(3), 1–43. https://doi.org/10.51325/ ijbeg.v2i3.44

Using Virtual Reality and Augmented Reality with ICT Tools for Enhancing Quality in the Changing Academic Environment in COVID-19 Pandemic: An Empirical Study

M. Raja and G. G. Lakshmi Priya

1 Introduction

ICT (Information and Communication Technologies) denotes Information Commu-nication Technologies, which embrace any type of technology for handling and communicating information. ICT tools have a distinctive significance within the education system as well as for implementing social transactions. The uses of ICT tools have also brought significant improvements in the ways teachers and other prominent education institutions work. The approach for ICT within processes of teaching and learning and the institutional strategies have a positive impact on the education system. In simple words, these educational institutions can quickly fulfil their job roles in an orderly manner and satisfy themselves using ICT tools to promote a favourable learning environment. Teachers and students can quickly render good collaboration within these favourable learning surroundings to achieve their academic goals. The use of multimedia tools makes this learning environment in classrooms more suitable and lively according to the needs of students.

In educational institutions, at different levels, individuals are expected to imple-ment various functions and tasks. Inputting different roles and tasks into operation, these individuals are also expected to use ICT. With the help of ICT, individuals can generate awareness amongst students apart from augmenting their knowledge level in different aspects. When educators look at enriching the methods of teaching and learning and the instructional methods, they use ICT. On the contrary, when the students have to complete their projects and assignments, they use ICT tools

M. Raja · G. G. Lakshmi Priya (✉)
Department of Multimedia, VIT School of Design, Vellore Institute of Technology, Vellore, Tamil Nadu, India
e-mail: lakshmipriya.gg@vit.ac.in

M. Raja
e-mail: raja.m@vit.ac.in

for carrying out their tasks in an organised manner and as per the expectations. Talking about Augmented Reality and Immersive Virtual Reality under ICT for education, their role in online learning and combined environment is at the initial stage. Many efforts are being made for framing core constraints and affordability, whereas prospective future developments have been outlined.

Within processes of traditional learning, a lot of emphases is laid on the contents. The syllabus and the contents are considered to be significant. Teachers need to equip themselves properly before coming to class to deliver their lectures. Using the ICT tools helps promote independent learning while considering alternative learning theories [1]. Augmented Reality and Virtual Reality are considered to be the two sides of a coin. Both of them aim at extending the sensorial surroundings of the individuals by mediating reality with the help of technology. The former depends mainly on the alternative set up to experience, and the latter improvise the current elements along with an additional layer of meaning. Recently, such technologies have started flourishing.

VR is related strongly to the growth and development of digital simulations and personal computers. Heilig Morton initiated the thought behind alternative reality for inhibiting. The mechanical machinery allows the spectators to enjoy sort movies and have multiple senses like smell, touch, sight, and hearing. Since then, computational developments support the creation of a rapidly advanced technology-mediated environment. A study explores the way of combining the sensory input to the virtual set-up. VR has helped in spreading awareness and has also been reformulated across different steps of innovation. Its traits and features are changing continuously, and the socio-economic attributes have also accepted its scope and definition. In the past three decades, the virtual world and immersive VR have emerged to be the most trending topics [2].

This paper aims to give insight into the impact of the COVID-19 pandemic on education, the benefits of ICT, VR and AR, and the literature review on similar topics. In addition to this, we ascertain the significance of using VR and AR with ICT tools to improve the quality of academic environments through data set collected with the help of a structured questionnaire and discussed the results.

1.1 Impact of COVID-19 on Education

The times that no one could have imagined. The epidemic times that sweep the world, leaving no place unaffected. COVID-19 has undoubtedly made an indelible mark on the world of education. By the end of March 2020, the virus had spread throughout India, forcing the closure of the majority of schools, colleges, and institutions.

Though the COVID-19 epidemic had many harmful impacts on education, there was also a beneficial consequence that might propel the education system and its approaches. The epidemic has paved the way for new techniques of spreading knowledge throughout the world. It was extremely difficult for India since many people

Fig. 1 The student
population in India

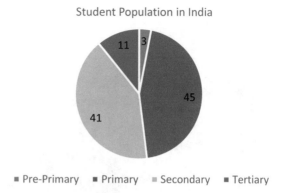

Student Population in India

■ Pre-Primary   ■ Primary   ■ Secondary   ■ Tertiary

live in places without internet access, while others attend less equipped government-run institutions. Several attempts were undertaken to continue teaching and learning using online techniques, but it was impossible to make it available to everyone.

According to UNESCO, approximately 1.5 billion school and college students were stranded at home in the middle of the raging epidemic, accounting for about 90 percent of the world student population as of March 2020. India tops the globe in school enrollment with 250 million kids and has one of the most extensive networks of universities and colleges. The Indian education system is extremely diverse, with over fifteen lakh schools and fifty thousand universities and colleges supported by over three hundred twenty million youthful and passionate students. The overall student population in India is represented in Fig. 1, based on the data obtained from UNESCO.

COVID-19 has increased the use of digital technology in education delivery. Education institutions shifted toward blended learning and pushed teachers and students to become tech-savvy. Smart technology, online, webinars, virtual meeting room, teleconferencing, digital tests, and evaluations became commonplace phenomena that would otherwise have been defined or would have come into practical use a decade or more later. There was an unprecedented collaboration among all stakeholders in education, including administrators, instructors, students, parents, and firms developing software for creative information transmission.

1.2 Role of ICT, VR, and AR During Pandemic

Technology is continuously expanding; education has embraced ICT and now provides handy solutions to assist people in improving their knowledge, education, and literacy level. An e-learning platform allows for quick access to knowledge and skills enhancement at any time and from any location. It offers a framework where individuals may self-guided create a personalised bundle relating to major

Fig. 2 Remote classroom using ICT

topic areas. E-Learning is a critical component of ICT. E-Learning is the use of electronic technology to access educational content other than in a traditional classroom. These courses are seizing the moment and attempting to fill the academic hole left by educational institutions' shut down due to the coronavirus outbreak.

During the epidemic, digital services and products became the saviour, and demand for new technological developments increased dramatically. The worldwide EdTech sector is growing at a rate of 17–25% each year. Because of the shutdown, it had an additional impetus to develop. Trends that were relevant before the present circumstance are gaining traction. Seamless classes are just a dream without ICT tools (Fig. 2).

The capacity of AR and VR to build remote, experiential learning environments offers a feasible alternative for governments and institutions to consider at this vital moment. The current state of AR and VR technology provides remote collaborative classes and huge numbers of students to congregate and study while limiting the danger of exposure.

Augmented Reality (AR) is a technology that cleverly blends virtual data with the actual environment. It employs a wide range of technical approaches, including multimedia, 3D modeling, real-time tracking and recognition, intelligent interaction, and sensing. Texts, photos, three-dimensional3D models, music, movies, and other virtual information are simulated and applied in the actual environment (Fig. 3)—the two types of information supplement one another, resulting in the improvement of the real world.

Computers, electronic data, and simulation technologies are all part of Virtual Reality (VR) technology. Its fundamental idea is that a computer generates a virtual world to give individuals the sensation of being immersed in it. VR helps students immerse into the content and learn without the distraction of the outside world (Fig. 4).

Fig. 3 Remote learning using AR

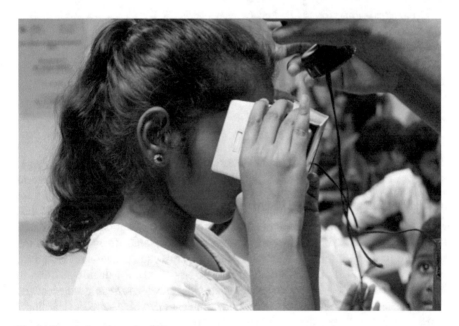

Fig. 4 Remote learning using VR

The applications of ICT, VR, and AR are enormous. These technologies have a greater potential to increase education standards, and hence, most institutes are coming forward to adopt them. Some of the key benefits of ICT, VR, and AR are highlighted in Fig. 5.

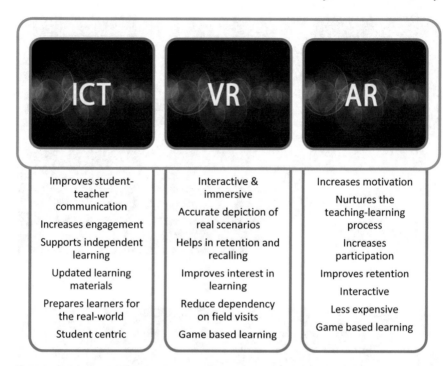

Fig. 5 Advantages of ICT, VR and AR

2 Literature Review

The new and advanced ICT tools can educate the teachers in large quality and better quality. A mix of the old ICT tools for broadening the coverage while getting access to new ICT tools which promote communication needs to offer cost efficiency for the education of teachers. A nationwide network of educating the community centers is equipped properly with computer labs and broadband connections. If trained staff members get access to online distance courses, it will help them in a big way. Also, the provision for support needs to be established amongst the educational institutions at different levels [3]. The ICT tools are expected to make a valuable addition to education and support efficient pedagogy to provide knowledge amongst the leaders. It is supposed to augment the process of communication, which supports learning.

Further, ICT also becomes quite pervasive computer-aided equipment. It's integrated into each aspect of the process of learning within educational institutions. Thus, the students and teachers cannot get the desired academic results, but they can also easily enhance the education management system [4]. While the implementation measures for promoting good quality education with the help of ICT tools happen, there are a few aspects that should be considered. These are defining the learning objectives is the first one. The education institutions at different levels have some learning objectives and goals which they expect to achieve. One of the main objectives of learning of these educational institutions is ensuring that learning and teaching are organised to cause the effective growth of the students. Students should easily augment their understanding and knowledge to accomplish their professional and personal goals. The efficiency of learning and teaching methods and the educational programs may be assessed within the structure of the learning goals. These can be defined as particular skills, knowledge, behavioural traits, and individuals' abilities [5].

Within educational institutions, individuals should have sufficient knowledge that is implemented. They include organisation, planning, staffing, directing, controlling, and coordinating. When the ICT tools need to be well integrated for making enhancements in education quality, the individuals have to plan the approaches and the methods. Organising helps promote the tasks and activities of organising according to their requirements. Directing is about directing the activities as well as functions that are according to educational objectives and goals. Staffing may be referred to the selection and recruitment of qualified, trained, and well-experienced individuals. Such individuals need to acquire an efficient understanding of performing their job duties. Coordinating is known as the coordination of different programs, approaches, and tasks. Controlling is the term used for controlling the resources and maintaining amiable work-environment conditions. It is important to ensure that the resources are used in a well-organised and appropriate manner [6].

When ICT is being integrated within the curriculum and the methods of instruction, the instructors and teachers are expected to work collaboratively with the students. Meetings are properly organised wherein matters are discussed and give one another suggestions and ideas. The directors and the principals should play the lead roles. It's their responsibility to make sure that the teachers and the staff members of the admin department are equipped well. They properly use technology. They also organise training programs and workshops for the staff members and the teachers. Thus, it helps them in augmenting their understanding and knowledge [7]. Once teachers acquire sufficient knowledge, they need to implement it in learning and teaching methods and instructions strategies. They also need to make sure that the technologies are used operationally for improving the learning amongst students and for helping them achieve their objectives and goals. That is why teachers should have the requisite leadership skills so that they may use them for promoting their well-being and also for their student's well-being.

When it comes to Virtual Reality and Augmented reality, they are considered to be the two faces of a coin. Both of them aim at extending the sensory environments of the individual. They do it by mediating reality with the help of technology. Virtual reality depends on an alternative set-up for experiencing, and Augmented Reality helps improve the current elements with some extra steps of meaning. Recently, all such technologies have started flourishing [8].

While the common VR may be experienced with the help of a screen, IVR— Immersive VR is about taking away forward through enrichment of the feeling of being present there. As it has been seen by Lorenzo, Lledo, and Pomares, in comparison to the conventional VR environment, in the IVR system, students are surrounded by an environment, and they receive stimuli constantly and have a possibility of interacting with it too.

While IVR and AR share their origin and premise, AR has taken a completely different path. Rather than concentrating on the divergent set-up, it depends on improving the real elements with the help of technology. AR may be interpreted as the view of the physical environment where the components are well integrated with the computer-aided sensory inputs. The main objective is seeing and experiencing the real world along with different virtual objects without losing control of the sense of reality.

VR is connected strongly with the development and growth of digital simulations and personal computers. Heilig Morton initiated the idea behind an alternate reality for inhabiting the DIEH7C7 installation way back in 1962. The mechanical machinery allowed the spectators to enjoy short movies and have multiple senses like smell, hearing, sight, and touch. Since then, computational advancements have supported the creation of advanced technological environments. The milestones of the path were the first head-mounted display device which Sutherland Ivan and Sproull Bob developed in the year 1968 [9]. Researches have explored the right way of combining the sensory inputs to the virtual set-up in the 1980s. Virtual reality has been spreading and reformulating itself with the help of several innovations. Its traits and features are constantly changing, and the socio-economic factors have also accentuated its scope and definition. In the past three decades, the immersive virtual world and reality have emerged as trending topics. Virtual reality is concerned with the rise of the internet, outlining the shared virtual environment where users interact. The multi-user virtual environment, the massive multipliers online, and the virtual world set-up are some of the glaring instances of this phenomenon. It highlights the social dimensions of Virtual Reality.

The virtual space has become quite a popular and sophisticated tool for distance learning, cooperation, and social relations [10]. Compared to the conventional Virtual Reality environment, in the IVR system, the students are surrounded by the environment and constantly receive stimuli and have the possibility of interacting with it. The first CAVE, a room with computer-aided elements that users visualised with HMDs or special glasses, represented the milestone in this regard. Multiple strategies have been adopted for obtaining this outcome, from considering the entire sight rotation to the reproduction of vibrations and the environment sounds. Devices such as HTC Vive, Oculus Rift, PlayStation, etc., lead to these innovations. However, it is argued that immersive is fluid and contextual attribution that depends on the technological trends and the expectations. It was yesterday, it's not today, and today, it won't be the same tomorrow [11].

The graph shown in Fig. 6 obtained from the http://www.statista.com website shows an enormous amount of growth in the sale of smartphones in 2021. Over thousand five hundred million unit sales are recorded during the first quarter of the year, and the overall sales are expected to grow at least double fold by the end of the year 2021. This shows the need for smartphones by people across the globe for various activities. Notably, during the COVID-19 pandemic, smartphone usage has increased due to the closure of schools and classes being conducted remotely.

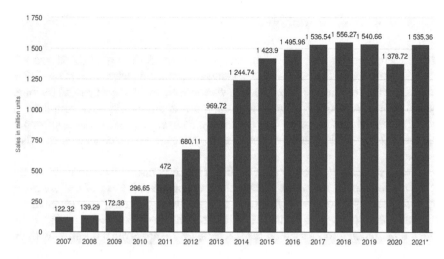

Fig. 6 Smartphone sales worldwide between 2007 and 2021

India holds the second position with over five hundred million active internet users, out of which 2/3rd of the user population are daily users. These statistics will rocket to multiple folds as the internet has become an inevitable part of our daily lives during the pandemic. Online classes are a dream without internet connectivity. Remote video conferencing tools have gained a lot of demand, and Zoom is one such ICT that is topping the list of video conferencing tools in terms of usage and revenue, as shown in Fig. 7.

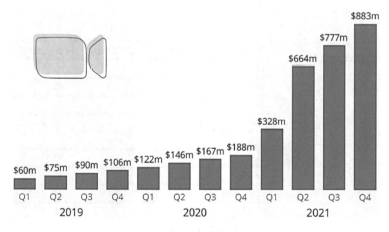

Fig. 7 Zoom's quarterly revenue projection during COVID-19 pandemic

3   Research Methodology

The present study is descriptive in which the reasons for using VR, AR, and ICT tools for improving the quality of academic environments have been studied. The sample size of the study is 140. The data was collected with the help of a structured questionnaire (Appendix A) on a five-point scale. Analysis was done with the help of the mean values and t-test. The descriptive analysis approach was employed in the research by sending google form questionnaires to the teachers to obtain their responses.

Table 1 presents the demographic profile of the respondents. There are 49% males and 51% females in the study. Among the respondents, 25% are of the age group up to 35 years, 46% of 35–50 years and 29% are above 50 years. The percentage of respondents who are post-graduate is 53%, and PhD is 47%. To identify the online video conferencing tools used by the teachers, they were given a question with the following options: Zoom Meetings, Google Meet, Microsoft Teams, GoToMeeting, and Others. The obtained responses show that the respondents widely use zoom Meetings, thus correlating with the quarterly revenue projections of zoom during the COVID-19 pandemic, as juxtaposed in Fig. 7. Followed by zoom meetings, Google Meet is the most preferred platform, next to that is Microsoft Teams, followed by GoToMeeting. About 3.6% of the respondents use conferencing platforms other than the four mentioned in the question, as shown in Fig. 8.

The qualifications and experience of each respondent differ, and hence the researchers also posted a question to seek if they are aware of VR and AR technologies. As shown in Fig. 9, the recorded responses show that only 37.14% (52 respondents) have experienced or used VR or AR technologies. 60.72% (85 respondents) have an idea of the technology but haven't experienced it, and 2.14% (3

Variables	Number of respondents	Percentage age
Gender		
Male	69	49
Female	71	51
Total	140	100
Age		
Upto 35	35	25
35–50	64	46
Above 50	41	29
Total	140	100
Educational qualification		
PG	74	53
Ph.D.	66	47
Total	140	100

Table 1 Demographic profile of the respondents

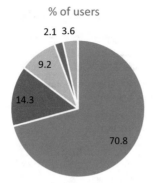

% of users

2.1 3.6

9.2

14.3

70.8

■ Zoom Meetings ■ Google Meet ■ Microsoft Teams ■ GoToMeeting ■ Others

Fig. 8 Most preferred conferencing tool

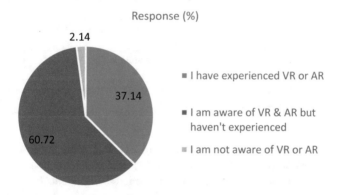

Response (%)

2.14

37.14

60.72

■ I have experienced VR or AR

■ I am aware of VR & AR but haven't experienced

■ I am not aware of VR or AR

Fig. 9 Respondents' experience with VR and AR

respondents) have never heard of VR and AR. To better understand the VR and AR technologies, the users must experience it rather than reading about it because teleportation happens only during technology [12].

Table 2 shows respondents' opinions on the role of ICT as a tool for enhancing quality in the changing academic environment. It is observed that Virtual Reality (VR) and Augmented Reality (AR) makes education more interesting for students is; the most important role of ICT as a tool for Enhancing Quality in the Changing Academic Environment with the mean value of 4.45. Virtual Reality (VR) and Augmented Reality (AR) depend mainly on the development of IT infrastructure (4.42). ICT has great potential in revamping the education system (4.39). It is essential to make the entire education system ICT savvy for a better tomorrow (4.36), and ICT technologies such as Virtual Reality (VR) and Augmented Reality (AR) give a close look of reality to students through technology (4.33).

Further, ICT tools help students in completing their assignments in a much better way (4.27). The role of the ICT tools in education can contribute significantly to

Table 2 Mean value of the role of VR, AR and ICT as a tool for enhancing quality in the changing academic environment

S. no.	Questions	Mean score
1	ICT tools help in improving the quality of education	3.84
2	ICT tools generate awareness amongst students in an easy and interesting manner	4.12
3	ICT tools help students in completing their assignments in a much better way	4.27
4	ICT has great potential in revamping the education system	4.39
5	It is not a difficult process for teachers and students in rural areas to acquaint themselves with ICT tools	3.56
6	ICT technologies such as Virtual Reality (VR) and Augmented Reality (AR) give a close look at reality to students through technology	4.33
7	Virtual Reality (VR) and Augmented Reality (AR) depends mainly on the development of IT infrastructure	4.42
8	Virtual Reality (VR) and Augmented Reality (AR) makes education more interesting for students	4.45
9	It is essential to make the entire education system ICT savvy for a better tomorrow	4.36
10	The role of the ICT tools in education can contribute significantly to shaping the future of a country's economy	4.24

shaping the future of a country's economy (4.24), ICT tools generate awareness amongst students easily and interestingly (4.12), and ICT tools help in improving the quality of education (3.84) were also considered important. Further, It is not difficult for teachers and students in the rural areas to acquaint themselves with ICT tools with the mean value of 3.56, which was not considered an important ICT role.

To learn more about teachers' perspectives on ICT, researchers asked open-ended questions that allowed them to freely express their thoughts on integrating VR and AR with ICT. Based on these responses, as shown in Fig. 10, the researchers understand the application of integrating VR and AR with ICT for students. The researchers divided the response descriptions into six categories based on these findings. Each identified category is considered to be a function of integrating VR and AR with ICT as listed below:

- Improves the quality of teaching-learning
- It helps to understand abstract concepts easily
- Fosters interaction between student and teacher
- Fosters interaction between student and teaching material
- VR, AR & ICT are essential tools for the learning community.

Table 3 shows the results of the t-test. It is found from the table that the significance value for all the statements except 1 is below 0.05. Hence all the statements regarding

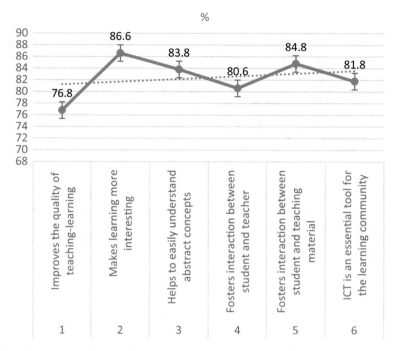

Fig. 10 Functions of VR and AR with ICT in education

Table 3 Mean Value. T-value and significance of the Role of VR, AR and ICT as a tool for Enhancing Quality in the Changing Academic Environment

S. no.	Questions	Mean score	t-value	Sig.
1	ICT tools help in improving the quality of education	3.84	4.209	0.000
2	ICT tools generate awareness amongst students in an easy and interesting manner	4.12	7.219	0.000
3	ICT tools help students in completing their assignments in a much better way	4.27	8.743	0.000
4	ICT has great potential in revamping the education system	4.39	8.631	0.000
5	It is not a difficult process for teachers and students in rural areas to acquaint themselves with ICT tools	3.56	0.611	0.271
6	ICT technologies such as Virtual Reality (VR) and Augmented Reality (AR) give a close look at reality to students through technology	4.33	9.527	0.000
7	Virtual Reality (VR) and Augmented Reality (AR) depend mainly on the development of IT infrastructure	4.42	10.971	0.000

(continued)

Table 3 (continued)

S. no.	Questions	Mean score	t-value	Sig.
8	Virtual Reality (VR) and Augmented Reality (AR) makes education more interesting for students	4.45	12.530	0.000
9	It is important to make the entire education system ICT savvy for a better tomorrow	4.36	10.309	0.000
10	The role of the ICT tools in education can contribute significantly to shaping the future of a country's economy	4.24	8.928	0.000

the Role of ICT as a tool for Enhancing Quality in the Changing Academic Environment are significant. However, for the 5th statement, it is not difficult for teachers and students in the rural areas to acquaint themselves with ICT tools; the significance value is above 0.05; hence, the role of ICT highlighted in the statement is not significant.

4 Conclusion

The use of the ICT tools has been making a significant contribution to bringing improvements in the overall quality of education at different levels. The individuals have this viewpoint mostly that progress can be easily made within learning and teaching, governance, instructional strategies, and the administrative functions when you implement the ICT tools. To make efficient use of the ICT tools VR and AR, individuals continuously s need to augment their abilities constantly. ICT tools act as the agents of change within learning and improve the management of education. Students from rural background should be given resources and opportunities to make use of technological advancements. Especially, technologies like Virtual Reality and Augmented Reality have conquered a sturdy position in the ICT arena. Appropriate use of these technologies will improvise the quality of education.

Appendix A

This research aims to ascertain the significance of virtual reality (VR) and augmented reality (AR) use with ICT tools for improving the quality of academic environments. All responses to this questionnaire will be held and treated in strict confidence and used only for research purposes. Thank you very much for completing the questionnaire, and your contribution is very much appreciated.

Gender: • Male • Female • Transgender

Age: • Upto 34 • 35–49 • 50 and above

Educational Qualification: • Undergraduate (UG) • Postgraduate (PG) • Doctorate (Ph.D.)

VR & AR experience: • I have experienced VR or AR • I am aware of VR/AR but haven't used • I am not aware of VR or AR.

1. **ICT tools help in improving the quality of education**

   • Strongly Disagree • Disagree • Neither Agree nor Disagree • Agree • Strongly Agree.

2. **ICT tools generate awareness amongst students in an easy and interesting manner**

   • Strongly Disagree • Disagree • Neither Agree nor Disagree • Agree • Strongly Agree.

3. **ICT tools help students in completing their assignments in a much better way**

   • Strongly Disagree • Disagree • Neither Agree nor Disagree • Agree • Strongly Agree.

4. **ICT has great potential in revamping the education system**

   • Strongly Disagree • Disagree • Neither Agree nor Disagree • Agree • Strongly Agree.

5. **It is not a difficult process for teachers and students in rural areas to acquaint themselves with ICT tools**

   • Strongly Disagree • Disagree • Neither Agree nor Disagree • Agree • Strongly Agree.

6. **ICT technologies such as Virtual Reality (VR) and Augmented Reality (AR) give a close look at reality to students through technology**

   • Strongly Disagree • Disagree • Neither Agree nor Disagree • Agree • Strongly Agree.

7. **Virtual Reality (VR) and Augmented Reality (AR) depend mainly on the development of IT infrastructure**

   • Strongly Disagree • Disagree • Neither Agree nor Disagree • Agree • Strongly Agree.

8. **Virtual Reality (VR) and Augmented Reality (AR) makes education more interesting for students**

   • Strongly Disagree • Disagree • Neither Agree nor Disagree • Agree • Strongly Agree.

9. **It is important to make the entire education system ICT savvy for a better tomorrow**

   • Strongly Disagree • Disagree • Neither Agree nor Disagree • Agree • Strongly Agree.

10. **The role of the ICT tools in education can contribute significantly to shaping the future of a country's economy**

   • Strongly Disagree • Disagree • Neither Agree nor Disagree • Agree • Strongly Agree.

References

1. Ben Guida, D. (2020). Augmented reality and virtual reality: A 360° immersion into western history of architecture. *International Journal of Emerging Trends in Engineering Research, 8,* 6051–6055.
2. Kerr, J., & Lawson, G. (2019). Augmented reality in design education: Landscape architecture studies as ar experience [J]. *International Journal of Art & Design Education, 5*(4), 3–6.
3. Liesa-Orús, M., Latorre-Cosculluela, C., Vázquez-Toledo, S., & Sierra-Sánchez, V. (2020). The technological challenge facing higher education professors: Perceptions of ICT tools for developing 21st century skills. *Sustainability, 12,* 5339.
4. Olimova, D. Z. (2020). The effectiveness of the implementation of ICT in the learning process. *European Scholar Journal (ESJ), 1*(4), 9–10.
5. Mohammed, A., & Santosh, S. (2019). The use of ICT tools in English language teaching and learning: A literature review. *5*(2), 29–32.
6. Kundu, A., Bej, T., & Dey, K. N. (2020). An empirical study on the correlation between teacher efficacy and ICT infrastructure. *International Journal of Information and Learning Technology, 37*(4), 213–238. https://doi.org/10.1108/IJILT-04-2020-0050
7. Yung, R., Lattimore, C. K., & Potter, L. E. (2021). Virtual Reality and tourism marketing: Conceptualising a framework on presence, emotion, and intention. *Current Issues in Tourism, 24*(11), 1505–1525.
8. Huang, K. T., Ball, C., Francis, J., Ratan, R., Boumis, J., & Fordham, J. (2019). Augmented versus virtual reality in education: An exploratory study examining science knowledge retention when using augmented reality/virtual reality mobile applications. *22*(2), 105–109.
9. Naresh, R. (2020). Education after COVID-19 crisis based on ICT tools. *31*(37), 464–468.
10. Flavián, C., Ibáñez-Sánchez, S., & Orús, C. (2021). Impacts of technological embodiment through Virtual Reality on potential guests' emotions and engagement. *Journal of Hospitality Marketing & Management, 30*(1), 1–20.
11. Graziano, T., & Privitera, D. (2020). Cultural heritage, tourist attractiveness and Augmented Reality: Insights from Italy. *Journal of Heritage Tourism, 15*(6), 666–679.
12. Surovaya, E., Prayag, G., Yung, R., & Khoo-Lattimore, C. (2020). Telepresent or not? Virtual reality, service perceptions, emotions and post-consumption behaviors. *Anatolia, 31*(4), 620–635.

An Empirical Study on the Effectiveness of Online Teaching and Learning Outcomes with Regard to LSRW Skills in COVID-19 Pandemic

P. Jayakumar⬛, S. Suman Rajest⬛, and B. R. Aravind⬛

1 Introduction

Language change in contemporary time is thought to be fueled largely by digital technology, which is regarded among the most important causes of linguistic change. As a result, the legacy of English Language education has undergone significant transformation in the previous decade, particularly to the amazing advent of the internet as an educational tool. Undoubtedly, sophisticated technology has proven to be a viable alternative to conventional ways of instructional teaching. However, it is quite fascinating to see whether or not contemporary instructors agree with the introduction of smart teaching methods into modern ones [1].

The widespread pandemic of Coronavirus disease 2019 (Covid-19) that began in the year 2020 and has spread across the globe seems to have had a significant impact on human activity. Almost every government in the countries affected by such a global epidemic has already called on its citizens to help combat the disease by using social distancing. As an outcome, the activities of formal education have increased. The educational setting has shifted from face-to-face instruction inside the classroom to online education. The use of digital technology to extend teaching outside of the classroom is known as online learning. This, of course, piques the interest of ELT scholars, who want to learn more about the effectiveness of digital English language learning outside of the classroom [2].

P. Jayakumar
St. Peter's Institute of Higher Education and Research, Chennai, Tamil Nadu, India

S. Suman Rajest (✉)
Department of English, Vels Institute of Science, Technology and Advanced Studies (VISTAS), Chennai, Tamil Nadu, India

B. R. Aravind
School of Social Sciences and Languages (SSL), Vellore Institute of Technology, Chennai, Tamil Nadu, India

483

Due to the extreme rapid advancement of technology in recent years, English language teaching (ELT) practitioners have only recently begun researching the approaches in which language learners practice and learn English outside of the classroom environment via readily available social media technology and internet materials [3]. Lee [4] defines this situation as informal digital learning of English (IDLE) that is identified as an independent entity, self-directed learning process through the use of a wide range of digital technologies such as mobile phones, social networking sites, and internet forums to understand and practice the English language.

In the digital era, teaching-learning is nothing without the effective use of technology. According to Oliver [5], online technologies influence the students' learning process and learning outcomes, and it also supports improving the learning opportunities. The use of the blended learning method creates a positive classroom climate in the process of teaching. It creates the interest among the students to learn the knowledge and skills mentioned in their outcomes. Tayebinik and Puteh [6] state, "Blended learning is more favourable to the students, and it also offers many advantages in learning platforms." In addition to traditional classrooms, online technologies are a supplement to achieve the students' learning outcomes. The participation of the students is more active when we use online technologies among the students. Eryilmaz [7] suggests, "Students have expressed that they learn more effectively in a blended learning environment." In the present century, students feel more comfortable learning English through online classes. In addition to the class lectures, the students show their interest in learning through digital sources.

Maher [8] recommended in his study that learning outcomes and their use have been undergone much criticism, but there is also some considerable gain to all stakeholders through learning outcomes. Finally, he concluded, "It is widely recognized that learning outcomes can enhance the educational process." Therefore, all stakeholders must have participated when creating a curriculum for the students. Learning objectives and learning outcomes are the two major areas in preparing a curriculum that must fulfil employment needs. First, this study has taken the satisfaction survey from the students to know whether they are aware of the learning objectives and outcomes. Secondly, the tertiary level students are attending classes online where the study evaluates the students are getting the learning outcomes. This study also tries to know the difference between traditional and online classes by asking few questions in the satisfaction survey presented in the following headings.

2 Literature Review

The continuous development of information and communication technology (ICT) has improved both education and transformed the way the field itself develops. Although powerful software and programs are now commonly used in education, and mobile devices such as tablet computers, personal digital assistants (PDAs), and laptops are becoming more common in the classroom, digital technologies may improve education. The use of digital technology in English teaching is also quite significant nowadays.

According to Salmon [9], a five-stage model for wholly online teaching and learning was first developed. This concept illustrates how to engage students online to provide them with learning experiences via online projects that he called e-activities and to move through several training phases at an even pace. In the first stage, a prerequisite course is established, and participants are inducted. In the second stage, students establish their internet accounts. In the third stage, participants' information is communicated. In the fourth stage, course-related conversations are initiated. Finally, real reflection and personal development are accomplished in the achievement of course goals.

Shih [10] conducted a study to ascertain Taiwanese EFL learners' online learning strategies. And to that end, it was discovered that learners who were more successfully employed a wider variety of techniques and utilized metacognitive and cognitive strategies more frequently than less successful learners.

Sharpe and Benfield [11] examined students' experiences with online education at Oxford Brokes University. They identified several similar features in students' online learning experiences and suggested implications such as the interpersonal reaction of the student experience and concerns about time and time management.

Puzziferro [12] investigated the association between students' self-regulated learning practices and their online learning results. The most often used techniques were effort regulation, time management and study environment, while peer learning and assistance seeking were the least often utilized. Additionally, it was shown that two online learning methodologies could predict students' grades: time and learning environment. Students who handled their time properly and studied in a conducive atmosphere were more likely to succeed in online courses.

Dixson [13] stated that online instructional activities or learner involvement are as successful as face-to-face education. It depends on cooperative learning and instructional presence from the teacher.

Miyazoe and Anderson [14] discovered that forum discussions and blogs on Moodle platforms are useful instructional tools to help students learn writing skills. This argument concluded that the study's significant encouraging findings regarding user forums and blogs to improve writing skills leaves little doubt about the value of these instructional resources in future research.

Learning a language takes a look at the effect of WhatsApp on learning English in Kuwait [15]. Alsaleem [16] discovered that by using WhatsApp as an authentic platform to improve writing, vocabulary, word choice, and communication in English, students could satisfy the language course's linguistic, grammatical, and functional objectives.

Gasmi [17] investigates the effects, possible advantages, and limits of WhatsApp to help Omani students strengthen their writing abilities. Finally, Allagui [18] investigates WhatsApp's usefulness in improving undergraduate students' fundamental EFL writing abilities at the university level, discovering that students' willingness to write is enhanced while using WhatsApp.

Jayakumar and Ajit [19] conducted a study on learning grammar through an Android application. The study recommended that the Android app made a remarkable improvement in teaching sentence structure among the students. They also suggested that technology helped the learners to learn the sentence structure better and faster. Bhuvaneswari et al. [20] made a study on online classroom pedagogy and digital learning. This study also recommended that students got improvement through online technologies.

Aparicio et al. [21] define online learning in two broad categories: learning and technology. Learning is defined as the cognitive process of acquiring information, while technology is defined as the instrument that facilitates acquiring information.

Furthermore, Aravind and Rajasekaran [22–24] conducted studies that found that modalities like TED talks and WordNet are very beneficial in vocabulary learning and improvement. Because vocabulary skills may be segmented, they lend themselves to developing educational approaches utilizing sophisticated technology.

Teachers and students are increasingly transitioning away from the conventional chalk-and-talk teaching process toward the contemporary, or technology-integrated, language classroom [25]. As a result of the integration of technological advancements into language classrooms, teaching and learning activities will become more participatory, as technology allows for two-way interactions; students will no longer be passive recipients of the educational process but will have the opportunity to provide input to their teaching staff as a result of this interconnection [26]. More impactful instructional and linguistic methods would result inside the English language classrooms due to this approach.

Similarly, Tsou et al. [27] asserted that digital narration might help students enhance their reading, writing, listening, speaking abilities, and overall academic success. In addition, using technology may make the learning process simpler and much more comfortable since learning may be completed more quickly and without much difficulty.

According to Jacobs [26], technological advances have the "ability to be continuously connected and share and exchange ideas and information across time and space using a wide variety of modalities". Isisag [28] students' educational techniques have to be relevant and appropriate in our real activities—which includes the implementation of inclusive education to give students new and distinctive learning opportunities –for them to be effective [29, 30].

3 Significance of the Study

The present study emphases the integration of technology-assisted language learning. In particular, the readability of the learning atmosphere has been replaced by online platforms due to the COVID-19 pandemic. Hence, engaging the learners were much larger and more comprehensive than the traditional classroom setting. As a result of exploring technological tools for language educators, the accessibility of innovative tools unquestionably improves the teaching-learning process. On the other hand, both instructors and learners will benefit from the timelessness and completeness of the objectivity of online learning. Furthermore, the implications of technological tools for pedagogical curricula can be discussed and designed.

4 Statement of the Problem

There is no doubt that a teacher cannot be replaced by anything. Due to the use of Information and Communication Technology in teaching and learning, it is inevitable to use Digital Technologies to make their teaching more effective. Despite having a problem-based curriculum, project-based curriculum, and student-centred curriculum, there are still failures in attaining the students' learning outcomes. To obtain learning outcomes in English language teaching, the institute is getting responses from all stakeholders. However, the employment needs of the students is still a question mark. Learner centric approach plays a vital role to achieve learning outcomes faster and easier. The learning environment is a major factor for the learning outputs. These factors influenced the researchers to do more about the topic and arrived at the present topic. Assuming the research problem, the study involved in a survey method to evaluate whether online classes meet the students' employment requirements and learning outcome.

5 Needs of the Study

As far as NAAC concern, acquiring learning outcomes are the principal part of any program. The program and its courses will be meaningless unless the students get the learning outcomes. The present study tells why learning outcomes are important in the following points.

- To strengthen the students' knowledge and skills in their course.
- To enable students to know the necessary areas in the course and program.
- To evaluate whether the students' learning outcomes fulfil the workplace needs.
- To focus the teaching as a learner-centred education.
- To enhance their higher-order thinking skills using Bloom's Taxonomy levels.
- To emphasize the objective of the course or program to the learners.

6 Research Objectives

The current study focuses on the efficacy of technology-enhanced learning outcomes in the COVID-19 pandemic. Hence, the objectives of the study are:

- To study the efficiency of online teaching, particularly for ESL learners

- To compare the impact of online learning through Technology Assisted Language Learning (TALL) in an ELL (English Language Learning) context
- To evaluate the learning outcomes of LSRW skills for tertiary level students through online teaching.

7 Research Questions

Based on the objectives of the study, this research is sought to answer the following research questions:

- Are technologies being used to support learning outcomes and teaching in tertiary level education?
- How is the use of Technology Assisted Language Learning impacting the higher education teaching-learning process?
- Is there a significant difference in the impact of using technology-based teaching to learn LSRW skills?

8 Research Methodology

The study collected the data through a structured questionnaire using a survey method in research. The questionnaire contained 15 questions. This is about the students' satisfaction survey on learning outcomes in English classrooms through online classes. The total number of students who participated in the study was 540 students who studied English as their Language subject in their program. The study is used to collect the data assisting a simple random sampling method. With the collected responses from the target audience, the data were analyzed qualitatively and quantitatively; for the present study, percentage analysis was employed to evaluate the students' satisfaction survey.

9 Analysis and Interpretation

Figure 1 shows that 8% of the students marked Strongly Disagree as their satisfaction, ten as Disagree, 14 as Neutral, 42 as Agree, and 26 as Strongly Agree.

Figure 2 shows that 4% of the students marked Strongly Disagree as their satisfaction, seven as Disagree, ten as Neutral, 51 as Agree, and 28 as Strongly Agree.

Figure 3 shows that 7% of the students marked Strongly Disagree as their satisfaction, 12 as Disagree, ten as Neutral, 22 as Agree, and 49 as Strongly Agree.

Figure 4 shows that 14% of the students marked Strongly Disagree as their satisfaction, eight as Disagree, 12 as Neutral, 18 as Agree, and 48 as Strongly Agree.

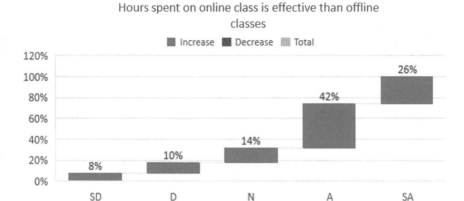

Fig. 1 Hours spent on online class is effective than offline classes

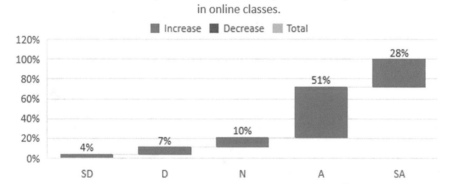

Fig. 2 Sometimes, I lose interest when I spend too many hours in online classes

Figure 5 shows that 5% of the students marked Strongly Disagree as their satisfaction, 11 as Disagree, 17 as Neutral, 44 as Agree, and 23 as Strongly Agree.

Figure 6 shows that 7% of the students marked Strongly Disagree as their satisfaction, 11 as Disagree, 14 as Neutral, 48 as Agree, and 20 as Strongly Agree.

Figure 7 shows that 5% of the students marked Strongly Disagree as their satisfaction, six as Disagree, 15 as Neutral, 20 as Agree, and 53 as Strongly Agree.

Figure 8 shows that 9% of the students marked Strongly Disagree as their satisfaction, 11 as Disagree, 16 as Neutral, 18 as Agree, and 46 as Strongly Agree.

Figure 9 shows that 8% of the students marked Strongly Disagree as their satisfaction, ten as Disagree, 42 as Neutral, 14 as Agree, and 26 as Strongly Agree.

Figure 10 shows that 4% of the students marked Strongly Disagree as their satisfaction, eight as Disagree, 18 as Neutral, 49 as Agree, and 21 as Strongly Agree.

Figure 11 shows that 7% of the students marked Strongly Disagree as their satisfaction, ten as Disagree, 13 as Neutral, 52 as Agree, and 18 as Strongly Agree.

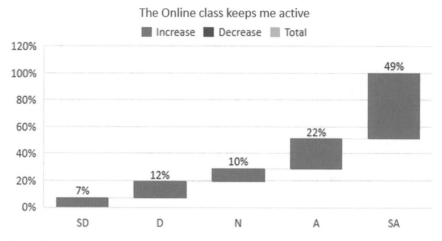

Fig. 3 The online class keeps me active

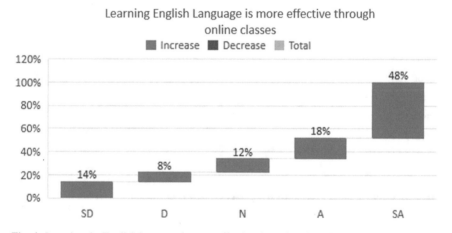

Fig. 4 Learning the English language is more effective through online classes

Figure 12 shows that 23% of the students marked Strongly Disagree as their satisfaction, 13 as Disagree, 47 as Neutral, ten as Agree, and seven as Strongly Agree.

Figure 13 shows that 8% of the students marked Strongly Disagree as their satisfaction, ten as Disagree, 14 as Neutral, 42 as Agree, and 26 as Strongly Agree.

Figure 14 shows that 9% of the students marked Strongly Disagree as their satisfaction, 12 as Disagree, 15 as Neutral, 43 as Agree, and 21 as Strongly Agree.

Figure 15 shows that 8% of the students marked Strongly Disagree as their satisfaction, 16 as Disagree, ten as Neutral, 14 as Agree, and 52 as Strongly Agree.

It is understood from Table 1 that students responded to 15 questions, and a percentage analysis was made to understand the feedback of 540 students. Based on

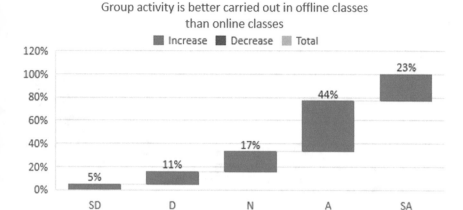

Fig. 5 Group activity is better carried out in offline classes than online classes

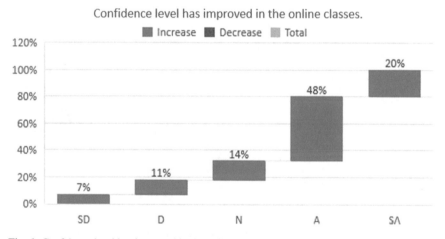

Fig. 6 Confidence level has improved in the online classes

the available evidence, the researchers presented the figures through the tables. The table shows that students responded to the higher amount of percentage Agree (42%) in statement 1. Followed by statement 2 is Agree (51%). Statement 3 is Strongly Agree (49%). Statement 4 is Strongly Agree (48%). Statement 5 is Agree (44%). Statement 6 is Agree (48%). Statement 7 is Strongly Agree (53%). Statement 8 is Strongly Agree (46%). Statement 9 is Neutral (42%). Statement 10 is Agree (49%). Statement 11 is Agree (52%). Statement 12 is Neutral (47%). Statement 13 is Agree (42%). Statement 14 is Agree (43%). Finally, statement 15 is Strongly Agree (52%).

For 5 statements, the table shows Strongly Agree, whereas it shows Agree to 8 statements and Neutral to 2 statements. The way the students think about that the online classes are effective. The analysis identified that the students accepted the

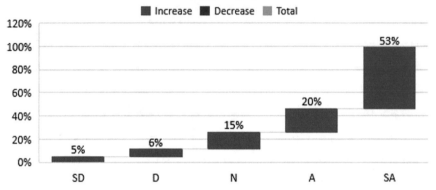

Fig. 7 Effective planning of the classes by the professor in the online teaching is highly remarkable

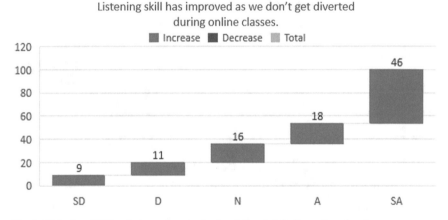

Fig. 8 Listening skill has improved as we don't get diverted during online classes

online teaching in the Pandemic Covid 19. It is also identified from the analysis that their learning outcomes are fully reached to the students. Undoubtedly, students' attendance is so high in online classes. Their satisfaction level is more, and their learning outcomes are attained towards 15 statements.

This study presented the response of the students through the graph for better understanding. The following graphs state the language skills viz. Listening, Speaking, Reading, and writing is essential for the second language learner and grammatical skills that help write and speak English accurately and fluently.

There are many studies conducted on LSRW skills using different methods and approaches still; there is a gap in attaining the learning outcomes concerning LSRW skills which are considered less. This study is made to evaluate the effectiveness of

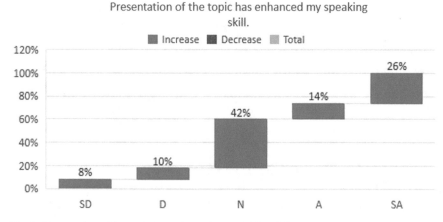

Fig. 9 Presentation of the topic has enhanced my speaking skill

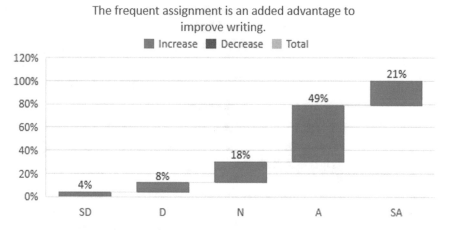

Fig. 10 The frequent assignment is an added advantage to improve writing

online classes and learning outcomes. Based on the available evidence, it is understood from the study that online classes are more effective than offline classes. Using Technology Assisted Language Learning (TALL) makes students comfortable attaining their learning outcomes easier and better. By seeing the responses, it is believed that students show their interest in learning English through online classes due to the digital technologies. Students are familiar with using digital technologies like ICT tools, Web 2.0 tools, and other mobile apps through the study. The students are more satisfied with the online classes. The responses are more positive from the students' side. They feel better in attending online classes.

The study found that students felt the online classes more effective than the offline classes through the percentage analysis. Moreover, the online classes kept them active. As a result, the listening skill was improved. The confidence level

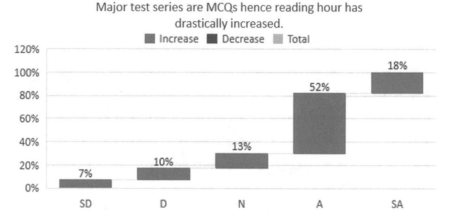

Fig. 11 Major test series are MCQs; hence reading hour has drastically increased

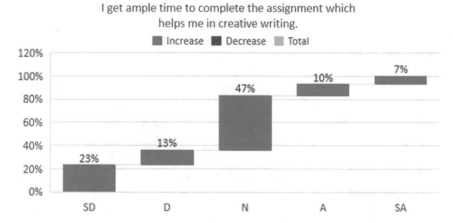

Fig. 12 I get ample time to complete the assignment, which helps me in creative writing

was also increased for the students while attending online classes. Students felt that there was no diversion because of online classes. Through presentations of the topic, students enhanced their speaking skills in the online classes. The frequent assignments were another area where students felt an added advantage to improve their writing skills. Reading hour has drastically increased using PPTs, class notes, and test series. The students stated that assignments made them write and speak sentences without grammatical errors. It also helped them in their creative writing.

Regarding attendance, they did not face any shortfall. On the contrary, audiovisual teaching attracted them to attend the class regularly without fail. Therefore they can attain the learning outcomes in general English classrooms.

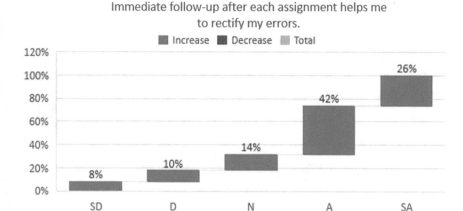

Fig. 13 Immediate follow-up after each assignment helps me to rectify my errors

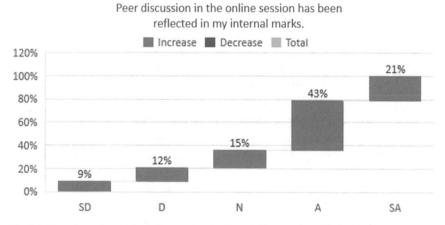

Fig. 14 Peer discussion in the online session has been reflected in my internal marks

The paper discusses some of the major findings. This study mainly focused on online classes and learning outcomes of the tertiary level students about LSRW skills. From statements 8, 9, 10, and 11 in Table 2, online classes created a platform for the students to enhance their LSRW skills. Students' listening skills have improved a lot by attending regular English classes and lectures with visuals given a better understanding of the subject taught. At the same time, students are also allowed to develop their speaking skill using seminar topics and other activities related to the subject. Students are motivated to read poems, novels, and dramas. By reading activities in the classrooms, students' reading skill has drastically increased. Assignments and tests are the major roles in developing their writing skill. Besides LSRW skills, this study also found some learning outcomes gained by the students in attending online classes. The students marked a higher percentage to creative writing and error-free

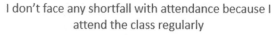

I don't face any shortfall with attendance because I
attend the class regularly

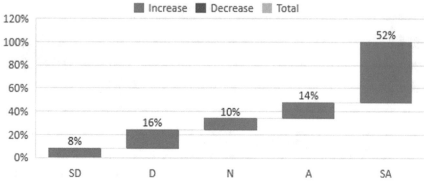

Fig. 15 I don't face any shortfall with attendance because I attend the class regularly

Table 1 Students' satisfaction survey on online classes and learning outcomes

S. no.	Statements
1	Hours spent on online class is effective than offline classes
2	Sometimes, I lose interest when I spend too many hours in online classes
3	The online class keeps me active
4	Learning the English language is more effective through online classes
5	Group activity is better carried out in offline classes than online classes
6	Confidence level has improved in the online classes
7	Effective planning of the classes by the professor in the online teaching is highly remarkable
8	Listening skill has improved as we don't get diverted during online classes
9	Presentation of the topic has enhanced my speaking skill
10	The frequent assignment is an added advantage to improve writing
11	Major test series are MCQs; hence reading hour has drastically increased
12	I get ample time to complete the assignment, which helps me in creative writing
13	Immediate follow-up after each assignment helps me to rectify my errors
14	Peer discussion in the online session has been reflected in my internal marks
15	I don't face any shortfall with attendance because I attend the class regularly

sentences. They said that online classes were effective, and classes kept them active throughout the class timings. Discussion and group activities made students work collaboratively. Hence, the study has proven from the percentage analysis that the students effectively attained the learning outcomes through online classes.

Table 2 Skill Set performances and disadvantages

Skill set	Reasons for better performance	Disadvantages
Listening	Students could perform well because of a. Less distraction b. Closer attention because of the availability of headphones c. Teacher interaction is not disturbed as everyone else is muted	Attention deficiency disorder may affect listening, and the non-availability of non-verbal cues may lead to miscommunication of the message
Reading	Students could a. Get peer reviews due to easy online accessibility b. Read ebooks that can be downloaded online	Online reading may cause tiredness, and that may lead to disinterest in reading
Writing	Doing more assignments are possible as the platform is user friendly	Students may use spell check and other online tools that may dissipate their honesty
Speaking	The component stage freight is completely ignored	Students may tend to hide videos and speak, which would result in improper speaking or just end up reading the material
Grammar	Students may feel less conscious because of less peer pressure through discouraging comments, body language, intimidation and eye-contact	Students may fail to learn grammar in an inductive approach

10 Conclusion

In English Language Teaching (ELT), teaching was done using many methods and approaches. Due to the emerging teaching technologies, the faculty and students are trained to use them effectively in their classrooms. However, they find it difficult to attain the learning outcomes employing online technologies. They are exposed to use online technologies, but they fail to evaluate the course's learning outcomes. By answering the questionnaire of the present study, the students will be able to evaluate their learning outcomes concerning LSRW skills of their own. The online classes create interest in learning among the students and improve the learners' position in evaluating the learning outcomes. The study recommends that online classes help to complete the subject and conduct the test within the subject credits. It also gives satisfaction to the students in attaining the learning outcomes.

Conflict of Interest No conflict of interest exists.

References

1. Abbasova, M., & Mammadov, N. (2019). The role of digital technology in English language teaching in Azerbaijan. *International Journal of English Linguistics, Canadian Center of Science and Education, 9*(2), 364–372. https://doi.org/10.5539/ijel.v9n2p364
2. Nugroho, A., & Atmojo, A. E. P. (2020). Digital learning of English beyond classroom: EFL learners' perception and teaching activities. *JEELS (Journal of English Education and Linguistics Studies), 7*(2), 219–243.
3. Sundqvist, P., & Sylvén, L. K. (2016). Extramural English in teaching and learning: From theory and research to practice. *Springer.* https://doi.org/10.1057/978-1-137-46048-6
4. Lee, J. S. (2019). Informal digital learning of English and second language vocabulary outcomes: Can quantity conquer quality? *British Journal of Educational Technology, 50*(2), 767–778. https://doi.org/10.1111/bjet.12599
5. Oliver, R. (1999). Exploring strategies for online teaching and learning. *Distance Education, 20*(2), 240–254.
6. Tayebinik, M., & Puteh, M. (2013). *Blended learning or E-learning?* arXiv:1306.4085.
7. Eryilmaz, M. (2015). The effectiveness of blended learning environments. *Contemporary Issues in Education Research (CIER), 8*(4), 251–256.
8. Maher, A. (2004). Learning outcomes in higher education: Implications for curriculum design and student learning. *Journal of Hospitality, Leisure, Sport and Tourism Education, 3*(2), 46–54.
9. Salmon, G. (2004). *E-moderating: The key to teaching & learning online* (2nd edn). New York: Routledge Flamer. UKOU. http://www.open.ac.uk/.
10. Shih, Y. C. D. (2005). Taiwanese EFL learners' online language learning strategies. In *Fifth EEE International Conference on Advanced Learning Technologies, 2005. ICALT 2005* (pp. 1042–1046). IEEE.
11. Sharpe, R., & Benfield, G. (2005). The student experience of E-learning in higher education: A review of the literature. *Brookes eJournal of Learning and Teaching, 1*(3), 1–9.
12. Puzziferro, M. (2008). Online technologies self-efficacy and self-regulated learning as predictors of final grade and satisfaction in college-level online courses. *The American Journal of Distance Education, 22*(2), 72–89.
13. Dixson, M. D. (2012). Creating effective student engagement in online courses: What do students find engaging? *Journal of the Scholarship of Teaching and Learning, 10*(2), 1–13.
14. Miyazoea, T., & Anderson, T. (2012). Discuss, reflect, and collaborate: A qualitative analysis of forum, blog and wiki use in an EFL blended learning course. *Procedia-Social and Behavioural Sciences, 34*, 146–152. https://doi.org/10.1016/j.sbspro.2012.02.030
15. Salem, A. A. M. (2013). The impact of technology (BBM and WhatsApp applications) on English linguistics in Kuwait. *International Journal of Applied Linguistics and English Literature, 2*(4), 65–69.
16. Alsaleem, B. I. A. (2013). The effect of "WhatsApp" electronic dialogue journaling on improving writing vocabulary word choice and voice of EFL undergraduate Saudi students. *Arab World English Journal, 4*(3), 224–237.
17. Gasmi, A. (2014). Mobile assisted language learning: Potential and limitations of using 'Whatsapp' messenger to enhance students writing skills. In *INTED2014 Proceedings* (pp. 7243–7243). IATED.
18. Allagui, B. (2014). Writing through WhatsApp: An evaluation of students writing Performance. *International Journal of Mobile Learning and Organisation, 8*(3–4), 216–231.
19. Jayakumar, P., & Ajit, I. (2016). Android app: An instrument in clearing Lacuna of English grammar through teaching 500 sentence structures with reference to the verb eat. *Man in India, 96*(4), 1187–1195.
20. Bhuvaneswari, G., Swami, M., & Jayakumar, P. (2020). Online classroom pedagogy: Perspectives of undergraduate students towards digital learning. *International Journal of Advanced Science and Technology, 29*(04), 6680–6687.
21. Aparicio, M., Bacao, F., & Olivera, T. (2016). An e-learning Theoretical Framework. *Journal of Educational Technology Systems, 19*(1), 292–307.

22. Aravind, B. R., & Rajasekaran, V. (2018). Advanced technological modality to explore ESL learners' vocabulary knowledge through social strategies. *Journal of Advanced Research in Dynamical and Control Systems, 10*(10), 250–256.

23. Aravind, B. R., & Rajasekaran, V. (2019). Improving noun hyponyms using select TED wordlist through WordNet: An investigational study. *Universal Journal of Educational Research, 7*(11), 2495–2500. https://doi.org/10.13189/ujer.2019.071129.

24. Aravind, B. R., & Rajasekaran, V. (2019). Technological modality to influence persuasive and argumentative vocabulary for effective communication with reference to selected TED talk videos. *International Journal of Recent Technology and Engineering, 7*(5), 165–170.

25. Lacina, J. (2004). Promoting language acquisitions: Technology and English language learners. *Childhood Education, 81*(2), 113–115.

26. Jacobs, G. E. (2010). *Writing instruction for generation 2.0* .Plymouth, England: Rowman and Littlefield Education.

27. Tsou, W., Wang, W., & Tzeng, Y. (2006). Applying a multimedia storytelling website in foreign language learning. *Computers & Education, 47*, 17–28.

28. Isisag, K. U. (2012). The positive effects of integrating ICT in foreign language teaching. In *International Conference Proceedings. ICT for Language Learning.*

29. Suman Rajest, S., Suresh, P. (2018). Impact of 21st century's different heads of learning skills for students and teachers. *International Journal of Multidisciplinary Research and Development, V*(IV), 170–178.

30. Rajest, S. S., & Suresh, P. (2019). An analysis of psychological aspects in student-centered learning activities and different methods. *Journal of International Pharmaceutical Research, 46*(01), 165–172.

ICT Virtual Multimedia Learning Tools/Affordances: The Case of Narrow Listening to YouTube Multimedia-Based Comprehensible Input for the Development of ESL Learners' Oral Fluency

Azzam Alobaid

1 Research Overview

Information and Communication Technology (ICT) is defined as new multimedia technologies which are used for creating, displaying, storing, manipulating, and exchanging information [1]. ICTs relate to the field of EFL/ESL learning and teaching through the use of computers, hardware and software, smartphones, gadgets, networks, internet, website, e-mail, television, radio, and other computer and network-based technologies. Various types of ICTs can be integrated to benefit language learners from their capacity for the creation, enhancement and optimization of better multimedia language learning environments for EFL/ESL learning and teaching [2]. "Multimedia represents a potentially powerful learning technology [3]," and research is constantly needed to identify how and which aspects of Multimedia Instruction (MI) create favorable learning conditions for people. Multimedia learning environment combines more than one media type like texts, sounds, videos and animations usually technology-based for enhancing understanding or memorization [4]. Multimedia learning-based research unveiled positive impact on L2 learners' performance owing to the increase in the use and exposure to ICT multimedia tools which provide multimedia elements for language learners in the learning environments [5]. YouTube is a web-based ICT multimedia learning tool widely known and integrated in L2 teaching/learning for its ever-increasing affordances [2].

Multimedia learning tools/environments are becoming more and more indispensable in the twenty-first century classrooms and have attracted researchers

A. Alobaid (✉)
Centre for Linguistics, School of Languages, Literature and Culture Studies, J.N.U., New Delhi 110067, India
e-mail: azzam14_llh@jnu.ac.in

© The Author(s), under exclusive license to Springer Nature Switzerland AG 2022
A. Hamdan et al. (eds.), *Technologies, Artificial Intelligence and the Future of Learning Post-COVID-19*, Studies in Computational Intelligence 1019,
https://doi.org/10.1007/978-3-030-93921-2_28

more than before due to their potentially positive effects and cognitive advantages [6–8]. Several studies reported advantages like a greater depth of language processing, focusing learners' attention, reinforcing the acquisition of vocabulary through multiple modalities, prompting long-term content retention [2, 5, 9, 10].

Given the multiple beneficial multimedia learning effects/cognitive advantages for ESL learners (i.e., increasing learners' comprehension of the spoken language input and attention to the (new) target language input (i.e., chunks), processing more information in their working memory, and retrieving these information/intake (i.e., chunks) when needed from their long-term memory [2]), this work proposes the use of YouTube as a multimedia learning tool for its potential multimedia learning effects/cognitive advantages and examines its possible role in creating and enhancing a multimedia learning environment and impact on the development of ESL learners' English speech fluency. This multimedia learning environment can provide and enhance what is known as the comprehensible input of the target language (by way of captioned videos) required for learners' language proficiency development[14]; ESL learners' engagement with and exposure to (by way of narrow listening technique [16]) the provided and enhanced multimedia- based comprehensible input may increase their gain of the target language chunks, which may in their turn, improve learners' L2 oral fluency.

Recent studies have shown a huge interest in ICT multimedia technology tools/affordances such as YouTube for language education purposes. Several studies explored the development of ESL/EFL speaking skill and the effectiveness of the overall use of YouTube as an ICT multimedia tool for enhancing the learners' speaking experience. Such studies reported that the use of the ICT-based multimedia technology such as YouTube can enhance learners' receptive (i.e., aural) and productive (i.e., oral and writing) language skills [11–13].

One of the main affordances is that ICTs such as YouTube can provide plenty of engaging and authentic multimedia materials, which can be a rich source of comprehensible input of the target language required for the learners' language proficiency development [14], for L2 learners to listen to and help them build their speaking proficiency [5]. Narrowly listening to such L2 input, when presented through multiple modalities simultaneously (aural and written) by way of captioned videos, can be responsible for the development of learners' oral fluency. In this respect, this work proposes the use of YouTube closed captions to enhance such L2 input for ESL learners, i.e., to make it much more comprehensible and chunkable. Within multimedia learning environments research, the role and advantages of captions for the development of ESL/EFL skills have received a special interest. Previous studies concluded that captions can be useful and facilitative to L2 learners due to their positive role or functions, namely improving general comprehension and figuring out meaning via the unpacking of language chunks, inducing higher and deeper levels of processing through focusing attention, improving language encoding and enhancing acquisition of vocabulary through various formats and ultimately improving retrieval and long-term retention of newly acquired language input [9, 10, 15]. Narrow Listening (NL) (seen as a powerful learning technique for the development of L2 learners' speaking proficiency and hence employed in this study) is "in accord with key second language acquisition and teaching principles highlighted in the literature, namely Comprehensible Input Hypothesis [14]. NL provides a great deal of input which is made comprehensible through repeated listening, self-selection of samples, and familiarity with topics" [16, 17]. This research proposes yet another way (i.e., captions) which has two substantially potential multimedia learning effects and cognitive advantages; both of which may lead to a more fluent L2 speech.

First, captions can make the target language input even more comprehensible while adopting the narrow listening technique to learn a given language. It was found that listening to a video two times, first time with and second with no captioning, can reduce listener's anxiety, activate selective and global listening strategies and promote automaticity in processing of L2 input [9]. Intralingual subtitles have potential value in helping the learner to not only better comprehend authentic linguistic input but also to produce comprehensible communicative output [18]. Having the opportunity to see and control subtitles, as opposed to not having that opportunity, results in both better comprehension and subsequent better productive use of the foreign language. Their study showed that allowing fifth- semester college students of French the possibility of seeing and controlling subtitles may increase their performance on video-based oral communicative practice tasks with multimedia courseware.

Second, captions may aid in chunk learning (seen as major factor of speech fluency) as some set phrases and expressions ease students' cognitive stress [19], free their attention capacity [20] and speed up their speaking processing [21]. In a previous study [22], it was concluded that chunk learning proved to be beneficial for Chinese EFL learners' speaking fluency as they needed to store large vocabulary in their memory and recall them automatically in speaking. By means of such automaticity, learners reduced the time pressure and cognitive load and thus increased their speaking fluency [23–25].

What follows are the main research questions to be asked and addressed in this study.

RQ.1 Does the oral fluency of ESL learners improve after the implementation of YouTube for the improvement of ESL oral fluency, given that learners engage with and are frequently exposed to (by way of the narrow listening technique) the provided and enhanced multimedia comprehensible input on YouTube?

RQ.2 What is the correlation between ESL learners' engagement with and frequent exposure to the proposed ICT-based multimedia learning tool (i.e., YouTube) for the development of L2 oral fluency and their oral fluency improvement, if any?

RQ.3 What is the role and impact of the frequent use of YouTube on the development of ESL learners' oral fluency?

The null (H_0) and alternative (H_a) hypotheses set in this work are:

H_0 Engagement with and exposure to (by way of narrow listening technique) the provided and enhanced multimedia comprehensible input (by way of captioned videos) on YouTube cannot increase ESL learners' gain of L2 chunks and thus their oral fluency.

H_a Engagement with and exposure to (by way of narrow listening technique) the provided and enhanced multimedia comprehensible input (by way of captioned videos) on YouTube can increase ESL learners' gain of L2 chunks and thus their oral fluency.

To the best of the author's knowledge, no research has yet covered these essential areas/gaps related to the use of YouTube affordances such as video captions with regard to their role and impact on ESL learners' oral fluency. Therefore, this work takes a further step and proposes using YouTube as an ICT multimedia learning tool to provide and enhance (by way of captioned videos) a multimedia-based comprehensible input for L2 learners; it's through learners' 'narrow listening[1]' (as a technique) to this multimedia-based comprehensible input, they will likely gain more L2 chunks and thus improve their ESL oral fluency over time.

In this work, speech fluency concerns the learner's capacity to produce language in real time without undue pausing or hesitation [20]. It was also defined as the learner's capacity to get across communicative intent without too much hesitation and too many pauses to cause barriers or a breakdown in communication [26]. Fluency can be measured using different metrics of quantitative and qualitative nature like speech rate, length of run, pause length, silence, false starts, repetitions and reformulations (see the Materials and Methods Sect. 3 for more details on the quantitative and qualitative metrics in this work). With regard to the concept of language chunks and the problem of oral fluency, researchers pointed to different aspects of fluency such as pausing and hesitation, automaticity and conversational speed and ready-made chunks [27–30]. Formulaic chunks which can increase speech rate and conversational flow as a significant aspect of fluent conversation. He argued that "chunks by their nature, are retrieved whole; they are not created anew each time; they are part of that automaticity which enables effortless accuracy [and fluency]" [28]. This linkage between oral fluency and the language learners' need to increase their L2 chunks is clearly of paramount significance, and thus underscored in this study.

2   Motivation of the Study

This section describes the interrelated selected theoretical concepts/techniques which incentivized the current study.

Listening and speaking constitute the two elements of oral language, and the existence of listening skills obliges speaking, but in terms of language acquisition listening is prior to speaking (i.e., listening forms the basis for speaking) [31]. Therefore, for successful communication, it is inevitable to combine listening and speaking in L2 education [31]. Given this close relationship between listening and speaking, this work suggests and employs the narrow listening technique as a source of L2 aural input with the aim of promoting ESL learners speaking or oral fluency. The

---

[1] In this work, narrow listening is used synonymously with (daily) frequent and highly engaged exposure to the target language input (in this case the multimedia-based comprehensible input).

value of narrow listening is stressed as a source of input beyond listening abilities per se for the development of pronunciation and fluency (outputs) [30]. Narrow listening can be seen as another subset of extensive listening because both require learners to receive a massive amount of aural [comprehensible] input [32].

Comprehensible input is "critical for second/foreign language acquisition [14]. Evidence shows that aural comprehensible input plays a key role in the early stages of this process" [33]. This suggests the essential role of listening (i.e., getting input) for speaking (i.e., producing output) a foreign/second language. In this regard, while listening to uncontrolled casual conversations may be too difficult for beginning and intermediate foreign language students to comprehend (not to mention to reproduce), narrow listening offers them a valuable and rewarding alternative, i.e., to make aural L2 input more comprehensible for learners [17]. Previous studies [17] showed that L2 learners found narrow listening as a technique to be "interesting, very helpful in improving listening comprehension, fluency, and vocabulary, and in increasing their confidence". This present study extends previous research by employing narrow listening as a learning technique in a multimedia learning environment/framework to take advantage of this successful technique and the multiple positive advantages of multimedia learning environment (see the Research Overview Sect. 1 for these advantages) provided by ICT tools like YouTube (i.e., captioned videos) for the development of the speech fluency of ESL learners in particular.

The motivation behind choosing to investigate the speech fluency skill is because it is one of the major characteristics of communicative competence [34] and is an important indicator for progressing in language learning [35] and one of the conditions which ensure the success in communication [36].

ICT-based multimedia learning tools like YouTube, which is one of many web-based ICT multimedia learning tools widely known and used by ESL/EFL learners, was employed in our study because numerous studies have supported the idea that overall student motivation and engagement in learning is enhanced by the implementation of instructional technology [37, 38]. Technology engages students behaviourally (more effort and time spent participating in learning activities); emotionally (positively impacting attitudes and interests towards learning); and cognitively (mental investment to comprehend content) [39]. Also, ICTs such as YouTube can provide plenty of engaging and authentic multimedia language learning materials; they can be a rich source of comprehensible input for L2 learners to listen to [5].

This research is novel in its application of an ICT widely popular platform, namely YouTube while adopting a collective methodology of the most debatbly famous and powerful Second Language Acquisition (SLA) theories within the framework of the cognitive theory of multimedia learning for the development of ESL learners' oral fluency.

3 Theoretical Framework

This study falls within the Cognitive Theory of Multimedia Learning, Comprehensible Input, Narrow Listening and Chunk Learning.

3.1 Cognitive Theory of Multimedia Learning

New knowledge can be learnt more effectively by way of electronic instructional materials displayed in verbal and visual modes [3].

Multimedia Principle. When learners are presented with words and pictures rather than words alone, learning is expected to be more effective. Multimedia includes multiple modalities of verbal (texts, narrations, spoken words/sounds) and visual (i.e., videos, photographs, pictures, animations and illustrations) representations presented together [40, 41]. Learners are likely to "process more information in their working memory and retrieve them when required from the long-term memory, for their learning can be supported when they depend on different sources to comprehend, which will also reduce the burden of relying on a single source" [42, 43].

3.2 Multimedia-Based Comprehensible Input: Rather Than Just Any Input

Comprehensible input is a language input which can be comprehended by listeners although not entirely understanding all of its words and structures. It is the crucial component for language acquisition, which can ideally be provided in an anxiety-free environment, containing messages that learners really want to hear [16]. Research into multimedia instruction "provided learners with rich exposure to the L2 input in meaning-based tasks built upon different media features" [5, 44]. In this work, the multimedia-based comprehensible input can be defined as the target language input which can be made comprehensible to L2 learners in a multimedia learning environment. In this study, L2 input provided on YouTube is made more comprehensible by way of captioned videos.

3.3 Narrow Listening and Chunk Learning

Narrow Listening technique [16] suggests that "students listen, as many times as they like and at their leisure, to several brief tape-recordings or conversations of proficient speakers discussing a topic in which the students themselves have an interest. Students change topics gradually, which allows them to expand their language competence at their own pace". Learners would benefit more if they study a 2 min recording in detail than they would to listen to large amounts of language superficially, e.g., 60 min program just one time. Through this technique a great deal of input is made comprehensible through repeated listening, self-selection of samples, and familiarity with topics [17]. The concept of narrow listening vis a vis the improvement/acquisition of L2 input applies to our use of video and audio (i.e., multimedia learning) in ESL/EFL contexts. More specifically, today's ICT technologies allow us to do more narrow listening and detailed analysis because the multimedia of video and audio are so much more accessible than they used to be. With respect to the possible acquisition of the target language chunks and narrow listening, it is during this detailed analysis (while narrowly listening) that students become consciously aware of how native speakers use their language, including chunks such as collocations, fixed expressions, formulaic utterances, sentence starters, verb patterns, idioms, catch/short phrases, etc. [45].

As suggested by [46], "learning by chunking is a cognitive process that recodes information into meaningful groups, called chunks, to increase learning efficiency/capacity. Chunks of information are generally composed of familiar or meaningful sets of information that are recalled together". From a psycholinguistic perspective, learners are able to decrease the amount of information that must be held in working memory by increasing the amount of information per chunk. They reported that chunking extends the ability to recognize or recall information (i.e., L2 chunks) or perform tasks [i.e., speaking] at a later time.

This work argues that ESL learners' engagement with and exposure to (by way of the narrow listening technique, [16]) the provided and enhanced multimedia-based comprehensible input can potentially increase their gain of the target language chunks and ultimately improve their L2 oral fluency over time. In this respect, this work aims at the task of chunking more of L2 input which can be made easier (in terms of comprehension, attention, analysis and recall) and more meaningful if L2 input is made more comprehensible. For this aim, the provided and enhanced multimedia-based comprehensible input (by way of captioned videos) can come into play and be handy for learners, i.e., L2 input can be made more comprehensible and chunkable. In effect, their L2 output will likely be more fluent. "Native speakers tend to use a great deal of prefabricated chunks of language which are not composed each time by the rules of syntax" [47, 48]. Reference [49] highlighted the fact that "many chunks are as frequent as or more frequent than the single word items which appear in core vocabulary". If clusters/chunks are so important to language use and are so widespread in discourse, they should be paid special attention. The reason why chunks are so widespread is because they can be processed more quickly and the mind can store these ready-made chunks in the long-term memory to be used in language production [48]. Research on chunks is based on the assumption that native speakers use plenty of chunks in their everyday language and they are considered as fluent speakers of language [48, 50]. Reference [9] reported that captions can be helpful "in the process of language analysis or decomposition. Because word boundaries cannot always be determined in fast running speech, but in writing words can be made more discrete, allowing the learners to unpack the speech in a more meaningful way". They quoted [51] who stressed that "learning to understand a language involves parsing the speech stream into chunks which reliably mark meaning." In a similar manner, this study contends that captions can make L2 aural input (especially fast running speech) more comprehensible and chunkable. With this aim, the study stresses the use and significance of captions to aid ESL learners whilst they narrowly listen to a given YouTube video.

To connect the dots among the above-mentioned theories of L2 learning, this work proposes the use of ICT-based multimedia learning tools like YouTube captioned-videos to make L2 input not only more comprehensible but also more chunkable for learners as they narrowly listen to it, which may very likely result in a more fluent L2 speech due to the increase of learners L2 intake which can accrue over time, given that these learners are engaging with and frequently exposed to such L2 multi-mediated input.

4 Materials and Methods

To examine whether and how far the widely used ICT multimedia learning tools like YouTube can actually help ESL learners improve their oral fluency over time, 14 bilingual Arabic ESL learners aged 15 were invited to take part in this experiment. In this respect, this small-scale correlational study employed a number of metrics (n.17) of both quantitative and qualitative nature to explore the performance of the same group (i.e., treatment group) of learners using a One-Group Pretest-Posttest Design $(O_1 \ X_1 \ O_2)$ [52]. This study was conducted at the Iraqi school, New Delhi, India. These participants were excited and keen to improve their English speech proficiency. Before the study began, the researcher clearly explained the objectives and significance of this research to these participants. The fully informed participants agreed to take part in this study which had four objectives.

The first objective of this study was to determine if the speech fluency of ESL learners improves over time due to the implementation of the proposed ICT-based multimedia learning tool, namely YouTube for the improvement of their speech fluency, given that they are engaging with and frequently exposed to ICT multimedia during their L2 learning process. In this regard, different outcomes were measured by the repeated measures test $(O_1 \ X_1 \ O_2)$, which was conducted before and after the implementation of YouTube as an ICT multimedia learning tool for the improvement of ESL learners' speech fluency (Table 1). This design can determine the potential of intervention (i.e., (X_1) by calculating the difference between the initial evaluation of the dependent variables before the intervention (i.e., O_1) and the second evaluation of the dependent variables after the intervention (i.e., O_2). The dependent variables in this study were the metrics (n.17) used to gauge the likely learners' speech fluency progress (see below in this section for information about these metrics), if any, over time. In this regard, the learners' responses (n. 84 in total)

Table 1 Summary table of Wilcoxon matched-pairs ranked test results: baseline and end line of learners' speech performance (before as compared to) after 5 months of exposure to YouTube: statistically significant differences based on (p. value < 0.5)

Speech fluency metrics	Baseline test: mean (sd)	End line test: mean (sd)	(End line-score)-(Baseline-score)			Wilcoxon signed rank			Effect size (Cohen's d value)
			Neg. rank	Posit. rank	Ties	Total	Z	P-Value	
Speech rate	41.23 (16.67)	64.86 (24.88)	0	14	0	14	−3.3	0.001	−0.62
Articulation rate	2668 (1045)	4084 (1546)	0	14	0	14	−3.3	0.001	−0.62
Number of silent pauses (SPs)	4.64 (3.23)	1 (1.11)	13	0	1	14	−3.19	0.001	−0.60
Mean length of SPs	4.25 (1.40)	2.38 (2.16)	11	0	3	14	−2.94	0.003	−0.56
Number of filled pauses	18.29 (10.69)	6 (4.47)	14	0	0	14	−3.3	0.001	−0.62
Number of false starts	3.43 (2.50)	1.50 (1.61)	10	2	2	14	−2.45	0.01	−0.46
Number of replacements	1.64 (1.69)	1 (0.96)	6	3	5	14	−1.38	0.168	−0.26
Number of reformulations	2.79 (2.36)	1.93 (1.82)	9	3	2	14	−1.19	0.233	−0.22
Number of Repetitions	7.43 (4.78)	3.86 (2.71)	10	2	2	14	−2.48	0.01	−0.47
Efficiency/effortlessness	4.07 (1.21)	6.07 (1.64)	0	14	0	14	−3.34	0.001	−0.63
Lexical retrieval/resource	4.21 (0.97)	5.64 (1.22)	0	14	0	14	−3.4	0.001	−0.64
Incidence of errors per AS-unit	0.72 (0.37)	0.63 (0.26)	10	3	1	14	−1.57	0.116	−0.30
Lexical diversity (VOCD)	43.50 (6.08)	45.54 (12.44)	8	6	0	14	−0.22	0.826	−0.04
Lexical diversity (MTLD)	29.11 (5.22)	31.56 (9.35)	6	8	0	14	−0.722	0.47	−0.14
Lexical density	45.72 (2.91)	46.74 (3.45)	7	7	0	14	−0.847	0.397	−0.16

(continued)

Table 1 (continued)

Speech fluency metrics	Baseline test: mean (sd)	End line test: mean (sd)	(End line-score)-(Baseline-score)			Wilcoxon signed rank			Effect size (Cohen's d value)
			Neg. rank	Posit. rank	Ties	Total	Z	P-Value	
Mean length of T. unit	9.70 (1.12)	8.92 (0.82)	11	3	0	14	−2.166	0.03	−0.41
Number of clauses per T. unit	1.26 (0.16)	1.18 (0.10)	10	4	0	14	−2.27	0.023	−0.43

to the given oral test samples[2] (*n.* 6 in total) were scored to provide a measure of the participants' performance on the speaking practice tasks, which helped answering the first research question. The speaking test tasks comprised International English Language Testing System (IELTS)-based speaking tasks, taking into consideration all the criteria and standards while administering this test in class and analysing its outcomes (i.e., participants' scores of this test) by four linguists.

The second objective was to determine the correlation between ESL learners' engagement with and frequent exposure to the proposed ICT-based multimedia learning tool and their speech fluency improvement, if any, after adopting this tool for the improvement of ESL speech fluency. To address this objective, different outcomes were set by the questionnaire questions (Qs. 4–29, Appendix A).

The third objective of this study was to compare[3] ICT multimedia learning tools with the non-ICT multimedia learning tools/environments for each of the afore-mentioned variables (i.e., engagement rate, frequent exposure to language learning tools/environments, i.e., these were the two independent variables employed in this study, see below in this section for more information). Different outcomes were set by the questionnaire questions (Qs. 4–29, Appendix A).

The final objective was to determine the potential role and impact of the frequent use of ICT multimedia learning tools like YouTube on the development of ESL learners' oral fluency. This was done through exploring the participants' personal perspectives on the use of affordances of ICT virtual multimedia learning tools/environments for the development of their L2 oral fluency. Different outcomes were set by the questionnaire closed-ended question in this regard (Q.30, Table 6).

The self-report questionnaire[4] (*n.* 30 in total, Appendix A) was conducted online at the end of this study, i.e., after 5 months of the implementation and actual use of the provided and enhanced multimedia learning environment (by way of YouTube captioned videos), which was proposed for the development of learners' L2 speech proficiency. Participants were to fill out this questionnaire which collected information about their daily digital (i.e., ICT-based learning/Online) and non-digital (i.e., Non-ICT-based learning/Offline) literacy practices with regard to their engagement rate, frequent exposure to each of these potentially helpful language learning tools in their learning environment [5] (Questions 4–29) and their personal perspectives towards ICT multimedia learning tools such as YouTube and its impact on their English speech fluency development [2] (Q. 30). Participants considered the suggested BBC 6 min English YouTube channel videos (as a sample) while answering

---

[2] Participants were given three oral test samples before and three after the experiment to set the baseline and end line in this study. The test speaking tasks included 1. introduction and interview (4–5 min) 2. individual long turn (3–4 min) 3. two-way discussion (4–5 min).

[3] This comparison was basically done to entertain other variables (AKA confounding factors as suggested by [5]) which might be producing some effect besides the potential YouTube potential effect in the likely development of learners' speech fluency of the participants, and thus the author could reasonably make a more insightful and valid judgment.

[4] This questionnaire was adapted from [2, 5] and adjusted to fit with the context and main objectives in this study. For this aim, the author followed the guidelines set by [53] for designing and analysing surveys in SLA research.

the questionnaire. Basically, this questionnaire focused on learners' personal experience and potential benefits of YouTube captioned videos used for ESL speech fluency development to identify and record the role and impact of YouTube on ESL speaking fluency progress. The survey (30 questions) was comprised of Likert and closed-ended items. The questionnaire items (Qs. 4–29, Appendix A) provided data for part of the quantitative analysis part of this study (i.e., correlation analysis). The questionnaire item (Q.30, Table 6) provided data for the qualitative analysis part. As for the questionnaire findings, all participants' responses (*n*. 14) were downloaded and tabulated into Microsoft Excel sheets and subsequently exported to SPSS version 21.0 for statistical analysis purposes.

The quantitative analysis part of this work involved a large number of metrics (*n*.17) of both quantitative and qualitative nature which were employed to detect the progress of 14 bilingual Arabic ESL learners' speech fluency. These metrics, which are widely cited in most recent SLA literature of speech fluency, were employed as the dependent variables in the analysis of the participants' L2 progress in terms of speech fluency and the correlational analysis in this study. These metrics were the following: (1) *Speech rate* which was measured by the total number of syllables produced in a given speech sample and then divided by the amount of total time required to produce the speech sample, including pause time, expressed in minutes (note that all frequency measures in this study were standardized by expressing them per minute [54]). All syllables were included in the count, including repair phenomena and filled pauses [55]. (2) *Articulation rate* which was measured by the mean number of syllables produced per minute over the total amount of time spent speaking when producing the speech sample following [55]. (3) *Number of silent pauses (SPs)* which was measured by the total number of pauses divided by the total amount of time spent speaking expressed in seconds and then was multiplied by 60 following [55]. (4) *Mean length of SPs* was calculated by dividing the total length of pauses whose duration is 0.4 seconds or longer by the total number of pauses whose duration is 0.4 seconds or longer [56]. (5) *Number of filled pauses (FPs)* which was measured by counting the number of all non-lexicalized FPs (e.g., um, uh, er) [56]. (6) *Number of false starts* which was measured by counting the number of rejected sounds, words or longer utterances that were cut off in a speech sample [56]. (7) *Number of replacements* which was measured by counting the number of words that were replaced with other words without additional modifications [56]. (8) *Number of reformulations* which was measured by counting the number of utterances longer than one word that were repeated with some modification [56]. (9) *Number of repetitions* which was measured by counting the number of words or longer stretches of speech that were repeated without modification. When a word was repeated more than once, the additional repetitions were counted as separate instances of repetition [56]. (10) *Efficiency/Effortlessness* which was measured according to the four kinds of patterns qualifying L2 efficiency and effortlessness, namely continuity, overall spontaneous language processing, naturalness and overall organization of speech [57]. (11) *Lexical retrieval,* expressed by three subcategories, namely the ease and difficulty of lexical retrieval in speech processing (i.e., closely related to the pausing phenomena), lexical encoding (i.e., searching for words and good structures) and the communication strategies used to overcome problems with lexical access such as self-correction or as fillers or paraphrasing [57].(12) *Incidence of errors per AS. Unit* which was measured by dividing the number of errors by the number of AS-units [58]. (13) *Lexical diversity (VOCD)* is a lexical diversity assessment tool producing an output index that is referred to as VOCD [59]. (14) *Lexical diversity (MTLD)* is another lexical diversity assessment tool producing an output index that is referred to as MTLD [60]. The author analysed the lexical diversity metric (both VOCD and MTLD) using the online softwareText Inspector (85). (15) *Lexical density* which was measured by dividing the number of lexical words by the total number of words multiplied by 100 [61]. The author used the online software Textalyzer(86) in this regard. (16) *Mean length of T. unit* which was measured by dividing the number of words by the total number of T. units in a speech sample [62]. (17) *Average number of clauses per T. unit* which was measured by dividing the number of clauses by the number of T. units in a speech sample [63].

To account for the effectiveness of the proposed method in this study, the author employed two independent variables in the correlational analysis part (i.e., engagement rate and frequent exposure to language learning tools/environments, including YouTube) not only due to their huge reputation and significance in SLA literature as essential factors/prerequisites for receptive and productive skills in foreign/second language learning but also as two essentially inherent criteria to operationalize the narrow listening technique adopted in this study for the development of learners' oral fluency. Learners' engagement factor/variable involved both observable behavior (i.e., involvement/participation and time spent on task) [64] and emotional or affective aspects into their conceptualization of engagement (i.e., enjoyment or interest) [65]. These definitions work well for the aim and context of the present study. Participants' engagement rate was assessed using a five-point Likert scale (i.e., 1-point was defined as extremely poor engagement and 5-points as extremely high engagement). The other independent factor (i.e., the learners' time range of exposure) was practically interpreted as the learners' frequent contact with the target language they are trying to learn, either generally or with specific language points[66]. Participants' range of daily (i.e., frequent) exposure time was defined by the amount of time (i.e., hours) per day using each of these tools for learning English; it was measured on a five-point Likert scale. Time range of exposure was coded as the following: 0 h= null exposure, 1 h = one hour, 2 h = two hours, 3 h = three hours, 4 h + = four hours and above. In this respect, learners' speaking tasks were rated by four experts and the inter-rater reliability analysis for the total speaking tasks (*n*. 84) showed Cronbach's alpha $\alpha = 0.80$. This correlational study hypothesizes that ESL learners who engage with, expose themselves more to (by way of narrow listening) and leverage from the provided and enhanced multimedia-based comprehensible input (by way of captioned videos) on YouTube can increase their gain of the target language chunks and ultimately improve their oral fluency over time.

Also, the author ran three statistical tests to analyze the datasets obtained in this study. The quantitative analysis part involved Wilcoxon and Spearman correlation.

First, Wilcoxon test was used to process the data elicited from the learners' test scores of the speaking tasks and thus to be able to account for any potential variation before and after the experiment (see Wilcoxon test results in Table 1). This test was suitable as the same group of participants were to be examined at two different points of time, which was the case in this work [67]. In this study, the critical Z value for a 95% confidence interval was used for hypothesis testing. For a 2-tailed test (which was the case in this study) if $Z < -1.96$ or $Z > +1.96$, the null hypothesis in this study can be rejected.

Second, Spearman correlation test was used to explore the strength and direction of the relationship between the variables in this study, i.e., between the language learning tools available to this group of learners with respect to the learners' engagement rate with and daily active exposure time to each of these language learning tools, including ICT multimedia learning tools like YouTube (i.e., this represents the independent variables for this correlational analysis whose data were elicited from learners' responses to questionnaire items (Qs 4–29, Appendix A) and the learners' L2 speech fluency progress after the experiment, if any, (i.e., this represents the dependent variables for this correlational analysis whose data were the learners' scores of the post-test speaking tasks provided in Table 1). As for the strength of relationship between these variables, the critical value for Spearman |r| was set at $0.05 = 0.464$. This was the significance level for a two-tailed test which can be found in the critical values table of the Spearman's ranked correlation coefficient (r_s) [68]. The degree of freedom df $= N - 2$, df $= 14 - 2 = 12$. Correlation coefficient values below the 0. 464 were considered insignificant and those above the 0. 464 were significant. The higher the r value or closer to |+1/−1|, the stronger the correlation is between any two of the concerned variables. Consequently, the better the learners' oral fluency is. Regarding the direction (i.e., positive $= +$ve/negative $= -$ve) of relationship between these variables, the oral fluency metrics (dependent variables) in the correlation table (Tables 2, 3, 4 and 5), namely *speech rate, articulation rate, efficiency/effortlessness, lexical retrieval/resource, lexical diversity (VOCD), lexical diversity (MTLD), lexical density, mean length of T.unit, number of clauses per T.unit* should be positively correlated with the independent variables as these dependent variables dealt with desirable aspects of oral fluency. However, the other oral fluency metrics, namely *number of silent pauses, mean length of silent pauses, number of filled pauses, number of false starts, number of replacements, number of reformulations, number of repetitions, incidence of errors per AS-unit* should be negatively correlated with the independent variables since these dependent variables dealt with undesirable aspects of oral fluency, i.e., hesitation phenomenon and errors which can be found in learners' oral speech; the lower the number of hesitation phenomenon and errors mean the higher oral fluency is. The results yielded from this analysis were used to answer the second research question in this study (see test results in Tables 2, 3, 4 and 5).

Third, the qualitative analysis part was based on Content analysis, which was technically employed to analyze the data elicited from the participants' responses to the questionnaire item (Q30, see test results in Table 6). This test was run using one closed-ended question (Q30, Table 6). This question concerns these ESL learners' personal perspectives about affordances of ICT virtual multimedia learning tools/environments for the development of their L2 speech fluency. More specifically,

Table 2 Spearman's correlation coefficients: rate of learners' engagement with ICT-based learning/online

Spearman's correlation (Rho)		Monomedium-based input (text/speech)		Multimedia-based input (text and speech)				
		Reading (books)	Listening (songs)	YouTube BBC 6 min English	Video games	Songs	Films	Audio books
Speech fluency metrics	Speech rate	Correlation coefficient (rs)	0.129	−0.096	0.518	−0.179	−0.311	0.165	0.147
		Sig. (p)	0.660	0.745	0.058	0.541	0.279	0.573	0.617
	Articulation rate	Correlation Coefficient (rs)	0.162	−0.124	0.501	−0.103	−0.371	0.053	0.102
		Sig. (p)	0.581	0.672	0.068	0.726	0.192	0.858	0.729
	Number of silent pauses (SPs)	Correlation Coefficient (rs)	−0.322	−0.020	0.073	0.282	−0.100	−0.143	−0.070
		Sig. (p)	0.262	0.945	0.805	0.329	0.734	0.627	0.812
	Mean length of SPs	Correlation Coefficient (rs)	−0.599*	−0.068	0.214	0.416	−0.106	−0.218	0.147
		Sig. (p)	0.024	0.818	0.463	0.139	0.717	0.455	0.617
	Number of filled pauses	Correlation Coefficient (rs)	0.157	−0.371	0.139	0.412	−0.619*	−0.310	0.041
		Sig. (p)	0.592	0.191	0.636	0.143	0.018	0.280	0.891

(continued)

Table 2 (continued)

Spearman's correlation (Rho)

		Monomedium-based input (text/speech)		Multimedia-based input (text and speech)					
		Reading (books)	Listening (songs)	YouTube BBC 6 min English	Video games	Songs	Films	Audio books
Number of false starts	Correlation Coefficient (rs)	−0.167	−0.158	0.025	0.081	0.005	−0.576*	−0.146
	Sig. (p)	0.568	0.589	0.931	0.783	0.987	0.031	0.617
Number of replacements	Correlation Coefficient (rs)	0.360	−0.181	−0.403	−0.322	0.160	0.061	−0.244
	Sig. (p)	0.206	0.535	0.153	0.261	0.585	0.837	0.401
Number of reformulations	Correlation Coefficient (rs)	0.208	−0.371	−0.080	0.128	−0.268	−0.477	−0.016
	Sig. (p)	0.475	0.191	0.785	0.664	0.354	0.085	0.958
Number of repetitions	Correlation Coefficient (rs)	−0.137	0.005	0.064	−0.037	0.002	0.178	0.318
	Sig. (p)	0.640	0.987	0.828	0.899	0.994	0.542	0.268

(continued)

Table 2 (continued)

Spearman's correlation (Rho)		Monomedium-based input (text/speech)		Multimedia-based input (text and speech)				
		Reading (books)	Listening (songs)	YouTube BBC 6 min English	Video games	Songs	Films	Audio books
Efficiency/effortlessness	Correlation Coefficient (rs)	−0.172	0.103	0.861**	−0.276	−0.008	0.156	0.334
	Sig. (p)	0.556	0.725	0.000	0.339	0.978	0.595	0.242
Lexical retrieval/resource	Correlation Coefficient (rs)	−0.454	0.035	0.813**	0.036	−0.049	0.014	0.561*
	Sig. (p)	0.103	0.906	0.000	0.904	0.867	0.961	0.037
Incidence of errors per AS-unit	Correlation Coefficient (rs)	0.415	−0.402	−0.686**	0.064	−0.251	−0.124	−0.201
	Sig. (p)	0.140	0.154	0.007	0.829	0.386	0.673	0.490
Lexical diversity (VOCD)	Correlation Coefficient (rs)	−0.253	0.124	0.811**	0.027	−0.009	0.135	0.286
	Sig. (p)	0.383	0.672	0.000	0.927	0.975	0.645	0.321

(continued)

Table 2 (continued)

Spearman's correlation (Rho)

		Monomedium-based input (text/speech)		Multimedia-based input (text and speech)				
		Reading (books)	Listening (songs)	YouTube BBC 6 min English	Video games	Songs	Films	Audio books
Lexical diversity (MTLD)	Correlation Coefficient (rs)	−0.255	0.201	0.732**	−0.047	0.136	0.043	0.222
	Sig. (p)	0.378	0.491	0.003	0.874	0.643	0.885	0.445
Lexical density	Correlation Coefficient (rs)	0.129	0.153	−0.315	0.201	−0.182	0.260	−0.144
	Sig. (p)	0.660	0.601	0.273	0.491	0.533	0.369	0.623
Mean length of T. unit	Correlation Coefficient (rs)	0.380	−0.388	−0.384	0.174	−0.497	−0.140	−0.073
	Sig. (p)	0.180	0.170	0.176	0.551	0.071	0.632	0.803
Number of clauses per T. unit	Correlation Coefficient (rs)	−0.435	−0.029	0.705**	0.045	0.216	−0.281	0.274
	Sig. (p)	0.120	0.922	0.005	0.877	0.458	0.330	0.343

*Correlation is significant at the 0.05 level (2-tailed)
**Correlation is significant at the 0.01 level (2-tailed)

Table 3 Spearman's correlation coefficients: time range of learners' exposure to ICT-based learning/online

Spearman's correlation (Rho)		Monomedium-based input (text/speech)		Multimedia-based input (text and speech)					
		Reading (books)	Listening (songs)	YouTube BBC 6 min English	Video games	Songs	Films	Audio books
Speech fluency metrics								
Speech rate	Correlation Coefficient (rs)	−0.405	−0.231	0.565[*]	−0.227	0.367	−0.540[*]	0.203
	Sig. (p)	0.151	0.426	0.035	0.434	0.197	0.046	0.486
Articulation rate	Correlation Coefficient (rs)	−0.405	−0.257	0.549[*]	−0.163	0.367	−0.540[*]	0.203
	Sig. (p)	0.151	0.374	0.042	0.578	0.197	0.046	0.486
Number of silent pauses (SPs)	Correlation Coefficient (rs)	0.267	−0.074	0.034	0.297	−0.410	0.547[*]	−0.469
	Sig. (p)	0.355	0.801	0.907	0.302	0.145	0.043	0.091
Mean length of SPs	Correlation Coefficient (rs)	0.220	−0.131	0.277	0.376	−0.282	0.563[*]	−0.563[*]
	Sig. (p)	0.449	0.654	0.338	0.185	0.329	0.036	0.036
Number of filled pauses	Correlation Coefficient (rs)	−0.587[*]	−0.381	0.205	0.398	0.348	−0.326	0.224
	Sig. (p)	0.027	0.179	0.482	0.158	0.223	0.255	0.442

(continued)

Table 3 (continued)

Spearman's correlation (Rho)

		Monomedium-based input (text/speech)		Multimedia-based input (text and speech)				
		Reading (books)	Listening (songs)	YouTube BBC 6 min English	Video games	Songs	Films	Audio books
Number of false starts	Correlation Coefficient (rs)	−0.261	0.123	0.221	−0.005	0.111	0.602[*]	−0.324
	Sig. (p)	0.367	0.674	0.448	0.986	0.704	0.023	0.258
Number of replacements	Correlation Coefficient (rs)	−0.213	0.408	−0.316	−0.359	−0.023	0.159	0.409
	Sig. (p)	0.464	0.148	0.271	0.208	0.939	0.587	0.147
Number of reformulations	Correlation Coefficient (rs)	−0.207	−0.163	0.185	0.020	0.110	0.110	−0.019
	Sig. (p)	0.478	0.577	0.526	0.945	0.707	0.707	0.949
Number of repetitions	Correlation Coefficient (rs)	0.155	−0.252	0.194	−0.068	0.000	−0.066	−0.169
	Sig. (p)	0.598	0.386	0.506	0.818	1.000	0.823	0.563

(continued)

Table 3 (continued)

Spearman's correlation (Rho)

		Monomedium-based input (text/speech)		Multimedia-based input (text and speech)				
		Reading (books)	Listening (songs)	YouTube BBC 6 min English	Video games	Songs	Films	Audio books
Efficiency/effortlessness	Correlation Coefficient (rs)	−0.286	0.082	0.799**	−0.315	0.267	−0.422	0.247
	Sig. (p)	0.321	0.782	0.001	0.272	0.357	0.133	0.394
Lexical retrieval/resource	Correlation Coefficient (rs)	−0.211	0.087	0.748**	−0.008	0.270	−0.292	0.096
	Sig. (p)	0.469	0.768	0.002	0.979	0.351	0.311	0.743
Incidence of errors per AS-unit	Correlation Coefficient (rs)	−0.051	0.055	−0.771**	0.153	−0.065	0.259	−0.056
	Sig. (p)	0.863	0.853	0.001	0.601	0.826	0.371	0.850
Lexical diversity (VOCD)	Correlation Coefficient (rs)	−0.152	0.068	0.761**	−0.040	0.194	−0.324	0.277
	Sig. (p)	0.604	0.818	0.002	0.893	0.506	0.259	0.337

(continued)

Table 3 (continued)

Spearman's correlation (Rho)

		Monomedium-based input (text/speech)		Multimedia-based input (text and speech)				
		Reading (books)	Listening (songs)	YouTube BBC 6 min English	Video games	Songs	Films	Audio books
Lexical diversity (MTLD)	Correlation Coefficient (rs)	−0.152	0.177	0.714**	−0.119	0.151	−0.281	0.314
	Sig. (p)	0.604	0.545	0.004	0.686	0.606	0.331	0.274
Lexical density	Correlation Coefficient (rs)	0.253	−0.387	−0.414	0.272	−0.065	−0.065	−0.277
	Sig. (p)	0.382	0.171	0.141	0.347	0.826	0.826	0.337
Mean length of T. unit	Correlation Coefficient (rs)	−0.203	−0.072	−0.521	0.237	0.195	−0.238	−0.019
	Sig. (p)	0.487	0.808	0.056	0.414	0.505	0.413	0.950
Number of clauses per T. unit	Correlation Coefficient (rs)	−0.025	0.134	0.732**	−0.027	0.065	0.195	0.186
	Sig. (p)	0.931	0.647	0.003	0.926	0.825	0.504	0.525

*Correlation is significant at the 0.05 level (2-tailed)
**Correlation is significant at the 0.01 level (2-tailed)

Table 4 Spearman's correlation coefficients: rate of learners' engagement with non-ICT-based learning/offline

Spearman's correlation (Rho)			Monomedium-based input (text/speech)			Multimedia-based input (text and speech)		
			Reading (books)	Listening (songs)	Video games	Formal schooling	Films	Audio books
Speech fluency metrics	Speech rate	Correlation Coefficient (rs)	0.140	−0.182	−0.179	−0.092	0.400	0.067
		Sig. (p)	0.634	0.553	0.541	0.753	0.156	0.820
	Articulation rate	Correlation Coefficient (rs)	0.054	−0.264	−0.103	−0.018	0.477	0.002
		Sig. (p)	0.856	0.384	0.726	0.950	0.084	0.994
	Number of silent pauses (SPs)	Correlation Coefficient (rs)	−0.194	−0.373	0.282	−0.117	−0.346	−0.366
		Sig. (p)	0.506	0.209	0.329	0.690	0.225	0.198
	Mean length of SPs	Correlation Coefficient (rs)	−0.291	−0.537	0.416	−0.161	−0.419	−0.543*
		Sig. (p)	0.312	0.058	0.139	0.583	0.135	0.045
	Number of filled pauses	Correlation Coefficient (rs)	0.175	−0.669*	0.412	0.317	0.615*	−0.315
		Sig. (p)	0.550	0.012	0.143	0.270	0.019	0.273
	Number of false starts	Correlation Coefficient (rs)	−0.087	−0.337	0.081	−0.172	0.052	−0.131
		Sig. (p)	0.769	0.261	0.783	0.557	0.859	0.655
	Number of replacements	Correlation Coefficient (rs)	0.206	0.145	−0.322	−0.409	0.256	0.109
		Sig. (p)	0.480	0.637	0.261	0.147	0.377	0.709

(continued)

Table 4 (continued)

Spearman's correlation (Rho)

		Monomedium-based input (text/speech)			Multimedia-based input (text and speech)		
		Reading (books)	Listening (songs)	Video games	Formal schooling	Films	Audio books
Number of reformulations	Correlation Coefficient (rs)	−0.278	−0.482	0.128	0.151	0.197	−0.217
	Sig. (p)	0.335	0.095	0.664	0.606	0.500	0.455
Number of repetitions	Correlation Coefficient (rs)	−0.114	−0.102	−0.037	−0.245	−0.309	−0.068
	Sig. (p)	0.698	0.740	0.899	0.399	0.282	0.817
Efficiency/effortlessness	Correlation Coefficient (rs)	0.121	−0.169	−0.276	0.019	0.536[*]	0.017
	Sig. (p)	0.680	0.582	0.339	0.949	0.048	0.955
Lexical retrieval/resource	Correlation Coefficient (rs)	0.110	−0.376	0.036	0.154	0.444	−0.258
	Sig. (p)	0.707	0.206	0.904	0.599	0.112	0.374
Incidence of errors per AS–unit	Correlation Coefficient (rs)	0.328	0.205	0.064	−0.130	−0.089	0.095
	Sig. (p)	0.253	0.501	0.829	0.659	0.761	0.747
Lexical diversity (VOCD)	Correlation Coefficient (rs)	0.005	−0.258	0.027	0.314	0.379	−0.095
	Sig. (p)	0.987	0.395	0.927	0.274	0.182	0.747
Lexical diversity (MTLD)	Correlation Coefficient (rs)	−0.063	−0.223	−0.047	0.351	0.461	−0.095

(continued)

Table 4 (continued)

Spearman's correlation (Rho)			Monomedium-based input (text/speech)			Multimedia-based input (text and speech)		
			Reading (books)	Listening (songs)	Video games	Formal schooling	Films	Audio books
		Sig. (p)	0.831	0.465	0.874	0.218	0.097	0.747
	Lexical density	Correlation Coefficient (rs)	0.100	0.141	0.201	−0.166	−0.439	0.136
		Sig. (p)	0.733	0.647	0.491	0.570	0.116	0.642
	Mean length of T. unit	Correlation Coefficient (rs)	0.322	0.076	0.174	0.204	0.196	0.012
		Sig. (p)	0.262	0.804	0.551	0.485	0.503	0.969
	Number of clauses per T. unit	Correlation Coefficient (rs)	−0.403	−0.235	0.045	0.130	0.218	−0.165
		Sig. (p)	0.153	0.440	0.877	0.658	0.455	0.574

*Correlation is significant at the 0.05 level (2-tailed)

**Correlation is significant at the 0.01 level (2-tailed)

Table 5 Spearman's correlation coefficients: time range of learners' exposure to non-ICT-based learning/offline

Spearman correlation (Rho)		Monomedium-based input (text/speech)			Multimedia-based input (text and speech)		
		Reading (books)	Listening (songs)	Video games	Formal schooling	Films	Audio books
Speech fluency metrics	Speech rate						
	Correlation Coefficient (rs)	0.518	−0.310	0.034	0.034	0.378	−0.447
	Sig. (p)	0.058	0.281	0.907	0.907	0.182	0.109
	Articulation rate						
	Correlation Coefficient (rs)	0.453	−0.241	0.103	0.034	0.378	−0.447
	Sig. (p)	0.104	0.407	0.726	0.907	0.182	0.109
	Number of silent pauses (SPs)						
	Correlation Coefficient (rs)	0.077	−0.291	−0.291	−0.291	0.073	0.073
	Sig. (p)	0.795	0.313	0.313	0.313	0.805	0.805
	Mean length of SPs						
	Correlation Coefficient (rs)	0.185	−0.299	−0.299	−0.150	0.150	0.150
	Sig. (p)	0.527	0.299	0.299	0.610	0.610	0.610
	Number of filled pauses						
	Correlation Coefficient (rs)	0.122	0.069	−0.104	−0.243	0.069	−0.451
	Sig. (p)	0.677	0.814	0.723	0.403	0.814	0.106
	Number of false starts						
	Correlation Coefficient (rs)	0.147	0.178	−0.036	−0.178	0.462	−0.320
	Sig. (p)	0.615	0.544	0.904	0.544	0.097	0.265
	Number of replacements						
	Correlation Coefficient (rs)	−0.226	0.036	0.471	−0.326	0.326	0.036
	Sig. (p)	0.436	0.902	0.089	0.256	0.256	0.902

(continued)

Table 5 (continued)

Spearman correlation (Rho)		Monomedium-based input (text/speech)			Multimedia-based input (text and speech)		
		Reading (books)	Listening (songs)	Video games	Formal schooling	Films	Audio books
Number of reformulations	Correlation Coefficient (rs)	−0.046	0.281	0.387	−0.105	0.281	−0.387
	Sig. (p)	0.876	0.330	0.172	0.720	0.330	0.172
Number of repetitions	Correlation Coefficient (rs)	0.097	−0.210	0.070	0.210	0.245	−0.035
	Sig. (p)	0.742	0.471	0.812	0.471	0.398	0.905
Efficiency/effortlessness	Correlation Coefficient (rs)	0.718**	−0.177	−0.354	0.035	0.354	−0.354
	Sig. (p)	0.004	0.545	0.215	0.904	0.215	0.215
Lexical retrieval/resource	Correlation Coefficient (rs)	0.579*	−0.143	−0.394	0.143	0.322	−0.143
	Sig. (p)	0.030	0.625	0.163	0.625	0.261	0.625
Incidence of errors per AS-unit	Correlation Coefficient (rs)	−0.595*	−0.103	0.379	−0.172	0.310	0.103
	Sig. (p)	0.025	0.725	0.182	0.556	0.281	0.725
Lexical diversity (VOCD)	Correlation Coefficient (rs)	0.561*	−0.172	−0.241	0.103	−0.034	−0.310
	Sig. (p)	0.037	0.557	0.407	0.726	0.907	0.281
Lexical diversity (MTLD)	Correlation Coefficient (rs)	0.540*	0.103	−0.310	0.034	−0.172	−0.241

(continued)

Table 5 (continued)

Spearman correlation (Rho)		Monomedium-based input (text/speech)			Multimedia-based input (text and speech)		
		Reading (books)	Listening (songs)	Video games	Formal schooling	Films	Audio books
Lexical density	Sig. (p)	0.046	0.726	0.281	0.907	0.557	0.407
	Correlation Coefficient (rs)	−0.303	−0.447	0.172	0.241	−0.034	0.034
Mean length of T. unit	Sig. (p)	0.293	0.109	0.557	0.407	0.907	0.907
	Correlation Coefficient (rs)	−0.403	−0.103	0.448	0.034	0.172	0.103
Number of clauses per T. unit	Sig. (p)	0.153	0.725	0.108	0.907	0.556	0.725
	Correlation Coefficient (rs)	0.335	0.242	−0.449	0.069	0.138	−0.069
	Sig. (p)	0.241	0.405	0.108	0.815	0.638	0.815

*Correlation is significant at the 0.05 level (2-tailed)
**Correlation is significant at the 0.01 level (2-tailed)

Table 6 ESL learners' personal perspectives about affordances of ICT virtual multimedia learning tools/environments for the development of their L2 speech fluency

Q30. Do you think the use of YouTube-based multimedia videos (by way of captions) made L2 input more comprehensible and chunkable and thus improved your English-speaking proficiency more effectively over time?	(n) % responses	Pedagogical implications
a- Listening to captioned videos makes learning English easier as they aid my comprehension of YouTube video materials; this makes the new L2 aural input more comprehensible for me. Yes No	(13) 92% (1) 8%	This demonstrates the helpful role of captioning to make new L2 aural input more intelligible for L2 learners.
b- Listening to captioned videos supports my attention to the new L2 aural input; this helps me gain more L2 input (i.e., chunks) such as lexical phrases, set phrases, fixed phrases/expressions, new vocabulary words; these language chunks can ultimately improve my overall English language use in speaking. Yes No	(12) 85% (2) 15%	This demonstrates how captions are helpful for learners to gain more of the aural input (i.e., language chunks) which may be employed in other speaking tasks.
c- Listening to captioned videos helps me analyze and recognize the new target language input (i.e., chunks) more clearly; this can facilitate the retention of the newly learnt/acquired language chunks and the overall process of listening and speaking. Yes No	(13) 92% (1) 8%	This demonstrates that captions can be utilised to help learners analyze the target language, draw their attention and improve retrieval of acquired language input and ultimately contributing to the development of their overall L2 speaking fluency.
d- Listening to captioned videos helps with an in-depth analysis of the target language chunks (such as collocations, fixed expressions, formulaic utterances, sentence starters, verb patterns, idioms, catch/short phrases, etc.) because the multimedia of visual and auditory materials makes the L2 input more accessible and comprehensible and ultimately more reproduceable. Yes No	(12) 85% (?) 15%	The use of captioned videos allows learners to do easier "narrow listening" and more detailed analysis of the new L2 input, i.e., chunks.

(continued)

Table 6 (continued)

e- Listening to captioned videos helps me see and parse structural/lexical patterns or chunks in the videos, which assists me in remembering and learning from the patterns presented in subsequent communicative speaking tasks. Yes No	(12) 85% (2) 15%	This shows the positive role of listening to captioned videos in fostering the acquisition of L2 chunks.
f- The use of captioned videos in English speaking classes can create and enhance a multimedia learning environment (given their potential positive effects and cognitive advantages) for different learning styles; they are especially supportive for the visual and/or auditory ESL learner as far as the L2 comprehensible input is concerned. Yes No	(14) 100% —	This demonstrates the assistive and positive role of captions to cater for different learning styles of ESL learners.
g- Listening to captioned videos helps me reinforce previous L2 knowledge, including language chunks, and practice speaking (processing) of what is taken in through the caption. Yes No	(14) 100% —	This demonstrates the ancillary role of video captions to instil previously acquired knowledge and practice (i.e., process) the likely new incoming knowledge.
h- The availability of new, interesting, useful information, language materials and topics on YouTube makes it easier, more engaging for narrow listening; ultimately, this makes me more active and better able in speaking classes; i.e., food for thought. Yes No	(13) 92% (1) 8%	This shows that the use of YouTube makes it easy to learn and develop speech proficiency over time due to the availability of new multimedia L2 input in every listening/speaking class.
i- Listening narrowly to the multimedia-based comprehensible input on YouTube helps me engage more with and chuck-learn more of L2 input and thus helps me improve my speaking skill. Yes No	(13) 92% (1) 8%	This demonstrated how narrow listening to YouTube-multimedia based comprehensible input can engage ESL learners and increase their gain of language chunks.

Note Adapted from *ICT multimedia learning affordances: role and impact on ESL learners' writing accuracy development*, by Azzam Alobaid. Retrieved from https://doi.org/10.1016/j.heliyon.2021.e07517 Copyright 2021, Heliyon

they were asked how they thought the use of YouTube-based multimedia videos (by way of captions) made L2 input more comprehensible and chunkable and thus improved your English-speaking proficiency more effectively. Participants' actual responses signified the potential role and impact of the frequent use of YouTube-based multimedia videos on the learners' ESL oral fluency development through indirectly exploring their personal experience in terms of the benefits gained from the actual use and potential advantages of the YouTube-based multimedia videos, especially when used for ESL oral fluency development. Participants were required to tick (if they agreed or not with) the given statements according to their actual experience. These statements were structured around the main research questions in this study and are widely discussed in recent literature of SLA which relate to the adopted theories in this study for the development of ESL learners' speaking proficiency.

5 Procedures

This research adopted the narrow listening technique/approach [69] for the development of learners speaking fluency while learners listened to the YouTube multimedia-based comprehensible L2 input. In terms of the learning materials, three different videos about different topics (mainly pedagogical rather than authentic or spontaneous) were nominated towards the end of every English-speaking class, and participants had to single out one of them (of personal interest and which can be familiar to them) to be the subject matter for the next speaking class. In this study, a great deal of L2 input could be made comprehensible through (1) repeated listening, (2) self-selection of samples, and (3) familiarity with topics as suggested by [16] as cited in [17] and also through (i.e., captions) which has two substantially potential multimedia learning effects and cognitive advantages, namely to make L2 input more comprehensible and chunkable for learners; both of which may lead to a more fluent L2 speech. Giving learners autonomy to select (i.e., selection of language materials can be according to the same author, the same genre, the same title, the same linguistic level) their own materials promotes greater motivation [30]. Topical familiarity has been found to be helpful for comprehension. From listening fluency perspective, if learners are provided with or possess good background knowledge over a topic and are familiar with topical vocabulary, they are more likely to be able to process the input more efficiently and eventually become fluent [32]. These self-selected videos were supposed to be brief, i.e., between 1 and 6 min. In this regard, BBC 6 min English YouTube videos were used in these speaking classes. Basically, participants were requested to make use of the captions of videos to understand (i.e., make comprehensible) the selected topic of a given video on their own at home while they narrowly listened to a given YouTube video. The idea behind narrowly listening to captioned videos (for as many times as required) prior to speaking tasks is to make the new L2 input all the more comprehensible by virtue of YouTube multi-mediated L2 input. While repeating the task of extensive listening (while optionally looking at video captions), learners are not only expected to receive a massive amount of aural input [32] but also to absorb (i.e., comprehend) more L2 input, recycle, consolidate and retain newly acquired L2 input for later productive tasks, i.e., speaking fluency. At this point, there is a potential that learners may well pick up a lot of new language chunks. Actually, learners were requested to note down those new language chunks (by definition language chunks are words that always go together, such as fixed collocations, or that commonly do, such as certain grammatical structures that follow rules) they could learn/pick up as they simultaneously listened to and watched the captioned videos. Examples of such chunks involved lexical phrases, set phrases, fixed phrases/expressions, new vocabulary words. Learners would use their knowledge of chunks to help them predict meaning and therefore be able to process language in real time. Areas of work and focus included idioms, collocations and verb patterns as types of chunks. Learners were encouraged to identify and record chunks (e.g., lexical and grammatical) when they find them in these videos. Afterwards, learners were constantly encouraged to try to make use of the new target language input (i.e., chunks) (potentially gained from the same given video) in the speaking tasks given to them. Speaking tasks based on narrow listening, which were structured around these videos content (i.e., video-based topical discussions), involved various speaking activities (i.e., getting/discussing general idea, reacting to the content, using the newly learnt L2 input in similar communicative contexts, summaries). Additionally, learners were encouraged to note down their comments and inquiries about the topic for open discussion in the next speaking class. Participants had to do this on their own at home for every speaking task.

At school, the selected video with captions turned on was viewed again by all learners and the teacher[5] at the beginning of the speaking class. Afterwards, learners were requested to bring up for open discussion their comments, inquires/doubts, the lexical and grammatical chunks they identified and recorded on their own and use them while discussing the overall idea of the new topic (video-based topical discussion). Following the narrow listening technique guidelines in speaking classes, there can be a different focus each time learners would narrowly listen to some video such as general comprehension, practice on listening skills, vocabulary words and language chunks.

The instructor would use the captions for their multimedia effects. Furthermore, the captions would serve other educational functions which may enrich the acquisition of the new L2 input such as when referring to/highlighting interesting language chunks. Also, the instructor may point out some language chunks with the help of captions if these chunks are found enriching for the learners' speaking experience. Once again, participants were reminded of the facilitative role of captioning which can aid them when needed (mainly to make the L2 input comprehensible) and most importantly to start using the newly acquired/learnt L2 input (i.e., chunks) in their English-speaking tasks in and out of the language classes. Last, when learners

---

[5] In this study, the author took a complete role of English language teacher in these ESL speaking classes.

become familiar with one theme or topic in a video, they can move on to the next one (i.e., new video).

6 Research Findings and Discussion

In investigating ICT multimedia learning potential role and effect in ESL learning/teaching context, our present study proposed YouTube as an ICT-based multimedia learning tool for the improvement of ESL learners' oral fluency, given that learners engage with and are frequently exposed to (by way of the narrow listening technique) the provided and enhanced multimedia comprehensible input through the suggested ICT tool, i.e., YouTube. Findings provide pedagogical insights and enable ESL learners/teachers to identify more effective methods for the development of oral fluency in a multimedia learning environment like YouTube. What can be construed from our study findings is that ESL learners' engagement with and frequent exposure to (by way of the narrow listening technique [16]) the provided and enhanced multimedia-based comprehensible input on YouTube can ultimately improve their L2 oral fluency over time. This study yielded a number of observations and findings which supported the effectiveness of using YouTube as an ICT multimedia learning tool for the development of ESL learners' oral fluency over time. The obtained results are further illustrated, discussed and interpreted in light of relevant previous research below.

Regarding the first research question, findings of this study (Table 1) demonstrated statistically significant differences in some ($n.9$) but not all of the quantitative and qualitative aspects of the learners' L2 oral fluency after 5 months of actual exposure to the provided and enhanced YouTube-based multimedia comprehensible input which was proposed for the development of ESL learners' L2 speech proficiency. As shown by Wilcoxon Matched-Pairs Ranked Test results (Table 1), the post-test scores were statistically significantly higher than pre-test scores in terms of **Speech rate** $Z = -3.3$, $p = .001$, $d = -0.62$; **Articulation rate** $Z = -3.3$, $p = .001$, $d = -0.62$; **Number of silent pauses** (SPs) $Z = -3.19$, $p = .001$, $d = -0.60$; **Mean length of SPs** $Z = -2.94$, $p=.003$, $d = -0.56$; **Number of filled pauses** $Z = -3.3$, $p= .001$, $d = -0.62$; **Number of false starts** $Z = -2.45$, $p = .01$, $d = -0.46$; **Number of Repetitions** $Z = -2.48$, $p=.01$, $d = -0.47$; **Efficiency/Effortlessness** $Z = -3.34$, $p = .001$, $d = -0.63$; **Lexical retrieval/resource** $Z = -3.4$, $p=.001$, $d = -0.64$.

The above results (Table 1) demonstrated statistically significant differences between the medians of this group of learners before and after the implementation of YouTube for the improvement of ESL oral fluency as shown by the concerned p values (significance level $p \leq 0.05$) of this test. Fig. 1 is a graphic representation of the developed oral fluency metrics, demonstrating the learners' test scores before as compared to after 5 months of exposure to YouTube. Also, given that the concerned Z scores in this test were below -1.96, this suggested that the above null hypothesis set in this work can be rejected in favour of the alternative hypothesis. Moreover, the effect size of this test (expressed by Cohen's d values) exceeded Cohen's [70]

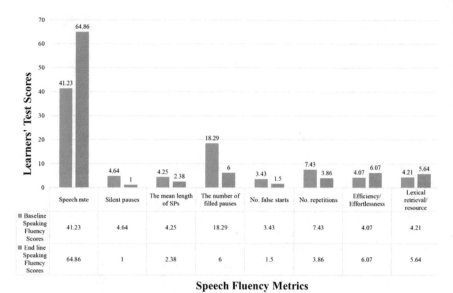

Fig. 1 Baseline versus end line test scores of learners (before as compared to) after 5 months of exposure to YouTube: statistically significant differences based on (*p*. value < 0.5)

convention for a medium effect (*d* = 0.50). However, other aspects of learners' speech fluency (*n*. 6) developed slightly but were statistically insignificant, namely *No. of replacements, No. of reformulations, Incidence of errors per AS. Unit, Lexical diversity (VOCD), Lexical diversity (MTLD), Lexical density*. Also, only two aspects of learners' oral fluency did not demonstrate any improvement in this study, namely *Mean length of T. unit and Number of clauses per T. unit.*

Based on the significance values (*n*. 9) of this test in this study, findings (Table 1) indicated that, by and large, statistically significant differences/positive effects were noticed after of the implementation of YouTube for the improvement of ESL learners' oral fluency. More specifically, YouTube as an ICT multimedia learning tool could provide and enhance the multimedia-based comprehensible input (by way of YouTube captioned videos) required for learners' L2 speech proficiency development. Over 5 months, learners in these English-speaking classes were encouraged to narrowly listen to the self-selected YouTube captioned videos in the manner suggested and described above (see the Procedures Sect. 5). Thus, the resultant effect in the development of the English speaking fluency of this group of learners may arguably be attributable to the positive and enhancing multimedia effects and cognitive advantages provided by such ICT multimedia learning tools as YouTube. More specifically, the proposed technique of narrowly listening to YouTube captioned videos, which was proposed to help learners improve their L2 oral fluency, seemed to have helped learners in many possible ways which could have probably led to a more fluent speech of English of this group of ESL learners. The present study may help explain how narrow listening to captioned videos may have led to a more fluent oral speech of ESL learners.

Initially, in terms of comprehensibility of the multimedia L2 input, it helped learners down the process of language analysis of the target language input. When listening to (fast) running speech, learners may not always be able to determine word boundaries of speech samples. Therefore, captioning could make L2 input more discrete; this enabled learners to unpack the aural speech samples in a more comprehensible way [9]. Comprehension (while listening) of the L2 input is an essential stage of second/foreign language learning. In this respect, [51] stresses that "learning/understanding language input involves parsing the speech stream into chunks which can reliably mark meaning." In this sense, captions support learners while they listen as they can see and are better "able to then parse structural patterns or language chunks in the videos, which may assist them in (not only figuring out meaning but also) remembering and learning from the patterns presented" [9] (as implied in the multimedia principle, Sect. 3).

Second, listening to captioned videos seems to have made processing L2 input in chunks all the more easier, which may have reduced the burden on the listeners to process individual bits and link form and meaning, and thus chunking extended the learners' ability to recognize, recall information or perform actual communicative tasks at a later time [46]. The provided and enhanced multimedia-based comprehensible input can ultimately improve their L2 oral fluency over time through the potential learning gain/acquisition of such language chunks. Generally, "processing L2 information in chunks reduces the burden on the learner to process individual bits and link form and meaning" [9]. Research evidence about processing of language chunks for learners demonstrates an advantage of formulaic over nonformulaic sequences [48]. This study may probably lend some explanation on why captioning could induce a faster and greater depth of L2 processing of such formulaic sequences or chunks [5, 9].

Third, listening to captioned videos seems to have helped learners with regard to vocabulary words gain. Vocabulary is important for EFL/ESL learners, just like the significance of bricks to a building [25]. When the receptive vocabulary is rather limited, learners can hardly put the "receptive vocabulary knowledge into productive use" [71], so it is essential for L2 learners to store a wealth of vocabulary in their long-term memory. Moreover, the ability to quickly recall words from one's mind may affect the speaking fluency [72]. Therefore, learners need to able to have a fast access to the words and expressions in their mind while speaking so that their speaking fluency can be enhanced [25]. Similarly, the use of captions induced "better vocabulary test scores in the aural mode, especially when videos were shown twice, once with captions and once without [9]." This may suggest some gain of the target language vocabulary (i.e., chunks) owing to the potential of captions, and thus learners can become more fluent speakers.

Regarding the second research question, findings of this study (Tables 2, 3, 4 and 5) were statistically significant. These findings demonstrated a range of statistically significant correlations (moderate and strong correlations) between the development of ESL learners' L2 oral fluency seen in (Table 1) and learners' engagement with and frequent exposure over 5 months to the proposed ICT-based multimedia learning tool, namely YouTube for the development of L2 oral fluency. Moreover, findings (Tables 2 and 3) showed linear relationships between the concerned variables in this work (see Fig. 2).

What follows are the correlation test findings given in more details and divided into two main categories—ICT-based learning versus Non-ICT-based learning.

6.1 ICT-Based Learning

The correlation coefficient findings (Table 2) seen between the participants' engagement rate with ICT-based mono-medium and multimedia learning tools and the learners' speech fluency improvement (Table 1) demonstrated various values, the majority of which were insignificant, especially in the mono-medium learning environments. However, the significant correlation values in terms of both strength (i.e., positive/negative association) and direction (i.e., linearity) were those in the multimedia environments of *YouTube, songs, films and audio books*. These significant correlations can be found in (Table 2). These findings revealed that the correlations of learners' engagement were significant on the multimedia rather than the mono-medium side. More specifically, there were more significant and stronger correlations between most of the oral fluency metrics used in this work and the multimedia of *YouTube* than other multimedia such as *songs, films or audio books*. This may indicate that learners were engaging with the multimedia of YouTube more and to a greater extent than the rest of other learning sources in their online learning environment, and that the multimedia (i.e., text and speech) were preferred over the mono-medium (i.e., either text or speech) environments for the development of their L2 speech proficiency.

The correlation coefficient findings (Table 3) seen between participants' time range of exposure to ICT-based mono-medium and multimedia learning tools and their oral fluency improvement (Table 1) were insignificant except for *YouTube, films and audio books*. These findings (Table 3) seem to confirm the above findings about the learners' engagement rate (Table 2) that while learners were giving far greater amounts of their learning time to the multimedia leaning environments, they gave little or no time in the mono-medium learning environment. This may indicate that learners were more engaged with or inclined towards the multimedia learning environments than the mono-medium environments for learning and improving their L2 speaking proficiency online. More importantly, compared to other ICT-based multimedia learning tools, YouTube was favoured more than other multimedia learning tools as far as improving their L2 speech proficiency is concerned.

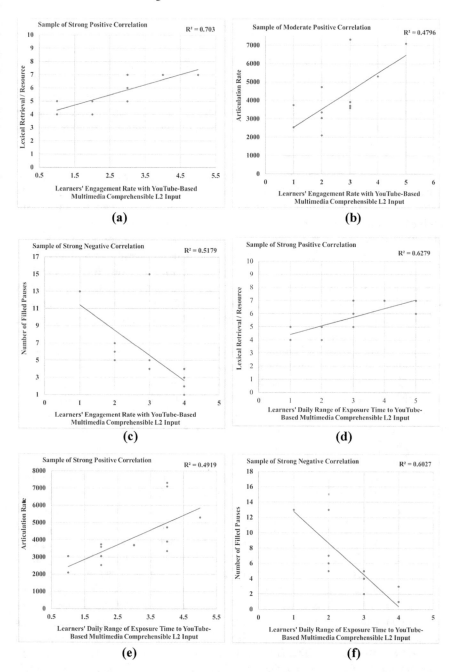

Fig. 2 Samples of positive and negative correlations. **a** Lexical retrieval and resource versus engagement. **b** Articulation rate versus engagement. **c** Number of filled pauses versus engagement. **d** Lexical retrieval and resource versus time. **e** Articulation rate versus time. **f** Number of filled pauses versus time

6.2   Non-ICT-Based Learning

The correlation coefficients seen in (Table 4) between the participants' engagement rate with the Non-ICT-based mono-medium and multimedia learning tools and the participants' oral fluency improvement (Table 1)demonstrated insignificant correlations across almost all the tools except for three moderate correlation values, namely *listening (songs), films and audio books*. These insignificant findings may suggest that learners were not engaging their learning in the Non-ICT-based mono-medium and multimedia learning environments at all.

The correlation coefficient findings seen in (Table 5) between the participants' time range of exposure to the Non-ICT-based mono-medium and multimedia learning tools and their oral fluency improvement (Table 1)were insignificant for most of the tools except for *reading books*. Whilst this may indicate that learners gave a good deal of their learning time to *reading books* in the Non-ICT-based mono-medium leaning environment, these correlation findings (Table 5) do not conform with findings (Table 4) which revealed that learners were not enjoying their learning time as much as they did in the ICT-based multimedia learning environment (see Table 2). This may suggest that learners were more inclined to the ICT-based multimedia learning environments than the Non-ICT-based mono-medium environments for improving their L2 speaking proficiency.

Taken together, the correlational findings about the ICT and Non-ICT-based mono-medium/multimedia learning tools with regard to the learners' high engagement rate with and frequent exposure to each of these tools indicated that a strong correlation exists between the development of oral fluency of this group of ESL learners over 5 months and the engagement with and frequent exposure (by way of narrow listening technique) to ICT-based multimedia rather than mono-medium learning tools, namely YouTube, which was proposed as an ICT multimedia learning tool to provide and enhance (by way of YouTube captioned videos) the multimedia-based comprehensible input of the target language (in this case English) required for learners' L2 speech proficiency (i.e., oral fluency) development. To put this into perspective, compared to other potential language learning tools in the learners' environment, ICT-based multimedia learning tools such as YouTube seems to be more engaging, and hence learners tended to spend far more of their learning time using it on daily basis for improving their speech proficiency than other multimedia learning tools. Consequently, this could have led to a more effective learning or greater gain in their L2 oral fluency; this may suggest the usefulness of narrow listening to ICTs such as YouTube which may have induced a more fluent use of oral language in some (but not all) aspects of speech fluency over time as learners could with the help of ICT-based multimedia learning (i.e., potential multimedia learning effects and cognitive advantages of captioned videos per se) handle their listening activity for improving their speaking fluency more efficiently. This study's findings lend some support and validate results of previous correlational studies. [73] found that ICTs can make it easier for learners to access language materials, stressing an existing correlation between ESL learning and the use of multimedia materials in a computer-enhanced language learning scenarios, signifying a positive impact on learning behaviour/outcomes [such as more fluent oral output] with increased motivation. [74] highlighted L2 learners' personal experience about the effectiveness of captioned video usage more than other (captioned) materials in an L2 learning context. Generally, these learners had a better experience with captioned than non-captioned L2 materials, for they felt that their listening was improving along with captions use. [75] analyzed the need for pedagogical usage of subtitles for EFL learning with the goal of improving their oral skills. [75] found that captioning assists learners to improve their aural comprehension skills, provides them with different types of support (visual, textual, and technological) for language development and also helps learners observe creditable input and produce perceptible output.

Based on the advantageous assumptions of the narrow listening technique (reflected through learners' engagement with and frequent exposure to ICT and Non-ICT learning tools, including YouTube) and the potential role of multimedia-based comprehensible input, generally, correlational findings of this study supported and added value to the conclusion about our first research question and thus drive the following concluding remarks. There can be a positive role and effect attributable to ESL learners' engagement with and frequent exposure (by way of narrow listening technique) to ICT-based multimedia rather than mono-medium learning tools such as YouTube, which was proposed as an ICT multimedia learning tool to provide and enhance (by way of YouTube captioned videos)the multimedia-based comprehensible input of the target language (in this case English) required for learners' L2 speech proficiency (i.e., oral fluency) development.

From a statistical perspective, although this study's findings about bivariate correlations between two or more variables seem significant, such as the case of linear relationships in this work between language learners' engagement rate and time range of exposure vis-à-vis the oral fluency improvement, this does not imply causation. Rather, a reasonable degree of association can be established implying a degree of strength and linearity among the concerned variables (See Fig. 2).

Regarding the third research question, the results (Table 6) supported and added value to the quantitative findings (Tables 1, 2, 3, 4 and 5) and the hypothesis in this study. More specifically, learners' responses (given in numbers and percentages in Table 6) brought more evidence about the potentially positive role and impact of the frequent use of ICT multimedia learning tools like YouTube on the development of ESL learners' oral fluency. This was clearly indicated by the participants' personal experience in terms of the benefits gained from the actual use (by way of narrow listening technique) and potential advantages of the provided and enhanced multimedia-based comprehensible input (by way of captioned videos) on YouTube which could increase their gain of the target language chunks and thus improve their speech fluency over time. For example, YouTube captioned videos which were basically proposed (to make L2 input comprehensible and chunkable) for their multimedia learning effects and cognitive advantages seemed to have aided learners in many ways as they narrowly listened to the target language in hand (in this case English) available in the self-selected YouTube videos. First, captions were helpful to make videos content and language materials (i.e., new L2 input) much more intelligible/comprehensible for L2 learners, especially when they listened to/were introduced to new L2 aural input. Comprehension of new L2 input has always been considered to be a prerequisite in the process of L2 acquisition/learning [14, 76]. In terms of aural comprehension, listening to captioned videos helped learners "not only attend to or identify discrete word boundaries more clearly, but also facilitated their aural comprehension of the L2 input in the aural mode as captions are unaffected by variations in accent and/or audio quality that can adversely affect aural comprehension" [77]. Second, listening to captioned videos supported and enhanced learners' attention to (new) target language aural input and helped them more clearly notice the language chunks (as implied by multimedia principle, Sect. 3); this helped them gain more of the aural L2 input (i.e., chunks) such as lexical phrases, set phrases, fixed phrases/expressions, new vocabulary words; these language chunks could ultimately improve learners' overall English language use in speaking as long as such language chunks could have been rightly employed in subsequent speaking tasks. Some instances of the language chunks picked up by this group of learners were *at the end of the day, you see what I mean, that's what/why, more or less, you can never tell.* These instances of language chunks were frequently used by the presenters in the self-selected videos on YouTube. Similarly, [5] demonstrated some improvement in ESL learners' writing quality and quantity post the integration of YouTube owing to some new language chunks picked by learners from the videos by way of chunk learning. [5] noted that these chunks were used frequently in these videos. Third, listening to captioned videos helped learners analyze and recognize (i.e., process) more information about the new target language aural input (i.e., chunks) in the working memory more clearly and easily; this could facilitate the retention of the newly learnt/acquired language chunks and the overall process of listening and speaking. This kind of in-depth analysis of the target language chunks (such as collocations, fixed expressions, formulaic utterances, sentence starters, verb patterns, idioms, catch/short phrases, etc.) can arguably be due to the multimedia of visual and auditory materials which make the L2 input more accessible and comprehensible and ultimately more chunkable. It is through this kind of detailed analysis that learners become consciously aware of how native speakers use their language. In other words, listening to captioned videos can help learners "see and parse structural/lexical patterns or chunks in the videos, which assist them in remembering (i.e., retaining these information whenever required from the long-term memory as implied in the multimedia principle, learning from the patterns presented and applying this new knowledge (i.e., chunks)" [9] in subsequent communicative speaking tasks. In this sense, video captions may aid in chunking of the L2 aural input. Fifth, the use of captioned videos in English speaking classes can create and enhance a multimedia learning environment (given their potential positive effects and cognitive advantages) for different learning styles; they are especially supportive for the visual and/or auditory ESL learner as far as the L2 comprehensible input is concerned. Sixth, this study found that listening to captioned videos could help learners reinforce previous L2 knowledge including language chunks and practice speaking (processing) of what is taken in through the captions. This demonstrates the ancillary role of video captions to instil previously acquired knowledge and practice (i.e., process) the likely new incoming knowledge. The above results seem to conform with previous research by [9] about the potential of captions use because they result in a greater depth of processing by focusing attention, reinforce the acquisition of vocabulary through multiple modalities, and allow learners to determine meaning through the unpacking of language chunks. From a cognitive perspective, the employment of nontypical input (i.e., video captioning) can facilitate processing language in chunks; this can be beneficial to overall L2 learning as the processing demands are reduced. As a result, due to the captioning effect which helps in chunk learning [78] "learners can analyze and break down the aural input into meaningful constituent structures" [79]. In light of our study, we believe that narrow listening to the multimedia-based comprehensible input on YouTube helped ESL learners engage with and chuck-learn more of L2 input. Similarly, it was reported by [17, 80] that students enjoyed the natural, off-the-cuff conversations of narrow listening. Many indicated that they liked narrow listening as it exposed them to French as it is really spoken while giving them control over the topic.

Arguably, the resultant positive effect or progress on the part of the learners' speaking fluency could be attributable to the multimedia learning effects and cognitive advantages of multimedia learning environments such as YouTube, provided that learners are engaging with and frequently exposed to its multi-mediated input. The author suggested the narrow listening technique as one effective way of engaging with and exposing learners to the claimed multimedia learning environment (see the Framework Sect. 3 which supports our claim, the reliability and validity of this technique). In this respect, we found that the availability of new and engaging language materials and topics (especially those rich in L2 chunks) on YouTube is of paramount significance since this can make it easier and more engaging for learners to do narrow listening and take advantage of that for developing their oral fluency.

Ultimately, this could drive a positive potential and effect for these learners that they have become more active and better able in ESL listening/speaking classes i.e., fluent as it can be seen in (Table 1), where the quantitative results have shown the actual participants' development of oral fluency. The outcomes of this research are in line with previous research revealing that listening while providing control (i.e., narrow listening) of the learning process stimulates students to speak and helps to improve their speaking skills [81]. That said, this study recommends narrow listening for its significant value as a source of [comprehensible input] beyond listening abilities per se for the development of oral fluency, performing a dialogue and free production [30]. Finally, findings of this study support previous claims and the strong belief about the use of captioned videos in L2 learning as a supplement to teaching lexical chunks explicitly [82], and our belief about the use of captioned videos as an aid to make L2 input much more comprehensible and chunkable.

7 Limitations

First, given that the small sample size is a limitation in this study, our research findings should be taken with some caution although they seem to be positive and significant, confirming nine of the seventeen speech proficiency (i.e.,fluency) measurements used in this study. Despite that these findings might seem generally promising, further research and more data are suggested to validate such findings in future studies. Second, due to the learners' limited attentional capacity [83], it was not realistic to focus on speaking fluency and accuracy at once. Therefore, the meaning could be mainly focused on during the speaking tasks, and then the language forms may be emphasized afterwards. Third, the effectiveness of the comprehensible input relates to major concerns over the uneasy task to identify each learner's L2 level. This can be equally challenging (if not impossible) to be determined for ESL teachers and learners and thus provide the right L2 input accordingly. In this regard, individual differences need to be entertained as the application of rules or techniques in any language classroom will likely be limited owing to these differences [84].

8 Concluding Remarks and Recommendations

The use of ICT-based multimedia learning tools like YouTube captioned-videos could make (new) L2 aural input not only more comprehensible but also more chunkable for learners as they narrowly listened to it, which may very likely have resulted in a more fluent L2 speech or oral output due to the increase of learners' L2 intake which could have accrued over time, given that these learners are engaging with and frequently exposed to such L2 multi-mediated input. It was concluded that ICT tools like YouTube can be helpful for ESL learners and thus recommended for optimization of speaking fluency. Therefore, instructors may encourage the use of ICTs in and out of ESL classrooms due to their multimedia/cognitive advantages for the development of ESL receptive and productive skills such as listening and speaking, respectively. Also, instructors/learners are encouraged to increase time on task by narrowly listening to captioned videos at leisure or while moving around by bus or sitting in the park, etc. Learners need to focus on the context of speech to remember vocabulary words, language chunks and the various tones; this may well induce the increase of L2 intake. Learners need to maximize opportunities for interaction while using the newly learnt L2 input in relevant communicative situations; this is the key towards fluency [76]. Finally, to enhance L2 students' learning effectiveness, especially while using ICT-based multimedia learning tools like YouTube, language instructors should guide their students on how to select and leverage appropriate materials for themselves.

Acknowledgements The author acknowledges that the abstract in its original form was presented at the Early Careers 15th LangUE Conference in Language and Linguistics at the University of Essex, UK, 10th June 2021. By extension, this extended abstract was presented at the 6th Saarbrücken

International Conference on Foreign Language Teaching at Saarland University of Applied Sciences, Germany, 27–29 October 2021.

Appendix A

Questions 4–10 Related to the Rate of Learners' Engagement with ICT-Based Learning/Online

Q.4 Rate the extent to which reading books online (Mono-medium: text/speech) made or did not make learning and improving English speaking engaging.

1. Not engaging
2. Fairly engaging
3. Engaging
4. Very engaging
5. Extremely engaging

Q.5 Rate the extent to which listening to songs online (Mono-medium: text/speech) made or did not make learning and improving English speaking engaging.

1. Not engaging
2. Fairly engaging
3. Engaging
4. Very engaging
5. Extremely engaging

Q.6 Rate the extent to which watching YouTube BBC 6 min English online (Multi-media: text and speech) made or did not make learning and improving English speaking engaging.

1. Not engaging
2. Fairly engaging
3. Engaging
4. Very engaging
5. Extremely engaging

Q.7 Rate the extent to which video games online (Multimedia: text and speech) made or did not make learning and improving English speaking engaging.

1. Not engaging
2. Fairly engaging
3. Engaging
4. Very engaging
5. Extremely engaging

Q.8 Rate the extent to which watching songs online (Multimedia: text and speech) made or did not make learning and improving English speaking engaging.

1. Not engaging
2. Fairly engaging
3. Engaging
4. Very engaging
5. Extremely engaging

Q.9 Rate the extent to which watching films online (Multimedia: text and speech) made or did not make learning and improving English speaking engaging.

1. Not engaging
2. Fairly engaging
3. Engaging
4. Very engaging
5. Extremely engaging

Q.10 Rate the extent to which reading audio books online (Multimedia: text and speech) made or did not make learning and improving English speaking engaging.

1. Not engaging
2. Fairly engaging
3. Engaging
4. Very engaging
5. Extremely engaging

Questions 11–17 Related to the Time Range of Learners' Exposure to ICT-Based Learning/Online

Q.11 How much time did you spend on reading books online (Mono-medium: text/speech) per day for learning and improving English speaking?

1. 0 h
2. 1 h
3. 2 h
4. 3 h
5. 4 h +

Q.12 How much time did you spend on listening to songs online (Mono-medium: text/speech) per day for learning and improving English speaking?

1. 0 h
2. 1 h
3. 2 h
4. 3 h
5. 4 h +

Q.13 How much time did you spend on watching YouTube BBC 6 min English online (Multimedia: text and speech) per day for learning and improving English speaking?

1. 0 h
2. 1 h
3. 2 h
4. 3 h
5. 4 h +

Q.14 How much time did you spend on video games online (Multimedia: text and speech) per day for learning and improving English speaking?

1. 0 h
2. 1 h
3. 2 h
4. 3 h
5. 4 h +

Q.15 How much time did you spend on watching songs online (Multimedia: text and speech) per day for learning and improving English speaking?

1. 0 h
2. 1 h
3. 2 h
4. 3 h
5. 4 h +

Q.16 How much time did you spend on watching films online (Multimedia: text and speech) per day for learning and improving English speaking?

1. 0 h
2. 1 h
3. 2 h
4. 3 h
5. 4 h +

Q.17 How much time did you spend on reading audio books online (Multimedia: text and speech) per day for learning and improving English speaking?

1. 0 h
2. 1 h
3. 2 h
4. 3 h
5. 4 h+

Questions 18–23 Related to the Rate of Learners' Engagement with Non-ICT-Based Learning/Offline

Q.18 Rate the extent to which reading books offline (Mono-medium: text/speech) made or did not make learning and improving English speaking engaging.

1. Not engaging
2. Fairly engaging

3.   Engaging
4.   Very engaging
5.   Extremely engaging

Q.19 Rate the extent to which listening to songs offline (Mono-medium: text/speech) made or did not make learning and improving English speaking engaging.

1.   Not engaging
2.   Fairly engaging
3.   Engaging
4.   Very engaging
5.   Extremely engaging

Q.20 Rate the extent to which video games offline (Mono-medium: text/speech) made or did not make learning and improving English speaking engaging.

1.   Not engaging
2.   Fairly engaging
3.   Engaging
4.   Very engaging
5.   Extremely engaging

Q.21 Rate the extent to which formal schooling offline (Multimedia: text and speech) made or did not make learning and improving English speaking engaging.

1.   Not engaging
2.   Fairly engaging
3.   Engaging
4.   Very engaging
5.   Extremely engaging

Q.22 Rate the extent to which watching films offline (Multimedia: text and speech) made or did not make learning and improving English speaking engaging.

1.   Not engaging
2.   Fairly engaging
3.   Engaging
4.   Very engaging
5.   Extremely engaging

Q.23 Rate the extent to which reading audio books offline (Multimedia: text and speech) made or did not make learning and improving English speaking engaging.

1.   Not engaging
2.   Fairly engaging
3.   Engaging
4.   Very engaging
5.   Extremely engaging

Questions 24–29 Related to the Time Range of Learners' Exposure to Non-ICT-Based Learning/Offline

Q.24 How much time did you spend on reading books offline (Mono-medium: text/speech) per day for learning and improving English speaking?

1. 0 h
2. 1 h
3. 2 h
4. 3 h
5. 4h +

Q.25 How much time did you spend on listening to songs offline (Mono-medium: text/speech) per day for learning and improving English speaking?

1. 0 h
2. 1 h
3. 2 h
4. 3 h
5. 4h +

Q.26 How much time did you spend on ideo games offline (Mono-medium: text/speech) per day for learning and improving English speaking?

1. 0 h
2. 1 h
3. 2 h
4. 3 h
5. 4 h +

Q.27 How much time did you spend on formal schooling offline (Multimedia: text and speech) per day for learning and improving English speaking?

1. 0 h
2. 1 h
3. 2 h
4. 3 h
5. 4 h+

Q.28 How much time did you spend on watching films offline (Multimedia: text and speech) per day for learning and improving English speaking?

1. 0 h
2. 1 h
3. 2 h
4. 3 h
5. 4 h +

Q.29 How much time did you spend on reading audio books offline (Multimedia: text and speech) per day for learning and improving English speaking?

1. 0 h
2. 1 h
3. 2 h
4. 3 h
5. 4 h +

References [85, 86] are given in the list but not cited in the text. Please cite them in text or delete them from the list.dear editor(s), kindly check my response to your query in text and bibliography

References

1. Zhou, Z. (2018). Second language learning in the technology-mediated environments. *Asian Education Studies, 3*, 18. https://doi.org/10.20849/aes.v3i1.307.
2. Alobaid, A. (2021). ICT multimedia learning affordances: role and impact on ESL learners' writing accu racy development. *Heliyon, 7*(7), e07517.
3. Mayer, R. E. (2005). Cognitive theory of multimedia learning. *The Cambridge Handbook of Multimedia Learning, 41*, 31–48.
4. Guan, N., Song, J., & Li, D. (2018). On the advantages of computer multimedia-aided English teaching. *Procedia Computer Science, 131*, 727–732.
5. Alobaid, A. (2020). Smart multimedia learning of ICT: Role and impact on language learners' writing fluency—YouTube online English learning resources as an example. *Smart Learning Environment, 7*, 24. https://doi.org/10.1186/s40561-020-00134-7
6. Olasina, G. (2017). An evaluation of educational values of YouTube videos for academic writing. *The African Journal of Information Systems, 9*(4), 2.
7. Kabooha, R., & Elyas, T. (2018). The effects of YouTube in multimedia instruction for vocabulary learning: Perceptions of EFL students and teachers. *English Language Teaching, 11*(2), 72–81.
8. Zhou, Q., Lee, C. S., Sin, S. C. J., Lin, S., Hu, H., & Ismail, M. F. F. B. (2020). Understanding the use of YouTube as a learning resource: A social cognitive perspective. *ASLIB Journal of Information Management.*
9. Winke, P., Gass, S., & Sydorenko, T. (2010). *The effects of captioning videos used for foreign language listening activities.* Michigan State University. Language Learning & Technology.
10. Kovacs, G., & Miller, R. C. (2014). Smart subtitles for vocabulary learning. In *Proceedings of the SIGCHI Conference on Human Factors in Computing Systems* (pp. 853–862).
11. Alvarez-Marinelli, H., Blanco, M., Lara-Alecio, R., Irby, B. J., Tong, F., Stanley, K., & Fan, Y. (2016). Computer assisted English language learning in Costa Rican elementary schools: An ex Perimental study. *Computer Assisted Language Learning, 29*(1), 103–126.
12. Kelsen, B. (2009). Teaching EFL to the iGeneration: A survey of using YouTube as supplementary material with college EFL students in Taiwan. *Call-EJ Online, 10*(2), 1–18.
13. Malhiwsky, D. R. (2010). Student achievement using web 2.0 technologies: A mixed methods study. *Open Access Theses and Dissertations from the College of Education and Human Sciences, 58.*
14. Krashen, S. D. (1985). *The input hypothesis: Issues and implications.* Addison-Wesley Longman Limited.
15. Sadiku, A. (2018). The role of subtitled movies on students' vocabulary development. *International Journal of Sciences: Basic and Applied Research, 42*, 212–221.
16. Krashen, S. D. (1996). The case for narrow listening. *System, 24*(1), 97–100.
17. Dupuy, B. C. (1999). Narrow listening: An alternative way to develop and enhance listening com prehension in students of French as a foreign language. *System, 27*(3), 351–361.

18. Borrás, I., & Lafayette, R. C. (1994). Effects of multimedia courseware subtitling on the speaking performance of college students of French. *The Modern Language Journal, 78*(1), 61–75.
19. Lewis, M. (1993). *The lexical approach: The state of ELT and a way forward*. Language Teaching Publications.
20. Skehan, P. (1996). Second language acquisition research and task-based instruction. In J. Willis, & D. Willis (Eds.), *Challenge and change in language teaching* (pp. 17–30). Oxford: Heinemann.
21. Goh, C. C. M. (2003). Speech as a psycholinguistic process: The missing link in oral lessons. *REACT, 22*(1), 31–41.
22. McLaughlin, B., & Heredia, R. (1996). Information-processing approaches to research on second language acquisition and use. In W. C. Ritchie, & T. K. Bhatia (Eds.), *Handbook of second language acquisition* (pp. 213–228). San Diego: Academic Press.
23. Wood, D. (2004). An empirical investigation into the facilitating role of automatized lexical phrases in second language fluency development. *Journal of Language and Learning, 2*(1), 27–50.
24. Zhou, J. Y., & Wang, X. F. (2007). Chunking-an effective approach to vocabulary teaching and learning in college classrooms. *CELEA Journal, 30*(3), 79–84.
25. Wang, Z. (2014). Developing accuracy and fluency in spoken English of Chinese EFL learn ERS. *English Language Teaching, 7*(2), 110–118.
26. Arevart, S., & Nation, P. (1991). Fluency improvement in a second language. *RELC journal, 22*(1), 84–94.
27. Ellis, R., & Barkhuizen, G. P. (2005). *Analysing learner language*. Oxford University Press.
28. McCarthy, M. (2005). Fluency and confluence: What fluent speakers do. *The Language Teacher, 29*(6), 26–38.
29. Zafarghandi, M., Tahriri, A., & Dobahri Bandari, M. (2015). The impact of teaching chunks on speaking fluency of Iranian EFL learners. *Iranian Journal of English for Academic Purposes, 4*(1), 46–59.
30. Tsang, A. (2019). Effects of narrow listening on ESL learners' pronunciation and fluency: An 'MP3 flood'programme turning mundane homework into an engaging hobby. *Language Teaching Research, 1362168819894487*.
31. Demir, S. (2017). An evaluation of oral language: The relationship between listening, speaking and self-efficacy. *Universal Journal of Educational Research, 5*(9), 1457–1467.
32. Chang, A. C. (2017). Narrow listening: A subset of extensive listening. *TLEMC (Teaching and Learning English in Multicultural Contexts), 1*(1).
33. Winitz, H., & Reeds, J.: *Comprehension and problem solving as strategies for language train ing*. De Gruyter Mouton.
34. Shahini, G., & Shahamirian, F. (2017). Improving English speaking fluency: The role of six factors. *Advances in Language and Literary Studies, 8*(6), 100–104.
35. Chambers, F. (1997). What do we mean by fluency? *System, 25*(4), 535–544.
36. Gorkaltseva, E., Gozhin, A., & Nagel, O. (2015). Enhancing oral fluency as a linguodidactic issue. *Procedia—Social and Behavioral Sciences, 206*, 141–147.
37. Mo, S. (2011). Evidence on instructional technology and student engagement in an auditing course. *Academy of Educational Leadership Journal, 15*(4), 149.
38. Schindler, L. A., Burkholder, G. J., Morad, O. A., & Marsh, C. (2017). Computer-based technology and student engagement: A critical review of the literature. *International Journal of Educa tional Technology in Higher Education, 14*(1), 25.
39. D'Angelo, C. (2018). The impact of technology: Student engagement and success. *Technology and the Curriculum: Summer 2018*.
40. Butcher, K. R. (2014). The multimedia principle. *The Cambridge Handbook of Multimedia Learning, 2*, 174–205.
41. Kanellopoulou, C., Kermanidis, & Giannakoulopoulos, A. (2019). The dual-coding and multimedia learning theories: Film subtitles as a vocabulary teaching tool. *Education Sciences, 9*, 210. https://doi.org/10.3390/educsci9030210.

42. Mayer, R. E. (2014). The Cambridge handbook of multimedia learning (2nd edn), *Cambridge handbooks in psychology*. Cambridge: Cambridge University Press.
43. Sweller, J. (2005). Implications of cognitive load theory for multimedia learning. In R. Mayer (Ed.), *The Cambridge handbook of multimedia learning* (pp. 19–30). Cambridge, UK: Cambridge University Press.
44. Plass, J. L., & Jones, L. (2005). Multimedia learning in second language acquisition. *The Cambridge handbook of multimedia learning* (pp. 467–488).
45. Lewis, M., Gough, C., Martínez, R., Powell, M., Marks, J., Woolard, G. C., & Ribisch, K. H. (1997). *Implementing the lexical approach: Putting theory into practice* (Vol. 3, No. 1, pp. 223–232). Hove: Language Teaching Publications.
46. Fountain S. B., & Doyle K. E. (2012). Learning by Chunking. In: N. M. Seel (Eds.), *Encyclopedia of the sciences of learning*. Boston, MA: Springer.
47. Pawley, A., & Syder, F. (1983). Two puzzles for linguistic theory: Nativelike selection and nativelike fluency. In J. Richards & R. Schmidt (Eds.), *Language and communication* (pp. 191–227). Longman.
48. Conklin, K., & Schmitt, N. (2008). Formulaic sequences: Are they processed more quickly than nonformulaic language by native and nonnative speakers? *Applied linguistics, 29*(1), 72–89.
49. McCarthy, M., & Carter, R. (2002). This that and the other: Multiword clusters in spoken English as visible patterns of interaction. *The Irish Yearbook of Applied Linguistics, 21*, 30–52.
50. Boers, F., Eyckmans, J., Kappel, J., Stengers, H., & Demecheleer, M. (2006). Formulaic sequences and perceived oral proficiency: Putting a lexical approach to the test. *Language Teaching Research, 10*(3), 245–261.
51. Ellis, N. C. (2003). Constructions, chunking, and connectionism: The emergence of second language structure. In C. J. Doughty & M. H. Long (Eds.), *The handbook of second language acquisition*. Malden, MA: Blackwell.
52. Zmyslinski-Seelig, A. N. (2017). Related measures. In M. Allen (Ed.), *The SAGE encyclopedia of communication research methods*. Sage Publications. https://www.bit.ly/2Qbm1zW.
53. Dörnyei, Z., & Csizér, K. (2012). How to design and analyze surveys in second language acquisition research. *Research Methods in Second Language Acquisition: A Practical Guide, 1*, 74–94.
54. Kormos, J. (2014). *Speech production and second language acquisition*. Routledge.
55. Riggenbach, H. (1991). Toward an understanding of fluency: A microanalysis of nonnative speaker conversations. *Discourse Processes, 14*(4), 423–441.
56. Peltonen, P., & Lintunen, P. (2016). Integrating quantitative and qualitative approaches in L2 fluency analysis: A study of Finnish-speaking and Swedish-speaking learners of English at two school levels. *European Journal of Applied Linguistics, 4*(2), 209–238.
57. Préfontaine, Y., & Kormos, J. (2016). A qualitative analysis of perceptions of fluency in second language French. *International Review of Applied Linguistics in Language Teaching, 54*(2), 151–169.
58. Takiguchi, H. (2003). A study of the development of speaking skills within the framework of flu ency, accuracy and complexity among Japanese EFL junior high school students. Un published master's thesis, Joetsu University of Education.
59. McKee, G., Malvern, D., & Richards, B. (2000). Measuring vocabulary diversity using dedicated 469 software. *Literary and Linguistic Computing, 15*(3), 323–338.
60. McCarthy, P. M., & Jarvis, S. (2010). MTLD, vocd-D, and HD-D: A validation study of sophisti cated approaches to lexical diversity assessment. *Behavior Research Methods, 42*(2), 381–392.
61. Ure, J. (1971). Lexical density and register differentiation. In G. Perren, J. L. M. Trim (Eds.), *Applications of linguistics. Selected papers of the second international congress of applied linguistics*, Cambridge 1969.
62. Szmrecsanyi, B. (2004). On operationalizing syntactic complexity. In *Le poids des mots. Proceedings of the 7th international conference on textual data statistical analysis*. Louvain-la-Neuve (Vol. 2, pp. 1032–1039).
63. Larsen-Freeman, D. (2006). The emergence of complexity, fluency, and accuracy in the oral and written production of five Chinese learners of English. *Applied Linguistics, 27*(4), 590–619.

64. Brophy, J. (1983). Conceptualizing student motivation. *Educational Psychologist Journal, 18*, 200–215.
65. Connell, J. P. (1990). Context, self, and action: A motivational analysis of self-system processes across the life span. *The Self in Transition: Infancy to Childhood, 8*, 61–97.
66. British Council, https://www.teachingenglish.org.uk/article/exposure. Accessed 15 May 2021.
67. Scheff, S. W. (2016). *Fundamental statistical principles for the neurobiologist: A survival guide.* Academic Press.
68. Zar, J. H. (1984). *Biostatistical analysis* (2nd ed.). Englewood Cliffs: Prentice-Hall.
69. Rodrigo, V. (2003). Narrow listening and audio library: The transitional stage in the process of de veloping listening comprehension in a foreign language. *Mextesol Journal: Mexican Association for English Teachers, 27*, 11–28.
70. Cohen, J. (2013). *Statistical power analysis for the behavioral sciences.* Academic Press.
71. Nation, I. S. P. (2005). Teaching and learning vocabulary. In *Handbook of research in second language teaching and learning* (pp. 605–620). Routledge.
72. Carter, R. (2001). Vocabulary. In R. Carter, & D. Nunan (Eds.), *The Cambridge guide to teaching English to speakers of other languages* (pp. 42–47). Cambridge: Cambridge University Press. https://doi.org/10.1017/CBO9780511667206.007.
73. Izquierdo, J., Simard, D., & Garza, M. G. (2015). Multimedia instruction & language learning atti tudes: a study with university students. *Revista Electrónica de Investigación Educativa, 17*(2), 101–115. http://redie.uabc.mx/vol17no2/contents-izqsimard.html.
74. Pujolà, J. T. (2002). CALLing for help: Researching language learning strategies using help facilities in a web-based multimedia.
75. Talaván, N.: Subtitling as a task and subtitles as support: Pedagogical applications. In *New insights into audiovisual translation and media accessibility. Brill Rodopi.*
76. Zhang, S. (2009). The role of input, interaction and output in the development of oral fluency. *English Language Teaching, 2*(4), 91–100.
77. Vanderplank, R. (1993). A very verbal medium: Language learning through closed captions. *TESOL Journal, 3*(1), 10–14.
78. Vanderplank, R. (1990). Paying attention to the words: Practical and theoretical problems in watch ing television programmes with uni-lingual (CEEFAX) sub-titles. *System, 18*(2), 221–234.
79. Winke, P., Gass, S., & Sydorenko, T. (2013). Factors influencing the use of captions by foreign language learners: An eye-tracking study. *The Modern Language Journal, 97*(1), 254–275.
80. Alzaneen, R., & Mahmoud, A. (2019). The role of management information systems in strengthening the administrative governance in ministry of education and higher education in Gaza. *International Journal of Business Ethics and Governance, 2*(3), 1–43. https://doi.org/10.51325/ijbeg.v2i3.44
81. Kondreteva, İ. G., Safina, M. S., & Valeev, A. A. (2016). Listening as a method of learning a foreign language at the non-language faculty of the university. *International Journal of Environmental & Science Education, 11*(6), 1049–1058.
82. Teng, (Mark) F. (2019). The effects of video caption types and advance organizers on incidental L2 collocation learning. *Computers & Education, 142.*
83. Willis, J. (2005). Introduction: Aims and explorations into tasks and task-based teaching. In C. Edwards, & J. Willis (Eds.), *Teachers exploring tasks in English language teaching* (pp. 1–12). Hampshire: Palgrave Macmillan.
84. Alahmadi, N. S. (2019). The role of input in second language acquisition: An overview of four theories. *Bulletin of Advanced English Studies (BAES), 3*(2), 70.
85. Text Inspector, https://www.textinspector.com. Accessed 20 Jan 2021.
86. Textalyzer, https://seoscout.com/tools/text-analyzer. Accessed 06 Jan 2022.

A Comprehensive Review of Blockchain Technology and Its Related Aspects in Higher Education

Bahaa Razia and Bahaa Awwad

1 Introduction

The use of developing technologies, such as the Internet and the World Wide Web, has improved significantly in the higher education sector. Web-based apps are widely used to improve connectivity, share knowledge, increase collaboration, and stimulate active learning. The use of Blockchain in education is still new, and the number of blockchain-based solutions available is currently restricted. Nonetheless, blockchain technology has the potential to give a number of advantages and opportunities. There are anticipated to be 2.7 billion students globally by 2025, with little more than 500 million already enrolled [1]. Blockchain technology was first developed in 2008 for the Bitcoin digital payment system [2]. This innovative technology has been developed and applied in a variety of companies, educational sectors, and research organizations around the world, and it has become the topic of extensive research in a variety of fields [3–5]. The basic goal of Blockchain is to eliminate the need for a "trusted" central authority to mediate transactions between diverse parties. In the blockchain, centralisation is seen as a critical factor that could lead to security issues, as well as cost and other issues such as being a single point of failure. The decentralised nature of blockchain technology, on the other hand, can help to raise the level of trust between parties in a system and help to eliminate the need for a trusted third or external party to conduct transactions between them. Each transaction that occurs between several parties can be stored as a distributed ledger using blockchain technology. The digital education technologies make it possible to assess how well a student has learned new knowledge and abilities, as well as to quickly rectify the learning process; education becomes more customised and adaptable [6] (Fig. 1).

Smart agreements (e.g., contracts) between several parties (executable code) can also be executed using blockchain technology [7]. These smart contracts have the

B. Razia · B. Awwad (✉)
Palestine Technical University - Kadoorie, Tulkrem, Palestine

© The Author(s), under exclusive license to Springer Nature Switzerland AG 2022   553
A. Hamdan et al. (eds.), *Technologies, Artificial Intelligence and the Future of Learning Post-COVID-19*, Studies in Computational Intelligence 1019,
https://doi.org/10.1007/978-3-030-93921-2_29

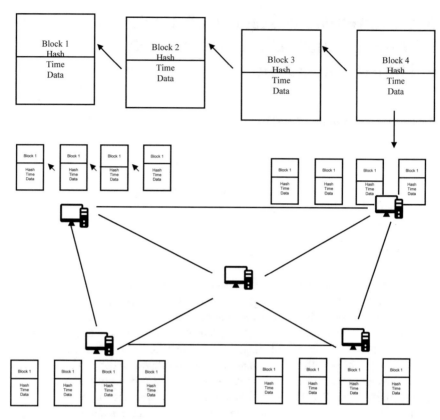

Fig. 1 Key elements of blockchain network

potential to improve the efficacy of blockchain solutions and enable the development of distributed apps for a variety of objectives in a variety of industries. Smart contracts can be utilised in the education sector to provide flexible distributed solutions that benefit all members of an online learning system, including teachers, administrators, and students.

Students and educational institutions, for example, may be able to engage into more individualised digital agreements that include deadlines, assignment requirements, and grading criteria [4]. There are certain characteristics of blockchain technology that make it worthwhile to investigate more in order to improve educational sectors. Immutability, reliability, availability, and trust are some of these characteristics. Immutability: The data recorded on the blockchain is deemed tamper-proof due to the cryptography that secures and ties blocks together, as well as the chronological order in which data is kept. Reliability: The network's decentralised design assures that it is more efficient in terms of service than centralised networks. There is no central authority, and it is possible that it will fail. Information transparency is becoming increasingly important. Highly transparent, decentralised data storage can be established using blockchain technology. Data is shared more efficiently, stored closer together, and retrieved readily by data owners because to the distributed nature of blockchain technology. Trust: In this case, blockchain technology can eliminate the requirement for third-party service providers to facilitate communication between parties. As a result, the primary goal of this study is to examine the research aspects that have been examined in the domain of education in terms of smart contract and blockchain technology, as well as to identify critical issues that need to be investigated in future projects. As a result, in order to reach this goal, a systematic analysis strategy is chosen [7]. Several academic papers and articles were carefully examined using the scientific database in order to determine the most relevant studies linked to blockchain applications in the education sector. These procedures aid in the gathering of useful data that will be utilised to better understand the potential obstacles and issues associated with blockchain technology in future research.

2 Literature Review

2.1 Blockchain Technology Overview

This section contains information on blockchain technology and its associated aspects. Furthermore, this part includes examples of existing educational blockchain projects as well as a basic description of smart contracts. Blockchain technology is defined as a distributed database that preserves transactions between diverse parties in an immutable, safe, and secure manner. Blockchain is a peer-to-peer (P2P) network that allows peers (nodes) to collaborate to administer the block and transaction exchange network.

All users in the peer-to-peer network must store their own data on the blockchain while synchronising all of their blocks with other users' data using a consensus process [8] (Filatov 2020). The consensus is defined by the longest chain that is accepted by the majority of peer nodes. As a result, a third trusted party isn't required because the participants can interact and send transactions directly among themselves.

A cryptographic hash identifies each block, and each block refers to the previous block in the chain, forming a blockchain. Each block contains many transactions, and the maximum size of each block varies depending on the blockchain platform. The blocks in the chain cannot be modified and are immutable, preventing double-spending [9], [10, 11].

The first blockchain generation was cryptocurrency, which is a digital money based on peer-to-peer networks and cryptography. One of the ideas included into blockchain technology is mining, which is the practice of linking transaction data to the public ledger of prior blockchain transactions. The mining process requires a miner on the blockchain network to gather transactions, generate a new transaction block, and perform mathematical methods to validate (e.g., verify) the new block and link it to the chain of previous blocks (Fig. 2).

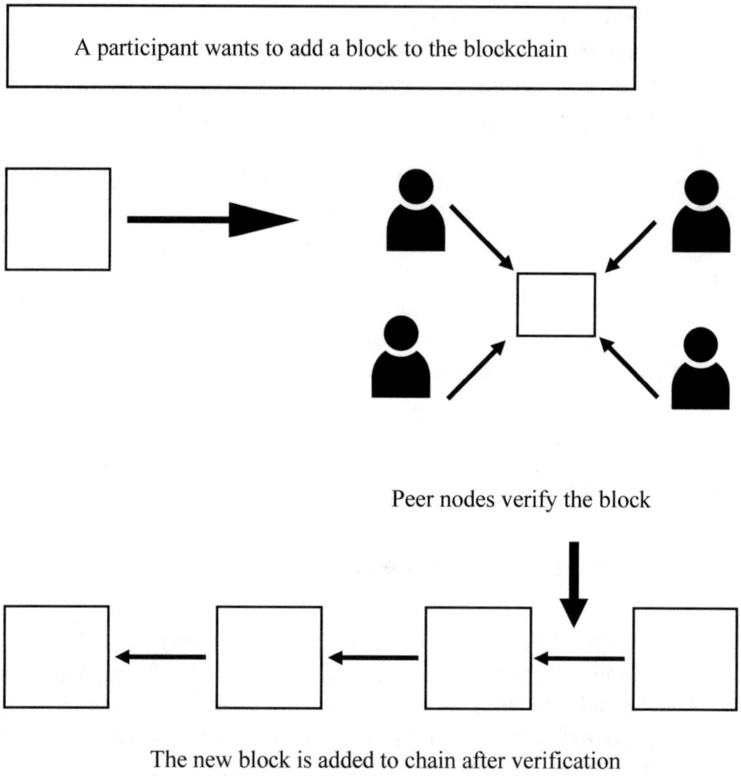

A participant wants to add a block to the blockchain

Peer nodes verify the block

The new block is added to chain after verification

Fig. 2 After verification, a new block is added to the blockchain

The freshly produced blocks will be tested and verified by the other miners' nodes in the blockchain network using a consensus process. The second blockchain generation, according to [4], has manifested itself in the form of Ethereum. This makes it easier to build and implement distributed applications. The Ethereum blockchain allows smart contracts to be built on top of it, which has given researchers and practitioners a wonderful opportunity to incorporate blockchain characteristics into other industries. In general, there are two types of blockchain: private and public blockchain [4, 11]. Participants in a public blockchain (for example, Ethereum) are able to access and communicate in the network. In contrast, only members with rights can join and communicate in a private blockchain network (such as Ripple).

2.2 Smart Contracts

A smart contract is a stateful computer program comprising event-condition-action that is executed by two or more parties who do not have tacit trust in each other

[12]. To clarify, it is a self-executed code used to impose conditions and roles between two or more players [2], [13]. The implementation of a smart contract using blockchain-related technology not only reduces third-party expenses during the transaction phase, but it also improves transaction stability and security. A smart contract can be either centralised or decentralised, and can be deployed in either a centralised or decentralised setting to operate off-chain or on blockchain [12], [14].

2.3 Blockchain Technology and Its Potential Applications in Higher Education

In recent years, the role of blockchain technology and its applications in the education sector has gotten a lot of attention across a lot of fields. Blockchain and smart contract technologies, as well as its connected elements, are heavily involved in a variety of disciplines, including education, in various ways and shapes. However, until now, education blockchains have mostly been used to store certificates and grades, with little emphasis paid to the use of smart contracts and blockchains to build a learning process infrastructure. The following are the most recent smart contract and blockchain uses in education. Other applications for blockchain and smart contracts in education include distributed file storage, online learning, student evaluations, payments, financing, digital rights protection, and identity management (see Table 1).

Table 1 Smart contracts and blockchain technology applications in education

Fields	Explanation	Example
Digital certificate	These applications are designed to have better control on the certificates received by students and to minimise dependency on needing third party intermediaries (e.g., universities and employers) for holding, checking and validating qualifications and credentials of students	Open blockchain [4] and the Blockcerts project [15]
Support services	These applications are intended to create a special Bitcoin-based cryptocurrency for the regulation of the educational products and services industry, (e.g. support services, participation in online courses, regulated studies)	Edgecoin [16]
Earnings	Such implementations connect learning to earnings. The blockchain technology is used in this case to store the studying hours or teaching and not the digital currency	The Ledger project [17]

2.4 Covid-19 and Its Impact

Students' learning will be severely disrupted, assessments will be disrupted, and public tests will be cancelled or substituted for qualifications as a result of the global lockdown of educational institutions. Education is moving online on an unprecedented and unproven scale. Because student assessments are now available online, there is a lot of trial and error and uncertainty for everyone involved. Many assessments have been postponed, while others have been canceled. Importantly, these disruptions will not only be a short-term problem; they will also have long-term consequences for the affected cohorts, resulting in increased inequality. In higher education, several institutions and colleges are replacing traditional assessments with online grading systems. Assessments will almost probably have a bigger measurement error than regular times because this is a new area for both teachers and pupils. Employers rank candidates according on their educational qualifications, such as degree categories and grade point averages [18]. The matching efficiency for new graduates in the job market may be diminished as a result of increasing noise in candidates' signals, resulting in slower pay growth and higher job separation rates. This is costly for both the individual and society as a whole (Fredriksson et al. 2018).

3 Research Methodology

In order to reach the research's goal and explore various uses of blockchain, particularly in the education sector, a systematic mapping approach was developed and utilised in this study [19]. Below is a detailed description of the method employed (Fig. 3). A systematic mapping study entails a number of steps that include locating, categorising, and analyzing previous relevant studies on the research topic (Eaganathan et al. 2019).

Defining research objectives, examining relevant articles, scanning publications, detecting key phrases in abstracts, data extraction and classifications, and finally data analysis and mapping are the main procedural steps of the systematic mapping

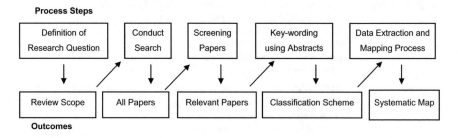

Fig. 3 The systematic mapping procedure

approach. Each procedure has a result that leads to the final result, which aids in the creation of the systematic map (see Fig. 3).

3.1 Definition of Research Questions (Scope of the Research)

The main goal of a thorough mapping analysis is to describe an area of study and identify the quantity, kind, and findings of research that are accessible within it. A additional goal may be to explain the venues where the field's literature has been written. Table 1 shows the questions that have been identified for this study in order to fulfil the research goal.

3.2 Conducting Primary Study Research

Using search terms and perusing relevant journal publications and conference proceedings, a large number of papers and articles were identified in this phase of the process. It is critical to arrange search strings in terms of comparison, demographics, result, and intervention in order to produce successful search strings [20]. To help restrict the scope of the study, terms like "college" and "blockchain" were utilised. In addition, well-known scientific databases and search engines, including as scopes, springer, science direct, ResearchGate, and science open, were utilised to perform the search. These scientific databases and search engines index high-quality and high-impact publications in the fields of information technology and education. The study was limited to high-quality and prominent journal papers that were discovered in books, periodicals, international journals, seminars, and conference proceedings. Papers are screened for exclusion and inclusion. Exclusion and inclusion criteria are used to eliminate publications that are unrelated to the study questions. As a consequence, this study used the stated search criteria to filter relevant articles and their outcomes in blockchain technology, particularly in the sector of education [21]. The first step is to filter out irrelevant articles and journal publications based on their titles. In the event that the paper's title was ambiguous, the abstract was investigated and examined. Non-English papers, grey literature (e.g., working papers, government documents), duplicates, and publications with no complete text were also eliminated.

3.3 Abstract Key-Wording (Classification Scheme)

Using the keyboarding criteria outlined, all relevant papers and journal publications were categorised at this step [21]. All potential keywords have been selected based on the abstracts of the publications and their contributions. As a result, the image below depicts the methical approach that was used to construct and define the

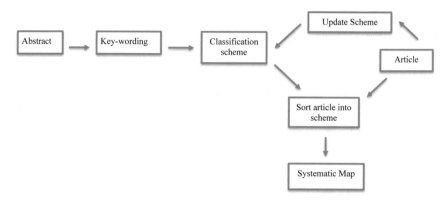

Fig. 4 The development of a categorisation scheme

categorisation scheme, as well as export information. This method aids in decreasing the time necessary to design the categorisation scheme and ensuring that the system takes previous research into account (Fig. 4).

3.4 Data Extraction and Mapping of Selected Studies (Systematic Map)

The real data extraction occurs during this stage. This procedure was designed to gather the information needed to answer the research questions in this study. As a result, Table 2 provides the defined criteria that were used to include specific components for assessing the papers. This criteria have been piloted in a subset of publications before being expanded to analyse the remaining papers in order to extract information. When extracting data, the categorisation scheme begins to evolve, which involves merging and splitting current categories as well as introducing new categories. An Excel table was utilised to document the data extraction procedure. To summarize, the following procedures were used to review the selected papers:

Title is the title of the selected papers or journal publications.
Author(s) refers to the research author's name (s).
Paper types include journal articles, chapters, papers, and conference papers.
Paper topic is related to the paper's subject and topic area.

Table 2 This study's research questions

Number	Research question
1	Which journals include papers on blockchain in the domain of higher education?
2	What are the challenges that face blockchain and its applications in higher education?

The year the work was published is referred to as the publication date.

Publication location refers to the location, nation, or organization of a journal, organization, or conference.

Paper purpose refers to the paper's principal goal.

Application implementation refers to the study's possible implementation.

Challenges refer to the study's potential and actual challenges.

4 Results

This research has resulted in 123 scholarly papers. However, 69 of them were irrelevant and were thus eliminated based on the criteria of the first phase of the screening procedure. There are two reasons why articles are omitted. The first reason is connected to the scope of the research, which is concerned with blockchain technology and the sector of higher education. As a consequence, all items that were not directly related to higher education were ignored. The second reason is that publications that merely addressed broad elements of blockchain and its pertinent parts were excluded. Following that, 22 more papers were discarded as duplicates. As a result, 32 papers were critically evaluated and analyzed in order to achieve a systematic mapping method. The following is a breakdown of these 32 articles by year of publication: In 2017, there were four articles, eleven in 2018, fifteen in 2019, and seven in 2020. All of the articles chosen were published after 2016, indicating that this is a fresh and innovative study topic. It is important to note that the number of published papers on this issue appears to be increasing year after year, indicating that there is a rising interest in the uses of blockchain in higher education and their connected features.

It is evident from Fig. 5 the geographical distribution of the articles chosen and evaluated for this study. This distribution of papers across nine nations demonstrates that worldwide academic attention has been devoted to the implementation of blockchain technology in higher education and its related issues. The most publications are produced in the United States by various colleges or businesses. Japan has the second most papers published in this subject. The remaining pieces were published in different countries.

Figure 6 shows that the bulk of these selected papers (46%) have been published in journal articles. 34% were published in conference proceedings, 14% were published in books, and the remaining 6% were published in periodicals. The great majority of papers in this research are from the Institute of Electrical and Electronics Engineers. The world's biggest technical professional organization for the advancement of technology (IEEE) provides a diverse variety of high-quality publications that enable technology professionals to share technical skills and information. Other publications dealt with technical and pedagogical issues. Table 3 summarizes the potential obstacles that may be encountered while using blockchain technology in higher education, particularly in universities.

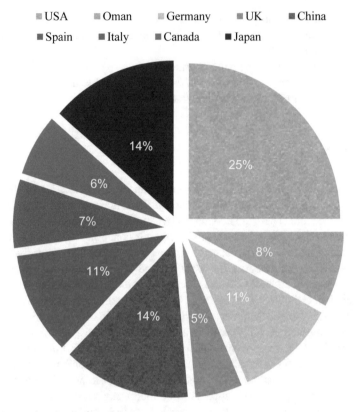

Fig. 5 Geographic distribution of chosen articles

Fig. 6 Publications are distributed based on their kind

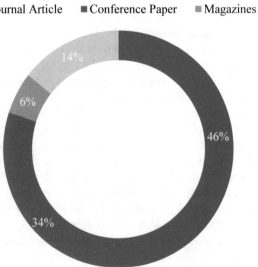

Table 3 The use of blockchain technology in education

Number	Value proposition	Example	References
1	Enhancing and motivating lifelong learning	BitDegree (tracking educational achievement and rewarding the parties engaged), ODEM (interaction between students and academic professionals), Open source framework	Ghaffar and Hussain [22] and Lam and Dongol [23]
2	Certificates and identity management	Digital Credentials Consortium, Blockcerts, Open Source University	Chen et al. [4], [15]
3	Credit transfer	Storing records and transferring student credit across universities	Shen and Xiao [16]
4	Admissions	Sharing admission procedures to allow students to apply for universities	Mori and Miwa [24]
5	Student exams and assessments	Production of exams and assessments	Gesto and Gohari [25, 26]
6	Parchment	Generating certificates and assessing qualifications and processing programmes	Parchment [27]

Table 4 Blockchain's problems in higher education

Number	Challenge	References
1	Motivation	Ghaffar and Hussain [22], Liu et al. [23], Liyuan et al. [29]
2	Scalability	Forment et al. [21], Juričić et al. [30] and Turkanovic´ et al. [31]
3	Blockchain usability	Han et al. [32], Mitchell et al. [33] and Liu et al. [23]
4	Immutability	Ghaffar and Hussain [22], Mitchell et al. [33] and Vidal et al. [34]
5	Privacy and security	Arndt and Guercio [35] and Han et al. [32]
6	Market option	Mori and Miwa [24, 36]
7	Innovation	Juričić et al. [30], Liu et al. [23] and Shen and Xiao [16]
8	Legal	Finck [37] and Shah et al. [38]

Table 4 represents many blockchain uses, particularly in the higher education sector. In this regard, it has been suggested that many conventional online learning materials dissatisfy students and instructors, which has a negative impact on successfully engaging learners [28]. As a consequence, the OpenSource blockchain architecture was investigated as a future platform that would assist learners in obtaining enhancements to present learning and teaching elements. Another significant sector is associated with a platform known as On-Demand Education Marketplace (ODEM). This platform allows students to engage with academic experts while also allowing and empowering them to trade certified credentials on the blockchain in a safe digital environment. The ODEM platform delivers excellent solutions for a variety of parties, including educational institutions, employers, students, and academic experts, by linking these parties without the need of middlemen. Another example of improving and rewarding lifelong learning is BitDegree, an online network that gives users with learning rewards for completing technical courses or meeting learning objectives, such as tokenised scholarships.

Some of the main benefits for learners in the Digital Credentials Consortium network include: obtaining credentials digitally, not having to pay for transcripts, securely sharing student records, and having access to a streamlined procedure for issuing multiple credentials to a single source of learners. Blockcerts is a platform that allows educational institutions to use blockchain accreditations within their curricula. Learners' journeys on the Blockcerts platform are meant to be simple and brief. Students may maintain, share, and receive credentials in this environment in a safe and efficient manner. Furthermore, adding an open source university platform would help users find future suggested employment opportunities and allow organizations to utilise the site to find qualified individuals.

In terms of exams and assessments, educational institutions may use blockchain technology to manage student assessments and exams. These management methods contribute to the creation of safe information that may be kept in a blockchain database (e.g., student attendance, participation reports, exam score, exam details, assessment evaluation criteria). In this scenario, blockchain technology assures that participants' data is trustworthy and genuine. As a consequence, organizational confidence and trust in the learning and teaching process in higher education will be attained.

In terms of credit transfer, organizations may securely store and transfer individuals' credit between institutions without the use of a third party. Another application is admission, in which colleges assist students and facilitate them to apply to numerous higher education institutions by sharing admissions procedures and associated features. Finally, parchment is regarded as a format that aids in the assessment of credentials, the generation of certificates, and the processing of academy programs. This enables learners, professors, and academic institutions to seek and exchange their credentials, as well as have a debate about academic achievement in a safe and easy setting.

5 Discussion

Based on the publishing trend, the application of blockchain technology for higher education has been addressed with rising interest. Because there have been few studies in this field, there is a need for more study on the potential applications of blockchain innovation in various industries, including education. As a result, 32 research articles were chosen and critically analyzed for this study in order to answer the research objective and research questions. The purpose of this study was to offer a systematic review in order to discover and analyse relevant blockchain technology research related to higher education. The article focuses on two major topics: examining cutting-edge blockchain-based apps created for educational purposes, and summarising the potential challenges and study gaps that require further attention in future initiatives.

RQ1: The first question of this study focused on finding blockchain-related publications and research articles, particularly in the sector of higher education. While several blockchain apps have been developed for educational reasons, just a handful have been used by organizations and stakeholders. These apps are categorised as follows: improving and inspiring lifelong learning, student evaluation, certificate and identity management, review papers, credit transfer, and admissions. The initial theme is associated with a specific application aimed at enhancing and motivating lifelong learning. This subject encompassed all journal papers dealing with lifetime learning records in blockchain, particularly in higher education. This platform enables learners to transfer their learning records from one institution to another in a secure and verified manner. The second subject is student evaluation, and numerous studies have created a platform to use blockchain technology for online coursework and examinations in high school. As a consequence, Ref. [33] created a system called dAppER that may be used to create test papers. This method allows for transparency that is suitable for viewing by external examiners and auditors, as well as some insights into how quality assurance frameworks in general are supplemented by the development of decentralised apps. This technology comprises a robotic quality assurance framework for the creation of research papers and their assessment schedules, as well as blockchain, which secures the checks and maintains the audits' immutable and trustworthy leader. Reference [16] created a system that uses blockchain technology to validate and trace students' replies. The third topic is linked to credit transfer. This blockchain feature enables educational establishments to transfer student credits and finished course data. By introducing a system for storing student papers and transcripts, as well as an electronic credit transfer technique [39]. Students will be able to transfer their finished course and related credits to some other university via this application. Turkanovic et al. [31] have also created EduCTX, a worldwide credit network for higher education that can be utilised for credit accumulation and transfer and is beneficial to both students and universities. The fourth is concerned with certificate and identity management. These apps are intended to provide better control over the certificates acquired by students and to reduce reliance on third-party intermediaries (e.g., universities and employers) for holding, verifying, and certifying students' qualifications and credentials [4]. The sixth topic is associated with the parchment application, which aids in the assessment of credentials, the generation of certificates, and the processing of academy programs. This enables learners, professors, and academic institutions to seek and exchange their credentials, as well as have a debate about academic achievement in a safe and easy setting. The last theme is about admissions registrations. Mori and Miwa [24], for example, established a digital application framework for university admission that aids in the organization of research documents and e-port folioing on a blockchain using smart contracts. Furthermore, [22] presented a framework for HEC and PEC that is utilised to check applicants' records in a short and secure manner, allowing students to apply for university admissions on a single platform.

RQ2: The second question of this study cantered on identifying the obstacles that blockchain technology and its implementation confront, particularly in the realm of higher education. While blockchains have many helpful educational uses, practitioners continue to encounter a variety of issues while utilising this educational technology. As a result, several difficulties were discovered in the papers studied. Motivation, blockchain usability, scalability, security, immutability, legal concerns, market alternatives, and innovation are among the obstacles. The first obstacle is a lack of motivation, which makes it difficult for stakeholders to deploy blockchain technology applications, particularly in the education sector. As a result, more research is needed to help users in adopting blockchain technology and its related applications. As a result, several academics have designed and developed functioning blockchain systems in order to increase the use of this technology in the field of education [22, 29, 40, 41]. The second problem is scalability, which may have an influence on the access latency of blockchain applications owing to an increase in the number of users, assistance, and big data sets. According to Ref. [21], the direction and implications of blockchain technology and its related features are difficult to anticipate, given probable future modifications and new scalability implementation techniques. The third difficulty is the usefulness of blockchain. Because of the technology's ambiguous language, this topic remains a major issue in the field of education. Furthermore, in order to preserve security, practitioners may be required to deal with sophisticated and numerous settings, such as recovery tools and primary keys. It is also essential to remember that adopting a peer-to-peer network has distinct criteria that might make access to blockchains difficult for end users. As a result, the usability of blockchain may be increased through application design interfaces that make it simple for people who lack technological expertise to utilise blockchain [42, 43]. As a result, further research on blockchain usability for individuals is envisaged. Furthermore, by developing simple specifications and implementing well-designed interfaces, the adoption of blockchain technology and its related features may be increased in a variety of areas, particularly education. The fourth problem is connected to immutability, which has been widely employed in blockchains and is regarded as a crucial property of blockchain and its related features. The immutability of the blockchain protects the data by making it impossible for the data recorded to be changed. Nonetheless, immutability is regarded as a key problem when implementing blockchain technology in the sphere of education (e.g., degree revocation). Reference [34] confirmed that educational degrees saved using blockchain technology are safe and cannot be changed owing to the blockchain technology's immutability. The immutability problem is closely related to several areas where student information must be safely documented and maintained. Admissions, evaluations, examinations, and credential verification are examples of these categories. The fifth problem is one of privacy and security, in which student identities must be secured by connecting actual and pseudonymous identities [16]. As a result, it is critical to ensure that the data stored on the network can be used safely by users [32, 44–46]. Other challenges are related to legal issues, where education providers face multiple challenges (e.g., ambiguous legislation, legislator perception of centralised and decentralised data) in ensuring that their products comply with data protection laws, including the General Data Protection Regulation (GDPR). When gathering and analyzing data, decision makers face a market choice issue. This is due to a lack of systematic and independent methods for validating the obtained data. Dealing with legal concerns becomes even more difficult as a result of the difficulty in handling these obstacles. Nonetheless, there are a variety of approaches that education providers may use to find new answers to these issues. For example, it is feasible to restrict or even prohibit personal data from being kept on the blockchain, as well as carefully examine whether the blockchain is used to fulfil the needs of a legal organization or to utilise permissioned blockchains with stricter usage rules.

6 Conclusion

Blockchain is a distributed ledger system that aids in the distribution of consensus methods and utilises cryptographic techniques in terms of traceability, decentralisation, and immutability. Blockchain and smart contracts offer features that can help with a wide range of innovative advances in higher education. Aside from managing certificates and evaluating achievements, blockchain technology may be applied in the sphere of education in a number of novel ways. Blockchain technology has the potential to be utilised by educators and learners in a variety of applications, including the design and implementation of learning activities, formative evaluation, and monitoring the whole learning and teaching process. Despite this, there have been little research on this topic. As a result, investigating the prospects for integrating blockchain technology and smart contracts to increase learner engagement and achievement is difficult.

This study seeks to address this issue by investigating existing studies of blockchain and its related features for education, as well as identifying academic challenges connected with blockchain implementation, particularly in the field of education. The analysis' conclusions have the potential to assist future academics in investigating and coping with new issues. Several blockchain apps have been developed for educational purposes, but just a handful have been used by organizations and stakeholders. These apps are categorised as follows: improving and inspiring lifelong learning, student evaluation, certificate and identity management, review papers, credit transfer, and admissions. While blockchains have many helpful educational uses, practitioners continue to encounter a variety of issues while utilising this educational technology.

As a result, several difficulties were discovered in the papers studied. Motivation, blockchain usability, scalability, security, immutability, legal concerns, market alternatives, and innovation are among the obstacles. In terms of covid-19, blockchain can aid in the development of a future education system that makes greater use of blended learning methods to reach all students at their level and provide more personalised teaching. To extend the use of digitalisation in various industries, including higher education, educational institutions must facilitate learning continuity. This will assist higher education institutions in recovering from the possible effects of Covid-19 by adopting a more flexible, secure, and resilient strategy.

Acknowledgements Special thanks to Palestine Technical University for their continued support.

References

1. Spies, M., & Brothers, P. (2020). Education in 2030. Five scenarios for the future of learning and talent [report] Holoniq.com [Electronic resource]. https://www.holoniq.com/wp-content/uploads/2020/01/HolonIQ-Education-in-2030.pdf.
2. Antonopoulos, A. M. (2017). *Mastering bitcoin: programming the open blockchain,* 2nd edn. O'Reilly Media, Inc.
3. Arndt, T. (2018). Empowering university students with blockchain- based transcripts. In *Cognition and Exploratory Learning in the Digital Age (CELDA)* 399.
4. Chen, G., Xu, B., Lu, M., & Chen, N.-S. (2018). Exploring blockchain technology and its potential applications for education. *Smart Learning Environments, 5,* 1.
5. Walport, M. 2016. Distributed ledger technology: Beyond blockchain. UK Government Office for Science. 1.
6. Rudich, K. (2020). Ed Clark, Universitet Sent-Tomasa: "U segodnyashnikh molodykh lyudey uzhe est' tsifrovaya lichnost' naryadu s real'noy" [Ed Clark, University of St. Thomas: "Present-Day Young People Have Already Obtained a Digital Personality in Addition to Their Real One"]. [Electronic resource]. https://hightech.fm/2020/01/20/ed-clark [in Russian].
7. Buterin, V. (2014). A next-generation smart contract and decentralised application platform. *Etherum,* 1–36.
8. Jirgensons, M., & Kapenieks, J. (2018). Blockchain and the future of digital learning credential assessment and management. *Journal of Teacher Education for Sustainability, 20*(1), 145–156.
9. Chohan, U. W. A., (2017). *History of bitcoin.* Available at SSRN: https://ssrn.com/abstract=3047875 or https://doi.org/10.2139/ssrn.3047875.
10. Andersen, S. C., & Nielsen, H. S. (2019). Learning from performance information. *Journal of Public Administration Research and Theory.*
11. Razia, B., Larkham, P., & Thurairajah, N. (2019). Risk assessment and risk engagement in the construction industry within conflict zones. In *Proceedings of the 35th Annual ARCOM Conference, Leeds, UK.* https://nrl.northumbria.ac.uk/id/eprint/41102/1/RISK_ASSESSMENT_AND_RISK_ENGAGEMENT.pdf.
12. Molina-Jimenez, C., Solaiman, E., Sfyrakis, I., Ng I., & Crowcroft, J. (2019). On and off-blockchain enforcement of smart contracts. In: G. Mencagli et al. (eds) Euro-Par 2018: *Parallel Processing Workshops. Euro-Par 2018. Lecture Notes in Computer Science* (Vol 11339).

13. Rani, S. (2020). *Are centralised cryptocurrency regulations the answer? Three countries; three different directions, 45 Brook.*
14. Solaiman, E., Wike, T., & Sfyrakis, I. (2020). Implementation and evaluation of smart contracts using a hybrid on and off-blockchain architecture. In: *Concurrency and computation: practice and experience.* Wiley Online Library.
15. Curmi, A., & Inguanez, F. (2020). *Blockchain based certificate verification platform.* Institute of Information & Communication Technology, MCAST IICT.
16. Shen, H., & Xiao, Y. (2018) Research on online quiz scheme based on double-layer consortium blockchain. In *2018 9th International Conference on Information Technology in Medicine and Education (ITME),* pp. 956–960.
17. Wood, G. (2014). *Ethereum: A decentralised transaction ledger* (pp.1–10).
18. Piopiunik, M., Schwerdt, G., Simon, L., & Woessman, L. (2020). Skills, signals, and employability: An experimental investigation. *European Economic Review, 123,* 103374.
19. Petersen, K., Feldt, R., Mujtaba, S., & Mattsson, M. (2008). *Systematic Mapping Studies in Software Engineering.*
20. Kitchenham, B. & Charters, S. (2007). *Guidelines for performing systematic literature reviews in software Engineering.* Technical Report EBSE-2007-01, School of Computer Science and Mathematics, Keele University.
21. Forment, M. A., Filvà, D. A., García-Peñalvo, F. J., Escudero, D. F., & Casañ, M. J. (2018). Learning analytics' privacy on the blockchain. *In Proceedings of the Sixth International Conference on Technological Ecosystems for Enhancing Multiculturality, 2018,* 294–298.
22. Ghaffar, A., & Hussain, M. (2019). BCEAP-A blockchain embedded academic paradigm to augment legacy education through application. In *Proceedings of the 3rd International Conference on Future Networks and Distributed Systems,* pp. 1–11.
23. Liu, L., Han, M., Zhou, Y., Parizi, R. M., & Korayem, M. (2020). Blockchain-based certification for education, employment, and skill with incentive mechanism. In *Blockchain cybersecurity, trust and privacy* (pp. 269–290). Springer.
24. Mori, K., & Miwa, H. (2019). Digital university admission application system with study documents using smart contracts on blockchain. *In International Conference on Intelligent Networking and Collaborative Systems, 2019,* 172–180.
25. Getso, M. A., & Johari, Z. (2017). The blockchain revolution and higher education. *International Journal of Information Systems and Engineering, 5*(1), 57–65. https://doi.org/10.24924/ijise/2017.04/v5.iss1/57.65.
26. Wu, Z., He, T., Mao, C., & Huang, C. (2020). Exam paper generation based on performance prediction of student group. *Information Sciences, 532,* 72–90.
27. Parchment. (2021). Turn Credentials into opportunities. https://www.parchment.com. Accessed 05 Mai 2021.
28. Devine, P. (2015). Blockchain learning: Can crypto-currency methods be appropriated to enhance online learning? http://oro.open.ac.uk/44966/. Accessed 11 Jan 2021.
29. Liyuan, L., Meng, H., Yiyun, Z., & Reza, P. (2019). E^2 C- Chain: a two-stage incentive education employment and skill certification blockchain. In *2019 IEEE International Conference on Blockchain (Blockchain),* pp. 140–147.
30. Juričic. (2019). Creating student's profile using blockchain technology. In *2019 42nd International Convention on Information and Communication Technology.xElectronics and Microelectronics (MIPRO), 2019,* 521–525.
31. Turkanovic. (2018). EduCTX: A blockchain-based higher education credit platform. *IEEE Access, 6*(2018), 5112–5127.
32. Han, M., Li, Z., He, J. (Selena), Wu, D., Xie, Y., & Baba, A. (2018). A novel blockchain-based education records verification solution. In *Proceedings of the 19th Annual SIG Conference on Information Technology Education (Fort Lauderdale, FL, 2018),* pp. 178–183.
33. Mitchell, I., Hara, S., and Sheriff, M. 2019. dAppER: decentralised application for examination review. In 2019 IEEE 12th International Conference on Global Security, Safety and Sustainability (ICGS3) (2019), 1–14.

34. Vidal, F., Gouveia, F., & Soares, C. (2019). Analysis of blockchain technology for higher education. In *2019 International Conference on Cyber-Enabled Distributed Computing and Knowledge Discovery (CyberC)*, pp. 28–33.
35. Arndt, T., & Guercio, A. (2020). Blockchain-based transcripts for mobile higher-education. *International Journal of Information and Education Technology, 10*, 2.
36. Bore, N., Karumba, S., Mutahi, J., Darnell, S. S., Wayua, C., & Weldemariam, K. (2017). Towards blockchain-enabled School Information Hub. In *Proceedings of the Ninth International Conference on Information and Communication Technologies and Development (Lahore, Pakistan, 2017), Article 19*.
37. Finck, M. (2019). Blockchain and the general data protection regulation—Can distributed ledgers be squared with European data protection law? Panel for the Future of Science and Technology, European Parliamentary Research Service. https://www.europarl.europa.eu/Reg Data/etudes/STUD/2019/634445/EPRS_STU(2019)634445_EN.pdf. Accessed 15 Jan 2021.
38. Shah, P., Forester, D., Berberich, M., & Raspé, C. (2019). Blockchain technology: Data privacy issues and potential mitigation strategies, Practical Law. https://www.davispolk.com/files/blo ckchain_technology_data_privacy_issues_and_potential_mitigation_strategies_w-021-8235. pdf. Accessed 15 Feb 2021.
39. Rivastava, G., Dwivedi, A., & Singh, R. (2018). *The future of blockchain technology*. Branson University, Issue 2(1).
40. Ali, A. Z., & Mohd, A. B. (2020). Effect of applying total quality management in improving the performance of Al-WAQF of ALBR societies in Saudi Arabia: A theoretical framework for "Deming's Model." *International Journal of Business Ethics and Governance, 3*(2), 12–32. https://doi.org/10.51325/ijbeg.v3i2.24
41. Zainal, M. M., & Hamdan, A. (2022). Artificial intelligence in healthcare and medical imaging: Role in fighting the spread of COVID-19. https://doi.org/10.1007/978-3-030-77302-1_10.
42. Alshurafat, H., Al, M. O., & Mansour, E. (2021). Strengths and weaknesses of forensic accounting: An implication on the socio-economic development. *Journal of Business and Socio-Economic Development, 1*(2), 85–105. https://doi.org/10.1108/JBSED-03-2021-0026
43. Kassim, E., & El Ukosh, A. (2020). Entrepreneurship in technical education colleges: applied research on university college of applied sciences graduates—Gaza. *International Journal of Business Ethics and Governance, 3*(3), 52–84. https://doi.org/10.51325/ijbeg.v3i3.49
44. Alzaneen, R., & Mahmoud, A. (2019). The role of management information systems in strengthening the administrative governance in ministry of education and higher education in Gaza. *International Journal of Business Ethics and Governance, 2*(3), 1–43. https://doi.org/10.51325/ijbeg.v2i3.44
45. Al, N., & Hamdan, A. (2021). Artificial intelligence and women empowerment in Bahrain. *Studies in Computational Intelligence, 2021*(954), 101–121.
46. Safari, K., Njoka, C., & Munkwa, M. G. (2021). Financial literacy and personal retirement planning: A socioeconomic approach. *Journal of Business and Socio-Economic Development, 1*(2), 185–197. https://doi.org/10.1108/JBSED-04-2021-0052
47. 2020. https://doi.org/10.1016/j.ins.2020.04.043.
48. Budhiraja, R., Budhiraja, S., & Rani, R. (2020). TUDoc chain-securing academic certificate digitally on blockchain. In S. Smys, R. Bestak, & A. Rocha (Eds), *Inventive Computation Technologies. ICICIT. Lecture Notes in Networks and Systems* (Vol. 98, pp. 150–160). https://doi.org/10.1007/978-3-030-33846-6_17.
49. Christensen, C. M. (2003). *The innovator's solution: Creating and sustaining successful growth*. Harvard Business Press. ISBN 978-1-57851-852-4.
50. Educators, W. (2018). *Higher education in an age of innovation, disruption, and anxiety*.
51. European Commission. (2016). eGovernment Benchmark 2016. A turning point for eGovernment development in Europe? Study carried out for the European Commission by Capgemini, IDC, Sogeti, and Politecnico di Milano.
52. Filvà, D. A., García-Peñalvo, F. J., Forment, M. A., Escudero, D. F., & Casañ, M. J. (2018). Privacy and identity management in learning analytics processes with blockchain. *In Proceedings of the Sixth International Conference on Technological Ecosystems for Enhancing Multiculturality, 2018*, 997–1003.

53. Grech, A., & Camilleri, A. F. (2017). *Blockchain in education*. Publications Office of the European Union.
54. Guo, J., Li, C., Zhang, G., Sun, Y., & Bie, R. (2020). Blockchain-enabled digital rights management for multimedia resources of online education. *Multimedia Tools Applications, 79*, 9735–9755. https://doi.org/10.1007/s11042-019-08059-1
55. Hanson, R. T., Staples, M. (2017). Distributed ledgers, scenarios for the Australian economy over the coming decades. Canberra. Commonwealth Scientific and Industrial Research Organisation.
56. Jing, L. (2019). Research on the application of blockchain technology in private university education. In *International Conference on Advanced Education and Management (ICAEM)*.
57. Klimonov, M., & Popova, V. (2020). Informatsionnaya sistema elektronnogo portfolio studentov na osnove tekhnologii blokcheyn [Blockchain-Based Information System of Students' Electronic Portfolio]. [Electronic resource]. https://drive.google.com/file/d/1T9_be9 MqN9dvbd5CUsWH69TY-gCwZAGp/view [in Russian].
58. Lam, T. Y., & Dongol, B. (2020). A blockchain-enabled e-learning platform. *Interactive Learning Environments., 2020*, 1–23.
59. Morabito, V. (2017). Business innovation through blockchain. *The B3 Perspective*. Springer.
60. Patel, K., & Das, M. L. (2020). Transcript management using blockchain enabled smart contracts. *In International Conference on Distributed Computing and Internet Technology, 2020*, 392–407.
61. Pinheiro, R., Young, M. (2017). The university as an adaptive resilient organisation: A complex systems perspective. In *Theory and Method in Higher Education Research* (pp. 119–136).
62. Li, Q., Zhang, X. (2017). Blockchain: Promoting openness and public credit of education by technology [J]. *Journal of Distance Education*.
63. Rashid, M., Deo, K., Prasad, D., Singh, K., Chand, S., & Assaf, M. (2019). TEduChain: A platform for crowdsourcing tertiary education fund using blockchain technology. [Electronic resource]. https://arxiv.org/pdf/1901.06327.pdf.
64. Razia, B., Thurairajah, N., & Larkham, P. (2017). Understanding delays in construction in conflict zones. In *International Research Conference: Shaping Tomorrow's Built Environment, Manchester, UK*. https://www.researchgate.net/publication/335661600_Understanding_Delays_in_Construction_in_Conflict_Zones.
65. Sahonero-Alvarez, G. (2018). Blockchain and peace engineering and its relationship to engineering education. In *2018 World Engineering Education Forum-Global Engineering Deans Council (WEEF-GEDC)*, pp. 1–6.
66. Sharples, M., & Domingue, J. (2016). The blockchain and kudos: A distributed system for educational record, reputation and reward, Cham.
67. Stachokas, G. (2019). *The role of the electronic resources librarian* (p. 176). Chandos Publishing.
68. Xu, Y., Zhao, S. I., Kong, 1, Y. Zheng, S. Zhang, 1, & Li, Q. (2017). *A high performance educational certificate blockchain with efficient query* (pp. 288–304). Springer.
69. Yan, Z. (2018). Research on the development and application of higher education based on blockchain technology [J]. *Chinese and Foreign Entrepreneurs*.
70. Yang M. X., Li X., Huanqing W., & Zhao K. (2017). Application model and realistic challenge of blockchain technology in education [J]. *Modern Distance Education Research*.
71. Yifu M, J. (2017). The demand analysis and technical framework of blockchain+education [J]. *China Educational Technology*.
72. Zhao, G., Di, B., & He, H. (2020). Design and implementation of the digital education transaction subject two-factor identity authentication system based on blockchain. In *2020 22nd International Conference on Advanced Communication Technology (ICACT)*, pp. 176–180.
73. Zhou, L., Zhang, L., & Zhao, Y. (2020). A scientometric review of blockchain research. *Information Systems and e-Business Management*. https://doi.org/10.1186/s40854-019-0147-z
74. Rahardja, U., Hidayanto, A. N., Hariguna, T., & Aini, Q. (2019). Design framework on tertiary education system in indonesia using blockchain technology. In *2019 7th International Conference on Cyber and IT Service Management, CITSM 2019* (5–8).

Insights into Experiences of E-learning as a Tool for Teaching During the COVID-19 Pandemic Among University Students

Noor Ulain Rizvi, Sunitha Prabhuram, and Cheshta Kapuria

1 Introduction

The coronavirus disease of 2019 (COVID-19) has ensued in extensive mortality and has revealed the frailties of economies globally. The first case of pneumonia caused by a novel coronavirus was diagnosed in December 2019. In the following month, China confirmed the person-to-person transmission nature of the virus. Apropos, the World Health Organisation (WHO) declared it to be a disease posing a risk to countries with inadequate health care systems, describing it as a Public Health Emergency of International Concern. After spreading to more than 114 countries, the contagious disease was declared a pandemic on March 11, 2020. The first case of COVID-19 in the United Arab Emirates (UAE) was identified on January 29, 2020. In UAE, till July 30 2021, there have been 677,801 confirmed cases of COVID-19, with 1,939 deaths reported to WHO. Worldwide, the national responses to curb the pandemic varied from complete or partial lockdowns to restrictions.

In UAE, the first restrictions were introduced at the end of March 2020 that enabled national sterilization drive to clean up streets and roads and boost public safety. The UAE closed all malls and markets for almost two months but reopened them, with precautions in place, in May 2020. Many governments took measures to avoid spreading the virus and ensure the continuity of the educational process, and universities worldwide adopted online learning. On similar grounds, the Ministry of Education restricted all in-person classes in schools and universities until the end of the academic year. This resulted in the postponement of opening campuses for the new semester. The students were shifted to "remote," "online," or "virtual" learning environments (referred to hereafter as remote learning).

N. U. Rizvi (✉) · S. Prabhuram
Manipal Academy of Higher Education, Academic City, Dubai, United Arab Emirates

C. Kapuria
University of Delhi, New Delhi, India

© The Author(s), under exclusive license to Springer Nature Switzerland AG 2022   573
A. Hamdan et al. (eds.), *Technologies, Artificial Intelligence and the Future of Learning Post-COVID-19*, Studies in Computational Intelligence 1019,
https://doi.org/10.1007/978-3-030-93921-2_30

The ongoing pandemic has had a profound influence on the well-being of learners and their educational attainment. At its peak, educational institutions' full and partial closures interrupted the education of around 1.6 billion learners worldwide [1]. This outbreak witnessed an upsurge in remote teaching. Educational institutions in the UAE were swift to initiate remote teaching on a national scale and for all grade levels. The expedient response of extensive migration from face-to-face to online-based learning has created educational experiences and implications that need to be preserved and leveraged in future.

In comparison to the traditional in-person learning technique, the remote learning experience requires better fundamental computer skills [2], the efficiency of human-human and human-machine interaction [3], as well as studying motivation [4]. Such learning platforms were considered an 'option' prior to the pandemic [5], but the unprecedented nature of the pandemic triggered the widespread usage to prevent an absolute closure of educational institutions and stunt the development of learners [6]. This paradigm shift due to the 'forced innovation' may have modified learners' perception of remote learning and could be different from the perception built-in studies before the pandemic.

This research attempts to gauge the learners' experiences and observations about remote learning while understanding its benefits and challenges. It also attempts to observe if the perception differs based on degree level across different specializations. Lastly, it provides concluding remarks on the future of education post-pandemic.

For this purpose, a detailed survey questionnaire was prepared, and the conclusions are based on responses of 538 university learners across different degree programs and specializations based in the UAE.

Thus, this research contributes to the development of the E-learning process. It provides information on the use of specific methods used to deliver the courses, the content of the course, and student's grievances, recommendations, and preferences for a particular teaching technique.

The following section describes the literature reviewed and identifies the research gaps. Section 3 explains the methodology used in this research, followed by a discussion of the findings in Sect. 4. Section 5 presents the concluding observations, future scope, and limitations.

2 Literature Review

The literature review has been subclassified into three sections: E-learning in higher education, benefits and shortcomings of it, current studies about the impact of the pandemic on education.

2.1 E-learning in Higher Education

Higher education is an ever-evolving and continuous transformation process; the universities must restructure the courses, programs, and pedagogy to keep up with learners' needs, requirements, and convenience. With the advent of technology, the internet and affordability, information technology and E-learning are increasingly observed as crucial factors leading to substantial investments by universities in these systems [7]. However, integrating innovative E-learning systems to reinforce learning and teaching remains challenging [8]. Due to its complex nature, E-learning has myriad definitions. Horton [9] defines it as using information and computer technologies and systems to build design learning experiences. It is described as using electronic media to provide distance learning solutions [10]. Concisely, E-learning transmits knowledge and education by employing various electronic devices [11].

Early evidence of distance education was illustrated in Isaac Pitman's endeavours of using post and shorthand to teach, engage and collaborate with learners [12]. The term E-learning was coined and began to be used in the mid-1990s [13]. It seems to be a natural progression of distance learning and teaching [14]. In a more convoluted and all-encompassing definition, E-learning may be a form of learning and teaching that integrates the electronic medium, which fosters the development, to make education and training qualitative [15]. It can also be perceived as a system for providing formal education or a network that enables disseminating information to a larger audience through electronic resources. Computers and the internet are essential in such transactions [16].This system offers a plethora of possibilities, sharing, uploading documents in different formats, and is web-based. It does not require the users to install additional tools, and the content can be download anytime without time constraints [17].

A unique element of E-learning, making it distinct from traditional or other methods, is its ability to focus on individual adjusted learning instead of instruction-focused [18]. Gallie and Joubert [19] mark this as a shift from a teacher-centred to a student-centred approach. In the traditional learning approach, the principal source of information is the faculty, who also assess. The quality of education is, thus, highly dependent on the instructor's knowledge and skills. Whereas in E-learning, other tools and systems are available for evaluation, and the learner has access to many documents available on the platform. The quality of education is dependent on the faculty's teaching style as well as their technological skills. The eight principles encompassing effective online teaching are the contact between faculty and student, collaborative learning, active learning, prompt feedback, clear communication by faculty about expectations from learners, encouragement to learners for allocating more time to tasks, application of technology, and diversified learning [20].

The literature reviewed more likely favours the effectiveness and usefulness of E-learning on the learner's performance. In a study, teachers trust the potential of E-learning in terms of improved collaborations, communication with students, increased flexibility, a better understanding of lectures, and enhanced educational process [21]. Lochner [22] demonstrated that the use of E-learning along with traditional teaching style increased the learners' engagement in lectures. Even with proven advantages of E-learning over traditional learning, there is a stigma attached [23]. Learners considered the courses taught online as less valuable than those taught in class [24]. Blended learning, a combination of face-to-face and online classes, is a relatively acceptable medium [25].

2.2 E-learning: Benefits and Shortcomings

If used effectively to integrate teaching and learning processes, E-learning can positively influence collaboration and performance in the educational process. The effectiveness of E-learning is determined by three critical elements, namely, student, technology, and institution [26]. Institution refers to teachers equipped with skills to use tools to create a learning environment for learners and innovatively get students closer and secure their attention. Students may feel isolated due to the absence of physical interaction; in this case, the institution must establish connections and relations.

The study by Navarro [27] concluded that learners subject to E-learning could better assimilate knowledge compared to learners exposed to the traditional way. It was especially effective for shy, slow learning and easily intimidated learners who generally do not express themselves in the classroom.

The recognition of E-learning can also be attributed to its ability to be flexible in delivering lectures and ease in accessing information [28]. It is also embedded with personalizing and adapting courses as per the learner's need [16]. It removes the constraint of space and time, enables access to various informational content, learns at own pace, facilitates collaboration, motivates interaction with peers, and exchanges viewpoints and ideas [29]. It saves them time and money, which would otherwise be spent on travelling [30]. From the benefits stated above, learners consider 'accessibility' the most significant benefit [31].

Indeed E-learning has many benefits but is marred with shortcomings as well. Learners may get distracted easily and lose focus [32]. With the attention span of an average millennial or a GenZ to be less than that of a goldfish, these seem to be a matter of concern [33].

Feeling of isolation due to the inability to meet peers physically and decreased motivation are other problems associated with it [34]. Apart from human behaviour, technological disturbances include loss of internet connectivity, slow devices, and system errors [32]. Physical problems related to eyesight and body posture arises due to increased screen time and reduced outdoor activities [35].

2.3 Remote Learning During a Pandemic

Conspicuously, the abrupt migration to online-based teaching and learning triggered by the suddenness of the pandemic in education cannot be precisely termed as 'online learning' or 'E-learning'. The 'emergency remote teaching' as a response to the pandemic cannot be meaningfully compared to the well-planned online learning experience [23]. This precipitous way to shift online by institutions at once could create an incorrect perception of online learning as not a strong alternative [23]. It takes approximately 6–9 months to develop a fully online course, as it includes planning, preparation, and development time before being delivered.

COVID-19 has rekindled the interest of researchers and academia towards the perception of online learning. Research [36] investigated the perception of the learner's enrolled in the English study program of UKI Toraja and found a favourable attitude towards it. They appreciated its usefulness and helpfulness in the time of crisis. Learners from Southeast University in China felt that faculties should know how to modify the previously taught content into a more online friendly environment and not simply transfer to online. They suggested that more projects and assignments must be included [2]. Some authors [37] identified seven crucial characteristics that are the foundation of online education. These factors optimize learning in exceptional circumstances, like the current pandemic. They include management and development of infrastructure needed for the internet, user-friendly tools that aid in assimilation and understanding of information, availability of diverse and interactive electronic resources, communities through the social network to reduce isolation, usage of effective techniques, collaborative work among institutions and lastly, platforms that assist learners and faculties to learn about recent policies followed at universities and by government.

Preliminary research [38] is conducted to report the pandemic's impact on education based on a survey of 31 countries. They report that the pandemic has further aggravated inequity and the digital divide. There is a need to shift to formative assessment and use proctoring services to control cheating and dishonesty.

A research [39] studies the impact using a verified social media handle via Instagram, involving followers from 95 countries and concluded that active-learning methods have a constructive impact on remote learning. Research [40] was conducted on 435 students across universities in Egypt and highlighted the cost efficiencies remote learning can have.

As observed from the literature reviewed, learners' perception was collected from experiments where the E-learning platforms complement the traditional learning practices. However, the studies pertaining to the exclusive usage of E-learning, as observed during the pandemic, are rather limited. This research focuses on understanding the perception of learners in UAE based on their experiences of exclusive remote learning through some part of the year 2020. It is worthy to note that prior to the pandemic, the education system in UAE seldom used E-learning platforms. The usage was limited to some faculties using Learning Management System (LMS) to upload course materials. Many faculties and learners were astounded and overwhelmed with the impromptu shift from a traditional education environment to an exclusive remote one. The expedient response of extensive migration from face-to-face to online-based learning has created educational experiences and implications that need to be preserved and leveraged in future. Hence, this research attempts to document the learners' experiences during the COVID-19 pandemic and gauge their expectations, readiness, and observations about remote learning. It also gauges their efficacy in engaging in remote classes based on the degree level across different specializations. These questions seem imperative as they are not yet very well understood and researched, keeping in mind the contemporary nature of the topic.

3 Methodology

3.1 Research Instrument, Scales, Identification of the Sample Population and Method of Analysis

The survey primarily builds on the three research questions. Five experts validated the structure of the questionnaire. A pilot study was conducted on 15 Manipal Academy of Higher Education (Dubai campus) students, which served as a robust pre-test. The questionnaire is designed to elicit specific responses (rather than general views), owing to the reduced attention span of millennials and GenZ. Many questions use a three-point Likert scale to measure attitudes and allow for a degree of opinion [41]. The questionnaire was administered to university learners across different degree programs and specializations based in the UAE who have been exposed to remote teaching in the past year. Data is analyzed using mean and frequency distribution. The total number of responses varies question-wise as respondents could choose not to reply.

3.2 Data Collection

A Google form was created, and the link was shared using emails, Linked In, and other social media platforms. The study restricted responses to only one response from each respondent. The participants in the study received information at the beginning of the questionnaire about the purpose of the survey and informed consent. The total responses received were 569, out of which 538 responses were used in the analysis. The email addresses were not collected to respect anonymity and confidentiality. The average time needed to answer the questionnaire was 15 min.

Respondents' profile The respondents are primarily university learners across different degree programs (Bachelors and Masters) and specializations (B.Com., B.B.A, B.Tech., B.Arch., B.A., B.Des., B.Sc.) based in the UAE. The cohort is

Table 1 Sociodemographic characteristics of respondents

Variable	Category	Count	Percentage
Gender	Female	293	54
	Male	245	46
Degree	Under-graduate	327	61
	Post-graduate	211	39
Country residing in (during the pandemic)	UAE	482	90
	Others:	56	10
Family members staying in the house, including you	1–3	84	16
	4–7	361	67
	More than 7	93	17
Most family members work from home	Yes	368	68
	No	170	32
Internet access	Sufficient	384	71
	Moderately sufficient	91	17
	Insufficient	63	12
Major stream of education	Commerce	257	48
	Science	223	41
	Arts	58	11

of diverse nationalities belonging to India, Pakistan, Sri Lanka, Africa, Egypt, the United Kingdom, among the major ones. 245 (46%) of the learners were male and 293 female (54%). 327 (61%) were in the Bachelor education programs, and the others, 211 (39%), were in the Master education programs. 10% of the respondents were living (stuck) outside UAE during the last year; hence, their experiences may differ from the remaining 90% residing in UAE (Table 1). The learners are mostly majoring in Commerce (48%), science (41%), and Arts (11%). Most of the learners claimed their internet access to be sufficient (71%); however, 12% (mainly those not residing in UAE) complained it was insufficient. The average household size is in line with the national average, between 4 and 7 members (67%), with 68% of students reporting many family members working from home.

4 Results

This section presents the data collected from learners who experienced remote teaching in the past year. For easier comprehension and unambiguous visual presentation, all the figures have been rounded off to the nearest whole number.

4.1 Learners' Experiences and Observations About Remote Learning

The device used to access the lectures has an impact on the experience of a learner. Simple touch-based devices may be beneficial for younger students; however, university students need devices with more computing power, storage, and an environment filled with innovative technologies which assist them in completing their assignments and projects. During the survey, approximately 70% of students reported using personal laptops and desktops, whereas 30% used mobiles and tablets (Table 2). The screen size of these devices is much smaller compared to an average laptop/desktop, making it challenging to browse and view content, thus influencing the learning experience [42].

After having experienced remote teaching for almost three semesters, 27% still have not adapted to this mode well, whereas 11% have inhibitions with this mode altogether (Table 3). Students also believed their attendance would have been better had the classes been conducted like before (Table 4), highlighting the inefficiencies that faculties, universities, and technological innovations need to be dealt with. Students' reasons for preferring in-person and synchronous remote classes emphasize the desire for social interaction and echo the research on the importance of social presence in remote courses.

A study [43] outlined Social Presence Theory in depicting students' perceptions of each other as real in different means of telecommunications. These ideas translate directly to questions surrounding remote education and pedagogy regarding educational design in networked learning, where connection across learners and instructors improves learning outcomes, especially with "Human-Human interaction" [44–46].

Table 2 Devices used

Devices used to access distance education lessons	Frequency	Percentage
Personal laptops	248	49
Desktop computers	106	21
Mobile devices/tablets	149	30
	503	100

Table 3 Adjustment to remote learning

Did you adjust to the new teaching mode?	Frequency	Percentage
Yes, swiftly	124	23
Yes, after initial difficulties	205	39
Have not adjusted well	146	27
Do not want to adjust	56	11
	531	100

Table 4 Attendance in remote and traditional modes

	How often did you attend the remote learning sessions?		How often would you have attended the classes if they were held on campus? (Assume no COVID-19 situation)	
	Frequency	Percentage	Frequency	Percentage
Regularly	135	25	283	55
Sometimes	374	70	219	42
Never	23	4	15	3
	532	100	517	100

Table 5 Courses undertaken

What courses did you study in the past year?	Frequency	Percentage
Core subjects only	405	81
Extracurricular subjects (music, physical education etc.)	35	7
Both	57	11
	497	100

81% of the respondents reported being taught mostly core subjects, with only 11% stating that the focus was on both core subjects and extracurricular subjects (Table 5). Extracurricular subjects and activities are imperative for overall development and employability skills development (47, 48). It also distinguishes students from competition based on skills that are usually not mentioned on their resumes due to their limited experience [49].

Perhaps, the attendance issue as discussed above can be resolved if the remote learning environment is embedded with increased engagement, peer-to-peer interaction, and encouraging student participation (Fig. 1). This question had students put forward their options if they believed the list provided was not exhaustive. Students' felt inclusion of co-curricular activities, changes in evaluation/assessment patterns, comprehensive database of online resources can prove beneficial in increasing the capabilities of e-learning. More than 60% emphasized adapting the course structure and pedagogy to suit the online environment. Students, especially those not residing in the UAE during the period, recommended a solution to mitigate the impact students face with unstable connections. They proposed a combination of pre-recorded videos, live sessions, and increased remote participation through projects and quizzes. A few also recommended the provision of teaching platforms with subtitles and playback functions.

The students reported that their universities provided live lessons and the recording of the same for future reference (Fig. 2).

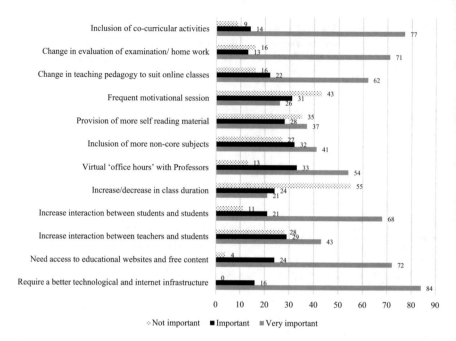

Fig. 1 Factors enhancing remote learning possibilities

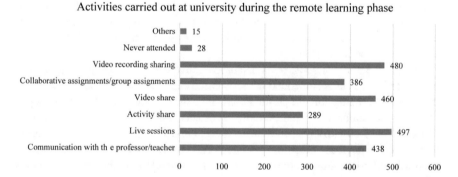

Fig. 2 Activities carried out by the university

E-learning supported by online tools does not replace teachers, who are the key to its success. However, online teaching methods require a unique set of soft and hard skills and a bank of suitable digital lesson materials. Students opined that digital socializing is an important skill to be focused upon in an online class. Other skills, namely, self-learning, adaptation, and research, are also relevant (Fig. 3).

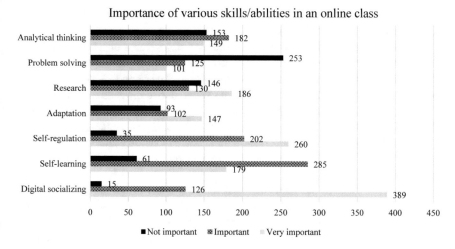

Fig. 3 Skills/abilities required in an online class

4.2 Benefits and Challenges of Remote Learning

Students stated that remote learning induced by the pandemic had challenges but showcased certain positive outcomes as well; they recommended specific approaches that can be used to support their learning (discussed in Sect. 4.1). Among the benefits identified from the literature, the ease of attending classes from any location and savings of travelling cost, i.e., reduction in limitation caused by geographical boundaries, is envisaged as the most important benefit (Table 6). However, since most universities did not alter their fee structure, it did not lead to significant savings in terms of fees paid.

The hierarchy of challenges in online learning mode may differ from the challenges faced during the pandemic enforced remote learning. Among the challenges, lack of concentration in the lectures, using other social media apps (79%), physical discomfort due to increased screentime (62%), and lack of motivation (52%)

Table 6 Benefits of remote learning

Benefits of online learning	Yes	No
Saves commuting (travelling) time	86	14
Limitation of physical space is broken (geographical boundaries)	92	8
Control over learning (you decide when and how to learn) that is, 'replay', 'fast forward' recordings, sit in bed etc	83	17
Students are not intimidated and can ask questions freely	62	38
Teachers can teach beyond traditional textbooks to include online resources	74	26
Online learning reduces financial costs	14	86

are the crucial ones. It implies a need to improve self-discipline, concentration, or technological innovations to restrict usage [2].

Usually, technical issues, poor connection and storage are considered challenges in developing countries; however, UAE relishes one of the highest download speeds. Thus, this was not seen as a challenge in the region. The online courses are meticulously designed for the said environment; however, due to the sudden shift, faculties could not (in some cases) appropriately adapt the content to the online environment. This was seen as a challenge by the majority (54%) of the respondents.

This section of results recommends that teachers modify the pace of remote teaching and account for the online environment that is starkly different from the in-person classroom.

This survey gave us the feedback that teachers must adapt the pace of online teaching to consider an environment completely different from a classroom. Presently, it is observed that the traditional classroom lessons are simply being duplicated online. There is a need to put more effort in innovating and designing the lessons that in turn increases the focus and attention span of students [2]; need to transform students' role from passive to engaged learners through greater interaction via presentations, discussions, etc. with the advent of 5G and artificial intelligence, this mode will become more challenging, but if appropriately harnessed, can be a boon.

However, the present research does not identify which active learning tools or a particular combination that works best. Deciphering this can be an interesting area for future research (Fig. 4).

4.3 Difference in Perception Based on Degree Level Across Different Specializations

As stated in Sect. 3.2, respondents are university learners across different degree programs (Bachelors and Masters) and specializations (Commerce, Business Administration, Technology, Architecture, Arts, Design, and Science).

Difference in opinion based on degree level: The survey responses suggest that expectations regarding remote learning readiness and academic achievement of advanced learners (post-graduates) were higher than the delivery. In their opinion, post-graduation courses usually involve more practical application than undergraduate courses that focus on building the foundation of subject matter. The course design of teaching a post-graduate course should be more thoughtful, scientific and give enough room for creativity and innovation. The final year undergraduate students and post-graduate students' internships were affected due to the abrupt nature of the unfolding events. The difference also arose in terms of the subjects students were majoring in; they are discussed below.

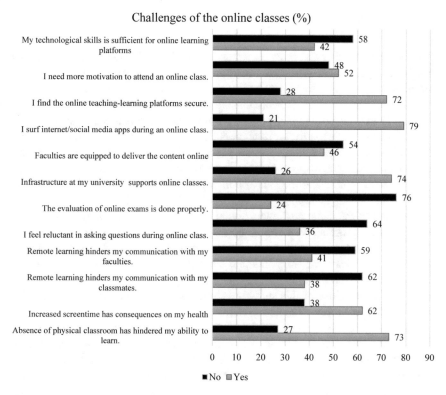

Fig. 4 Challenges of the online classes

Difference in opinion based on specialized field of study: Learners are more receptive to theoretical and numerical subjects during live lessons, implying such subjects (including social sciences) have transited swiftly to remote learning. These subjects have many databases and libraries online, including self-learning books, quizzes, activities, recorded lectures, games, and case studies. Students working on theoretical assignments and literature reviews largely remained unaffected. A periodic online meeting with the mentor allowed monitoring and support.

However, some subjects (especially in the science, engineering, and arts domain) that entail experiential learning do not translate well. The work of [2] also supports this conclusion. Subjects needing laboratory experiments, except those related directly to the pandemic, mainly were suspended or delayed; thus, learning was affected. Subjects involving field study, sampling, and involving working in groups on experiential experiments were deeply affected.

Even though the pandemic severely impacted the normal educational progress, the institutions must treat this as an opportunity to identify the deficiencies and expedite the much-needed reforms in online learning by incorporating state of the art technologies, innovations, course content and efficient management.

5 Conclusion

To summarise, the first and foremost robust conclusion is that most learners transited to remote learning swifty, or after initial difficulties. However, they felt the content and delivery needs modification to suit the remote environment. Hence, it is suggested that faculties must be provided with distance learning training periodically; cues can be taken from Singapore to make sure the transition to online is seamless if the need arises in the future. Secondly, the methodology for assessment must be modified along with technological innovation, especially in an online setup, to ensure the system has not tampered with and avoid instances of academic dishonesty. Based on the present experiences of remote learning, accessibility and flexibility are considered essential benefits, but the simple duplication of the content from traditional to remote learning environments is the major challenge. Social isolation is also a challenge that needs to be addressed appropriately. A similar conclusion was drawn by Gay [50] who stated that the current education system was too narrow and failed to embrace the humanity of faculties and learners; the pandemic and the subsequent social distancing has further increased the social isolation, which perhaps is not suitable for mental well-being. The conclusions of this study are similar to some extent to other researches based on other countries [39, 40, 51, 52].

It is worthy of mentioning that the pandemic-induced forced innovation did not give the faculties enough time to reconstruct the content, earlier designed for the in-person environment. Therefore, it is important to turn this emergency into an opportunity by promoting knowledge-sharing collaborations and resources internationally to lay the foundation of a global education network. There is a need for infrastructure to enable online learning, just like there is a well-developed infrastructure around the in-person education system, such as libraries, classrooms, hostels, career services, to name a few. The involvement witnessed by the local community to support students during the pandemic (initiatives such as #PandemicCamp, 'Education Uninterrupted', free psychological counselling sessions, among others) must be harnessed to build social cohesion.

After this forced 'trial run' of remote learning, the policymakers may survey the community, collect feedbacks, reflect on them and start a dialogue to build a future-proof education model. Opportunities for an inclusive educational system cater to people of determination, suffering from illness, and athletes who cannot study regularly due to the simultaneous practice sessions.

The study provides perspicacious insights to decipher an essential subject of educational attainment during the pandemic however suffers from certain limitations that can be addressed in the future. Firstly, the reliability of the results can be enhanced by including a much larger sample from the population. Open-ended questions and validation with empirical data may perhaps improve the findings and provide granularity to the discussion. It is often observed that males are technologically more proficient than females [51], whereas females can express their emotions better than their male counterparts in online forums [53]. Both these characteristics contribute to the success in online learning, and it will be interesting to observe the difference in perception about remote learning based on gender.

To conclude, the COVID-19 pandemic has offered an opportunity for investing in the education sector, to transform education by 'reinventing', 'reorienting', 'redefining' and 'reimagining' and "leave no one behind", as envisaged in the Sustainable Development Goals.

Declaration This paper was presented at Oxford Education Research Symposium held from August 5th–7th, 2021.

References

1. UNESCO. Education: From disruption to recovery. (2020). https://en.unesco.org/covid19/edu cationresponse.
2. Sun, L., Tang, Y., & Zuo, W. (2020). Coronavirus pushes education online. *Nature Materials, 19*(6), 687–687.
3. Cuadrado-García, M., Ruiz-Molina, M. E., & Montoro-Pons, J. D. (2010). Are there gender differences in e-learning use and assessment? Evidence from an interuniversity online project in Europe. In *Procedia-social and behavioral sciences* (pp. 367–71).
4. Hartnett, M. (2016). *The importance of motivation in online learning. In Motivation in online education* (pp. 5–32). Singaapore: Springer.
5. Abou, M., Taj-, I., Seddiek, N., El-, M. M., & Nosseir, A. (2014 June). E-learning and students' motivation: A research study on the effect of e-learning on higher education. *International Journal of Emerging Technologies in Learning, 9*(4), 20–26.
6. Coman, C., Țîru, L. G., Meseșan, L., Stanciu, C., & Bularca, M. C. (2020 Jan). Online teaching and learning in higher education during the coronavirus pandemic: Students' perspective. *Sustainability, 12*(24), 10367.
7. Popovici, A., & Mironov, C. (2015). Students' perception on using eLearning technologies. In *Procedia-social and behavioral sciences* (pp. 1514–1519).
8. Fischer, H., Heise, L., Heinz, M., Moebius, K., & Koehler, T. (2014). E-Learning trends and hypes in academic teaching. Methodology and findings of a trend study. In *International Association for Development of the Information Society.*
9. Horton, W. (2011). E-learning by design (13th Edn). Pfeier: San Francisco, CA, USA: John Wiley & Sons.
10. Engelbrecht, E. (2005). Adapting to changing expectations: Post-graduate students' experience of an e-learning tax program. *Computers & Education, 45*(2), 217–229.
11. Koohang, A., & Harman, K. (2005). Open source: A metaphor for e-learning. *Informing Science.*
12. Bezovski, Z., & Poorani, S. (2016). The evolution of e-learning and new trends. *Information and Knowledge Management, 6*(3), 50–57.
13. Lee, B. C., Yoon, J. O., & Lee, I. (2009). Learners' acceptance of e-learning in South Korea: Theories and results. *Computers & Education, 53*(4), 1320–1329.
14. Sangrà, A., Vlachopoulos, D., & Cabrera, N. (2012). Building an inclusive definition of e-learning: An approach to the conceptual framework. *International Review of Research in Open and Distributed Learning, 13*(2), 145–159.
15. Sangrà, M. A., Vlachopoulos, D., Cabrera, L. N., & Bravo, G. S. Towards and inclusive definition of e-learning.
16. Babu, G. S., & Sridevi, K. (2018). Importance of E-learning in Higher Education: A study. *International Journal of Research Culture Society., 2*(5), 1–8.

17. Raheem, B. R., & Khan, M. A. (2020). The role of E-learning in Covid-19 crisis. *International Journal of Creative Research Thoughts, 8*, 3135–3138.
18. Oye, N. D., Salleh, M., & Iahad, N. A. (2012). E-learning methodologies and tools. *International Journal of Advanced Computer Science and Applications, 3*(2).
19. Gallie, K., & Joubert, D. (2004). Paradigm shift: From traditional to online education. *Studies in Learning, Evaluation, Innovation and Development, 1*(1), 32–36.
20. Cheung, C., & Cable, J. (2017). Eight principles of effective online teaching: A decade-long lessons learned in project management education. *PM World Journal: A global resource for sharing knowledge in program and project management, 6*(7).
21. Burac, M.A., Fernandez, J. M., Cruz, M. M., & Cruz, J.D. (2019). Assessing the impact of e-learning system of higher education institution's instructors and students. *IOP Conference Series: Materials Science and Engineering, 482*(1), 012009.
22. Lochner, L., Wieser, H., Waldboth, S., & Mischo-Kelling, M. (2016). Combining traditional anatomy lectures with e-learning activities: How do students perceive their learning experience? *International journal of medical education., 7*, 69.
23. Hodges, C. B., Moore, S., Lockee, B. B., Trust, T., & Bond, M. A. (2020). The difference between emergency remote teaching and online learning. *Educause Review, 27*(1), 1–9.
24. Galy, E., Downey, C., & Johnson, J. (2011). The effect of using e-learning tools in online and campus-based classrooms on student performance. *Journal of Information Technology Education: Research, 10*(1), 209–230.
25. Tagoe, M. (2012). Students' perceptions on incorporating e-learning into teaching and learning at the University of Ghana. *International Journal of Education and Development Using ICT, 8*(1), 91–103.
26. Tham, C.M., & Werner, J. M. Designing and evaluating e-learning in higher education: A review and recommendations. *Journal of Leadership and Organizational Studies, 11*(2), 15–25.
27. Navarro, P., & Shoemaker, J. (2000). Performance and perceptions of distance learners in cyberspace. *American Journal of Distance Education, 14*(2), 15–35.
28. Bakia, M., Shear, L., Toyama, Y., & Lasseter, A. (2012). Understanding the implications of online learning for educational productivity. Office of Educational Technology, US Department of Education.
29. Arkorful, V., & Abaidoo, N. (2015). The role of e-learning, advantages and disadvantages of its adoption in higher education. *International Journal of Instructional Technology and Distance Learning., 12*(1), 29–42.
30. Cantoni, V., Cellario, M., & Porta, M. (2004). Perspectives and challenges in e-learning: Towards natural interaction paradigms. *Journal of Visual Languages and Computing, 15*(5), 333–345.
31. Al-, H. (2011). Faculty members and students perceptions of e-learning in the English department: A project evaluation. *Journal of Social Sciences, 7*(3), 291.
32. Sadeghi, M. (2019). A shift from classroom to distance learning: Advantages and limitations. *International Journal of Research in English Education, 4*(1), 80–88.
33. Gausby, A. (2015). Attention spans Canada: 21. https://dl.motamem.org/microsoft-attention-spans-research-report.pdf, Consumer Insights, Microsoft Canada; 2015.
34. Indira, D., & Sakshi, A. (2017). Online learning. *International Education and Research Journal, 3*(8), 32–34.
35. Nazarlou, M. M. (2013). Research on negative effect on e-learning. *International Journal of Mobile Network Communications and Telematics, 3*, 11–16.
36. Allo, M. D. (2020). Is the online learning good in the midst of Covid-19 pandemic? The case of EFL learners. *Journal Sinestesia, 10*(1), 1–10.
37. Huang, R. H., Liu, D. J., Tlili, A., Yang, J. F., & Wang, H. H. (2020). *Handbook on facilitating flexible learning during educational disruption: The Chinese experience in maintaining undisrupted learning in COVID-19 outbreak* (pp. 1–54). Beijing: Smart Learning Institute of Beijing Normal University.
38. Bozkurt, A., Jung, I., Xiao, J., Vladimirschi, V., Schuwer, R., Egorov, G., et al. (2020). A global outlook to the interruption of education due to COVID-19 pandemic: Navigating in a time of uncertainty and crisis. *Asian Journal of Distance Education, 15*(1), 1–26.

39. Nguyen, T., Netto, C. L., Wilkins, J. F., Bröker, P., & Vargas, E. (2021). Insights into students' experiences and perceptions of remote learning methods: From the COVID-19 pandemic to best practice for the future. *Frontiers in Education.*
40. El Said, G. R. (2021). How did the COVID-19 pandemic affect higher education learning experience? An empirical investigation of learners' academic performance at a university in a developing country. *Advances in human-computer interaction.*
41. Rizvi, N. U., Kashiramka, S., & Singh, S. (2021). Basel III in India: A double-edged sword. *Qualitative Research in Financial Markets.*
42. Maniar, N., Bennett, E., Hand, S., & Allan, G. (2008). The effect of mobile phone screen size on video based learning. *Journal of Software, 3*(4), 51–61.
43. Short, J., Williams, E., & Christie, B. (1976). The social psychology of telecommunications. Toronto; London; New York: Wiley.
44. Goodyear, P. (2002). Psychological foundations for networked learning. In *Networked learning: Perspectives and issues* (pp. 49–75). London: Springer.
45. Goodyear, P. (2005). Educational design and networked learning: Patterns, pattern languages and design practice. *Australasian Journal of Educational Technology, 21*(1), 82–101.
46. Tu, C. H. (2002). The measurement of social presence in an online learning environment. *International Journal on E-Learning, 1*(2), 34–45.
47. Baker, G., & Henson, D. (2010). Promoting employability skills development in a research-intensive university. Education+Training.
48. Clark, G., Marsden, R., Whyatt, J. D., Thompson, L., & Walker, M. (2015). "It's everything else you do…": Alumni views on extracurricular activities and employability. *Active Learning in Higher Education, 16*(2), 133–147.
49. Roulin, N., & Bangerter, A. (2013). Students' use of extracurricular activities for positional advantage in competitive job markets. *Journal of Education and Work, 26*(1), 21–47.
50. Gay, G. (2018). *Culturally responsive teaching: Theory, research, and practice. Multicultural education series.* New York: Teachers College Press.
51. Yawson, D. E., & Yamoah, F. A. (2021). Gender variability in E-learning utility essentials: Evidence from a multi-generational higher education cohort. *Computers in Human Behavior, 114*(106558).
52. Rizvi, N. U., & Prabhuram, S. (2021). Education in troubled times and the way forward: Learners' perspective. In *Proceedings of Oxford Education Research Symposium.* Oxford, United Kingdom.
53. Zhang, Y., Dang, Y., & Chen, H. (2013). Research note: Examining gender emotional differences in web forum communication. *Decision Support Systems, 55*, 851–860.

The Influence of Artificial Intelligence on Smart Learning: An Overview

Abdulsadek Hassan

1 Introduction

In recent years, the world has witnessed a resounding revolution in the field of Artificial Intelligence, and at the present time there is not without a field of employment of "Artificial Intelligence" applications, whether in Education, medicine, engineering, armament, manufacturing, investment, space sciences, communications, technology, cinema and art which places responsibilities on the shoulders of the Ministries of Education and universities together [1, p. 1795]. A massive development of its policies, curricula and strategies to keep pace with the modern artificial revolution, which was the spark of new areas for educators in the search for enriching the culture of Artificial Intelligence and including it theoretically and in practice in the different stages of education [2, p. 588].

From this, electronic curricula must be developed that simplify for students what Artificial Intelligence is, its objectives and concerns raised around it, and the dimensions of its application and uses in various fields of science and life [3, p. 25], so that all strategies are directed towards encouraging learners to solve current and future problems and create new projects depending on the assumptions of Artificial Intelligence knowledge, inside and outside university fences [4, p. 201].

Perhaps the education sector in some countries has the least share of the tremendous wave of change brought about by Artificial Intelligence systems in the past few years [5, p. 381]. This is due to the nature of the educational sector based on a large number of people—especially the educated side—which necessitates the decision-makers to follow certain educational policies. However, in recent years there has been a great boom in open e-learning platforms in which the world's largest universities [6, p. 331].

A. Hassan (✉)
Ahlia University, Manama, Kingdom of Bahrain

A. Hamdan et al. (eds.), *Technologies, Artificial Intelligence and the Future of Learning Post-COVID-19*, Studies in Computational Intelligence 1019,
https://doi.org/10.1007/978-3-030-93921-2_31

Technological progress will constitute an important leap in the educational field, as it will enable the improvement of the quality of education in the near future. Artificial Intelligence will also enable the study of learners 'behavior and work to help the students [7 , p. 12]. In other words, the ideal application of different pedagogical principles and respect for the multiple intelligences of the learner [8, p. 90].

Technology interferes better than humans in certain contexts, and will inevitably evolve to become more and more present in our lives in all fields, including the field of education, and here we may ask: Will Artificial Intelligence replace the lectures? [9, p. 1].

It is difficult to definitively answer this question at the present time, but most researchers believe that the role of the lecture will always be present, but it will differ in terms of its practical and educational value, to become more comprehensive, so that it will pay more attention to the social dimension that does not and will not be able to compensate for it. Universities for many learners remain human interaction and human contact [10, p. 15].

Universities and students are a great source of data as it is possible to design university systems capable of managing university and student data simultaneously and saving them in the form of huge databases [11, p. 581]. This big data can be used to train huge neural networks that can predict vulnerability at the individual student level and the shortage of material and human resources at the university level before it occurs [1, p. 1795]. Artificial Intelligence depends on data—a lot of it—so such algorithms will help university to make informed decisions about their institutions, which will increase the quality of educational outcomes and reduce the costs of these university [12, p. 3647]. As a simple example, data can be collected for preparing, using and retrieving books and the number of students from university over previous years, and then predicting the need for books in various universities in the coming year based on the expected number of students in all universities [13, p. 835].

Therefore, this study identifies the role of Artificial Intelligence on E-Learning system, especially during Coronavirus pandemic.

2   The Concept of Artificial Intelligence

In short, "Artificial Intelligence" is the behavior and certain characteristics of computer programs that make mimic human mental capabilities and patterns of work. The most important of these characteristics is the ability to learn, reason, and react [14, p. 161]. It is worth noting here that "Artificial Intelligence" programs differ from programming other information systems in that they have the ability to conclude on their own and realize what they should and should not do and not, not like the previous ones, by following pre-programmed steps. The use of this type of smart programs and mechanisms will not replace humans or lectures in particular, but rather it will greatly enhance human capabilities and contributions to the educational system [15, p. 415].

Artificial Intelligence science aims to develop systems that achieve a level of intelligence similar to or better than human intelligence. The AI applications are designed to imitate the actions of the human mind [16, p. 610]. In order for this to happen, it identified aspects of the superiority of human intelligence in the method of deduction and thinking and limited it to five points or steps: Categorization, Specific Rules, Heuristics, Past Experience, and Expectation [17, p. 705].

The goal is to place human knowledge inside the computer within what is known as knowledge bases, and then the computer can, through software tools, search for these rules, make comparison and analysis, in order to draw and conclude the best answers and solutions to various problems [18, p. 90]. This is similar to what a person does when he tries to solve new problems encountered in daily life, based on his previous experiences, through his expectations of possible results, and through skills in deduction and comparison between the best solutions available [19, p. 21].

3 Types of Artificial Intelligence

Researchers believed that the types of Artificial Intelligence can be divided into:

1. **Narrow Artificial Intelligence**: The easiest form of Artificial Intelligence, and its behavior in response to a situation, and it has a certain atmosphere of its own to work [20, p. 5442].
2. **Strong Artificial Intelligence**: It has the ability to make decisions based on gathering a lot of information (M. M. L 2017, p. 610).
3. **Artificial superintelligence**: Models that are under experience, and it has two types, the first seeks to understand human ideas, and has the ability to interact in a limited way, and the second has the ability to predict the feelings of others, and they are a highly intelligent generation, and are able to stand up to attacks electronic, they can be used in wars, and in medicine [17, p. 709].

There is an opinion that predicts the large spending on Artificial Intelligence in 2030, just as Artificial Intelligence has been used in the wars that took place on global dimensions [19], p. 19]. In spite of this, Artificial Intelligence can be reprogrammed, and make it aim to serve the hacker, which helps to cause serious damage, and it can be used to know many social and political information about a particular country, and it can currently predict natural disasters [21, p. 1]. Artificial Intelligence cannot distinguish between civilian and military personnel in a particular war, and armament systems must be able to cancel the electronic attack resulting from Artificial Intelligence, which is very difficult, and Artificial Intelligence has become an emerging technology that helps overcome global challenges [22, p. 137].

Recent attempts to introduce AI in universities have resulted in improvements in evaluating students before and during the educational process, placing students at appropriate levels for them, and increasing the efficiency of the classroom schedule [23, p. 1109]. These developments have allowed the introduction of different educational plans for a variety of students, but this sorting leads to consequences if the students' experiences or income level are not taken into consideration and if they are low-income or minority groups who suffer from poor preparation and lowering expectations about them [24, p. 4].

The spread of AI technology could also lure some regions to replace lectures with software, as is already happening in places like the Mississippi Delta. In light of the lecture shortage, these areas have turned into online platforms [25, p. 495]. However, this was reflected negatively on the students who suffered from the absence of trained human lectures who not only know the subjects, but also know and care about students [3, p. 26].

Technology software vendors have not helped solve this crisis. The educational landscape is now filled with virtual universities where modern technology companies have promised to reach hard-to-learn students as well as black, Latino or Hispanic students [26, p. 77], and to create efficiencies in under-funded areas. Instead, several startups were hit by scandals after they found that nearly 2,000 students last year had not been educated, and two online universities in Indiana were forced to close [27, p. 127].

But AI still has real benefits. For example, it can save lectures' effort that is wasted in time-consuming chores like correcting and evaluating homework [28, p. 90] but AI will not work if its purpose is to avoid the hard work resulting from searching for competent lectures. In order for the application of AI to progress, it must improve working conditions and increase job satisfaction for lectures. AI should reduce exhaustion and increase the desire to perform the job [29, p. 120]. But if technicians do not work with black lectures, they will not be able to know the conditions that must be changed until the maximum benefit from this technology is achieved [30, p. 75].

We must introduce different programmers into technology and include people of color in all aspects of AI development, while continuing to train lectures on its proper use and introduce regulations and laws to reduce discrimination methods in its application [31, p. 57].

AI technology will continue to disrupt institutions that have always existed, and the education system will suffer from these disruptions. But with supervision, these new systems can be used to produce satisfied lectures, excellent students, and finally—equality in classes [32, p. 111).

4 The Goals of Artificial Intelligence

Learning

One of the most important goals of Artificial Intelligence is the ability to learn by relying on the principle of trial and error [33, p. 156].

Problem Solving

In fact, the feature of problem-solving in this type of intelligence is a systematic method based on a series of procedures, which are relied upon to achieve a set of goals and solutions observed earlier and are divided into private and general solutions [34, p. 35].

Logic and Conclusion

Evolution plays an important role in employing intelligence in a machine, as it is based on the principle of scanning the surrounding environment by the sensory devices that it possesses, whether natural or artificial [35, p. 1825].

Building Perceptions

The secret of this goal lies in reaching the conclusions appropriate to the situation, and the inferences are divided into a deductive or induction and the decision is made accordingly [36, p. 74].

Pronunciation and Language

The concept of language is not just about pronunciation. But also, to the issuance of gestures, signals and reactions in specialized fields, just as modern robots can create dialogues with ease away from complexity [37, p. 649].

5   Advantages of Using Artificial Intelligence in Education

1. It uses non-digital coding, meaning that it is more complex than regular computers that rely on (only one and zero), which means the possibility of making complex decisions [22, p. 137], and its enormous potential that it can add to various fields of study. In addition, Artificial Intelligence programs have the ability to solve problems even with incomplete data, and they can even deal with contradictory and sometimes contradictory data [38, p. 76].
2. In addition to saving time and effort, it can contribute to providing an alternative reality for students, it accustoms students to confrontation and keeping pace with modern technology [39, p. 371].
3. Artificial Intelligence can contribute to presenting questions to students in a way that reveals the weaknesses, and the mental preparations of each student, in addition to following up and exploring the methods of learners [40, p. 39].
4. Artificial Intelligence helps students to choose the right questions. It is also a great space and a vent for them, as experiments have shown that students are more able to converse away from the lecture [41, p. 39].

6 The Impact of Artificial Intelligence on Learning

Artificial Intelligence aims to improve people's lives and double their thinking abilities through the way it interacts with them [37, p. 649], for example, if you search on the Internet for new books, your social media accounts will pick up results signals and feed you with a quantity of ads, suggesting several other nearby books shop [42].

Artificial Intelligence makes it is possible to apply these strategies in the classroom.

1. **Changing what and how we know**

   The focus of the teaching process is less on content and more on how knowledge is used. Artificial Intelligence can provide students with an education that is adapted to the needs of students, through which they can apply their knowledge in new ways [43, p. 117]. As a result, Artificial Intelligence is changing the curricula and methods of communicating information, and in the future, it is expected to rely on a self-directed method of education so that the means allow the student to lead the educational process [44, p. 281].

2. **Preparing students for future jobs**

   While AI allows classroom and lecture adjustments and exercises to change [45, p. 164], it is imperative to warn that today's students will take on different jobs in the future. It is difficult to prepare for an unpredictable situation, but teaching students' ways to think critically, solve and confront difficulties may be the best preparation they get [46, p. 17].

3. **Leaving administrative duties**

   The lecture has to do many things in a specific period, but what if conditions are available for him that makes him not do everything alone? [47, p. 123]Artificial Intelligence will do some of the lecturer's tasks, which will save him time, making him focus on what is most important, which is "direct work with students" [47, p. 5]

4. **Redefining lecture skills needed by successful education**

   Because the lecturer's role in the classroom will change as well as his duties will change, the lecture will not do the daily routines—such as recording attendance and absenteeism and distributing grades—AI will take over [48, p. 246].

7 Why Will the Need for Artificial Intelligence Systems Increase?

Classroom sizes are growing at a much faster rate than university budgets. This can limit individual interaction between students and lectures [49, p. 240]. Some experts argue that the pressure on classroom resources may be reducing proficiency at present in subjects such as mathematics [22, p. 137]. Also, the demand for cheap educational solutions that can supplement the student's time in the classroom makes the educational technology industry markets a paralyzed sector [50, p. 122]. It is striking that this sector, until recently, lacked the required development, which prompted both start-ups and companies that have many years in the industry do not find educational solutions that rely on the use of Artificial Intelligence to interact with students and see where they excel and where they need improvement [49, p. 240].

8 The Future of Artificial Intelligence

Humans' partnership with computers will continue to function because the human component has unique skills. Computers can predict a certain pattern using search algorithms and to collect data in abstract, objective, scientific ways [51, p. 64].

The human element brings emotion and human feeling to the classroom, so the lecture has an important role to play in achieving empathy for educational experiences [52, p. 225].

One of the goals of integrating Artificial Intelligence in the classroom is to use technology and benefit from its capabilities in the best possible way, which are based on academic and educational professional grounds [45, p. 164].

There is a distinguished experience that Finland provided to its citizens with the aim of introducing Artificial Intelligence services and trying to help them understand the best available capabilities and what are the expected future results based on the choices and decisions they make today [45, p. 165], and this initiative has contributed to educating citizens on the awareness of the extent to which the educational system benefits from Artificial Intelligence [53, p. 457].

9 How Will Artificial Intelligence Systems Change the Future of Education?

1. Automating basic educational activities such as student assessment and monitoring of these student grades using intelligence techniques business that uses a lot of lectures' time and effort [54, p. 408].

    Synthetic lecture can give evaluations to students on homework and monitoring grades and saving this time and effort and using it in interaction with the student or preparation lessons or the development of his skill is personally [55, p. 22].

    - Create personalized educational programs for each student in accordance with his needs and skills called adaptive educational programs [49, p. 241]. These programs respond to student needs and focus on topics that interests them, repeating topics that the student has not mastered, and helping the student to achieve better results [56, p. 45].

- Indicate the topics that need further clarification and study for improvement in courses where there are some gaps within the curriculum and not covered by the lecture's explanation and discovered by the computer through intelligence programs [40, p. 18].
- The artificial method by not all students answering a certain question is given in the program is a signal to the lecture that the explanation of this part of the curriculum has not been completed, and it also helps the student by using the program solves this question by giving them the missing information guide them to the correct answer [57, p. 302].

2. Artificial Intelligence programs make students get additional support from lectures especially with regard to high-level thinking and creativity skills, but manufacturers are currently working on developing this aspect in Artificial Intelligence systems and making them comparable to human lectures [58, p. 281).

3. Artificial Intelligence programs act as guides for students and lectures as they guide lectures to discuss topics on which students need training courses and give notes on how successful these courses are by monitoring student progress [10, p. 7].

4. Artificial Intelligence systems can change how they find information and interact with it such as some programs that already exist: Google and Amazon [59, p. 12].

5. This can be applied in the field of education also in the same way, which enables the future student to do the scientific research required of them is better and more interactive than today's students [60, p. 289].

6. Providing students with privacy in learning and assessing their level without the knowledge of their peers, which enhances confidence within himself and makes them accept to make mistakes sometimes in order to learn from their mistakes, they learn in a relatively judgment-free environment and AI provides support to students and solutions to improve their performance [55, p. 14].

7. In the future, Artificial Intelligence systems will change where students and lectures learn and how acquiring skills. The student can learn from anywhere in the world at any time by means of virtual reality technologies. and distance learning technologies that will have a great deal in the future [41, p. 22].

10   The Dimensions of Artificial Intelligence in Education

Many universities are keen to apply the dimensions of Artificial Intelligence to the fullest in order to reach the best experience in education, and among the most important applications of Artificial Intelligence in education:

Automated Grading

Artificial Intelligence applications can be employed in education by monitoring grades for students within the educational environment, so the robot or machine evaluates the student and his knowledge by analyzing his answers and providing feedback [58, p. 282]. Accordingly, appropriate personal training plans are drawn up for each student, in addition to informing the students of the grades obtained, and the use of this method is distinguished from error and favoritism completely [55, p. 24].

Feedback for Lectures

The feedback is the evaluation of students in relation to academic performance and what they have achieved, the feedback is one of the best applications of Artificial Intelligence in education and the most valuable source of information about student performance evaluation at all [19], p. 22], and this application is based on many new technologies such as chats with Artificial Intelligence robots and electronic or machine learning in addition to conduct conversations as is the case in interviews [53], p. 458], and resort to monitor the dimensions of the conversation and adapting it according to the answers provided by the student that reflect his personality, educational and intelligence level [9], p. 2].

Virtual Facilitators

The virtual medium is considered as a means of great benefit in terms of assisting students and providing them with accurate answers that students constantly need, and such an experiment was conducted and proved its worth at Georgia Institute of Technology by a robot supported by the system emanating from Artificial Intelligence [22], p. 138]. This robot was known as Jill Watson and is one of the applications of Artificial Intelligence in education [18], p. 91].

Campus Conversations

Bigdata-madesimple indicated that on-campus chats can be included within the applications of Artificial Intelligence in education, where electronic dialogue sessions are held between students and the robot for the purpose of obtaining aid related to their university matters) [61]. Whether it is the nature of the educational environment on campus or how to get to the lecture hall, find the parking lot, communicate with the faculty, and a lot of valuable information, in which the Artificial Intelligence will benefit the user [18], p. 92].

Personalized Learning

The importance of this application is in meeting that the needs of each separate (learner from his own students) [62], p. 4854]. Where it provides the learner with a series of educational programs that contribute to raise his efficiency in learning, and such applications help in identifying weaknesses of the learner and work to strengthen them through the educational curricula provided with them [9], p. 2], and the applications of Artificial Intelligence in education are characterized by their ability to adapt the needs of students, whether they were individual or group, regardless of complexity [63], p. 114].

Adaptive Learning

Adaptive learning is one of the most useful and important fields and applications of Artificial Intelligence in education, as this type of learning contributes to make remarkable progress through teaching students individually [5, p. 381], and adjustments are made to educational paths and curricula whenever the need arises, and a detailed report is provided to the lecture about materials that are difficult for the student to understand and assimilate [64].

Distance Learning (Proctoring)

Distance education is considered as one of the most prominent types of modern education, and this modern technology includes opportunities to take examinations remotely with the imposition of supervisory systems that are subject to Artificial Intelligence to monitor the student, and verify that there is no cheating, as it is a method by which the credibility and accuracy of the test is verified [62, p. 4854].

Helping Students with Special Needs

The applications of Artificial Intelligence in education are not limited to normal students, but they also meet the students with special needs as well and motivates them to adapt to the educational environment and assimilate educational materials and thus lead them towards the fullness of success, as well as increase the efficiency of students' social skills [65, p. 1323].

11 Disadvantages of Artificial Intelligence in Education

Among the most prominent defects and disadvantages that we may gain from applications of Artificial Intelligence in education:

1. The high cost of implementing Artificial Intelligence applications in education [4, p. 201].
2. An increase in the unemployment rate among the teaching staff [3, p. 25].
3. The possibility of hacking and self-replication of viruses that may infect robots [17, p. 706].
4. The classroom atmosphere is devoid of the spirit of cooperation and harmony that the lecture provides to the student [26, p. 78].
5. Boredom and unwillingness to learn on the part of students through engaging with a machine [36, p. 74].
6. Difficulty in using robots and dealing with them [42].
7. Inflicting a negative impact on human behavior as a result of restricting interaction with the machine [49, p. 241].

The Aims of Intelligence Artificial in Education:

The most prominent uses of this technology in Education were for the following goals:

1. **Specialization**: One of the main problems associated with university classes is explaining the curriculum in one form and template for all students without exception or taking into account the difference between students' cognitive and

study skills [55, p. 23]. Therefore, many leading companies in the field of Artificial Intelligence are working to solve this problem by providing an educational system that customizes the learning process according to the performance and skills of each individual student [60, p. 290].

These systems evaluate the performance and skills of students and based on the performance of each student and the strengths and weaknesses, appropriate lessons are determined for him in order to enhance his strengths and eliminate his weaknesses in relation to the curriculum [4, p. 202]. This system also helps lectures to accurately determine the level of their students and know what each student needs from the curriculum in order to increase success rates [4, p. 202], as many companies provide some programs that can conduct exercises and tests, determine grades, correct answers, and inform students of their performance in those tests [66].

2. **Training**: Artificial Intelligence is used to build websites and smart training programs that can identify and measure students' learning styles and methods, evaluate their knowledge, and then provide customized training according to what each student has obtained from the evaluation [30, p. 76].

3. **Grades**: Correcting tests and determining grades is the most difficult matter in the teaching process, and that this process takes a lot of time that can be saved in matters better for lectures such as planning lessons or developing lectures'skills [5, p. 383]. The companies have provided some programs that can conduct exercises and tests, determine grades, correct answers, and inform students of their performance in those tests [10, p. 8].

4. **Curriculum and teaching quality**: Artificial Intelligence can identify gaps in educational curricula and teaching, based on students'performance in tests and exercises [65, p. 1322]. For example, if many students solve a question incorrectly, the AI technology can identify the problem and the reason behind the students' inability to answer, which helps lectures better explain and focus on specific parts of the curriculum improvement [49, p. 241].

5. **Instant evaluation of students**: In the age of technology, students cannot live without computers and smartphones [10, p. 16], as they use social networking sites, chat programs and sometimes distance learning programs over the Internet all the time based on Artificial Intelligence technology [31, p. 58], by this technology the lecture can immediately evaluate students' cognitive and study skills, which helps them to develop their academic level [27, p. 128].

Test Systems

It is possible to use Artificial Intelligence techniques to change the regular test systems that evaluate students uniformly according to one test, which leads to injustice to the creative student greatly [39, p. 371], because these systems focus directly on model answers in traditional tests, and even if some universities try currently changing the methods of their tests to address the issue of innovation and reliance on abilities and creativity [48, p. 247], but the traditional view is still prevalent, and the tests are seen in their current way as more like memory tests [40, p. 19], in addition to the great burden that falls on the lecture, which is to correct a huge amount of tests annually—especially if these tests are in the form of essay questions—so AI systems can correct a large part of these tests, by translating words and studying patterns very accurately [41, p. 23].

Therefore, Artificial Intelligence systems can fill the gaps in individual differences between learners, and they can also "liberate" lectures from a large part of their responsibilities, which is reflected positively on the educational and research process at the same time, in universities [66], for example, there will be greater opportunities for professors to focus on output [18, p. 91]. At the same time, Artificial Intelligence systems will reduce a large part of the cost to universities—especially government— which will contribute to an increase in the number of seats within universities and institutes [4, p. 201].

The universities applied technology (smart observer) in the exams for some courses that were held remotely, which are used to supervise the implementation of electronic exams and monitor students while performing the test remotely and automatically to develop the electronic testing systems used at the university and adopt the latest technologies in modern electronic testing and monitoring systems [55, p. 24].

This technology enables a faculty member at the university to monitor students' performance while performing their tests with high accuracy, using modern digital means and artificial intelligence that can distinguish between natural and suspicious movements, so that the camera detects every movement that the student makes during the test and analyzes and classifies it [27, p. 128].

12 The Most Important Applications of Artificial Intelligence in Education

Artificial Intelligence is no longer a luxury increase in the field of education, as it has become in the developed countries of the world one of the pillars of educational development, and one of the most important ways to develop academic subjects [65, p. 1322], and among the most important applications of Artificial Intelligence in education are the following:

- Digital systems for universities in the sense of establishing data networks, through which large neural networks can be established, that can anticipate weaknesses and how to treat them for all students, as well as contribute to information management and address problems first, the most important of which is (Class Era) [18, p. 92].
- Contributing to the work of algorithms in the establishment of educational tools that work on reformulating educational curricula and crystallizing them in line with students' interest, in order to reach the shortest paths to deliver the study materials [3, p. 25].
- Developing students' abilities to communicate with human-like systems, which is the biggest motivator for them and is prepared and equipped to deal immediately

with humans in all linguistic and social situations, in order to help and enhance the ability to communicate and increase social skills [44, p. 282].

13   The Most Important Initiatives for Artificial Intelligence in the Field of Education

1. **Thinkster Math**: It is an application that mixes the mathematics curriculum and personal learning style, and this application monitors the mental processing of each student, which gradually reveals on the screen of the smart phone and presents the user with various problems according to his ability [67, p. 252], and as soon as the user writes about how he reached the answer. The application analyzes its work and determines why it was wrong or how it understood a specific aspect of problem solving [3, p. 26]. This application improves logical processing for all students by providing immediate and personal feedback [55, p. 24].

2. **Brainly**: It is a social networking site for classroom questions. The program allows learners to ask homework questions and receive automatic answers from their colleagues [18, p. 93]. This program helps to reach correct answers, and the program has a variety of experts in academic subjects working to create a classroom-like environment [66].

3. **Content Technologies, Inc**. Educators insert curriculum description into the content technology engine, and then content technology machines use algorithms to produce personalized books and study materials, based on the core concepts of the curriculum, and this initiative seeks to help publishers to create effective textbooks that enable each learner to obtain individualized learning [18, p. 92].

4. **Mika site**: This site at Carnegie Mellon University in America provides teaching tools based on Artificial Intelligence for busy learners and students who lack interest of a personal nature [63, p. 112]. Mika specializes in teaching undergraduate students to fill gaps in overcrowded classrooms. The application is guided by the learning process of each student and makes the learner aware of their daily progress, and lessons are adjusted according to each student's needs [62, p. 4855].

5. **Netex Learning**: This site allows lectures to design the curriculum through several digital devices [68, p. 79]. The site helps the most technically trainers to integrate interactive elements such as sound, image and self-assessment into their digital lesson planning, all in a virtual platform for personalized learning [14, p. 162].

    On this site, lectures can create student-specific material ready for publication on any digital platform while delivering videoconferences, digital discussions, personalized assignments, and educational analyzes and illustrates visual displays of each student's personal growth [66].

14   Characteristics of Artificial Intelligence-Based Education Programs

Before dealing with the characteristics of Artificial Intelligence-based in education programs, we will try to present the general characteristics of intelligence synthetic:

- The use of intelligence in solving the presented problems.
- The ability to think and perceive [37, p. 650].
- The ability to acquire and apply knowledge.
- The ability to learn and understand from previous experiences and experiences.
- The ability to use old experiences and employ them in new situations [30, p. 77].
- The ability to use trial and error to explore different matters.
- The ability to quickly respond to new situations and circumstances.
- The ability to deal with difficult and complex cases [24, p. 5].
- The ability to deal with ambiguous situations without information.
- The ability to distinguish the relative importance of the elements of the presented cases [26, p. 78].
- The ability to visualize, create, understand, and perceive visual matters.
- The ability to provide information to support administrative decisions [18, p. 91].

How Do Smart Education Systems Work?

Simulation methods are used and include more interactive learning environments that coerce learners into apply their learned knowledge and skills and thus constitute fertile environments for learners to retrieve and apply the knowledge and skills they have learned more interactively in a practical situation [64].

Characteristics can be identified that must be present in any smart learning program, Smart Teaching distinguishes from traditional teaching programs in six characteristics:

1. **Instantaneous generation of dialogue**: This feature is related to the program's ability to interact with the student in two directions, in traditional programs, the program raises a question and the student responds, but with this feature it became possible for the student to ask his question, in other hand, the program was presented in an interactive form, which made it possible to use the Socratic method of learning which was not available in learning by the computer before [63, p. 113].
2. **Semantic or cognitive networks**: the program is built in the form of knowledge network made up of facts, rules, and the relationships between them. Unlike the traditional program, which divides its content into screens organized in linear or branching form [55, p. 24]. In other words, the smart learning program contains two types of knowledge:

    - Knowledge related to the subject of the program being taught, and it changes according to the change of the program [62, p. 4854].

- Pedagogical knowledge, which is the knowledge related to the rules of teaching of the subject, and it is fixed for each field of specialization.

3. **Student model**: the program decides which information to provide to the learner in the next phase. The program must determine the student's previous knowledge and the student's current knowledge structure [34, p. 36]. This structure is formed by interacting with the student and analyzing his mistakes and this requires the existence of a system to diagnose the mistakes committed by the learner because of some errors independent of the content that may arise from indifference or haste in the answer [30, p. 76].

4. **Rules for diagnosing errors**: The smart teaching system must exploit the students, mistakes in order to correct some concepts. Also, student mistakes can be viewed as symptoms of misconceptions, and for misunderstandings to be diagnosed. The student must know his current state of knowledge, as well as his educational history [4, p. 203].

5. **Natural language processing**: one of the distinguishing features of the smart learning program is interaction through the user's natural language. The quality of the communication between the program and the learner improves dramatically if the program understands the natural language input of the student whether written or spoken [43, p. 118], it also adopts many features of the smart learning program, such as effective dialogue with students, diagnosing student errors on progress in natural language processing that is an area of Artificial Intelligence science [18, p. 91]. The main goal for natural language processing research is to make the communication between a computer and a human being naturally [58, p. 282], i.e., using human language such as Arabic and English. The natural language processing system in any program is divided into two parts: The first part is language comprehension, the second part aims to find ways to make the computer understand the instructions given to it in natural human language [24, p. 4]. It is the production of the natural language, and it aims to make the computer capable of producing a natural language similar to that which a person deals with in his life daily [4, p. 202].

6. **The ability to learn**: this refers to teaching applications through the smart computer that has the ability to change the behavior in teaching according to the behavior of the group of students interacting with them. It may appear to the program that the student is learning a specific topic with one strategy more than others, which leads the program to make it a priority within its teaching strategies [3, p. 25].

15   What Are the Factors Driving the Adoption of Artificial Intelligence?

There are three factors driving the development of AI across Smart Learning:

1. **Provides easy and affordable high-performance computing**: The abundance of education computing power that has enabled easy access to affordable, high-performance computing power. Prior to this development, the only computing environments available for AI were based and cost prohibitive [6, p. 331].
2. **Large amounts of data are available for learning**: Artificial Intelligence needs to learn with a lot of data to make correct predictions [58, p. 282]. The emergence of different tools for collecting disaggregated data, in addition to the ability of universities to store and process this data, whether structured or unstructured data, has enabled more universities to create Artificial Intelligence algorithms easily and cheaply [32, p112].
3. **Applied AI technology provides a competitive advantage**: Companies are increasingly recognizing the competitive advantage of applying AI insights to business goals and making them a business priority [30, p. 76]. For example, targeted recommendations provided by AI technology can help universities make better decisions faster. Many AI features and capabilities can reduce costs, reduce risk, and more [19, p. 19].

Components of Smart Education Systems

Much of the teaching method is based on presenting facts and concepts to learners and then making a test by means of questions, either oral or written [13, p. 835]. This method relies on exposing students to large amounts of information and test their ability to remember [16, p. 611]. It uses smart learning systems, where simulation and more interactive learning environments are forced learners to apply their learned knowledge and skills, they help learners to retrieve and apply knowledge and skills more effectively in scientific situations [19, p. 19]. Learning systems are made up Smartphone from the following basic components:

First—Module Expert

They contain curricula, teaching strategies, and information. The student's knowledge includes exams, questions [31, p. 57], and issues to be taught, which are more than just a representation of the data that it is a model or formulation of a method that represents a skilled person (expert). And it may contain what is called the expert system, which is specialized in providing solutions in areas requiring extraordinary experts [32, p. 112].

Second—Module Student

This unit includes information for all students and is concerned with following up the level of performance, and it gives indications of the student's level of understanding of the scientific subject [30, p. 78].

The ability to monitor errors, which reflects the true level of the learner and helps this unit to how to adapt the educational environment for the learner by analyzing the interaction that takes place between him and the smart system during the exams and solving problems [37, p. 650], and this is what the system is based on his decisions that reflect the student's unity:

1. The current state of knowledge of the student [44, p. 281].
2. The student's level of understanding in learning a specific lesson or subject.
3. The time that the student took to solve a specific problem.
4. The student's level of answering questions and the ability to recall the previous lessons [46, p. 17].
5. Measuring the student's educational behavior (The number of correct times that the student pursues to learn) [45, p. 165].

Third—Module Pedagogy

Based on the information received from the student unit. This unit defines the method of the constructive learning process. It makes educational decisions for each student according to his abilities and absorption and is implemented unit.

These decisions are at the appropriate time, and the methods of education are predetermined, which are:

1. **Teaching**: The system presents a specific lesson to the student within a plan of displaying contents lesson and desired goals [58, p. 282].
2. **Evaluation**: includes issues and questions and appropriate solutions to them [29, p. 121].
3. **Review**: reviews the student's answers and expertly explains concepts and completes student answers with all the information recorded on the system [22, p. 138].

Fourth—Module Explanation

This unit collects information from the specialist knowledge base (curriculum), as well as the student unit information in order to answer the student's questions and provide the appropriate explanation [4, p. 202].

This unit does the following:

1. Determine the contents of the explanation of the topic or the answer to the question [22, p. 137].
2. Determine the appropriate method for the explanation [44, p. 281].
3. Gathering information and arranging sentences to give coherent answers that can be understood [55, p. 23].

Fifth—Interface User Communication Unit

This unit controls the interactions between the student and the system through a group of outputs such as screen design, how to present the scientific material, and how to conduct the dialogue [69, p. 61].

Digital Universities in Education

Universities and students are a great source of data as it is possible to design universities systems capable of simultaneously managing university and student data and saving them in the form of huge databases [24, p. 5]. This big data can be used to train huge neural networks that can predict vulnerability at the individual student level and the shortage of material and human resources at the university level before it occurs. Artificial Intelligence depends on data—a lot of it—so such algorithms will help ministries and universities to make informed decisions about their institutions, which will increase the quality of educational outcomes and reduce the costs of these universities [4, p. 203]. As a simple example, data can be collected for preparing, using and retrieving books and the number of students from universities over previous years, and then predicting the need for books in various universities in the world in the coming year based on the expected number of students in all universities [18, p. 91]. Thus, the ideal quantity of books is sent to universities instead of the increase and decrease that occurs every year and causes students to receive their textbooks late to the middle of the semester sometimes [62, p. 4854].

It is expected that the classrooms will soon move from the traditional framework of learning to the use of a combination of robots and Artificial Intelligence designed as needed, and a large and increasing proportion of students will benefit from robots that are durable and flexible [55, p. 12], and classroom lectures will be freed from administrative matters and will devote themselves to the task of teaching. As for the classroom itself, the options for "specialized services according to needs" provided by Artificial Intelligence techniques would help improve students' enjoyment during classes and improve their grades at the same time [53, p. 457]. And well-trained robots can complement the role of experienced lectures in providing tutoring and extra lessons to strengthen and develop students' skills. This technology can solve the problems of the lack of lectures or the scarcity of qualified lectures in some fields. It will help the average lecture develop his abilities and fill any deficiency he has [49, p. 241].

But it must be noted here that Artificial Intelligence is not supposed to replace innate or natural intelligence [31, p. 59]. The purpose is not to replace the lecture in the classroom or to dispense with him completely, but for the human mind to work side by side with the artificial mind in an elaborate calculated combination. There is another problem that various technologies, software, methods, and Artificial Intelligence applications can also contribute to reducing their effects and is represented in this information explosion and the continuous technical and cognitive development 67, p. 251].

If the development of scientific curricula and the printing of textbooks is a long and complex process that may in turn take 5 years, then with Artificial Intelligence in educational devices and software will be able to extract the knowledge and skills required at a specific time, thus updating the lessons automatically and presenting them to the student in a manner that suits his needs and abilities [21, p. 1].

AI Specialization in the World:

The world's universities are currently racing to prepare advanced academic programs at the highest level of professionalism for new disciplines, which have become necessary in practical life, and are frequently used in marketing and employment operations, and in light of the implementation of the comprehensive development plan in the world, the focus is on education and teaching the new disciplines required on the job market [12, p. 3648].

The Artificial Intelligence specialization is at the forefront of the new majors that are being focused on in many universities around the world, which prepare distinguished academic programs that develop students' skills and capabilities and qualify them scientifically and practically for the requirements of the labor market [5, p. 382].

The specialization of Artificial Intelligence is one of the most important branches of basic computer science that focuses on simulating the human mind in its operations, by dealing with computer devices that have been elaborately manufactured and programmed to accommodate the surrounding environment and carry out the tasks of the human mind [56, p. 47].

Education Transformations in the Post-Corona Era

The world's education systems have witnessed unprecedented turmoil this year due to the Coronavirus pandemic, so most of the world's university closed their doors to more than 1.5 billion students, or more than 90% of the total number of students [62, p. 4853], according to recent figures issued by the UNESCO Institute for Statistics. Education experts have agreed that post-corona education will not be the same as before it [70, p. 201], especially with the emergence of a highly automated infrastructure using the data of the Fourth Industrial Revolution and Artificial Intelligence systems, and that there are expected changes that will be large and structural in education patterns, methods, directions, and policies [71, p. 0135].

The most important new patterns of digital structure in education are the following.

Distance education: This pattern has been used, in many countries of the world, as an alternative to traditional education since the outbreak of the pandemic. In terms of his positive results [63, p. 112].

E-learning: Which combines distance education and education in the classroom, through modern means and mechanisms of communication, including computers, networks and multiple media, combining sound, image, graphics, search mechanisms, and digital libraries [55, p. 23]. The greatest benefit is expected that this educational pattern will prevail in most educational institutions around the world during the foreseeable future, and one of its most important forms is what has become known as blended education, which combines technology-enhanced education with direct education (face to face) [61, 72.

Smart Learning: The trend is increasing towards adopting Artificial Intelligence techniques, in order to promote online education, adaptive learning software, and research tools that allow students to quickly interact, benefit from information, and acquire skills [62, p. 4853].

And the results of many academic studies have indicated that the use of adaptive learning benefits about the student's progress in his educational path, promotes active education, helps struggling students, and assesses the factors affecting student success [65, p. 1322]. However, the effective integration of these new technologies into the university curriculum requires good planning and the provision of the necessary resources [73].

The pattern of Artificial Intelligence is also linked to the use of robots in the field of education, as educational institutions' adoption of robots in teaching is increasing day after day, especially after the success of the experiment with robots that teach languages, as well as the teaching of some basic subjects, as is the case in China and some Scandinavian countries [62, p. 4854].

Shifts in University and Curricula

When the severity of the pandemic subsides, and countries begin to open their university to students—and this will often be in stages—there are significant transformations that will take place. However, these transformations will vary according to the financial capabilities and plans adopted by each country [68, p. 78, 74]. The basic education sector will witness new situations in many countries of the world in general, including:

- **Social distancing**: where access to the classroom will be graded, and the principle of social distancing will be carefully observed. No handshakes, no physical closeness. Friendships, social networks, and much more will remain suspended [75, p. 218].
- **Multiple working hours within the same university building**: the need for social distancing among students will impose fewer of them in the classroom. And then it will become necessary for educational institutions to operate two shifts, and perhaps three shifts every day, especially in overcrowded university. This will undoubtedly put more pressure on the faculty and administrative staff [76, p. 143].
- **Review of study abroad**: All forms of international education have been affected by the pandemic, and this will continue for at least some time. This effect extends to study abroad plans, training programs, and exchange of experiences [62, p. 4855].
- **Universities and support for Artificial Intelligence**: Although Artificial Intelligence has become a very widespread idea, and there are many who support the idea of integrating it and using it in various walks of life, and despite its promises to change the reality of learning forever [62, p. 4854].

Discussion

Smart learning depending on AI is an integrated concept on the use of technologies applications, and all the tools that scientific progress can provide to stimulate the learning process.

Information and communication technology is not an end in itself, but rather a goal that could lead to a qualitative shift in the way learning, and in the value of the teaching and learning process.

Smart learning is based on the judicious use of information technology for its supportive role in the learning process, but we must not forget that electronic equipment is a means that only facilitates and facilitates learning and will not replace education.

Smart learning programs mainly work on student success by activating and tightening the connection between three elements: school, home and infrastructure.

The success of the smart learning experience depends largely on clarity of vision and defining the role of all concerned parties, such as educational institutions, communication companies, community institutions, and even parents, and it depends on incrementalism when applied as well.

The repercussions of Coronavirus pandemic did not affect the work of university education thanks to artificial intelligence and its tools in many countries of the world, which enabled the continuity of education for all students, through distance learning, as teachers and professors were no longer able to go to their universities due to the repercussions of the virus, and proved that learning platforms are cognitive shields that has enabled educational materials at all levels, and has produced remarkable positive outcomes, since the outbreak of the Corona crisis. The virtual education system is characterized by advanced educational characteristics capable of evaluating students' levels, due to its reliance on artificial intelligence techniques adapted to the educational needs of learners, which contributes to bridging knowledge gaps and providing a pioneering educational experience.

Despite all the advantages of implementing smart learning programs, there are those who see that it has many drawbacks, perhaps the most prominent of which is the reduction of the lecture's role, which in turn means the reduction of the human relations aspect within the university.

References

1. Santos, J., Rodrigues, J. J., Casal, J., Saleem, K., & Denisov, V. (2018). Intelligent personal assistants based on internet of things approaches. *IEEE Systems Journal, 12*(2), 1793–1802.
2. Renz, A., Krishnaraja, S., & Gronau, E. (2020). Demystification of Artificial Intelligence in education—How much AI is really in the educational technology? *International Journal of Learning Analytics and Artificial Intelligence for Education, 2*(1), 4–30.
3. Lin, P. H., Wooders, A., Wang, J. T. Y., & Yuan, W. M. (2018). Artificial intelligence, the missing piece of online education? *IEEE Engineering Management Review, 46*(3), 25–28.
4. Al-Emran, M., Malik, S. I., & Al-Kabi, M. N. (2020). *Toward Social Internet of Things (SIoT): Enabling Technologies, Architectures and Applications*, pp. 197–209. Springer
5. Hussain, M., Zhu, W., Zhang, W., Abidi, S. M. R., & Ali, S. (2019). *Artificial Intelligence Review, 52*(1), 381–397.
6. Mostafa, L. (2015). Advanced intelligent system. In *1st International Conference on Advanced Intelligent System and Informatics (AISI2015), November 28–30, Beni Suef, Egypt*, pp. 329–339.

7. Liu, Z., Yang, C., R"udian, S., Liu, S., Zhao, L., & Wang, T. (2019). *Interactive learning environments* (pp. 1–30).
8. Sun, D., Mao, Y., Du, J., Xu, P., Zheng, Q., & Sun, H. (2019). *2019 Eighth International Conference on Educational Innovation through Technology (EITT) (IEEE)*, pp. 87–90.
9. Xing, W., & Du, D. (2019). The use of Artificial Intelligence (AI) in education. *Journal of Educational Computing Research, 57*(3), 1–11.
10. Stefan, A. D. P., & Sharon, K. (2017). Exploring the impact of Artificial Intelligence on teaching and learning in higher education. *Research and Practice in Technology Enhanced Learning, 1*, 3–13.
11. Roll, I., & Wylie, R. (2016). Evolution and revolution in Artificial Intelligence in education. *International Journal of Artificial Intelligence in Education, 26*(2), 582–599.
12. Strubell, E., Ganesh, A., & McCallum, A. (2019). Energy and policy considerations for deep learning in NLP. In *Proceedings of the 57th Annual Meeting of the Association for Computational Linguistics*, pp. 3645–3650.
13. El Naqa, I., Ruan, D., Valdes, G., Dekker, A., McNutt, T., Ge, Y., Wu, Q. J., Oh, J. H., Thor, M., Smith, W., et al. (2018). Machine learning and modeling: Data, validation, communication challenges. *Medical Physics, 45*(10), e834–e840.
14. Velarde Hermida, O., & Casas-Mas, B. (2019). An empirical review on the effects of ICT on the humanist thinking. *Observatorio (OBS*) Journal, 13*(1), 153–171.
15. León, G. C., & y Viña, S. M. (2017). La inteligencia artificial en la educación superior. *Oportunidades y Amenazas. INNOVA Research Journal, 2*(8), 412–422.
16. Cairns, M. M. L. (2017). Computers in education: The impact on schools and classrooms. *Life schools classrooms* (pp. 603–617). Springer.
17. Timms, M. J. (2016). Letting artificial intelligence in education out of the box: Educational cobots and smart classrooms. *International Journal of Artificial Intelligence Education, 26*(2), 701–712.
18. Fang, Y., Chen, P., Cai, G., Lau, F. C. M., Liew, S. C., & Han, G. (2019). Outagelimit-approaching channel coding for future wireless communications: Root-protograph low-density parity-check codes. *IEEE Vehicular Technology Magazine, 14*(2), 85–93.
19. Chassignol, M., Khoroshavin, A., Klimova, A., & Bilyatdinova, A. (2018). Artificial intelligence trends in education: A narrative overview. *Procedia Computer Science, 136*, 16–24.
20. Kim, Y., Soyata, T., & Behnagh, R. F. (2018). Towards emotionally aware AI smart classroom: Current issues and directions for engineering and education. *IEEE Access, 6*, 5308–5331.
21. Sharma, R. C., Kawachi, P., & Bozkurt, A. (2019). The landscape of Artificial Intelligence in open, online and distance education: Promises and concerns. *Asian Journal of Distance Education, 14*(2), 1–2.
22. Pokrivcakova, S. (2019). Preparing teachers for the application of AI-powered technologies in foreign language education. *Journal of Language and Cultural Education, 7*(3), 135–153.
23. Wartman, S. A., & Combs, C. D. (2018). Medical education must move from the information age to the age of Artificial Intelligence. *Academic Medicine, 93*(8), 1107–1109.
24. Sutton, H. (2019). Minimize online cheating through proctoring, consequences. *Recruiting Retaining Adult Learners, 21*(5), 1–5.
25. Crowe, D., LaPierre, M., & Kebritchi, M. (2020). Knowledge based artificial augmentation intelligence technology: Next step in academic instructional tools for distance learning. *TechTrends, 61*(5), 494–506.
26. Xiao, R., Xiao, H. M., & Shang, J. J. (2020). Artificial Intelligence and educational reform: Prospects, difficulties, and strategies. *China Educational Technology, 4*, 75–86.
27. Qi, H. Y., & Han, L. P. (2020). How to use the Internet to improve the quality of rural primary education and teaching. *Western China Quality Education, 6*, 127–128.
28. Jiang, F. Y. (2019). Challenges and changes of elementary education in the era of "Internet+." *Education Modernization, 49*, 90–91.
29. Cheng, M., & Wang, X. Y. (2020). Research on the quality evaluation of network teaching in the smart classroom. *Journal of Fujian Computer, 36*(2), 120–121.

30. Hsu, C. C., & Wang, T. I. (2018). Applying game mechanics and student-generated questions to an online puzzle-based game learning system to promote algorithmic thinking skills. *Computers & Education, 121*, 73–88.
31. Shang, Y. F. (2019). Classification evaluation system of network teaching quality based on mobile terminal. *Journal of Anyang Institute of Technology, 18*(6), 56–59.
32. Yu, Z. G. (2020). Using network resources to optimize primary school Chinese teaching. *Learning Weekly, 7*, 111–112.
33. Li, M., Su, Y. (2020). Evaluation of online teaching quality of basic education based on artificial intelligence. *International Journal of Engineering Pedagogy, 15*(16),147–161.
34. Salem, A. B. M. (2019). Computational intelligence in smart education and learning. In *Proceedings of the International Conference on Information and Communication Technology in Business and Education; University of Economics: Varna, Bullgaria*, pp. 30–40.
35. Salem, A. B. M., Nikitaeva, N. (2019). Knowledge engineering paradigms for smart education and smart learning systems. In *Proceedings of the 42nd International Convention of the MIPRO Croatian Society, Opatija, Croatia, 20–24 May*, pp. 1823–1826.
36. Voskoglou, M. G. (2019). An application of the "5 E's" Instructional treatment for teaching the concept of the fuzzy set. *Sumerianz Journal of Education, Linguistics and Literature, 2*, 73–76.
37. Bhatt, U., Xiang, A., Sharma, S., Weller, A., Taly, A., Jia, Y., Ghosh, J., Puri, R., Moura, J., & Eckersley, P. (2020). Explainable machine learning in deployment. In *Proceedings of the 2020 Conference on Fairness, Accountability, and Transparency (FAT* '20), Association for Computing Machinery*, pp. 648–657.
38. Agrawal, A., Gans J., & Goldfarb, A. (2018). *Prediction machines: The simple economics of Artificial Intelligence*. Harvard Business Review Press.
39. Rosert, E., & Sauer, F. (2019). Prohibiting autonomous weapons: Put human dignity first. *Global Policy, 10*, 370–375.
40. Zawacki-Richter, O., Marín, V. I., Bond, M., & Gouverneur, F. (2019). Systematic review of research on Artificial Intelligence applications in higher education–where are the educators? *International Journal of Educational Technology in Higher Education, 16*, 39.
41. Zawacki-Richter, O., Marín, V. I., Bond, M., & Gouverneur, F. (2019). Systematic review of research on Artificial Intelligence applications in higher education—Where are the educators? *International Journal of Educational Technology in Higher Education, 16*(39), 1–27.
42. Chai, C.S., Lin, P.Y., Jong, M. S. Y., Dai, Y., Chiu, T. K. F., & Huang, B. Y. (2020). Factors influencing students' behavioral intention to continue Artificial Intelligence learning. In *Proceedings of the International Symposium on Educational Technology, (ISET020), Bangkok, Thailand*, pp. 24–27.
43. Cautela, C., Mortati, M., Dell'Era, C., & Gastaldi, L. (2019). The impact of artificial intelligence on design thinking practice: insights from the ecosystem of startups. *Strategy Design Research Journal, 12*, 114–134.
44. Fryer, L. K., Nakao, K., & Thompson, A. (2019). Chatbot learning partners: Connecting learning experiences, interests and competence. *Computers in Human Behaviors, 93*, 279–289.
45. Li, H., Gobert, J., & Dickler, R. (2019). Evaluating the transfer of scaffolded inquiry: What sticks and does it last? In S. Isotani, E. Millán, A. Ogan, P. Hastings, B. McLaren, & R. Luckin (Eds.), *Artificial intelligence in education*(pp. 163–168). Cham: Springer.
46. Ma, Y., & Siau, K. L. (2018). Artificial intelligence impacts on higher education. In *MWAIS 2018 Proceedings. 42. Proceedings of the Thirteenth Midwest Association for Information Systems Conference, Saint Louis, Missouri May, 17–18*.
47. Villegas-Ch, W., Arias-navarrete, A., & Palacios-pacheco, X. (2020). Proposal of an architecture for the integration of a Chatbot with artificial intelligence in a smart campus for the improvement of learning. *Sustainability, 12*, 150.
48. Hill, J., Ford, W. R., & Farreras, I. G. (2015). Computers in Human Behavior Real conversations with Artificial Intelligence: A comparison between human—Human online conversations and human—Chatbot conversations. *Computers in Human Behavior, 49*, 245–250.

49. Markham, I. S., Mathieu, R. G., & Wray, B. A. (2020). Kanban setting through Artificial Intelligence: A comparative study of artificial neural networks and decision trees. *Integrated Manufacturing Systems, 11*, 239–246.
50. Peng, M.Y.-P., Tuan, S.-H., Liu, F.-C. (2017). Establishment of business intelligence and big data analysis for higher education. In *Proceedings of the International Conference on Business and Information Management, Beijing, China, 23–25 July*, pp. 121–125.
51. Duan, Y., Edwards, J. S., & Dwivedi, Y. K. (2019). Artificial Intelligence for decision making in the era of Big Data, Evolution, challenges and research agenda. *International Journal of Information Management, 48*, 63–71.
52. Goksel, N., & Bozkurt, A. (2019). Artificial Intelligence in education: current insights and future perspectives. In S. Sisman-Ugur, & G. Kurubacak (Eds.), *Handbook of research on learning in the age of transhumanism* (pp. 224–236). Hershey, PA: IGI Global.
53. Loftus, M., & Madden, M. G. (2020). A pedagogy of data and Artificial Intelligence for student subjectification. *Teaching in Higher Education, 25*, 456–475.
54. Malik, G., Tayal, D. K., & Vij, S. (2019). An analysis of the role of Artificial Intelligence in education and teaching. In P. Sa, S. Bakshi, I. Hatzilygeroudis, & M. Sahoo(Eds.), *Recent findings in intelligent computing techniques. advances in intelligent systems and computing* (pp. 407–417). Singapore: Springer.
55. Renz, A., Krishnaraja, S., & Gronau, E. (2020). Demystification of Artificial Intelligence in education–How much ai is really in the educational technology? *International Journal of Learning Analytics and Artificial Intelligence for Education (IJAI), 2*, 4–30.
56. Topol, E. J. (2019). High-performance medicine: The convergence of human and Artificial Intelligence. *Nature Medicine, 25*, 44–56.
57. Knox, J. (2020). Artificial Intelligence and education in China. *Learning, Media and Technology, 45*(3), 298–311.
58. Silander, C., & Stigmar, M. (2019). Individual growth or institutional development? Ideological perspectives on motives behind Swedish higher education teacher training. *Higher Education: The International Journal of Higher Education Research, 77*, 265–281.
59. Renz, A., & Hilbig, R. (2020). Prerequisites for Artificial Intelligence in further education: Identification of drivers, barriers, and business models of educational technology companies. *International Journal of Educational Technology in Higher Education, 17*(14), 2–21.
60. Perez, S., Massey-Allard, J., Butler, D., Ives, J., Bonn, D., Yee, N., & Roll, I. (2017). Identifying productive inquiry in virtual labs using sequence mining. *International Conference on Artificial Intelligence in Education* (pp. 287–298). Springer.
61. Xu, X., Jiang, X., Ma, C., Du, P., Li, X., Lv, S., Yu, L., Chen, Y., Su, J., Lang, G., Li, Y., Zhao, H., Xu, K., Ruan, L., & Wu, W. (2020). Deep learning system to screen coronavirus disease 2019 pneumonia, arXiv:2002.09334. http://arxiv.org/abs/2002.09334.
62. Di Vaio, A., Boccia, F., Landriani, L., & Palladino, R. (2020). Artifcial intelligence in the agri-food system: Rethinking sustainable business models in the COVID-19 scenario. *Sustainability, 12*(12), 4851–4872.
63. Butt, C., Gill, J., Chun, D., & Babu. B. A. (2019). Deep learning system to screen coronavirus disease 2019 pneumonia, Appl. Intell., 110–126.
64. Schmelzer, R. (2020) Artificial or human intelligence? *Companies Faking AI. Forbes, 4.*
65. Gupta, S. B., & Gupta, M. (2020). Technology and E-learning in higher education. *Technology, International Journal of Advanced Science and Technology, 29*(4), 1320–1325.
66. Mruthyunjaya, V., & Jankowski, C. (2020). Human-Augmented robotic intelligence (HARI) for human-robot interaction. In K. Arai, R. Bhatia, & S. Kapoor (Eds.), *Intelligent Systems and Computing, Proceedings of the Future Technologies Conference (FTC), San Francisco, CA, USA, 2019*, p. 1070. Springer: Cham, Switzerland.
67. Hassani, H., Silva, E. S., Unger, S., TajMazinani, M., & Feely, S. M. (2020). Artificial Intelligence (AI) or Intelligence Augmentation (IA): What Is the Future? *AI, 1*, 143–155.
68. Pravat Kumar Jena. (2020). Impact of covid-19 on higher education in India. *International Journal of Advanced Education and Research, 5*(3), 77–81.

69. Zheng, N. N., Liu, Z. Y., Ren, P. J., Ma, Y. Q., Chen, S. T., Yu, S. Y., Xue, J. R., Chen, B. D., & Wang, F. Y. (2017). Hybrid-augmented intelligence: Collaboration and cognition. *Frontiers of Information Technology and Electronic Engineering, 18,* 153–179.
70. Becker. (2019). Artificial Intelligence in medicine: What is it doing for us today? *Health Policy Technology, 8*(2), 198–205.
71. Naudé, W. (2020). Artificial Intelligence against COVID-19: An early review, IZA Inst. Labor Econ., Maastricht, The Netherlands, Tech. Rep.
72. Afana, A., Agha, E. L., & A. (2019). The role of organizational environment in enhancing managerial empowerment in Al-AQSA network for media and art production. *International Journal of Business Ethics and Governance, 2*(2), 30–63. https://doi.org/10.51325/ijbeg.v2i 2.40
73. Alshurafat, H., Al Shbail, M. O., & Mansour, E. (2021). Strengths and weaknesses of forensic accounting: An implication on the socio-economic development. *Journal of Business and Socio-economic Development, 1*(2), 85–105. https://doi.org/10.1108/JBSED-03-2021-0026
74. Alzaneen, R., & Mahmoud, A. (2019). The role of management information systems in strengthening the administrative governance in ministry of education and higher education in Gaza. *International Journal of Business Ethics and Governance, 2*(3), 1–43. https://doi.org/10.51325/ijbeg.v2i3.44
75. Pravat Ku Jena. (2020). Challenges and Opportunities created by Covid-19 for ODL: A case study of IGNOU. *International Journal for Innovative Research in Multidisciplinary Filed, 6*(5), 217–222.
76. Pravat Ku Jena. (2020). Impact of pandemic COVID-19 on education in India. *Purakala, 31*(46), 142–149.

The Governmental Support in Distance Education and Digital Learning Management Systems

The Government Support in Distance Education: Case of Bahrain

Layla Faisal Alhalwachi, Amira Karam, and Allam Hamdan

1 Introduction

The COVID-19 pandemic has conveyed the pivotal role of technology in combating the spread of the virus without the sacrifice of education and knowledge. Technology has resulted in any impartial reason for outside gatherings to desist as there is seemingly an application for a wide range of necessities such as: groceries, classrooms, entertainment, as well as accessing financial services and any governmental documents. This is a clear connotation that technology is no longer a simple luxury, but a necessity when considering crisis management.

The temporary shutdown of educational establishments has been the tactic employed by most governments as an attempt to damper the spread of the Covid-19 virus. This tactic has also consequently affected a large statistic of all current students worldwide.

Some countries were more vulnerable to the effects of the pandemic and may have more heavily disadvantaged groups than other countries, which is the reason UNESCO is striving to accomplish extra support to the groups that require it the most as a method to maintain the education continuity.

The Coronavirus pandemic has created the largest disruption in education systems around the world in history, affecting nearly 1.6 billion students in 190 countries and regions around the world, according to a report issued by the United Nations on the impact of the pandemic on education [1]. The Covid-19 virus has been the most substantial inhibition in worldwide education systems in human history. In accordance with a UN report on the effects of the pandemic regarding education, it has been stated that there have been approximately 1.6 billion affectees spanning

L. F. Alhalwachi · A. Hamdan
Ahlia University, Manama, Bahrain

A. Karam (✉)
Bahrain Institute for Banking and Finance BIBF, Manama, Bahrain

© The Author(s), under exclusive license to Springer Nature Switzerland AG 2022
A. Hamdan et al. (eds.), *Technologies, Artificial Intelligence and the Future of Learning Post-COVID-19*, Studies in Computational Intelligence 1019,
https://doi.org/10.1007/978-3-030-93921-2_32

190 individual countries globally [1]. On the other hand, the effects of this pandemic have not all been negative, for it has been an opportunity to resolve classroom density issues in schools as well as universities that are typically found in less economically developed countries. This issue has been observed as a roadblock to a successful learning environment in which the students may learn in maximum fruition. The improvised system of social learning has carried out the huge burden for solving this issue for classrooms worldwide. This issue would have otherwise been neglected, or not been achieved if the classrooms were to operate as normal.

Education in the Kingdom of Bahrain keeps pace with the development of technology before the crisis of Covid 19, in all universities, institutes and public and private schools, but not basically and at all times it is applied. After the crisis experienced by the world, especially in the Kingdom of Bahrain, all educational institutions relied mainly on technology in distance learning to continue the educational process, and various programs such as Microsoft teams and other modern programs were used to keep pace with the technological age.

2 Distance Education

The Corona pandemic forced many countries to start implementing and extending the concept of distance education [2]. As education has become one of the most affected sectors after the disruptions caused by the Corona pandemic. In the face of closures, educational institutions around the world are rapidly adapting to distance education in emergency situations. At a time when the United Nations UNESCO reported that 290 million students worldwide have stopped going to school due to the emerging corona virus [3]. The Corona crisis will lead to a change in the way the world views education, despite some of its disadvantages, distance education remains an alternative to traditional education in critical situations [4].

Distance education of all kinds has become a matter of concern to the most prestigious schools and universities, and education centres and trainers, with the tremendous development that the world is witnessing today, especially in the field of information technology and the expanding horizons of educational technology [5]. Distance education is basically a form of modern education in which the curricula are presented through the use of electronic media in the educational process without commitment to a specific time or place. As [6] defined it as a means to support the educational process and its transformation from indoctrination to creativity, developing thinking skills, solving problems, and providing teachers with skills to deal with modern technologies and interaction, which increases the expansion of the concept of self-education according to their previous experiences and skills. In addition, Marek et al. [7] defined distance education as one of the modes of education that has emerged in recent times due to the spread of the Internet and the development of educational methods and their suitability for the virtual world that provides an integrated educational environment for communication between students and professors via the Internet away from the obstacles of geographical distances.

2.1 Platforms of Distance Education and Its Impact

Distance education platforms are effective social learning environments to support teamwork and the exchange of opinions and experiences, as these social educational platforms intend to achieve permanent communication between their users without regard to place and time [8]. Where studies have shown that learners interact to a greater degree in educational materials and courses offered through educational platforms, so distance education platforms can constitute an information system that schools, universities and institutions can use in the educational process, either entirely through the Internet or by integrating it with the traditional method of education [9].

Distance education platforms have many positives as they clearly and significantly help in raising the level of students and scientific communities, as it provides alternatives for people who are unable to go to educational institutions under a circumstance that prevents them from doing so [10]. Also, these platforms work to employ students' abilities and even develop them instead of wasting and losing them. Moreover, it contributes to filling the gaps that may arise from the lack of teaching staff [11]. Furthermore, [12] mentioned that the online education platforms help students exchange views and ideas, which helps creative thinking, and enables teachers to create virtual classes for students, in addition to the ease of communication between the teacher and parents, which allows parents to follow the results of their children, since it helps teachers track the performance of their students in performing certain skills, and their progress. In addition, the distance education platforms have a role in providing learners with the necessary information skills for self-learning, developing creative thinking and making the learner more in control of the educational process and time management [13].

Furthermore, [14] showed that there has become an urgent educational necessity to use distance education platforms due to the many benefits and advantages they offer to the learner, and the reduction in the burdens that fall on the shoulders of the teacher, as well as an attempt to provide an effective educational climate that helps raise the efficiency of the educational process and achieve the overall quality and produce it in a good way. However, distance education platforms face many obstacles that stand in their way and limit its successes, such as the absence of real support from official institutions, the lack of competencies that are interested in developing self-education, and the absence of a culture of volunteering and initiative to produce educational materials [15].

2.2 Content Delivery of Distance Education

Distance education content refers to the information and knowledge contained in the educational material, which aims to achieve established educational goals [16]. To achieve success in any educational system, there are some means that teachers and students need in exchanging information, since distance education depends on a set of main tools that educational institutions of all kinds depend on, thus achieving the objectives of the educational process in conveying the necessary information from the teacher to the students, and then testing the students in the extent of their understanding of the information and knowledge they receive [17]. Therefore, distance education tools are selected by determining the appropriate type of media for each concept or part of the content, represented by static or animated images, video clips, and various fonts, which are determined according to the content and objectives of the program, and the characteristics of the learners [18].

As creating an appropriate environment for content delivery between students and teachers contributes to improving the outcomes of the learning process, as the multiplicity of communication channels develop the learning process and achieves its goals for students to obtain the greatest possible benefit from the information provided by teachers [19]. The distance education system is characterized by its complete dependence on the Internet to facilitate the educational process, as this type uses many means to complete the educational process that may range from e-mails, dialogue sessions, and live broadcast sessions, in addition to some references and printed materials in some cases, thus, this educational model is managed through an integrated virtual learning system that integrates technologies and multiple media in one place [6]. For instance, Edmodo platform is an interactive learning environment that employs Web 2 technology and combines the advantages of electronic content management systems with social networks Facebook, as it enables teachers to conduct electronic exams, distributes roles, and divides students into work groups, helps to exchange ideas and opinions between teachers and students, and shares scientific content, and allows parents to communicate with teachers and see the results of their children, which helps to achieve high-quality educational outcomes.

2.3 Distance Learning Effect from the Students' Perspective/Problems Associated with Online Teaching and Learning

With the start of adopting the distance education system in many countries of the world, there were many views on its feasibility and effectiveness [20]. As students and parents confirmed that the distance education system is the platform of the future and contains many features, most notably the interaction between the student and the teacher, as well as retrieving the lesson at any time, and also encourages parents to educate electronically [21]. In addition, for students who do not fit into the traditional classroom environment, distance education provides an exceptional opportunity because it provides the flexibility they need to succeed. On the other hand, [22] illustrated that many students believe that distance education cannot be a substitute for traditional education, whatever the circumstances, especially in scientific and technical subjects such as mathematics, biology, chemistry and other subjects that require practical and field applications.

Like all learning models, distance education suffers from some inherent problems, especially in the areas of isolation, support, technology, and discipline [23]. In the beginning, the distance education system is not simple and requires study and intelligence in implementation and application, so there is a need for qualified staff capable of managing this technical system [2, 24 25. In addition, [26] pointed out to the loss of the human factor in the distance education system, and the absence of effective dialogue and discussion, as many students are unable to express their ideas in writing, and they need direct verbal communication to express what they believe. Furthermore, [27] indicated the lack of an interactive study environment that raises the performance of the teacher and the response of the student, as the scientific material is often limited to the theoretical side and neglects the live experiences, especially in scientific subjects.

One of the most challenges facing distance education is the limited ability of educational institutions to establish large networks and provide large numbers of devices and equipment, in addition to its update, especially as the information and communication technologies are witnessing multiple developments and transformations, in a rapid and continuous manner, which makes it difficult to acquire various of these technologies [17, 28–30. Moreover, explained that there are a large number of current academic staff in developing countries who are unable to use digital technology in a way that enables them to teach through it. Also, [31] revealed that the academic achievement of students in distance education is very weak, compared to the traditional education system, this is in addition to the fact that learners' withdrawal from school and receiving lessons from a distance lose the sense of prestige and the usual order in school hours, as well as a loss of justice in evaluating learners and a decrease in the level of creativity, innovation and development.

2.4 Governmental Support at Distance Education During Covid-19

The Kingdom of Bahrain has succeeded in overcoming the exceptional circumstances and challenges imposed by the Coronavirus pandemic, placing the health and safety of its children at the forefront of its priorities by implementing the decision to suspend studies in all higher education institutions, public and private schools and kindergartens as of March 17 of the last year 2020, with continuation of the educational process through digital platforms and the work of the administrative and educational staff through a combination of formal and distance education, taking into account the precautionary measures, as it has an advanced and integrated system of communications and information technology.

The idea of distance education in the Kingdom of Bahrain was not a product of emergency conditions during the period of the Corona pandemic, as the various educational bodies in the Kingdom were ready for such crises. This is represented in training the learning community on various e-learning methods, creating electronic platforms and virtual classes, and preparing smart educational programs and lessons, so that the Bahraini learning community is ready to implement the idea of distance education. The Bahraini Ministry of Education has taken proactive steps at the regional level by investing in information and communication technologies and employing them for the service of education since 2005, by launching His Majesty King Hamad's project for future schools, the most important of which was the development of a suitable infrastructure that is the first of its kind in the region, the provision of e-learning platforms for schools and universities, the use of digital educational tools, in addition to the integration of technology in education. During the COVID-19 pandemic, the Bahraini government, has provided all communication channels between the student, the teacher and the parent through the educational portal, televised lessons, and live lessons through virtual classrooms and others, so as to ensure the continuation of the education process at a rate of 100% for all students using "Teams" at various levels.

To sum up, distance education is a distinct choice for those who know how to benefit from its advantages, and how to develop the educational learning process to suit it, because it will lead the educational future in the world, where the education process through the inverted classrooms will become the natural view of education. In the end, the Kingdom of Bahrain, with the leadership it has achieved for the world in the field of distance education, is able to benefit from the experience and develop it in line with the global scientific openness and educational contexts, so that it remains a pioneer in its field.

3 Research Methodology

Quantitative design is the approach that is used in this study in order to perform the research work and observe the support of distance education in Bahrain, by addressing the key study questions referred to in the introduction part.

3.1 Data Collection

The research collected quantitative data from online designed surveys. The survey questions were created by the researchers based on the experience and knowledge gained from the reviewed theories.

3.2 Study Population and Sampling

This study targeted students in Bahraini universities including public and private universities. According to statistics for higher education in Bahrain, the number of students in Bahraini universities is 38,113 students. This study has used 5% margin error and 95% confidence level to come up with 381 sample size as the minimum responses required. However, we received 486 responses which are above the minimum size of sample. All survey responses are valid for analysis and no exclusion has been made.

4 Data Analysis and Interpretation

This part presents and summarizes the findings from the methods used in this study, the quantitative data collected from survey questionnaires as mentioned earlier in research methodology. This part contains three statistical analyses presented in five parts. The first section is about "Demographic Analysis". The second section is "Descriptive Analysis" that presents quantitative data using two types of testing, frequency and percentage. The third section is "Path Analysis".

4.1 Demographic Analysis

Table 1 presents the demographic information collected, such as age, education level, etc. Table 1 introduces the population alongside with some important details in regard to the population. The questions that were asked to the population are included in an easy to assess format. The population size is also included in numbers as well as percentages.

Table 1 illustrates the demographic data of the participants whom 40% of the sample were males and 60% were females. There was age variety as well which enhanced the data collected as it is covering different age groups. 86% of the sample were university students and 12% were master's degree students where only 1% were diploma students.

The age group majority was from 17 to 21, mainly Bachelor degree students which is the main focus and aim of this research to analyze their feedback and views on the support they have received from different entities and processes whether national policies encourage and support distance education or not, whether labour market values distance graduates equally, the level of employers support distance learning...etc.

Figure 1 shows the viewpoint of the study sample about the support students in distance education from several parties. The students seem to have a main preference of studying at home, as observed with 62% of the sample usually preferring to access

Table 1 Demographic data

Question	Answer options	Frequency	Percent
Gender	Male	195	40
	Female	291	60
	Total	486	100
Age	From 17 to 21 years	279	57
	From 22 to 26 years	120	25
	From 27 to 31 years	39	8
	From 32 to 36 years	18	4
	Greater than 36 years	21	4
	Missing data	9	2
	Total	486	100
Do you have a job now?	Yes	369	76
	No	117	24
	Total	486	100
University	Public university	348	72
	Private university	138	28
	Total	486	100
What level are you studying?	Diploma degree	6	1
	Bachelor degree	420	86
	Master degree	60	12
	Total	486	100

lectures in their own abodes, while 13% only access it from coffee shops. Whereas, only 1% accessed classes from Public libraries.

4.2 Descriptive Analysis

A. Supporting of Distance Education

Table 2 which displays the values of the support of distance education introduces the spread of the sample's opinions regarding the amount of support provided to

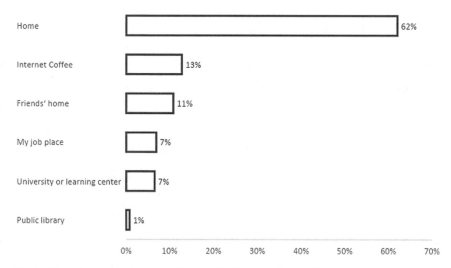

Fig. 1 Where do you usually use the Internet for distance education?

distance education. It is observed through the results that the population does not believe that there is strong support for distance education. Only between 1 and 6% would strongly agree that there are tools or support for those students that undergo distance education.

The trend of the answers follows through and is visible that the majority of the questionnaire answers provide us with the knowledge that people do not believe that online education is supported. The people that had the opinion that online education is not supported by sufficient materials, remained with this opinion as there is a very low standard deviation all throughout. The mean is approximately 3.5 ± 0.45, with on average the mean was chosen by 60–80% of the participants who answered the questionnaire.

B. Rating the Distance Education Support

The participants have rated the support of distance education from 1 to 10. The types of support, however, were more specific in this section of the questionnaire. The type of education which is believed to be the most supported is family support with the clear majority believing it to be a perfect 10 rating in terms of online learning support. Friends support came next with 222 responses with perfect 10. The government support is still highly recognized with 156 responses with perfect 10 where only 12 responses rated 1; Kingdom of Bahrain has proven to lead when it comes to providing Distance Education Support (refer to Table 3).

Figure 2 shows the type of education which is believed to be the most supported is family support with the clear majority believing it to be a perfect 10 rating in terms of online learning support. Friend's support came next with 222 responses with perfect 10. The least was the societal support. As it is shown in the graph, the family and government support took the lead as majority of the responses—highlighted in Table

Table 2 Supporting of distance education

Statements	Totally agree	Agree	Don't know	Disagree	Totally disagree	SD	Mean	Mean percent
National policies encourage and support distance education	9	36	114	195	129	0.97	3.83	77
	2%	7%	23%	40%	27%			
Labour market values distance graduates equally	15	69	210	135	57	0.96	3.31	66
	3%	14%	43%	28%	12%			
Employers support distance learning	12	51	171	165	87	0.98	3.54	71
	2%	10%	35%	34%	18%			
More people/companies involved in distance education	3	39	108	246	90	0.86	3.78	76
	1%	8%	22%	51%	19%			
Distance education institutions began to emphasize learner support	12	39	93	261	78	0.91	3.73	75
	2%	8%	19%	54%	16%			
The necessary infrastructure is available for distance education	21	60	87	222	93	1.06	3.63	73
	4%	12%	18%	46%	19%			
There is community and government support for poor families to obtain distance education	12	33	84	195	162	1.00	3.95	79
	2%	7%	17%	40%	33%			
There is community and government support for the special needs to obtain distance education	6	36	117	183	144	0.96	3.87	77
	1%	7%	24%	38%	30%			

2 and the graph—gave a perfect 10 to both family with 222 responses and government with 156.

4.3 Path Analysis

A.   Supporting of Distance Education Based on Gender

Table 3 Rating of distance education support

Distance education support	Rating									
	1	2	3	4	5	6	7	8	9	10
Family support	12	6	15	18	24	30	18	57	45	261
Governmental support	12	18	21	24	51	18	60	66	60	156
Societal support	24	15	15	24	51	54	66	75	39	123
Classmates support	12	15	12	36	30	42	57	90	69	123
University support	24	6	12	12	24	27	66	108	54	153
Friends support	21	12	9	12	15	27	33	57	78	222

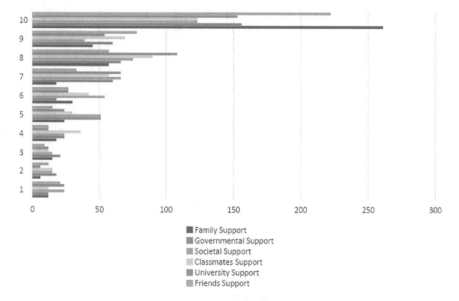

Fig. 2 Distance education support

The Paired Samples t Test or also sometimes referred to as the dependent samples test compares the means of two measurements taken from both genders with regard to the support of distance education as the dependent variable and a set of independent variables represented by national policies, labour market, employer's support, government support and supporting of special needs. The purpose of the test is to determine whether there is statistical evidence that the mean difference between paired observations from both males and females is significantly different from zero. The Paired Samples t Test is a parametric test.

Table 4 reflects that there are no differences in the male and female perspectives on supporting national policies and the labor market for distance education in Bahrain ($p\text{-value} > 0.05$).

Table 4 Paired samples test and Wilcoxon Z test

Support of distance education	Gender		Correlation		Paired samples test		Wilcoxon Z test	
	Male	Female	ρ	p-value	T-test	p-value	Z-test	p-value
National policies	3.831	3.823	0.006	0.929	0.051	0.959	−0.128	0.899
Labour market	3.246	3.354	0.033	0.644	−1.118	0.265	−1.030	0.303
Employer's support	3.338	3.682	−0.199	0.005	−3.143***	0.002	−3.215***	0.001
Government support	4.169	3.805	−0.059	0.414	3.609***	0.000	−3.764***	0.000
Supporting of special needs	4.123	3.703	0.026	0.713	4.590***	0.000	−4.239***	0.000

*** 1% level

While Table 4 reflects that there is a difference in the views of males and females in Bahrain about Employer's support, Government support and Supporting of special needs where the opinions of women in the study sample were higher than those of men with statistical significance.

In the above table, the t statistic appears to the left of the degrees of freedom. The values of 0.051, −1.118, and −3.143 are less than 1 for National policies, Labour market and Employer's support. Whereas the T value for government support and supporting of special needs is 3.609 and 4.590 indicating that the sample difference in means is almost 4 standard errors and 5 standard errors respectively to the right where the null hypotheses say the center of the sampling distribution is. Wilcoxon Rank-Sum produces a test statistic value (i.e., z-score), which is converted into a "p-value." A p-value is the probability that the null hypothesis—that both populations are the same—is true (Table 5). In other words, a lower p-value reflects a value that is more significantly different across populations.

B. Supporting of Distance Education Based on Age

The one-way analysis of variance (ANOVA) is used to determine whether there are any statistically significant differences between the means of the observed age groups with regard to the different dimensions mentioned above.

Table 5 ANOVA and Kruskal-Wallis tests

Age categories	Support of distance education				
	National policies	Labour market	Employer's support	Government support	Supporting of special needs
From 17 to 21 years	3.667	3.151	3.452	4.022	3.925
From 22 to 26 years	3.850	3.150	3.600	3.975	3.700
From 27 to 31 years	4.769	4.308	3.615	3.615	3.923
From 32 to 36 years	4.333	3.667	3.667	4.333	4.500
Greater than 36 years	3.714	4.000	4.143	3.286	3.429
ANOVA					
F-test	13.634***	18.954***	2.785**	4.514***	4.244***
Sig	0.000	0.000	0.026	0.001	0.002
Kruskal–Wallis test					
Chi-Square test	58.309***	65.010***	15.329***	26.669***	20.633***
p-value	0.000	0.000	0.004	0.000	0.000

*** 1% level

ANOVA (and Kruskal-Wallis) tests whether there is an age effect on the support of distance education. ANOVA (and Kruskal-Wallis) tests show that there is a statistically significant age effect on all five support of distance education variables (p-value <0.01) at 1% significance level. Looking at the mean values across age groups shows that higher age groups from 32 to 36 years had higher values of the support of distance education compared to the young age group from 17 to 21 years.

5 Conclusion and Recommendations

This paper discusses the analyses of the quantitative part of an online administered survey amongst learners who benefited from distance education during the COVID-19 pandemic. The survey attempted to measure the government support and the effectiveness of the new mode of learning through earners' satisfaction and reflection on their online learning experience and the support they have received from their family, friends, and classmates.

The recommendations are as follows:

- There was clear government support and respondents confirmed that in their feedback, yet clearer policies and procedures should be thought of, created, and communicated.
- The challenging issues when rapidly pivoting to distance teaching and learning should be tackled for the sake of ensuring the continuity of academic learning for students.
- There should be a mechanism for supporting the students who lack skills for independent study.
- Universities should ensure the continuity and integrity of the assessment of student learning.
- Last but not least, universities should ensure the well-being of students and of teachers.

Similarly, Handling the challenging barriers which include:

- Availability of technological infrastructure,
- Addressing student emotional well-being,
- Addressing the right balance between digital and screen free activities, and
- Managing the technological infrastructure.

References

1. United Nation Development Programme. COVID-19 and human development: Assessing the crisis, envisioning the recovery. 2020 UNDP.
2. Anderson, T., & Rivera-Vargas, P. (2020). A critical look at educational technology from a distance education perspective. *Digital Education Review, 37*, 208–229.
3. UNESCO. (2020). 290 Million students out of school due to COVID-19: UNESCO releases first global numbers and mobilizes response. UNESCO.
4. Al Lily, A. E., Ismail, A. F., Abunasser, F. M., & Alqahtani, R. H. A. (2020). Distance education as a response to pandemics: Coronavirus and Arab culture. *Technology in Society, 63*, 101317.
5. Agasisti, T., & Soncin, M. (2021). Higher education in troubled times: On the impact of Covid-19 in Italy. *Studies in Higher Education, 46*(1), 86–95.
6. Shearer, R. L., Aldemir, T., Hitchcock, J., Resig, J., Driver, J., & Kohler, M. (2020). What students want: A vision of a future online learning experience grounded in distance education theory. *American Journal of Distance Education, 34*(1), 36–52.
7. Marek, M. W., Chew, C. S., & Wu, W. C. V. (2021). Teacher experiences in converting classes to distance learning in the COVID-19 pandemic. *International Journal of Distance Education Technologies (IJDET), 19*(1), 40–60.
8. Ames, K., Harris, L. R., Dargusch, J., & Bloomfield, C. (2021). 'So you can make it fast or make it up': K–12 teachers' perspectives on technology's affordances and constraints when supporting distance education learning. *The Australian Educational Researcher, 48*(2), 359–376.
9. Zinovieva, I. S., Artemchuk, V. O., Iatsyshyn, A. V., Popov, O. O., Kovach, V. O., Iatsyshyn, A. V., ... & Radchenko, O. V. (2021). The use of online coding platforms as additional distance tools in programming education. *Journal of Physics: Conference Series, 1840*(1), 012029. IOP Publishing.
10. Ergüzen, A., Erdal, E., Ünver, M., & Özcan, A. (2021). Improving technological infrastructure of distance education through trustworthy platform-independent virtual software application pools. *Applied Sciences, 11*(3), 1214.
11. Vakaliuk, T. A., Spirin, O. M., Lobanchykova, N. M., Martseva, L. A., Novitska, I. V., & Kontsedailo, V. V. (2021). Features of distance learning of cloud technologies for the organization educational process in quarantine. *Journal of Physics: Conference Series, 1840*(1), 012051. IOP Publishing.
12. Fiş Erümit, S. (2021). The distance education process in K–12 schools during the pandemic period: evaluation of implementations in Turkey from the student perspective. *Technology, Pedagogy and Education*, 1–20.
13. Hash, P. M. (2021). Remote learning in school bands during the COVID-19 shutdown. *Journal of Research in Music Education, 68*(4), 381–397.
14. Ramírez, J. M., Hernández-, A. G., López, A. D., & Pérez-León, V. E. (2021). Measuring online teaching service quality in higher education in the COVID-19 environment. *International Journal of Environmental Research and Public Health, 18*(5), 2403.
15. Engzell, P., Frey, A., & Verhagen, M. D. (2021). Learning loss due to school closures during the COVID-19 pandemic. *Proceedings of the National Academy of Sciences, 118*(17).
16. Sobaih, A. E. E., Hasanein, A. M., & Abu, A. E. (2020). Responses to COVID-19 in higher education: Social media usage for sustaining formal academic communication in developing countries. *Sustainability, 12*(16), 6520.
17. Bhagat, S., & Kim, D. J. (2020). Higher education amidst COVID-19: Challenges and silver lining. *Information Systems Management, 37*(4), 366–371.
18. Parker, A. (2020). Interaction in distance education: The critical conversation. AACE Review (formerly AACE Journal), 13–17.
19. Moukarzel, S., Del Fresno, M., Bode, L., & Daly, A. J. (2020). Distance, diffusion and the role of social media in a time of COVID contagion. *Maternal & Child Nutrition, 16*(4).

20. Hebebci, M. T., Bertiz, Y., & Alan, S. (2020). Investigation of views of students and teachers on distance education practices during the Coronavirus (COVID-19) pandemic. *International Journal of Technology in Education and Science (IJTES), 4*(4), 267–282.
21. Zhu, X., & Liu, J. (2020). Education in and after Covid-19: Immediate responses and long-term visions. *Postdigital Science and Education, 2*(3), 695–699.
22. Tümen, S. (2020). College students' views on the pandemic distance education: A focus group discussion. *International Journal of Technology in Education and Science, 4*(4), 322–334.
23. Aboagye, E., Yawson, J. A., & Appiah, K. N. (2021). COVID-19 and E-learning: The challenges of students in tertiary institutions. *Social Education Research*, 1–8.
24. Alshurafat, H., Al, M. O., & Mansour, E. (2021). Strengths and weaknesses of forensic accounting: An implication on the socio-economic development. *Journal of Business and Socio-economic Development, 1*(2), 85–105. https://doi.org/10.1108/JBSED-03-2021-0026
25. Afana, A., Agha, E. L., & A. (2019). The role of organizational environment in enhancing managerial empowerment in Al-Aqsa network for media and art production. *International Journal of Business Ethics and Governance, 2*(2), 30–63. https://doi.org/10.51325/ijbeg.v2i2.40
26. Fatonia, N. A., Nurkhayatic, E., Nurdiawatid, E., Fidziahe, G. P., Adhag, S., Irawanh, A. P., ... & Azizik, E. (2020). University students online learning system during Covid-19 pandemic: Advantages, constraints and solutions. *Systematic Reviews in Pharmacy, 11*(7), 570–576.
27. Almazova, N., Krylova, E., Rubtsova, A., & Odinokaya, M. (2020). Challenges and opportunities for Russian higher education amid COVID-19: Teachers' perspective. *Education Sciences, 10*(12), 368.
28. Alzaneen, R., & Mahmoud, A. (2019). The role of management information systems in strengthening the administrative governance in ministry of education and higher education in Gaza. *International Journal of Business Ethics and Governance, 2*(3), 1–43. https://doi.org/10.51325/ijbeg.v2i3.44
29. Hasan Al-, M. (2019). Public governance and economic growth: Conceptual framework. *International Journal of Business Ethics and Governance, 2*(2), 1–14. https://doi.org/10.51325/ijbeg.v2i2.21
30. Safari, K., Njoka, C., & Munkwa, M. G. (2021). Financial literacy and personal retirement planning: A socioeconomic approach. *Journal of Business and Socio-economic Development, 1*(2), 185–197. https://doi.org/10.1108/JBSED-04-2021-0052
31. Gonzalez, T., De La Rubia, M. A., Hincz, K. P., Comas-Lopez, M., Subirats, L., Fort, S., & Sacha, G. M. (2020). Influence of COVID-19 confinement on students' performance in higher education. *PloS one, 15*(10).

Perception and Impact of Coronavirus (COVID-19) on People's and Lifestyle: E-learning, People, Well-Being, Quality of Life, Protective Measures and Government

Mukhtar AL-Hashimi and Hala AL-Sayed

1 Introduction

The appearance of the novel Coronavirus Disease 2019 (COVID-19) was initially observed in Wuhan, China, spread across the globe in more than 200 countries. The main causative factor of this disease is the Severe Acute Respiratory Distress Syndrome Coronavirus-2 (SARS-CoV-2). This virus is one type of zoonotic virus that belongs to the cluster of the beta-corona virus. The homolog of coronavirus is the SARS-CoV that has caused severe acute respiratory syndrome in 2002 as well. Unlike SARS, the transmissibility ratio of COVID-19 is higher; however, the morbidity and mortality rates are lower [1]. Significant health outcomes were also associated with Middle East Respiratory Syndrome (MERS-CoV) that had outbroken in 2012 [2].

Initially, the spread of COVID-19 was linked with the live seafood market present in the city of Wuhan [3]. However, it was evident later that the virus is spread through airborne transmission, droplets, and fecal–oral routes. The entry of Coronavirus within the cells is facilitated through the receptors of angiotensin-converting enzyme-2. The specific location of these receptors is in type II alveolar cells of the lungs, however, their broad expression is also observed at other multiple sites in the human body [4]. The incubation period of this virus is around 14 days, having a median of about four days. The significant symptoms experienced after contracting this virus include, fatigue, fever, dyspnea, dry cough, and muscle aches. Occasionally, this virus also results in watery diarrhea [5]. The severity of this disease is different in different cases with the variable course of the illness. The symptoms of this disease range from being asymptomatic, having mild signs or severe symptoms that may even require mechanical ventilation [6]. Humans have no pre-formed immunity against

M. AL-Hashimi (✉)
Ahlia University, Manama, Bahrain

H. AL-Sayed
Dr. Hala Al-Sayed Medical Center, Manama, Bahrain

this virus because it belongs to the family of zoonotic viruses. The best approach for protecting against this virus is implementing preventive strategies because there are limited treatment options.

The COVID-19 disaster is unparalleled, and the media around the world is awash with conflicting reports about this novel virus and its associated illness. However, everyone agrees that it is highly infectious, and people worldwide are turning to vaccination as the most successful strategy. Amid this deluge of media content, the public is still being fed a slew of false ideas. Since a person's experience, understanding, and memory have such a significant impact on their behavior, clarifying public expectations may be vital in deciding the outcome of COVID-19.

The lives and livelihoods of human beings have been turned upside down. As a result, actual economic costs are more varied and potentially more severe than those captured solely by financial indicators. They have negative consequences for people's mental well-being, a rise in the incidence of domestic violence, and a possible reduction in today's children's educational achievement [7, 8]. The virus and its economic consequences are affecting people from vulnerable groups and lower-income families, as they have been in recent financial crises and pandemics.

The remainder of the chapter is set out as follows. Firstly, it discusses maintaining the quality of life in the COVID-19 pandemic, followed by behavioral responses of people and the impact of COVID-19. Further, the chapter narrates about perceived threat and stress linked with COVID-19, along with discussing the emotional well-being of people about COVID-19. The chapter also discusses elderly people's perceptions of COVID-19, the overall perception of people about personal protective measures, and wearing masks. In the succeeding headings, the chapter discusses e-learning managerial perspectives of people, trust of government about COVID-19, and perceptions of people from a poor household. The chapter also focuses on discussing the lessons learned by people, organizations, and government from COVID-19 and the perception of COVID-19 across different cultures. Lastly, the chapter talks about the perception of immunization against COVID-19, followed by concluding remarks.

2   Impact of COVID-19 on E-learning

The educational institutions all around the world have shifted away from traditional learning techniques since the emergence of the COVID-19 virus and shifted towards online education. The traditional classroom setting has been abruptly replaced by electronic gadgets and online apps in the educational system [9]. Majority of the institutions have encouraged professors and students to use e-learning platforms for instructional reasons and to encourage students to study from their homes [10]. Faculty members were encouraged to submit study materials in the form of Power-Point presentations, PDFs, or Word documents, as well as alternative formats such as audio and video for uploading to online platforms. The instructional processes have been carried out via applications such as Zoom and Voov due to the virus's extensive dissemination. Students have been able to enrol in a variety of undergraduate and graduate courses over the internet. Academic demands have been met mostly via the use of programmes such as Google Classroom and Zoom.

During this pandemic, well-known businesses including as Google, Microsoft, and Zoom have provided educational establishments numerous elements of their products that might be useful in the field of education for free. For instance, the Microsoft team users were 750 on March 10th, which increased to 138,698 by March 24th [11]. On request, Zoom has increased the video calling time limitations in Italy, Japan, the United States, and China [12]. The world still needs far greater access to communication solutions like Zoom and Google Meet. As a result of the spread of the fatal COVID-19 virus, there has been a massive and abrupt change in the area of academics throughout the world. There has been a worldwide shift to online teaching and learning approaches [13]. To limit the spread of the virus and protect the safety of instructors and students, the traditional classroom environment has been replaced by digital methods.

Educators have broken down the instructional content into smaller modules to help pupils focus and comprehend the information better. In an audio-visual lecture, faculty members have reduced the tempo of their speech to draw their attention and allow them to take notes on important topics from the lectures and write down the essential information on the board. Inexperienced faculty members sought advice from online teaching assistants to ensure that each class's objectives and demands were met. Educators have changed their teaching methods by giving students innovative and skilled projects that can meet their learning needs and keep them engaged throughout online sessions. The professors combined online teaching approaches with offline self-study. They also engage students in conversation and offer comments on their assignments to promote their viewpoints. Students will not gain superficial, unclear, and fragmented knowledge using this method of instruction. Instead, students would engage in in-depth learning through a variety of dialogues. In conventional in-class teaching, insufficient pre-class study preparation, restricted involvement in class discussions, and inadequate discussion depth are all typical occurrences; similarly, these issues should not be neglected in online education [14, 15].

3 E-learning Managerial Perspectives of People

E-learning was rising at a rate of around 15.4% per year in educational institutions worldwide, with little uncertainty or burden on such institutions or students before the COVID-19 pandemic [16]. Due to global restriction initiatives to prevent the dissemination of COVID-19, educational institutions started offering the majority of their resources online, including lecturers and various examinations through multiple channels for over 60% of students worldwide. According to World Health Organization (WHO) reports, COVID-19 has been identified in over 216 countries, with millions of confirmed cases in some regions [3]. Several countries have taken precautionary steps such as school and university lockdowns and transitioning to complete

E-learning mode [16]. This move was taken in response to the WHO's clear recommendation that social distance guidelines be implemented to deter the dissemination of COVID-19. This lockout began in the middle of the spring semester, and both teachers and students were caught off guard.

However, several surveys have previously looked at crucial success factors (CSFs) in the education field from the viewpoints of both instructors and students to strengthen the E-learning environment in the future. Organizations should identify the most critical CSFs to accomplish to advance a project's goal. So far, these experiments have looked at the CSFs in E-learning at regular intervals. However, CSFs during the COVID-19 pandemic is likely to vary from CSFs in regular periods for a variety of factors. To begin with, the transition to E-learning was unplanned for all educational institutions during COVID-19. Not all organizations were able to make the transition seamlessly. Not all of them had already implemented E-learning, instead of those who had already provided E-learning and were planning and engaging in the process. Second, certain factors other than educational factors, such as political and health factors, affect the mechanism during COVID-19, creating an unusual circumstance. For instance, unlike during COVID-19, where students were under curfew, students can visit the library, attend tutoring sessions, and even go to locations with good internet access speeds if they do not have a good internet connection at home. Third, course content for classes taught through E-learning before COVID-19 was well-prepared compared to lessons not intended to be conducted through E-learning during COVID-19.

Focusing on disasters like the COVID-19 pandemic, the new thesis is essential because no recent analysis has looked at problems where the whole school infrastructure around the world has been disrupted. Many educational establishments have moved away from traditional classroom instruction and toward e-learning. Adoption of an e-learning system is not easy, and no one system can meet the needs of all disciplines and organizations around the world. The various potential methods and their essential attainment variables are examined in this study.

4   Maintaining Quality of Life in Pandemic COVID-19

Quality of life (QoL) has primarily been studied in research focused on non-communicable and chronic conditions over the last decade. It is characterized as a patient's overall subjective view of the impact of a disease or medical condition on various realms such as physical, social, psychological, and occupational functioning [17]. A previous study found substantial deterioration in HRQoL for general health realms, physical disabilities, and social functioning among survivors of the severe acute respiratory syndrome (SARS) six months after the pandemic began. The COVID-19 pandemic has harmed HRQoL, according to a new Moroccan report [18]. Implementing prevention strategies affects people's everyday lives and their overall functioning and well-being [19]. Previously, it has been observed that at the detrimental psychological consequences of quarantine after the SARS epidemic, and more recently, the word "coronaphobia" has been coined to explain stress and anxiety in general populations [20]. Experts in mental health have voiced concern about the effect of the COVID-19 pandemic on families' social functioning and well-being. Despite recent literature warnings about the broader psychological impact of mass quarantine to regulate COVID-19 spread on individuals' QoL [21], observational data on the effects of the COVID-19 pandemic in different countries is restricted.

In many other parts of the world, the COVID-19 epidemic was declared a public health emergency, with increased morbidity and mortality rates, resulting in a stressful experience for the entire population [22]. People's everyday habits have significantly been affected by a strict commitment to protective steps such as wearing facemasks, regular hand-washing, disinfecting objects, and, most importantly, keeping social distance and quarantine of contaminated persons [22]. There is increased reliance of people on cyber tools for academic and professional practices throughout the lockdown time in the home setting. According to some figures, devices such as smartphones, tablets, and notebooks grew by almost 20% during the lockout era in the United States. According to a study conducted in Saudi Arabia, students attending simulated classes during the initial stages of the pandemic faced low to high levels of stress [23]. Fear and anxiety related to the COVID-19 pandemic were moderate to high in a study conducted in China during the early pandemic season [24]. Individuals who have been quarantined are more likely to experience a wide variety of negative emotions, including anxiety, rage, remorse, and a sense of lack of power. People are more likely to feel less comfortable in their physical surroundings due to these reactions, and they may be more concerned with the washing of face and hands, disinfecting surfaces, and other health habits [25]. Some of the findings have shown that patients with chronic health problems, such as Parkinson's disease, breast cancer, and those who are undergoing surgery have a lower quality of life, especially in the physical and psychosocial aspects [26, 27].

The religious-dimension products, which assessed the impact of the pandemic on religious practices, clarified a minor variation in the QoL ranking. When it comes to the effects of the COVID-19 pandemic on different QoL dimensions, the conclusions are consistent with the results of the e-survey, which looked at QoL and society consistency during the COVID-19 pandemic. This study confirmed that the general public is concerned about financial circumstances, lack of confidence in governments, and the pandemic's effects on people's mental well-being. Male members of Saudi society were more likely to have lower QoL levels around these measurements due to the social structure and cultural beliefs. Extensively, males work outside of their homes, making them feel less secure. An online poll of the general population in Italy found that young people who had to work outside of their homes experienced higher levels of anxiety and tension [19]. People who worked in positions where they had to interact with others often, such as healthcare centers and transport stations, were more likely to be stressed because they were afraid of contracting COVID-19.

Males who were the primary breadwinners for their households were more likely to experience depression and distress due to conditions such as ill-health or work loss during the pandemic. Furthermore, social distancing constraints have restricted the social interactions of male participants, affecting their friendships. Moreover, the closing of mosques during the lockdown resulted in religious practices is prohibited. Before the COVID-19 pandemic, a comprehensive analysis report revealed the detrimental impact of social alienation on people's physical and psychological well-being [28]. Due to more restrictive physical controls and segregation as part of primary prevention steps during the COVID-19 pandemic, this risk increased [29]. According to a web-based study, people quarantined during the SARS outbreaks in 2004 indicated depression and alienation as a result of the decreased physical interaction with friends and family members. They were unable to go to stores to purchase food, groceries, or drugs due to their commitment to protective measures such as wearing a maintaining social distancing and wearing face masks, which increased their feelings of alienation.

The people with little or no teleworking experience experienced discomfort and depression due to the abrupt transition and new demands of this work world. These results suggest that corporate assistance should concentrate on delivering psychological support during times of crisis and reducing employee perceptions of workplace dissatisfaction and incompetence. Also, amid the crisis and widespread frustration at social levels during the COVID-19 pandemic, organizations may improve integrity among working individuals by appreciating their existing efforts and readiness to respond to and embrace new work demands.

5   Behavioral Responses of People and Impact of COVID-19

During the pandemic, psychological depression is one of the most common health issues in Saudi Arabia. Most people expressed concern about the pandemic's present global state, hearing that someone had tested positive for COVID-19 or died from it and received excessive detail about COVID-19 and travel limitations. It is significant to stop the spread of this deadly virus based on the general public's behavior. Owing to the religious and social norms and frequent hosting of high-profile international spiritual mass conferences, compliance with precautionary steps such as social distancing, face coverings, hand, and respiratory grooming, and avoiding crowds and poorly ventilated spaces is difficult [30].

Many nations implemented lockdowns in the early stages of the pandemic to slow the virus's spread. This was essential to mitigate deaths, avoid overburdening healthcare facilities, and earn time for setting up pandemic response systems to stop transmission after the lockout [31]. According to a study conducted in South Korea, most participants concluded that quarantine was unsuccessful in controlling MERS or even increased the risk of MERS spreading [32]. This may be due to the research time delay since the South Korean study was completed in 2015, resulting in a significant knowledge gap on quarantine relative to the current research.

The prevalence of preventive habits is reduced when people perceive themselves to be in a low-risk situation. During the H1N1 pandemic in Saudi Arabia, residents with low-risk perceptions were less likely to practice social distancing and activities like wearing masks and hand washing, and residents with a lower fear of SARS were less likely to practice cautious behaviors for protecting themselves [33]. It is found that when one's mood changed from pessimistic to optimistic, there is an increase in the mean score of overall behavioral reactions that includes preparedness, preventive actions, and self-quarantine behaviors. Even after controlling for all potential confounders, there was still an essential direct relationship between attitude and behavior. This behavior is likely to indicate a high level of knowledge about the epidemic, its severity, and mode of dissemination among those with a positive attitude toward the COVID pandemic, and this high level of expertise may be one of the reasons that lead to a high level of compliance with preventive behaviors.

In the prevention of any pandemic epidemic, socio-demographic characteristics play a critical role in influencing a community's beliefs and behaviors. Gender and age are the two most important variables influencing people's risk attitudes and willingness to take risks [34]. The female gender significantly affects amenability to precautionary steps, preparedness, self-quarantine practices, and overall behavior. Women are generally able to defy convention to protect their children's welfare and adopt the community's structure to their benefit.

6 Perceived Threat and Stress Associated with COVID-19 Pandemic

Evidence shows that paranoia raises tension and anxiety, potentially leading to mental disorders, although pandemics are traumatic. The interpretation of a threat to one's well-being is a critical concept in learning how people respond to challenging situations. Two primary constructs are defining the people's perception towards stress [35]:

- The disease's apparent seriousness of the adverse effects it has on one's life.
- The risk of developing the disease is known as perceived susceptibility.

This understanding of illness affects one's perception of stressful experiences and is influenced by previous experiences. The connection between threat perception and emotional and mood changes has been well established in the literature. Many studies, for example, have shown that psychiatric disorders are linked to a strong sense of a disease's danger [36].

The public's social media access to information about the pandemic, mixed with pervasive disinformation, has been a double-edged sword [37]. People nowadays obtain, interpret, and act on medical knowledge obtained from outdated social media outlets. This begs whether health awareness played a part in the social effects and anxiety felt during the pandemic. Despite significant healthcare advances in recent decades, decreased levels of health literacy are considered an important concern across the globe [38].

Health awareness is critical for predicting the social effects of a pandemic on the general public. A higher degree of health awareness was shown to be linked with PTSD symptoms negatively. This is in line with recent research on the connection between mental health and health literacy during this pandemic. According to one report, people with high health awareness were less likely to be stressed [39]. According to a survey conducted in the United States, people with low health literacy were less likely to recognize their vulnerability to the COVID-19 virus. People with poor health awareness are also less likely to acknowledge COVID-19 signs and are likely to show decreased preparedness for the pandemic [40]. More emphasis on health awareness in COVID-19 public health communications is critical, as it has consequences for both the pandemic's progression and the general population's well-being.

7   Emotional Well-Being of People About COVID-19

COVID-19 pandemic may be psychologically taxing because of the unpredictability of the situation, the complexity of whether the disease can be controlled, and the severity of the threat. Individuals and cultures are under enormous emotional strain due to these factors, as well as disinformation and exaggerated media attention [3]. In a Chinese survey of COVID-19's psychiatric effects, 26% said they had mild to moderate depression, 4.3% said they had extreme depression. More than a third said they had moderate to severe anxiety levels [41]. People of family and children are more likely to be stressed. COVID-19 has been linked to elevated levels of emotional tension in China, according to another report. Furthermore, Gavidia reports that during COVID-19, 88% of workers in an American survey said mildly to severe stress levels [42]. Likewise, during the MERS-CoV epidemic in Saudi Arabia and the severe acute respiratory syndrome (SARS) outbreak in the United States, healthcare staff reported feeling more worried and stressed.

During the COVID-19 pandemic, age, gender, marital status, and socioeconomic status are significant predictors of emotional well-being. Compared to the rest of the sample subjects, older adults, women, and married people had higher levels of psychological distress. Females, the elderly, and married people showed higher anxiety, depression, and behavioral/emotional power loss. The elderly and females likely view the pandemic as having little social influence, making them less able to control their impulses, leading to higher depression, anxiety, and psychological distress. Similar effects were seen in a study that looked at gender as an indicator of psychological illness, suggesting that men are slightly less stressed than women [41]. Sex, social support, and personal interactions were all variables in psychological distress in another study [43].

Natural disasters and pandemics elicit different societal responses in other societies. Psychiatric issues cause such emergencies in people that are more likely to have mental disorders. Older adults, pregnant females, foreign students, migrant workers, homeless people, and people with a history of mental disorders have all been listed as disadvantaged groups in the literature.

8   Elderly People and COVID-19 Pandemic

The risk of depression and anxiety is escalated among elderly individuals, considering their increased vulnerability to COVID-19. The perceived disadvantages in everyday life decreased as interventions were eased, and the reasons for this changed as the pandemic progressed. The perception of shortcomings resulting from the interventions was higher at the start of the pandemic and among younger people. The early stages of the pandemic and younger age were significant predictors of daily life limits. There may be a variety of explanations why older people thought the restrictions were less severe. For example, lower perceived weaknesses in the elderly may be explained by fewer commitments and more past threatening scenarios [44]. Other considerations, such as the pandemic process and age, played a minor role in how severe the respondents saw the interventions. The interventions were especially harsh in everyday life for men, educated people, people with depression, and people with mobility impairments, regardless of age or pandemic phase.

The most significant weakness was the lack of social interaction. After numerous interventions were eased, these remained the primary causes of presumed shortcomings in daily life. While presumed disadvantages have been linked to younger ages, the elderly are more vulnerable to the consequences of insecurity and a lack of social connections [45].

There is an indication of relevant fear towards COVID-19 infection based on the psychological burden side during the pandemic. In February 2020, a representative survey of people in Wuhan (the epicenter of China's COVID-19 outbreak) and Shanghai yielded similar findings. In this case, Wuhan had a slightly higher incidence of mild to extreme anxiety than Shanghai. Stress was linked to a perception of disease damage and apprehension about the accuracy of knowledge. With the substantial easing of steps in phase 3, this anxiety grew. It's worth noting that the respondents were older adults suffering from various neurological and internal illnesses. Many respondents claimed that they were at a higher risk of developing COVID-19 disease and hence fell into the risk category accounting for the high degree of anxiety. About half of the respondents were dismissive of the easing of sanctions, which is understandable. Insecurity and concerns about the future, on the other hand, declined throughout the pandemic. This is the best indication that the respondents responded to the unique demands of the situation and were more prepared than they were at the start of the pandemic since enough knowledge contributes to less psychological stress, such as fear.

Elderly people with underlying disorders are more vulnerable than most people because of their deteriorating cognitive capacity, impaired physical performance and physical health, and low immune function [46]. Furthermore, the aging population is vulnerable to anxiety when their cognitive abilities deteriorate, resulting in psychiatric dysfunction [47]. In the absence of successful symptomatic medications, the elderly population may pose more significant health risks in unexpected illnesses. Most nations nowadays have an aging population. Globally, the proportion of elderly people over 65 years old is expected to rise from 11% in 2019 to 16% in 2050.

9 Perception of People About Personal Protective Measures

The global health care system has been tested by the onslaught of an emerging infectious disease known as COVID-19. In the lack of proven therapeutics and the excellent vaccine, and with the number of new infections continuing to rise at an unprecedented pace worldwide, preventative measures are critical for breaking the virus's chain of transmission and controlling infection rates. Evidence shows that public knowledge and prevention behaviors are essential in disease management. As a result, health officials have taken crucial measures to ensure that reliable information is accurately disseminated to monitor and maintain public behavior.

According to the advice of WHO, the precautionary measures applied by the general public to control the COVID-19 pandemic include; avoid handshaking, wearing masks, washing hands, and avoid frequent touching face and ears [3]. If the public follows these measures carefully, the prevalence of illness in the general population will be reduced, and the risk of transmission will be reduced. This research took into account various social factors to further understand the factors that influence observed results and assess the general public's experience, comprehension, and attitudes in responding to and implementing proposed mitigation measures.

The vast majority of people showed a high level of understanding of hand hygiene and the importance of washing hands before touching their faces to prevent transmission. Hand-washing compliance is likely one of the most cost-effective ways to prevent the transmission of SARS-CoV2 in populations [48]. The people have a high understanding of the suggested hand-washing process and the risk of SARS-CoV2 transmission via polluted surfaces. Multiple advertising programs may have resulted in high levels of knowledge of COVID-19 transmission and protective steps to combat its dissemination.

The explanations for inadequate adherence to multiple preventative measures were looked at. Participants who did not stop handshake expressed apprehension due to feelings of contempt for others or the impression that such a measure will be unsuccessful in preventing SARS-CoV2 transmission. In many cultures, including Arab culture, the handshake is closely correlated with traditional values, and the problems associated with avoiding those rituals, while understandable, must be addressed by raising awareness [49]. To effectively monitor the spread of COVID-19 among populations, local health authorities should resolve specific knowledge gaps.

When returning home, the public was advised to wash their hands with soap for at least 40 s or with alcohol spray for at least 20 s. One out of every seven people in this survey said they disinfected their hands with soap and water or alcohol gel for less time than prescribed, which could lead to ineffective disinfection. When they returned home, most people said they did not wash their hands or did not follow the defined process.

10 Perception of People About Wearing Face Masks

Following the latest revision of the WHO and CDC's proposed techniques, universal masking was introduced, and it has since become a popular standard. Community-wide face masking is thought to help monitor COVID-19 by minimizing the spread of the virus by infectious saliva and respiratory droplets from people who have subclinical or minor infections [50]. According to studies conducted in China and Italy, 78% and 50–75% of people with positive molecular samples, respectively, were asymptomatic [51]. As a result, wearing masks in public by asymptomatic people was previously questioned and dismissed as ineffective. However, there is strong evidence that wearing a face mask decreases the likelihood of COVID-19 transmission by a significant amount. The use of a face mask in public is linked to a frequent decrease in COVID-19 transmission, which aids in the disease's prevention [52]. Even though healthcare providers and symptomatic patients consistently advocate the use of face masks, it is not recommended for the general public and the wider community [53]. Nonetheless, wearing a public mask is now widely recommended, particularly in places where group transmission is heavy. However, the use of face masks by active community members to minimize the risk of infectious respiratory infections is debatable.

N95 masks, medical masks, and non-medical fabric masks are the three styles of masks currently available. Medical masks are poorly fitting equipment used by health care staff and infected people to minimize the chance of infectious respiratory droplets being spread amongst people while sneezing or coughing. However, the safety rate in the phase of expiratory pollution ranges from 33 to 100%, depending on the type of face mask used. Cloth face masks, for example, have modest effectiveness in preventing disseminated respiratory infections caused by particles of the same size or smaller than COVID-19 particles [54]. As a result, many nations have made the wearing of face masks mandatory.

The rate of infection prevention will be maximized if the population adheres to face masking to a high degree. Individuals' compliance with wearing face masks during outbreaks can be hampered for a variety of purposes. The most significant of them being a lack of expertise, a misunderstanding, an unfavorable presence, and obstacles to enforcement. The evaluation of the community's compliance with wearing face masks necessitates gathering data on their experience, behaviors, and expectations and identifying and evaluating the obstacles to compliance. The main barriers experienced by people in wearing masks include; decreased perceived susceptibility to COVID-19, social and physical discomfort, and identity perceptions [55].

Despite the debate and the worldwide mask manufacturing sacristy, more countries recommend or require the use of face masks in public. Face masks worn in the community seemed to be appropriate even in the absence of other security measures. It was also found that wearing face masks in conjunction with social distancing can be very helpful in lowering COVID-19 risk in the population [56]. The rules for wearing masks vary significantly across countries. According to reports, the overall societal advantage of wearing a face mask depends on enforcement and early use

during delivery and hand hygiene [57]. Even though the participants in this study were mindful of the advantages of wearing face masks, the obstacles to doing so could have reduced their ability to comply.

The level of comfort in which people wore face masks seemed to influence their willingness to do so, particularly for extended periods [58]. N95 masks, on the other hand, seemed to be more painful, requiring tedious precautions. Despite the high enforcement and awareness of the value of wearing face masks and when and how to do so, there are some perceived societal obstacles to the procedure.

11 Trust of Government and Perception of COVID-19

Many countries replied towards COVID-19 by enacting varying degrees of stringent legislation to stop the spread of the disease [59]. However, since persistent non-compliance and overt demonstrations against more stringent policies have occurred, it is critical to consider the reasons that foster societal acceptance of such initiatives. How to successfully coordinate and ensure public health guidelines and containment strategies is an ongoing global concern. During the Ebola virus disease epidemic in Liberia in 2014, effective risk communication and health education boosted community support and participation, resulting in adopting critical preventive behaviors [60].

Another interesting result was how public confidence in government correspondence on COVID-19 affected risk-aversion behavior. Though family and relative guidance affected behavior, greater confidence in government contact had temporally complex correlations when health authorities advised this behavior. Earlier messaging had effectively persuaded the public that facemasks were only needed for people who were ill. Increased local spread and new research on the importance of pre-symptomatic infection and how facemasks can mitigate virus transmission led to the recommendation of mandatory facemask use in public areas [61]. This messaging was highly effective. It was associated with a significant increase in facemask usage beginning in week 14, well before fines for non-compliance were implemented in the second half of week 15.

As new research appears and risk perceptions shift in an emerging pandemic, changes to public records could be required. Government messaging, unlike recommendations from families and relatives, must be rationally calibrated. When there is evidence that transmission is growing, health officials may need to take preventive action, even though the public believes the risk of infection is minimal. And if public concerns remain, health officials can need to de-escalate more disruptive steps that are no longer necessary. During an epidemic, maintaining public confidence in health officials, both in their reaction and interactions, is critical to implement preventive behavior suitable for the situation.

When dealing with an epidemic that is already emerging, faith is a valuable asset. Although public perception of rising illnesses and deaths can influence specific behavior, policymakers should ideally initiate evidence-based general messaging strategies and initiatives ahead of those adverse outcomes. Such campaigns may be essential as they contradict popular belief, and well-executed behavioral cohort studies may help by predicting the interplay between expectations, confidence, and behavior.

12 Perception of People from Poor Household Towards COVID-19

The emergence of the COVID-19 pandemic in low- and middle-income countries raises serious questions about the readiness of these countries' health systems to deal with the epidemic as it spreads. With healthcare services still overburdened before the pandemic, it's becoming apparent that implementing the same policies seen in high-income countries in low- and middle-income countries will not be achievable [62]. Current guidelines place a heavy emphasis on hospital-based approaches, however, primary response measures might not be feasible, targeting shortages of hospital beds, personal protective equipment, current budget constraints, and ventilators. Furthermore, low- and middle-income countries must provide medical assistance to vulnerable communities, such as people and families living in poverty.

Households in resource-poor settings in low- and middle-income countries can lack familiar and credible sources of knowledge about disease etiology, making them ill-equipped to reduce the risk of infection during emerging outbreaks of COVID-19 pandemic [63]. It is essential to understand the perceptions and responses of the general public towards COVID-19 for effective pandemic responses in low- and middle-income countries. This is only possible through the evaluation of prevailing public health-related messages, along with different communicational strategies.

In areas where there are many COVID-19 outbreaks, the government must enact broad prevention steps, such as an increased population quarantine. Strict home quarantine, lockdowns in regions with positive COVID-19 cases, suspension of public transit, and air and sea traffic restrictions were among the steps taken [64].

Even though the people do not represent the general public, they represent a population segment with insufficient access to health care and are often excluded from social services. The use of health insurance varies greatly depending on one's social status [65]. Low-income households are often heavily affected by malnutrition. When combined with a shortage of access to health services, infant death rates in the lowest wealth quintile are six times greater than in the richest. During health threats and emergencies, these groups are often underestimated and underserved, thus, government agencies, multilateral organizations, and non-governmental organizations (NGOs) must properly direct public health response and coordination strategies [66–68].

Based on the demographics of the people, they got their content from conventional media outlets like television and radio rather than social media. Owing to the population, many residents do not own smartphones that can access social media, in addition to the unsatisfactory internet accessibility in these regions. This result emphasizes the importance of avoiding a sole reliance on social and digital media as knowledge-sharing mediums of public health engagement strategies [69, 70].

When it comes to preparing primary care response programs, it's essential to note the demonstrated lack of desire to seek medical help, even though COVID-19 symptoms are present. Mainly where there are no public health crises, unbalanced health care access is found in this demographic due to obstacles such as long drives to health providers, missing a day of income for commuting, the affordability of services, and the stigma associated with future infectious disease diagnosis [64, 71, 72]. Similar obstacles to accessing health care can apply to COVID-19 treatment if and when it is required in these households.

13   Perception of Muslim Community Members About COVID-19

While Muslims do not form a single homogeneous community, they will usually follow Islamic teaching's recommendations to uphold family kinship, care for and look after elderly family members, and pray or eat breakfast in the congregation [73]. Such rituals practiced by some Muslim communities include dining together (from a single plate) and exchanging utensils and intimate social behaviors such as kissing the hands or heads of elderly relatives. Muslims find the new covid-19 interventions of social distancing and self-isolation "foreign and ludicrous" because of these activities [73]. Because of various cultural influences, health authorities have identified Muslims as one of the at-risk populations/groups. The Muslims are to blame for the increased spread of covid-19 in the nation [74, 75].

This disparity may be explained by one of two factors. First, sure participants in our study believed they were at higher risk of catching the virus due to their work-related exposure. A new analysis in health enforcement in COVID-19 discovered that general anxiety/fear of covid-19 was linked to social distancing and grooming practices. Owing to low incomes, self-employment in jobs where working from home was not an alternative, and people's ineligibility for financial assistance, the economic need was also established as a driving factor contributing to increased virus exposure.

Second, members of the Muslim community are instilled with a sense of obligation due to their religion, which mandates that people have a responsibility to defend not only themselves but also others from injury. Any people in their group were not adhering to social distancing suggestions, according to the participants. It's difficult to explain why certain people followed social distancing laws and others did not.

The Muslims were told to separate themselves from others in their culture socially ran counter to the Muslim community's normative cultural and social traditions and values. Though adhering to social distancing rules was undeniably tricky, particularly during Ramadan and the Eid festival. Such that, the Muslims were not able to break the fast together in the traditional way. However, they felt that these deviations from normative traditions were entirely appropriate and by religious teachings. Evidence shows that such obligations, such as the requirement to respect authority authorities, are linked to increased compliance with COVID-19 limits. A sense of responsibility is instilled in children at a younger age [73, 76] (Table 1).

Table 1 Overview of reviewed studies

S. no	Category	Reviewed studies/articles	Conclusion
1	Maintaining quality of life in pandemic COVID-19	Haraldstad et al. [17]	Delivering psychological support need to be concentrated during times of crisis and reduce employee perceptions of workplace dissatisfaction and incompetence
		Mucci et al. [18]	
		Mazza et al. [19]	
		Lee et al. [20]	
		Rubin and Wessely [21]	
		Lee and You [22]	
		AlAteeq et al. [23]	
		Zhang and Ma [24]	
		Harper et al. [25]	
		Bargon et al. [26]	
		Greco et al. [27]	
		Mazza et al. [19]	
		Leigh-Hunt et al. [28]	
		Saltzman et al. [29]	
2	Behavioral responses of people and impact of COVID-19	Al-Hanawi et al. [30]	Psychological depression is one of the most common health issues
		Flaxman et al. [31]	
		Tomas et al. [32]	
		Winters et al. [33]	
		ul Haq et al. [34]	
3	Perceived threat and stress associated with COVID-19 pandemic	Broadbent et al. [35]	Paranoia raises tension and anxiety, potentially leading to mental disorders during pandemic
		Lai et al. [36]	
		Ioannidis [37]	
		Paakkari and Okan [38]	
		Nguyen [39]	
		Bailey et al. [40]	

(continued)

Table 1 (continued)

S. no	Category	Reviewed studies/articles	Conclusion
4	Emotional well-being of people about COVID-19	World Health Organization [3]	
		Yang [48]	
		'Abdulmun'im Al-Nasser [49]	
5	Elderly people and COVID-19 pandemic	Zhong et al. [44]	Elderly population may pose more significant health risks in unexpected illnesses in the absence of successful symptomatic medications
		Sun [45]	
		Paraskevis et al. [46]	
		Qian et al. [47]	
6	Impact of COVID-19 on E-Learning	Mnyanyi and Mbwette [9]	Impact of COVID-19 on E-Learning
7	Perception of people about wearing face masks	Cheng et al. [50]	Community-wide face masking is thought to help monitor COVID-19
		Day [51]	
		Lyu and Wehby [52]	
		Eikenberry [53]	
		Lima et al. [54]	
		Shelus et al. [55]	
		Li et al. [56]	
		MacIntyre and Chughtai [57]	
		Al Naam et al. [58]	
8	E-Learning managerial perspectives of people	Alqahtani and Rajkhan [16]	E-Learning managerial perspectives of people
9	Trust of government and perception of COVID-19	Lim et al. [59]	Trust of government and perception of COVID-19
10	Perception of people from poor household towards COVID-19	Bong et al. [62]	Perception of people from poor household towards COVID-19
11	Perception of Muslim community members about COVID-19	Hassan et al. [73]	Perception of Muslim community members about COVID-19
12	Lessons people, organization and governments learned from COVID-19	Malave and Elamin [77]	Increased prevalence of infections among the healthcare worker sharing information timely collaboration between government and different organizations spread of COVID-19
		Park et al. [78]	
		Kim et al. [79]	

(continued)

Table 1 (continued)

S. no	Category	Reviewed studies/articles	Conclusion
13	Perception of COVID-19 across different cultures	Rickard [80]	Perception of COVID-19 across different cultures
14	Perception about immunization against COVID-19	Lurie et al. [81]	Perception about immunization against COVID-19

14 Conclusion

It is important to recognize and enhance real information, understanding, and perception to prevent its dissemination because COVID-19 has become a global problem with no cure. Furthermore, the lack of epidemiological data on COVID-19 should be brought to the public's attention. Certainly, the researchers are in the process of identifying the properties of this new virus and they have little previous knowledge. Furthermore, even though most individuals afflicted with this infection have a moderate type of disease, the fact that we don't know precisely who it will affect and therefore cause death makes COVID19 a deadly disease. Only a handful of them has a thorough understanding of COVID-19. As a result, further awareness-raising efforts from responsible authorities are needed. Furthermore, this survey will assist healthcare agencies and accountable authorities in planning their public information campaigns at this crucial time. Additionally, the general public benefits and is well informed to read the perspectives of particular participants. Protective attitudes are linked to higher levels of intelligence, perceived seriousness, negative sentiment, and exposure to and confidence in official government media. The most important factor of defensive habits is official government correspondence.

References

1. Yi, Y., Lagniton, P. N., Ye, S., Li, E., & Xu, R. H. (2020). COVID-19: What has been learned and to be learned about the novel coronavirus disease. *International Journal of Biological Sciences, 16*(10), 1753.
2. Cherry, J. D., & Krogstad, P. (2004). Sars: The first pandemic of the 21st century. *Pediatric Research, 56*(1), 1–5.
3. World Health Organization. (2020). Mental health and psychosocial considerations during the COVID-19 outbreak, 18 March 2020 (No. WHO/2019-nCoV/MentalHealth/2020.1). World Health Organization.
4. Patel, A. B., & Verma, A. (2020). COVID-19 and angiotensin-converting enzyme inhibitors and angiotensin receptor blockers: What is the evidence? *JAMA, 323*(18), 1769–1770.

5. Qiu, H., Wu, J., Hong, L., Luo, Y., Song, Q., & Chen, D. (2020). Clinical and epidemiological features of 36 children with coronavirus disease 2019 (COVID-19) in Zhejiang, China: An observational cohort study. *The Lancet Infectious Diseases, 20*(6), 689–696.
6. Zhou, F., Yu, T., Du, R., Fan, G., Liu, Y., Liu, Z. & Cao, B. (2020). Clinical course and risk factors for mortality of adult inpatients with COVID-19 in Wuhan, China: A retrospective cohort study. *The Lancet, 395*(10229), 1054–1062.
7. Pancani, L., Marinucci, M., Aureli, N., & Riva, P. (2020). Forced social isolation and mental health: A study on 1006 Italians under COVID-19 quarantine.
8. Van Lancker, W., & Parolin, Z. (2020). COVID-19, school closures, and child poverty: A social crisis in the making. *The Lancet Public Health, 5*(5), e243–e244.
9. Mnyanyi, C. B., & Mbwette, T. S. (2009). *Open and distance learning in developing countries: The past, the present, and the future.* Dares salaam: Open University of Tanzania.
10. Cheng, H. N., Liu, Z., Sun, J., Liu, S., & Yang, Z. (2017). Unfolding online learning behavioral patterns and their temporal changes of college students in SPOCs. *Interactive Learning Environments, 25*(2), 176–188.
11. OECD. (2020). A framework to guide an education response to the COVID-19 pandemic of 2020, https://read.oecdilibrary.org/view/?ref=126_126988-t63lxosohs&title=A-framework-to-guide-aneducation-response-tothe-Covid-19-Pandemic-of-2020. Accessed 16 Jun 2020.
12. Molla, R., & VOX. (2020). Microsoft, Google, and Zoom are trying to keep up with demand for their now free work-from-home software, https://www.vox.com/recode/2020/3/11/21173449/microsoft-google-zoom-slackincreaseddemand-free-work-from-home-software. Accessed 16 Jun 2020.
13. Basilaia, G., Dgebuadze, M., Kantaria, M., & Chokhonelidze, G. (2020). Replacing the classic learning form at universities as an immediate response to the COVID-19 virus infection in Georgia. *International Journal for Research in Applied Science and Engineering Technology, 8*(3), 101–108.
14. Johnson, S., Bamber, D., Bountziouka, V., Clayton, S., Cragg, L., Gilmore, C., ... & Wharrad, H. J. (2019). Improving developmental and educational support for children born preterm: Evaluation of an e-learning resource for education professionals. *BMJ Open, 9*(6), e029720.
15. Su, C. H., Tzeng, G. H., & Hu, S. K. (2016). Cloud e-learning service strategies for improving e-learning innovation performance in a fuzzy environment by using a new hybrid fuzzy multiple attribute decision-making model. *Interactive Learning Environments, 24*(8), 1812–1835.
16. Alqahtani, A. Y., & Rajkhan, A. A. (2020). E-learning critical success factors during the covid-19 pandemic: A comprehensive analysis of e-learning managerial perspectives. *Education Sciences, 10*(9), 216.
17. Haraldstad, K., Wahl, A., Andenæs, R., Andersen, J. R., Andersen, M. H., Beisland, E., ... & Helseth, S. (2019). A systematic review of the quality of life research in medicine and health sciences. *Quality of Life Research, 28*(10), 2641–2650.
18. Mucci, F., Mucci, N., & Diolaiuti, F. (2020). Lockdown and isolation: Psychological aspects of COVID-19 pandemic in the general population. *Clinical Neuropsychiatry, 17*(2), 63–64.
19. Mazza, C., Ricci, E., Biondi, S., Colasanti, M., Ferracuti, S., Napoli, C., & Roma, P. (2020). A nationwide survey of psychological distress among Italian people during the COVID-19 pandemic: Immediate psychological responses and associated factors. *International Journal of Environmental Research and Public Health, 17*(9), 3165.
20. Lee, S. A., Jobe, M. C., Mathis, A. A., & Gibbons, J. A. (2020). Incremental validity of coronaphobia: Coronavirus anxiety explains depression, generalized anxiety, and death anxiety. *Journal of Anxiety Disorders, 74*, 102268.
21. Rubin, G. J., & Wessely, S. (2020). The psychological effects of quarantining a city. *BMJ, 368.*
22. Lee, M., & You, M. (2020). Psychological and behavioral responses in South Korea during the early stages of coronavirus disease 2019 (COVID-19). *International Journal of Environmental Research and Public Health, 17*(9), 2977.
23. AlAteeq, D. A., Aljhani, S., Althiyabi, I., & Majzoub, S. (2020). Mental health among healthcare providers during coronavirus disease (COVID-19) outbreak in Saudi Arabia. *Journal of Infection and Public Health, 13*(10), 1432–1437.

24. Zhang, Y., & Ma, Z. F. (2020). Impact of the COVID-19 pandemic on mental health and quality of life among local residents in Liaoning Province, China: A cross-sectional study. *International Journal of Environmental Research and Public Health, 17*(7), 2381.
25. Harper, C. A., Satchell, L. P., Fido, D., & Latzman, R. D. (2020). Functional fear predicts public health compliance in the COVID-19 pandemic. *International Journal of Mental Health and Addiction*, 1–14.
26. Bargon, C. A., Batenburg, M. C., van Stam, L. E., van der Molen, D. R. M., van Dam, I. E., van der Leij, F., … & Verkooijen, H. M. (2020). The impact of the COVID-19 pandemic on quality of life, physical and psychosocial well-being in breast cancer patients and survivors-A prospective, multicenter cohort study. medRxiv.
27. Greco, F., Altieri, V. M., Esperto, F., Mirone, V., & Scarpa, R. M. (2020). Impact of COVID-19 pandemic on health-related quality of life in uro-oncologic patients: What should we wait for? *Clinical Genitourinary Cancer*.
28. Leigh-Hunt, N., Bagguley, D., Bash, K., Turner, V., Turnbull, S., Valtorta, N., & Caan, W. (2017). An overview of systematic reviews on the public health consequences of social isolation and loneliness. *Public Health, 152*, 157–171.
29. Saltzman, L. Y., Hansel, T. C., & Bordnick, P. S. (2020). Loneliness, isolation, and social support factors in post-COVID-19 mental health. *Psychological Trauma: Theory, Research, Practice, and Policy*.
30. Al-Hanawi, M. K., Mwale, M. L., Alshareef, N., Qattan, A. M., Angawi, K., Almubark, R., & Alsharqi, O. (2020). Psychological distress amongst health workers and the general public during the COVID-19 pandemic in Saudi Arabia. *Risk Management and Health-Care Policy, 13*, 733.
31. Flaxman, S., Mishra, S., Gandy, A., Unwin, H. J. T., Mellan, T. A., Coupland, H., … & Bhatt, S. (2020). Estimating the effects of non-pharmaceutical interventions on COVID-19 in Europe. *Nature, 584*(7820), 257–261.
32. Tomas, M. E., Sunkesula, V. C., Kundrapu, S., Wilson, B. M., & Donskey, C. J. (2015). An intervention to reduce health care personnel hand contamination during care of patients with Clostridium difficile infection. *American Journal of Infection Control, 43*(12), 1366–1367.
33. Winters, M., Jalloh, M. F., Sengeh, P., Jalloh, M. B., Conteh, L., Bunnell, R., … & Nordenstedt, H. (2018). Risk communication and Ebola-specific knowledge and behavior during 2014–2015 outbreak, Sierra Leone. *Emerging Infectious Diseases, 24*(2), 336.
34. ul Haq, S., Shahbaz, P., & Boz, I. (2020) Knowledge, behavior, and precautionary measures related to the COVID-19 pandemic among the general public of Punjab province, Pakistan. *The Journal of Infection in Developing Countries, 14*(08), 823–835.
35. Broadbent, E., Wilkes, C., Koschwanez, H., Weinman, J., Norton, S., & Petrie, K. J. (2015). A systematic review and meta-analysis of the brief illness perception questionnaire. *Psychology & Health, 30*(11), 1361–1385.
36. Lai, J., Ma, S., Wang, Y., Cai, Z., Hu, J., Wei, N., … & Hu, S. (2020). Factors associated with mental health outcomes among health care workers exposed to coronavirus disease 2019. *JAMA Network Open, 3*(3), e203976.
37. Ioannidis, J. P. (2020). Coronavirus disease 2019: The harms of exaggerated information and non-evidence-based measures.
38. Paakkari, L., & Okan, O. (2020). COVID-19: Health literacy is an underestimated problem. *The Lancet Public Health, 5*(5), e249–e250.
39. Nguyen, H. C., Nguyen, M. H., Do, B. N., Tran, C. Q., Nguyen, T. T., Pham, K. M., & Duong, T. H. (2020). People with suspected COVID-19 symptoms were more likely depressed and had a lower health-related quality of life: The potential benefit of health literacy. *Journal of Clinical Medicine, 9*(4), 965.
40. Bailey, S. C., Serper, M., Opsasnick, L., Persell, S. D., O'Conor, R., Curtis, L. M., … & Wolf, M. S. (2020). Changes in COVID-19 knowledge, beliefs, behaviors, and preparedness among high-risk adults from the onset to the acceleration phase of the US outbreak. *Journal of General Internal Medicine, 35*(11), 3285–3292.

41. Wang, C., Pan, R., Wan, X., Tan, Y., Xu, L., Ho, C. S., & Ho, R. C. (2020). Immediate psychological responses and associated factors during the initial stage of the 2019 coronavirus disease (COVID-19) epidemic among the general population in China. *International Journal of Environmental Research and Public Health, 17*(5), 1729.

42. Gavidia, M. (2020). How has COVID-19 affected mental health, the severity of stress among employees? *American Journal of Managed Care.*

43. Brooks, S. K., Webster, R. K., Smith, L. E., Woodland, L., Wessely, S., Greenberg, N., & Rubin, G. J. (2020). The psychological impact of quarantine and how to reduce it: A rapid review of the evidence. *The Lancet.*

44. Zhong, Y., Liu, W., Lee, T. Y., Zhao, H., & Ji, J. (2021). Risk perception, knowledge, information sources, and emotional states among COVID-19 patients in Wuhan, China. *Nursing Outlook, 69*(1), 13–21.

45. Sun, D. Y. (2007). *Emergencies and behavioral decisions.* Beijing, China: Social Science Literature Press.

46. Paraskevis, D., Kostaki, E. G., Magiorkinis, G., Panayiotakopoulos, G., Sourvinos, G., & Tsiodras, S. (2020). Full-genome evolutionary analysis of the novel Coronavirus (2019-nCoV) rejects the hypothesis of emergence as a result of a recent recombination event. *Infection, Genetics, and Evolution, 79*, 104212.

47. Qian, Z. H. A. O., Caihong, H. U., Renjie, F. E. N. G., & Yuan, Y. A. N. G. (2020). Investigation of the mental health of patients with novel coronavirus pneumonia. *Chinese Journal of Neurology*, E003.

48. Yang, C. (2020). Does hand hygiene reduce SARS-CoV-2 transmission? *Graefe's Archive for Clinical and Experimental Ophthalmology, 258*(5), 1133–1134.

49. Al-Nasser, M. (1993). The social function of greetings in Arabic. *Zeitschrift für Arabische Linguistik*, 15–27.

50. Cheng, V. C. C., Wong, S. C., Chuang, V. W. M., So, S. Y. C., Chen, J. H. K., Sridhar, S., ... & Yuen, K. Y. (2020). The role of community-wide wearing of face mask for control of coronavirus disease 2019 (COVID-19) epidemic due to SARS-CoV-2. *Journal of Infection, 81*(1), 107–114.

51. Day, M. (2020). Covid-19: identifying and isolating asymptomatic people helped eliminate the virus in an Italian village. *BMJ: British Medical Journal, 368.*

52. Lyu, W., & Wehby, G. L. (2020). Community use of face masks and COVID-19: Evidence from a natural experiment of state mandates in the US: The study examines the impact on COVID-19 growth rates associated with state government mandates requiring face mask use in public. *Health Affairs, 39*(8), 1419–1425.

53. Eikenberry, S. E., Mancuso, M., Iboi, E., Phan, T., Eikenberry, K., Kuang, Y., ... & Gumel, A. B. (2020). To mask or not mask: Modeling the potential for face mask use by the general public to curtail the COVID-19 pandemic. *Infectious Disease Modelling, 5*, 293–308.

54. Lima, M. M. D. S., Cavalcante, F. M. L., Macêdo, T. S., Galindo Neto, N. M., Caetano, J. Á., & Barros, L. M. (2020). Cloth face masks to prevent Covid-19 and other respiratory infections. *Latin-American Journal of Sick, 28.*

55. Shelus, V. S., Frank, S. C., Lazard, A. J., Higgins, I. C., Pulido, M., Richter, A. P. C., ... & Hall, M. G. (2020). Motivations and barriers for the use of face coverings during the COVID-19 pandemic: Messaging insights from focus groups. *International Journal of Environmental Research and Public Health, 17*(24), 9298.

56. Li, T., Liu, Y., Li, M., Qian, X., & Dai, S. Y. (2020). Mask or no mask for COVID-19: A public health and market study. *PloS One, 15*(8), e0237691.

57. MacIntyre, C. R., & Chughtai, A. A. (2020). A rapid systematic review of the efficacy of face masks and respirators against coronaviruses and other respiratory transmissible viruses for the community, healthcare workers, and sick patients. *International Journal of Nursing Studies, 108*, 103629.

58. Al Naam, Y. A., Elsafi, S. H., Alkharraz, Z. S., Alfahad, O. A., Al-Jubran, K. M., & Al Zahrani, E. M. (2021). Community practice of using face masks for the prevention of COVID-19 in Saudi Arabia. *PloS One, 16*(2), e0247313.

59. Lim, V. W., Lim, R. L., Tan, Y. R., Soh, A. S., Tan, M. X., Othman, N. B., ... & Chen, M. I. (2021). Government trust, perceptions of COVID-19 and behavior change: Cohort surveys, Singapore. *Bulletin of the World Health Organization, 99*(2), 92.
60. World Health Organization. Novel coronavirus–China. Geneva: World Health Organization. Accessed 12 May 2020.
61. Ng, Y., Li, Z., Chua, Y. X., Chaw, W. L., Zhao, Z., Er, B., ... & Lee, V. J. (2020). Evaluation of the effectiveness of surveillance and containment measures for the first 100 patients with COVID-19 in Singapore—January 2–February 29, 2020. *Morbidity and Mortality Weekly Report, 69*(11), 307–311.
62. Bong, C. L., Brasher, C., Chikumba, E., McDougall, R., Mellin-Olsen, J., & Enright, A. (2020). The COVID-19 pandemic: Effects on low-and middle-income countries. *Anesthesia and Analgesia.*
63. Dodson, S., Good, S., & Osborne, R. (2015). Health literacy toolkit for low and middle-income countries: A series of information sheets to empower communities and strengthen health systems.
64. Lau, L. L., Hung, N., Go, D. J., Ferma, J., Choi, M., Dodd, W., & Wei, X. (2020). Knowledge, attitudes, and practices of COVID-19 among income-poor households in the Philippines: A cross-sectional study. *Journal of Global Health, 10*(1).
65. Philippine Statistics Authority. (2018). Philippine statistics authority. Retrieved from Philippine Statistics Authority Web site: https://psa.gov.ph/vegetable-root-crops-main/tomato.
66. Alshurafat, H., Al Shbail, M. O., & Mansour, E. (2021). Strengths and weaknesses of forensic accounting: An implication on the socio-economic development. *Journal of Business and Socio-economic Development, 1*(2), 85–105. https://doi.org/10.1108/JBSED-03-2021-0026.
67. Alzaneen, R., & Mahmoud, A. (2019). The role of management information systems in strengthening the administrative governance in ministry of education and higher education in Gaza. *International Journal of Business Ethics and Governance, 2*(3), 1–43. https://doi.org/10.51325/ijbeg.v2i3.44.
68. Smith, J. A., & Judd, J. (2020). COVID-19: Vulnerability and the power of privilege in a pandemic. *Health Promotion Journal of Australia, 31*(2), 158.
69. Afana, A., & EL Agha, A. (2019). The role of organizational environment in enhancing managerial empowerment in Al-Aqsa network for media and art production. *International Journal of Business Ethics and Governance, 2*(2), 30–63. https://doi.org/10.51325/ijbeg.v2i2.40.
70. Hasan Al-Naser, M. (2019). Public governance and economic growth: Conceptual framework. *International Journal of Business Ethics and Governance, 2*(2), 1–14. https://doi.org/10.51325/ijbeg.v2i2.21.
71. Aminova, M., & Marchi, E. (2021). The role of innovation on start-up failure vs. its success. *International Journal of Business Ethics and Governance, 4*(1), 41–72. https://doi.org/10.51325/ijbeg.v4i1.60.
72. Safari, K., Njoka, C., & Munkwa, M. G. (2021). Financial literacy and personal retirement planning: a socioeconomic approach. *Journal of Business and Socio-economic Development, 1*(2), 185–197. https://doi.org/10.1108/JBSED-04-2021-0052.
73. Hassan, S. M., Ring, A., Tahir, N., & Gabbay, M. (2021). How do Muslim community members perceive Covid-19 risk reduction recommendations-a UK qualitative study? *BMC Public Health, 21*(1), 1–14.
74. Buller, A. (2020). Why are British Muslims being falsely blamed for the spread of Covid-19? Arabian Business: https://www.arabianbusiness.com/politics-economics/451249-why-are-muslims-being-falselyblamed-for-the-spread-of-covid-19-in-britain. Accessed 20 Sep 2020.
75. Zainal, M. M., Hamdan, A., & Al Mubarak, M. (2021). Exploring the role of artificial intelligence in healthcare management and the challenge of coronavirus pandemic. *Internet of Things, 2021*, 243–260.
76. Alalwi, B., Mazzuchi, T., Hamdan, A., & Al Mubarak, M. (2021). Blockchain technology implications on supply chain management: A review of the literature. *Applications of Artificial Intelligence in Business, Education and Healthcare*, 23–38.

77. Malave, A., & Elamin, E. M. (2010). Severe acute respiratory syndrome (SARS): Lessons for future pandemics. *AMA Journal of Ethics, 12*(9), 719–725.
78. Park, J. E., Jung, S., & Kim, A. (2018). MERS transmission and risk factors: A systematic review. *BMC Public Health, 18*(1), 574.
79. Kim, K., Andrew, S. A., & Jung, K. (2017). Public health network structure and collaboration effectiveness during the 2015 MERS outbreak in South Korea: An institutional collective action framework. *International Journal of Environmental Research and Public Health, 14*(9), 1064.
80. Rickard, L. N. (2021). Pragmatic and (or) constitutive? On the foundations of contemporary risk communication research. *Risk Analysis, 41*(3), 466–479.
81. Lurie, N., Saville, M., Hatchett, R., & Halton, J. (2020). Developing Covid-19 vaccines at pandemic speed. *New England Journal of Medicine, 382*(21), 1969–1973.

The Impact of Digital Learning Management System on Students of Higher Education Institutions During Covid-19 Pandemic

K. Sivasubramanian⬢, K. P. Jaheer Mukthar⬢, V. Raju⬢, and Kolachina Srinivas⬢

1 Introduction

The Covid-19 virus health issue has become a global phenomenon. It also creates many other issues related to human life and society, including education, employment opportunities, industries, farm activities, etc. In this critical time, the higher education student could not go to colleges and universities. Because of this pandemic situation, digital online form e-learning or virtual learning started familiarized in education. The digital form of virtual teaching and learning process becomes the mandatory solution for the educational field. The Digital learning system is termed virtual and blended learning through digital platforms. There always be confusion among online learning, electronic learning and digital learning management systems. The digital type of learning is accompanied by modern technology and innovation to transform knowledge. The digital learning method is learning enabled by innovative technology that gives the learners few elements of regulation of time, place, and way [1]. In this various phenomenon time portrait, the learning is not restricted with any particular time frame. The power of the internet and electronic devices allowed the learners to adopt the insights of concepts at any time. The place denotes that the learning is not limited to one single place, like within the walls of a school or colleges. It is

K. Sivasubramanian (✉) · K. P. Jaheer Mukthar · V. Raju · K. Srinivas
Department of Economics, Kristu Jayanti College Autonomous, Bengaluru, India
e-mail: sivasubaramanian@kristujayanti.com

K. P. Jaheer Mukthar
e-mail: jaheer@kristujayanti.com

V. Raju
e-mail: raju.v@kristujayanti.com

K. Srinivas
KL Business School, KL Deemed to Be University, KLEF, Vaddeswaram, Guntur District, Andhra Pradesh, India

657

accessible anywhere in the world. Thirdly, the path refers to that there is a limited way of pedagogy by the instructor. The modern interactive, intelligent and adoptive technological process allows the learners to equip knowledge on their style and make it fruitful. Fourthly, the pace learning system is not limited to the pace of a whole classroom of the learners.

The student's digital learning system is the blend of technological adoption, digitalized content and teaching. In these three blends of the learning system, the technology is one of the prominent mechanisms and creates the platform for delivering subject content. It enables the learners to receive the equipped content from the digital portal. This includes the accessibility of the internet of things, unique software modules and hardware, such as the desktop computer, laptop and smart phones etc. The technology is termed as the innovative instrument, but not the content of the teaching material. Secondly, the digital form of content refers to the rich quality of academic study and teaching material. These uploaded materials on the digital learning management platform are used to deliver the lecture through the technology. It is not just the PDF form of study material or the PowerPoint presentation, but it is a form for students to engage and interact for a good learning experience. Thirdly, the teaching or instructing option that enables the students to receive the lecture from the educator. It is very important for the educators or the instructors in the digital learning management system. Technological advancement and innovation in teaching-learning pedagogies have changed teacher's role in the learning process. But it never abolishes the ultimate and unique requirement of the teacher.

There was an unadorned constraint on the availability of digitally skilled professionals [2]. In the world of digital learning practice, teachers could deliver a rich amount of personalized support and guidance to ensure learners are enriched with the content delivery and travel on the track across the whole course time. The present paper aims to study the impact of the digital learning management system on the students of higher education institutions. It is also connected with employability to lifetime employability, but it is often used as a new fortification in the labour market. Though employability becomes a slogan in the structural literature, no measurement can be revealed. The theoretical model provides a framework for forthcoming analytical research on employability. It will help determine the major factor that may impact labour market changeovers for people [3].

2 Literature Review

Digital learning in India arrives with technological and innovational growth and development. India is growing in the very fastest phase in digital learning in schools and universities. An exponential growth rate is found in the digital form of learning through technology in the nation. India plays a key role in encouraging and attracting enormous supply and demand of smart phone and electronic gadgets. With the highest population of 130 crores [4]. The dramatic usage of the internet has changed the nation's method and style of life, including rural and urban areas. Modernization, technological upliftment and innovation of new software led to change the lifestyle of Indians are moved towards online. This online transformation penetrated many sectors such as online shopping, digital information, digital banking, chatting online, gaming online, and digital learning. The perception and usage of innovation show the benefit of virtual learning. But parenthetically, this study report is limited to well-developed countries where the accessibility and affordability of technology and inventory support for virtual learning are not problems. Boolean logic of game learning online is becoming very simple and easy to understand [5]. Studies identified that all the stakeholders check the clear associations among the worldwide involvement and employability surrounded outcomes connected with the falsifying of networks, experiential learning process, language learning, and technical and soft skills [6]. Progression in internet and administrative structure has influenced the ways the information is moved and perceived. Today's super bandwidth internet helps, ranging from a data text format and audio-video streaming. Digital learning as a simulated learning atmosphere depends on the internet. The developing and developed countries that have been helping virtual learning through the digital learning platform got a positive future ahead [7]. The new age of digital learning methodology is entirely changing the old format of the teaching and learning process. In this area of e-learning, technology plays a key role in the implementation of digital platforms. Based on their analysis, higher educational institutions are effectively applying e-learning sources [8]. The atmosphere of higher education is sprouting across the globe, increasing the level of costs, reducing the budget potential of institutions, and the growing need for online education made the higher educational institutions implement the digital learning method to deliver the content. This results in the immediate adoption of the e-learning process for the benefit of students [9]. They made the statement that, undoubtedly, the online learning system continues to shine in many institutions. Because of this expectation, the government and educational institutions started focusing on implying online teaching to higher education students [10].

Digital learning is an advantage to students' world, and it could reach anywhere on the globe. It is very much helpful to the students across the world can learn anywhere and at any time. Shortly, the digitalized system of learning and teaching methodologies could save a lot of money and time. They suggested that the respective government must own the digital learning equipment such as portal, websites and applications so that all classes of the nation could have accessibility. Finally, their study concluded that e-learning is flexible for the students to carry out their study easily [11]. In expectation of good growth in e-learning, the professional agencies, government, and educational institutions are started focusing on applying and implementing digital learning tools. They also found that the operative and efficient way of implementation led to gain the greatest benefit and acquire the highest knowledge by the student's community. The developed and developing countries like India, the e-learning hike the level of education, quality of education, and accessibility of education to all have been ensured for the nation's development. E-learning and digital learning tools are proven to be highly adaptive. The digital learning methodology helps to increase the percentage of literacy among the total population in India.

They also added that, through the e-learning pedagogy, many opportunities could be obtained and speed up development in developing nations [12]. The e-learning substitutes often complement learning through an appropriate method, development, and content delivery of the particular course. He has clinched that all vital dimensions of digital learning such as instructor or teacher, learner, course content, technology, and situation have a significant relationship with the digital learning system [13]. India always being the essence of motivation on the foundation of traditional information across the world. It is a scientific technology, value, growth of sustainable medicine always sets the best place. In this scenario, the people's mindset irrespective of all the propels and bustle going on presently, the covid-19 pandemic carried the alteration in the teaching-learning model. The chief concern of the peoples is a disease and employment [14]. About 81% of the responses towards the online platform e-classes. It is also found that the most preferred virtual online platforms such as zoom, Google meet, and Cisco WebEx [15].

3 Objectives

i.   To visualize the significance of digital learning for the higher educational institutions and student in the study area.
ii. To find out the perception of the students from the higher education institutions in the study area.
iii. To evaluate the pros and cons of the digital learning management system tools
iv. To compare the classroom learning process and virtual learning process through digital portals.

4 Hypotheses

Ho1: There is an association between understanding or quality of learning of the concepts and gender of the students through an online learning platform.

Ho2: There is no association between understanding or quality of learning and teaching efficiency in online flip classes.

Ho3: There is a relationship between quality of learning and doubt clearance in e-learning.

Ho4: There is a correlation between learning quality and the personal attention of the course teacher.

Ho5: There is an association between good learning outcome and study material given by the course teacher.

Ho6: There is no association between quality learning outcome and issues of accessibility towards the online instruments. (Electronic Gadgets).

Ho7: There is no relationship between quality learning outcome and students' psychological issues due to the Covid-19 pandemic.

H08: There is no correlation between quality learning outcome.

5 Significance of the Research on Digital Learning Management System (DLMS)

The concept of e-learning has been divided into two categories such as academic learning and training. The academic use of digital learning is widely used across the world. It is also a growing phenomenon in developing countries as well, particularly in India. The net market worth of $247 million is constituting in the digital learning industry in India during the year 2016. This value is anticipated to expand to around $1.96 billion in 2021, an annual growth of 52%. The users registered for the digital learning system in colleges, universities, and e-education applications were estimated at 1.6 million in 2016. The number is expected to hike by 9.6 million in the year 2021. About 48% of the total population (15–40 age group) has a high aspiration with lesser income group. The impact of the covid-19 pandemic is playing a key role in the digital e-learning system in India [4].

6 Research Methodology

6.1 Research Design

An empirical method of research design is used to carry out the present research study. Thus, the study is completely based on observation and obtain the data through the direct survey approach. The information obtained through the direct collection of data from the respondents. Moreover, the collected data is synthesized and compared with the basic theory or with the hypothesis framed by the researcher for the research study. Hence, the data is collected from the sample respondent directly, the degree of validity, results, and uniqueness is stood with real experience gathered at the field by the researcher.

6.2   Data, Sample and Sampling Design

The direct survey method is applied through a structured questionnaire and distributed to the target samples in the study area. Chennai is selected as the study area to research the digital learning system. Because Chennai city has many private, aided, and government colleges and universities, the research would wish to identify the impact of the e-learning system and resources of higher education institutions in the city. This becomes the vital and valid reason for selecting Chennai as the study area to collect the sample and proceed with the research work. The purposive sampling technique is used to select sample respondents from the population universe. This type of sampling design is a kind of non-probability sampling approach. The observer or the data enumerator will be sure of their own decision while choosing the samples participation from the whole population. It is also be termed selective, subjective or judgmental sampling. The sample size is determined as 160 from the population universe in the study area. Students from various institutions, including private, aided, and Government College and university, have been chosen for sample response. Collected data has been tabulated and analyzed through SPSS statistical software tools. Inferences are made to determine the relationship between the variables. The following statistical tools have been used to analyze the data, such as correlation, multivariate analysis and factor analysis. All these statistical tools have been analyzed with the SPSS software package. The data further analyzed through a regression model by using the GRETL software.

7   Profile of the Study Area

Chennai city is the capital of Tamil Nadu state from south India and India's highest industrial and commercial hub. Chennai city is selected for the present research survey because of its uniqueness in various attributes such as employment opportunities, educational opportunities, urbanization and lifestyle change. The city is meant for more than 24 Indian based companies comprising turnover worth $1 billion, diversified with many industries, namely, software, automobile, health care, manufacturing and financial sector. Moreover, the economic condition of the city is placed as the third-highest per capita GDP in India [16].

8   Data Analysis and Results

The current shift in educational worker's market policy has occasioned universities being positioned under increasing heaviness to harvest employable graduates [17]. The corona virus epidemic created substantial challenges for the universal higher edification community [18]. After assessing some well-recognized web-based knowledge systems, E-Learning scheme in LIS education delivers free admission to professionals [19]. A multivariate test is applied to access the relationship among the variables such as understanding level of the concepts by the students through online classes, gender, teaching efficiency, doubt clearance by the teacher, personal attention provided by the course teachers, study materials uploaded on the respective MOODLE portals, the convenience of students, accessibility issues towards the digital classes, various psychological issues, the type of institution and internet-related technical issues.

The Table 1 emphasizes the constant variable of the understanding level of the students in the study area. The multivariate test design is carried out, and it is significant at 0.000 level with a 99.99% confidence level. The Pillai's Trace test value is 0994. The calculated f-value is 2323.149, with the degrees of freedom at 10. The probability value is at the 0.000 significance level for all the intercept measurements. The second intercept measurement is Wilk's Lambda; the calculated value is 0.006 with the f-test value of 2323.149. Degrees of freedom is at 10 at 0.000 significance level. The third and fourth measurement value calculated at 158.037, and the f-test value is 2323.149, with degrees of freedom at 10. The probability value at 0.000 level, the confidence level at 99.99%. The dependent variable is termed as the quality learning outcome with the various independent variable. The multivariate analysis executed to bring out the association between a dependent variable and independent variables. The independent variables are considered in this analysis are gender, teaching efficiency, doubt clearance, personal attention, availability of study

Table 1 Multivariate test

Effect		Value	F	Hypothesis df	Error df	Sig
Intercept	Pillai's Trace	0.994	2323.149[b]	10.000	147.000	0.000
	Wilks' Lambda	0.006	2323.149[b]	10.000	147.000	0.000
	Hotelling's Trace	158.037	2323.149[b]	10.000	147.000	0.000
	Roy's Largest root	158.037	2323.149[b]	10.000	147.000	0.000
Quality learning outcome	Pillai's Trace	0.264	1.435	30.000	447.000	0.066
	Wilks' Lambda	0.753	1.461	30.000	432.150	0.058
	Hotelling's Trace	0.306	1.486	30.000	437.000	0.050
	Roy's Largest Root	0.213	3.169[c]	10.000	149.000	0.001

[a] Design: Intercept + Understanding level
[b] Exact statistic
[c] The statistic is an upper bound on F that yields a lower bound on the significance level

materials, convenience, accessibility problems, psychological issues of the students, type of institution, and internet issues.

The model expresses the understanding level (Table 2), and the gender F test value is calculated at 1.763. Its probability is significant at 0.157, which is much higher than the acceptance level of 0.005. Hence the null hypothesis is accepted. The teaching efficiency also accepted due to the higher significance level. The doubt clearing aspect of the teacher is associated with the quality outcome of the student at 0.000 significance level. Personal attention, study material, students' psychological issues, and accessibility problems are also not significant at the 5% level. The type of institution and the internet issues are associated with the quality learning outcome of the student because the statistical p-value level is less than 0.005 significance level.

The correlation coefficient is also measured between the dependent and independent variables, and inferences are made to demonstrate the degree of relationship.

Table 2 Tests of between-subjects effects

Source		Type I sum of squares	Df	Mean square	F	Sig
Corrected model	Gender	1.230[a]	3	0.410	1.763	0.157
	Teaching efficiency	0.296[b]	3	0.099	0.099	0.960
	Doubt clearance	1.431[c]	3	5.477	5.416	0.000
	Personal attention	0.160[d]	3	4.053	0.621	0.602
	Study material	0.208[e]	3	0.069	0.502	0.681
	Convenience	0.040[f]	3	0.013	0.204	0.893
	Problems of access	1.116[g]	3	3.372	3.905	0.040
	Psychological issues	0.629[h]	3	0.210	1.224	0.303
	Type of institution	5.719[i]	3	1.906	4.692	0.004
	Internet issues	0.346[j]	3	0.115	4.206	0.000

[a] R Squared = 0.033 (Adjusted R Squared = 0.014)
[b] R Squared = 0.002 (Adjusted R Squared = −0.017)
[c] R Squared = 0.027 (Adjusted R Squared = 0.008)
[d] R Squared = 0.012 (Adjusted R Squared = −0.007)
[e] R Squared = 0.010 (Adjusted R Squared = −0.009)
[f] R Squared = 0.004 (Adjusted R Squared = −0.015)
[g] R Squared = 0.070 (Adjusted R Squared = 0.052)
[h] R Squared = 0.023 (Adjusted R Squared = 0.004)
[i] R Squared = 0.083 (Adjusted R Squared = 0.065)
[j] R Squared = 0.041 (Adjusted R Squared = 0.022)

The teaching efficiency is correlated with the students' learning outcome doubt clearance, personal attention, study material, and internet issues. Karl Pearson's coefficient of correlation is calculated at -0.034 value for teaching efficiency and learning outcome. It has a negative relation with each other. The other variable, doubt clearance value, is calculated at -0.021, which is also has a negative relation with teaching efficiency. The other three variables, such as personal attention, study material and internet-related issues, are also negatively correlated with Karl Pearson's calculated values of -0.136, -0.04 and -0.056. Like-wise, the quality of learning is measured with teaching efficiency, which is calculated at -0.034 value. It has a negative correlation with each other. The doubt clearance and study material negatively correlate with quality learning outcome with the calculated values of -0.008 and -0.053, respectively. Personal attention has a positive correlation at 0.004, the degree of correlation is less, but it is positive, as shown in Table 3.

The doubt clearance of the students with the respective course teacher is calculated at -0.021, which has a negative correlation coefficient with the teaching efficiency of the course taught through the e-learning method. The learning outcome is also negatively correlated with the dependent variable. Personal attention, study material and internet issues are positively correlated with the calculated correlation coefficient of 0.085, 0.09, and 0.047. To personal attention, the teaching efficiency, study material and internet related issues are negatively correlated. The calculated Karl Pearson's correlation coefficient is -0.136, -0.025 and -0.015, respectively, with the personal attention given by the course teacher. Personal attention is positively correlated with learning outcome and doubt clearance; the correlation coefficient is calculated at 0.004 and 0.085, respectively. The other variable is the study material, which negatively correlates with teaching efficiency, learning outcome of the students in higher education institutions, personal attention, and internet issues. The calculated correlation coefficients are -0.04, -0.053, -0.025 and -0.04, respectively, with study material. The internet issues faced by the students are positively correlated with learning outcome and doubt clearance with the calculated value of 0.162 and 0.047 respectively. It is negatively correlated with teaching efficiency, personal attention and study material with the calculated values of -0.056, -0.015 and -0.04.

Table 4 represents the ordinary least square; the understanding level was termed the dependent variable. The independent variables considered teaching efficiency, doubt clearance, personal attention, study material, convenience, the problem of access, the students' financial constraint, the student's psychological situation, and the internet issues of the students, as shown in Figs. 1 and 2.

Table 5 exhibits both independent and dependent variables coefficient and its confidence level at 95% confidence intervals. The understanding level (constant) calculated coefficient is 0.918666 with a confidence level at 95% intervals, as shown in Table 6.

The above diagram (Fig. 3) emphasizes the factors which determine the constant variable. The constant variable is the quality learning outcome. An important association between a high mark of exertion expectation and a high proportion of usage of the learning situation [20].

Table 3 The degree of correlations

Correlations

		Teaching efficiency	Learning outcome	Doubt clearance	Personal attention	Study material	Internet issues
Teaching efficiency	Pearson correlation	1	0.034	−0.021	−0.136	−0.04	−0.056
	Sig. (2-tailed)		0.666	0.795	0.087	0.62	0.479
	N	160	160	160	160	160	160
Quality learning outcome	Pearson correlation	−0.034	1	−0.088	0.004	−0.053	0.162[a]
	Sig. (2-tailed)	0.666		0.27	0.962	0.507	0.04
	N	160	160	160	160	160	160
Doubt clearance	Pearson correlation	−0.021	−0.088	1	0.085	0.09	0.047
	Sig. (2-tailed)	0.795	0.27		0.283	0.259	0.553
	N	160	160	160	160	160	160
Personal attention	Pearson correlation	−0.136	0.004	0.085	1	−0.025	−0.015
	Sig. (2-tailed)	0.087	0.962	0.283		0.75	0.855
	N	160	160	160	160	160	160
Study material	Pearson correlation	−0.04	−0.053	0.09	−0.025	1	−0.04
	Sig. (2-tailed)	0.62	0.507	0.259	0.75		0.62
	N	160	160	160	160	160	160
Internet issues	Pearson correlation	0.056	0.162[a]	0.047	−0.015	−0.04	1
	Sig. (2-tailed)	0.479	0.04	0.553	0.855	0.62	
	N	160	160	160	160	160	160

[a] Correlation is significant at the 0.05 level (2-tailed)

9 Impact of Covid-19 on Higher Education

The covid-19 pandemic has impacted the economic and societal recession by way of unemployment, the final year student severely affected due to non-availability of campus placements. The expenses increased for buying smartphone and laptops for online virtual classes. The impact would differ from case to case; the overall influence on higher learning is likely termed to be highly significant. The quality of

Table 4 The ordinary least square

Model: OLS, using observations 1–160

Dependent variable: understanding level

	Coefficient	Std. error	t-ratio	p-value
Const	0.918666303	0.541807015	1.69556	0.092043915
Teaching efficiency	−0.009705373	0.043043757	−0.225476893	0.821915208
Doubt clearance	−0.062402939	0.075606743	−0.825362091	0.41047676
Personal attention	0.004776601	0.147293251	0.032429188	0.974172897
Study material	−0.037774349	0.115846723	−0.326071797	0.744824499
Convenience	−0.094199089	0.166599059	−0.565423892	0.572630245
Problems of access	0.169761712	0.137491969	1.234702751	0.21887182
Financial constraint	−0.047340262	0.147171568	−0.321667174	0.748152657
Psychological issues	0.045504156	0.103430088	0.439950858	0.660605953
Internet issues	0.367226131	0.184340503	1.992107669	0.048174202
Mean dependent variable	1.20625	S.D. dependent variable	0.527207846	
Sum squared residual	41.91866748	S.E. of regression	0.528637667	
R-squared	0.051479734	Adjusted R-squared	−0.005431481	
F(9, 150)	0.904562196	P-value (F)	0.523011874	
Log-likelihood	−119.8747603	Akaike criterion	259.7495206	
Schwarz criterion	290.5012587	Hannan-Quinn	272.236736	

learning by the student is also declined due to non-direct teaching by the teachers. The covid-19 pandemic drastically affected the entire educational operations in India. Higher education is one of the most affected areas due to this pandemic situation. The eruption of the virus bound the lockdown of all the sectors of the economy, including the educational sector, particularly in higher education. All the higher learning institution were closed with the finish of all the educational events and generated many issues and challenges for all the stakeholders [21]. The distinct activities such as examinations, result publications, seminars, conferences, entrance tests, and competitive examinations conducted by the educational institutions were postponed. Most of the entrance test related higher education stands cancelled and postponed, which have created huge challenges for the career of student's community. The major challenges faced by the students and teachers in the teaching-learning process. Because they could not be able to present physically in the institution. The best way for the institutions to move towards electronic learning system to continue

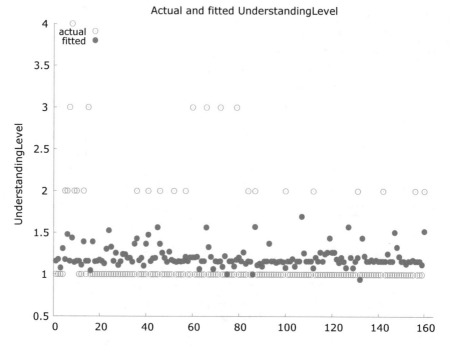

Fig. 1 Actual model fit for the constant variable

the teaching and learning process. But, the concern regarding the immediate implementation to have the continuity of syllabus and face this new pedagogical process. Though, the issues of digital platform and the practical problems on the students' alignment, the higher learning institutions have the greatest responsibility to support the students. The implementation process is time-consuming and cost consuming in buying the license for paid versions of the digital learning platform. The pandemic condition makes the compulsory adoption of the digital learning system. It allows the teachers to students to learn the innovative method of teaching-learning pedagogy. The higher educational institutions started directing webinars, orientation, induction and regular classes to support various digital learning platforms. The e-learning applications used in this process include zoom meeting, google meet, skype, YouTube live, Facebook lives, and Webex platforms. The digital initiative was taken to generate the operative and active virtual learning mode of teaching-learning to create awareness among the students for e-learning classes. The teachers use various platforms such as MOODLE, Digital Learning Management System (DLMS), Google classroom, WhatsApp, Google Drive and other platforms to share the important study materials [22]. The University Gants Commission (UGC) also has given an appropriate guideline for the proper conduct of online examinations to all the higher educational institutions in India [23]. Most of the higher education institutions made the changeover to online mode [24].

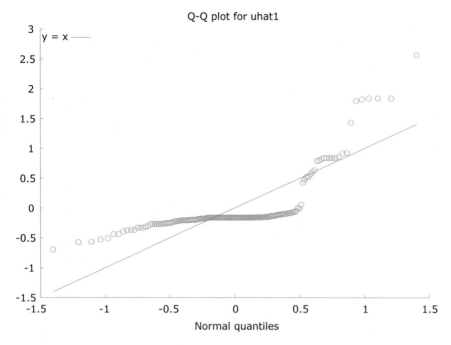

Fig. 2 Model plot for the constant variable

10 Pros and Cons of Digital Learning Approach

The covid-19 pandemic has both the positive and negative sides of influence on academic and research in the higher education sector. While checking the negative aspects of the virtual learning system, it has been impossible for the research and researchers to travel and work together locally and globally.

10.1 Pros of Virtual Learning System

i. Availability, accessibility and affordability of digital content at anytime and anywhere and on the go convenience.
ii. Wide range of clarity of the content based on the good resource materials through various formats such as expert talks, notes, video content, audio content, animations and blogs etc.,
iii. Personal interaction is highly possible without direct or primary contact with the teachers to save time and money for both learners, teachers, and the institutions.
iv. Many alternative and optional resources can be made available by the course teachers on the MOODLE portal, such as PowerPoint presentations, course lecture in both video and lesson form.

Table 5 Statistical confidence Levelt(150, 0.025) = 1.976

Variable	Coefficient	95 confidence intervals
Const	0.918666	(−0.151893, 1.98923)
Teaching efficiency	−0.00970537	(−0.0947558, 0.0753450)
Doubt clearance	−0.0624029	(−0.211795, 0.0869888)
Personal attention	0.00477660	(−0.286261, 0.295814)
Study material	−0.0377743	(−0.266677, 0.191128)
Convenience	−0.0941991	(−0.423383, 0.234985)
Problems of access	0.169762	(−0.101909, 0.441433)
Financial constraint	−0.0473403	(−0.338137, 0.243457)
Psychological issues	0.0455042	(−0.158864, 0.249872)
Internet issues	0.367226	(0.00298675, 0.731466)

v.   Online mode of teaching and learning much helpful in the process of evaluation on various stages. The students will get an immediate and faster response to their queries.

vi. Boon for both teachers, students and higher educational institutions during the Covid-19 pandemic situation.

10.2 Cons of Digital Learning Platform

i.   The one-to-one contact, presence of the teacher and direct personal guidance is missing phenomenon.

ii. There is no proper guidance and special assistance for the slow learners

iii. The poor and developing nations suffer from the non-availability of good and cheaper internet access, power back-ups and other basic infrastructural facilities to meet the virtual classes.

iv. There are no interactions among the students to students and discussion with teachers through face-to-face talk.

v.   The role and responsibility are limited towards the progression of students.

Table 6 White's test for heteroscedasticity OLS, using observations 1–160 dependent variable: uhat^2, omitted due to exact collinearity: 12 series

	Coefficient	Std. error	t-ratio	p-value
Const	1.64796	11.4450	0.1440	0.8858
Teaching efficiency	1.47625	1.38259	1.068	0.2878
Doubt clearance	0.817122	2.34686	0.3482	0.7283
Personal attention	0.0468350	3.89253	0.01203	0.9904
Study material	0.757272	3.09745	0.2445	0.8073
Convenience	−2.18750	3.91269	−0.5591	0.5772
Problems of access	−3.54389	6.55042	−0.5410	0.5895
Financial constraint	−1.66571	4.22563	−0.3942	0.6942
Psychological issue	−2.47946	2.33051	−1.064	0.2896
Internet issues	2.20091	6.71081	0.3280	0.7435
Sq teaching effectiveness	−0.110676	0.0899905	−1.230	0.2212
X2_X3	0.160425	0.175110	0.9161	0.3615
X2_X4	−0.429214	0.337919	−1.270	0.2065
X2_X5	−0.225115	0.221379	−1.017	0.3113
X2_X6	0.0334508	0.339902	0.09841	0.9218
X2_X7	−0.00448980	0.543076	−0.008267	0.9934
X2_X8	0.0781206	0.341079	0.2290	0.8192
X2_X9	−0.0632587	0.275895	−0.2293	0.8190
X2_X10	−0.0497964	0.607943	−0.08191	0.9349
sq_Doubt Clearance	−0.149795	0.132862	−1.127	0.2619
X3_X4	0.0828406	0.908753	0.09116	0.9275
X3_X5	−0.206910	0.512497	−0.4037	0.6871
X3_X6	0.146264	0.662670	0.2207	0.8257
X3_X7	−0.172453	0.528574	−0.3263	0.7448
X3_X8	−0.201504	0.799655	−0.2520	0.8015
X3_X9	0.0514162	0.359945	0.1428	0.8867
X4_X5	−0.0361340	0.951863	−0.03796	0.9698
X4_X7	0.724565	2.75929	0.2626	0.7933
X4_X8	0.117964	1.92985	0.06113	0.9514
X4_X10	−0.332794	2.42338	−0.1373	0.8910
X5_X6	0.0153000	1.02285	0.01496	0.9881
X5_X7	0.261828	0.565549	0.4630	0.6443
X5_X8	−0.0565862	1.04918	−0.05393	0.9571
X5_X9	0.579958	0.641777	0.9037	0.3680
X5_X10	−0.879768	1.07784	−0.8162	0.4160

(continued)

Table 6 (continued)

	Coefficient	Std. error	t-ratio	p-value
X6_X7	0.266253	1.28120	0.2078	0.8357
X6_X8	1.23034	1.38685	0.8871	0.3768
X6_X9	0.0663662	0.840137	0.07899	0.9372
X7_X9	1.56525	0.732309	2.137	0.0346 **
X7_X10	0.148264	1.23454	0.1201	0.9046
X8_X9	−0.0129698	0.749123	−0.01731	0.9862
X8_X10	0.322443	1.30016	0.2480	0.8046
X9_X10	−0.180272	1.05188	−0.1714	0.8642

Unadjusted R-squared = 0.180482, Test statistic: TR^2 = 28.877139, With p-value = P(Chi-square(42) > 28.877139) = 0.938340

vi. Study materials and lesson content are designed and uploaded on the Modular Object-Oriented Dynamic Learning Environments (MOODLE) by the course teacher, which is common for all learners (both slow and fast learners).

vii. The unexpected loss of internet connectivity, unsupported file formats and software are always the gaps in e-learning.

viii. There are issues such as no quality standards for e-learning materials and high cost of accessibility also persist.

11 Conclusion

The digital learning method is learning enabled by innovative technology that gives the learners few elements of regulation of time, place, and way. In this various phenomenon time portrait, the learning is not restricted with any particular time frame. The power of the internet and electronic devices allowed the learners to adopt the insights of concepts at any time. The place denotes that the learning is not limited to one single place like within the walls of a class in school or colleges. It is accessible anywhere in the world. Thirdly, the path refers to that there is a limited way of pedagogy by the instructor. The modern interactive, intelligent and adoptive technological process allows the learners to equip knowledge on their style and make it fruitful. Fourthly, the pace learning system is not limited to the pace of a whole classroom of the learners. The student's digital learning system is the blend of technological adoption, digitalized content and teaching. In these three blends of the learning system, the technology is one of the prominent mechanisms and creates the platform for delivering subject content. It enables the learners to receive the equipped content from the digital portal. This includes the accessibility of the internet of things, unique software modules and hardware, such as the desktop computer, laptop and smartphones etc. The technology is termed as the innovative instrument, but not the content of the teaching material. Secondly, the digital form of content refers to the

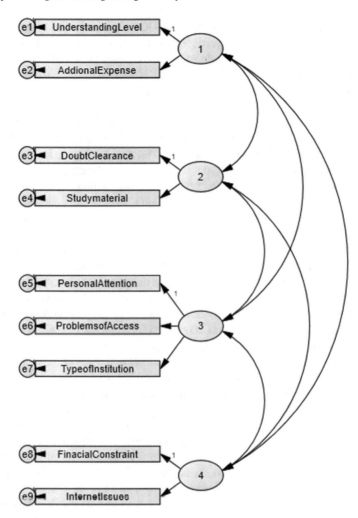

Fig. 3 Factor analysis test diagram

rich quality of academic study and teaching material. These uploaded materials on the digital learning management platform are used to deliver the lecture through the technology. The concept of e-learning has been divided into two categories: academic learning and training [25]. The academic use of digital learning is widely used across the world. It is also a growing phenomenon in developing countries as well, particularly in India. The net market worth of $247 million is constituting in the digital learning industry in India during the year 2016. This value is anticipated to expand to around $1.96 billion in 2021, an annual growth of 52%. A multivariate test is applied to access the relationship among the variables such as understanding level of the concepts by the students through online classes, gender, teaching efficiency, doubt clearance by the teacher, personal attention provided by the course teachers, study materials uploaded on the respective MOODLE portals, the convenience of students, accessibility issues towards the digital classes, various psychological issues, the type of institution and internet-related technical issues [4]. Internet instruction is soon to convert to the world's leading form [26]. Until freshly, the business situation branded by the biosphere in which countries were more than associated than ever before [27]. Recent information and statements kill time of e-learning plays a vibrant role in the progress of an individual and the forthcoming country [28]. Academic qualification is seen as an important measurement of their employability [29]. The impact would differ from case to case; the overall influence on higher learning is likely termed to be highly significant. The quality of learning by the student is also declined due to non-direct teaching by the teachers. Hence, the online e-learning system is rapidly growing in the educational sector, especially in the higher educational system [30]. The digital division will not be led to the prohibition of learners from poor and marginalized backgrounds [31].

The present technology has the probability of attaining Widespread quality education and progressing the learning outcomes. However, to release it's possible, digital technology has to be enhanced [32]. At the same time, the traditional institutions and course teachers viewed e-learning as a hazard [33]. About 1.3 billion students worldwide have not been able to attend a school or university [34]. Several higher education institutions have the technical skills required for the virtual learning process and maters in handling social media platforms [35]. The changes in learner's comportment towards the style and preference of specific degree programmes may become a substantial effect after the pandemic [36]. The outbreak of this pandemic required many nations to impose the lockdown. The impact of Covid can be visible in every compass of life to education, specifically in higher education. The nation's widest sector is adversely affected to a large extent. The teaching, learning and evaluation method entirely changed due to the pandemic. The developing nations like India, virtual learning recorded tremendous influence on the lives of those students who are less privileged and comes from marginalized sections of the nation. It had several effects on the conventional teaching and learning process of interaction between students and teachers. The students from technically deprived areas of learning become vulnerable. The course of education and the curriculum methods are changed. The changes made in terms of technological transformation [37, 38]. Pandemic has affected immensely on the educational pattern in India. Even though it caused many problems, numerous opportunities also arrive. The government of India and the distinct stakeholders of teaching-learning discovered the likelihood of all kinds of learning by digital innovation. India did not equip enough to reach the entire corners of the country through the virtual platforms. The precedence must be to employ virtual technological innovations to generate a beneficial situation for a mass young people of India. It is compulsory and necessary for higher educational institutions to equip their technical knowledge to face the virtual digital classes [39, 40]. Being the virtual learning contents are fundamentally different from the old learning method, there would be a requirement for preceding coaching for the student and teachers [41]. The world is fronting a predicament because of the Coronavirus pandemic situation, especially the developing nations like India. Most of the nations have been exaggerated due to the straight and the unintended belongings of the virus and survives of lots of publics have been altered, in numerous cases are irrecoverable. The pandemic situations influenced the higher learning sector as the main rolling standard of the educational sector—the students are involuntary to continue in their homelands to replace the risk of conceivable contagion and grief due to infection of the virus colleges to run during the pandemic situation. This has triggered almost all colleges to spread out to the digital route for running scheduled classes through digital modes. An added development has occurred in the cumulative admiration of pre-recorded access to online courses made available by various platforms. The covid-19 pandemic caused a tremendous downtrend in the global economy, and it is also responsible for the mass influence on higher educational learning [42]. The covid-19 pandemic was progressing very rapidly, more specifically in the developing nations, and it has impacted the research activities of the higher educational institutions [43–45]. On the whole, the economic and societal recession has been impacted by the covid-19 pandemic by way of unemployment, the final year student severely affected due to non-availability of campus placements. The expenses increased for buying smartphone and laptops for online virtual classes.

12 Policy Recommendations

Irrespective of the covid created numerous negative influences on learning, at the meantime, it brought additional dichotomy to the teaching and learning activities. Nevertheless, there are numerous setbacks of virtual learning, but it becomes mandatory for the learning system during this pandemic condition. The following vital policy recommendations have been given based on the present study.

i.   Both state and central government has to provide free high-speed internet services to all the higher learning students

ii.   The state government of Tamil Nadu must support financially to equip the state-owned colleges and universities to implement MOODLE platform for the benefit of students.

iii. All the online classes to be recorded and monitored by the institutions to have proper online etiquettes on the virtual model.

iv.   The respective parents also have to take cautious watching and monitor internet usage by their children while attending the virtual classes.

v.   The government and government-aided colleges and universities lack in implementing the virtual mode of classes due to the non-availability of funds. So, the state government's education department has to allocate a special fund to facilitate the virtual classes properly.

vi.   Most of the government college students did not have smartphone and computer for online classes. So, they must have to provide free smartphones for education purposes.

vii.   In this pandemic situation, the telecommunication companies and corporates could extend their support towards higher education students by providing free internet connectivity for students.

References

1. Florida Virtual School. Accessed at https://gosa.georgia.gov/about-us/what-digital learning. Accessed 17 May 2021.
2. Agarwal, E. S. (2013). E-learning: New trend in education and training. *International Journal of Advanced Research, 1*(8), 797–810.
3. Arnold, A. R. (2007). Self-perceived employability: Development and validation of a scale. *Personnel Review, 36*(1), 23–41.
4. Census of India. (2020). Census estimation. Government of India.
5. Weng, J.-F., Tseng, S.-S., & Lee, T.-J. (2010) Teaching Boolean logic through game rule tuning. *IEEE Transaction on Learning Technology, 3*(4).
6. Clarke, J. E. (2010). International experience and graduate employability: Stakeholders perceptions on the connection. *Higher Education, 59*(5), 599–613.
7. Sood, M., & Singh, V. (2014). E-learning: Usage among Indian students. *International Journal of Scientific and Engineering Research, 5*(4).
8. Arkorful, V., & Abaido, N. (2014). The role of E-learning, the advantages and disadvantages of its adoption in higher education. *International Journal of Education and Research, 2*(2).
9. Homavazir, Z. F. (2015) *Impact of E-learning on student learning and employability – A study in India.* Unpublished PhD thesis, Submitted to D. Y. Patil University, Mumbai.
10. Pande, D., Wadhai, V. M., & Thakre, V. M. (2016). Current trends of E-learning in India. *International Research Journal of Engineering and Technology, 3*(1).
11. Vivekananda, M., & Satish, R. U. V. N. (2017). Emerging trends in E-learning in India. *International Journal of Advances in Electronics and Computer Science, 4*(6), 1–5
12. Bhongade, D., & Yogesh, M. S. (2018). Prospect of E-learning in Indian higher education: Trends and issues. *International Journal of Current Engineering and Scientific Research, 5*(5). Technical Organization Research in India.
13. Selvaraj, C. (2019). The success of E-learning systems in management education in Chennai city- Using user's satisfaction approach. *The Online Journal of Distance Education and E-Learning, 7*(2).
14. Misra, M. (2020). Impact of covid-19 on education system. *Agrospheres: E-Newsletter, 1*(3), 17–18.
15. Mistral, S. P. (2007). E-learning in Mega Open University: Faculty attitude, barriers and motivators. *Educational Media International, 44*(4), 323–338.
16. Raghavan, T. C. A. S. (2015) Linking urban India to drive growth. *Live Mint.* Retrieved at: https://www.livemint.com/Opinion/VZ2ZW6DSRqTj7PIxbVLcIK/Linking-urban-India-to-drive-growth.html.
17. Bridgestock, R. (2009). The graduate attributes we've overlooked: Enhancing graduate employability through career management skills. *Higher Education Research & Development, 28*(1).
18. Crawford, J., Henderson., K. B., Rudolph, J., Malkawi, B., Glowatz, M., Burton, R., Magni, P. A., & Lam, S. (2020). Covid-19: Twenty countries higher education intra-period digital pedagogy responses. *Journal of Applied Learning & Teaching, 3*(1), 1–20
19. Imran, S. M. (2013). *Trends and issues of E-learning in education in India: A pragmatic perspective.* India: Aligarh Muslim University.
20. Keller, C., Hrastiski, S., & Carlsson, S. A. (2008). Students acceptance of E-learning environment: A comparative study in Sweden and Luthiana. In *Online Proceedings of 15th European Conference on Information Systems* (pp. 395–406).

21. Jena, P. K. (2020). Online learning during lockdown period for covid-19 in India. *International Journal of Multidisciplinary Educational Research, 9,* 5(8), 82–92.
22. Jena, P. K. (2020). Impact of covid-19 on higher education in India. *International Journal of Advanced Education and Research, 5*(3), 77–81.
23. University Grants Commission. (2020). UGC guideline on online examinations and academic calendar in view of covid-19 pandemic. Retrieved on June 2021.
24. Crawford, J., Henderson, K. B., Rudolph, J., Malkawi, B., Glowatz, M., Burton, R., Magni, P. A., & Lam, S. (2020). COVID-19: 20 countries' higher education intra-period digital pedagogy responses. *Journal of Applied Learning & Teaching, 3*(1), 1–20. https://doi.org/10.37074/jalt. 2020.3.1.7.
25. Marinoni, G., & Van't Land, H. (2020). The impact of COVID-19 on global higher education. *International Higher Education, 102.*
26. Markovie, M. R. (2010). Advantages and disadvantages of E-learning in comparison to traditional form of learning. *Annals of University of Petrosoni, Economics, 10*(2), 289–298.
27. Menon, P., Pande, V., Jadhav, S., & Agarkhedkar, S. (2020). Perception of faculty and students about E-learning about feasibility, acceptance and implementation in covid-crisis. *Research Square.*
28. Tubaishat, A., & Lansari, A. (2011). Are students are ready to adopt E-learning? A preliminary e-readiness study of a university in the Gulf region. *IJICT.*
29. Tomilson, M. (2008). The degree is not enough: Students perceptions of the role of higher education credentials for graduate work and employability. *British Journal of Sociology and Education, 29*(1).
30. Watanabe, K. (2005). A study on needs for E-learning – Through the analysis of national survey and case studies. *National Institute of Informatics, Progress in Informatics, 2,* 77–86.
31. Dalit Camera. (2020). 10th standard student succumbs to digital divide in Bengal. Retrieved July 23, 2020, from https://www.dalitcamera.com/10th-standard-studentcommits-suicide-digital-divide/.
32. The Indian Express. (2020). Digital divide may turn shift to online classes operational nightmare, warn experts. https://indianexpress.com/article/education/digitaldivide-may-turn-shift-to-online-classes-6448262/.
33. Kandri, S. E. (2020). How COVID-19 is driving a long-overdue revolution in education. The World Economic Forum COVID Action Platform. Retrieved July 23, 2020, from https://www.weforum.org/agenda/2020/05/how-covid-19-is-sparking-a-revolutionin-higher-education/.
34. McCarthy, N. (2020). COVID-19's staggering impact on global education. The World Economic Forum COVID Action Platform. Retrieved July 23, 2020, from https://www.weforum.org/agenda/2020/03/infographic-covid19-coronavirus-impactglobal-education-health schools/.
35. MILLER, M. D. (2016). Online learning: Does it work? In *Minds online: Teaching effectively with technology* (pp.19–41). Harvard University Press. https://www.hup.harvard.edu/catalog.php?isbn=9780674660021&content=toc.
36. Tripathi, S. K., & Amann, W. C. (2020). COVID-19 and higher education: Learning to unlearn to create education for the future. United Nations. May 4. Retrieved July 23, 2020, from https://academicimpact.un.org/content/covid-19-and-highereducation-learning-unlearn-create-education-future.
37. Tari, S., & Amonkar, G. (2021) Impact of covid on higher education in India. *Educational Resurgence Journal, 2*(5), 22–27.
38. Dhoot, P. (2020). Impact of covid-19 on higher education in India. *International Journal of Creative Research Thoughts, 8*(6), 22–29.
39. Rajest, S. S., & Suresh, P. (2018). The "Four Cs" education for 21st century's learners. *Research Guru Online Journal of Multidisciplinary Subjects, XII*(I), 888–900.
40. Rawal, M. (2021). An analysis of covid-19 impacts on Indian education system. *Educational Resurgence Journal, 2*(5), 35–40.
41. Koul, P. P., & Bapat, J. O. (2020). Impact of covid-19 on education sector in India. *Journal of Critical Reviews, 7*(11).

42. Rashid, S., & Yadav, S. S. (2020). Impact of covid-19 pandemic on higher education and research. *Indian Journal of Human Development*. Sage Publications.
43. Bowen, J. A. (2012). *Teaching naked: How moving technology out of your college classroom will improve student learning. Jossy-Bass.* Hoboken: John Wiley & Sons.
44. Miliszewska, I. (2007). It is fully on or partly off? The case of fully online provision of transnational education. *Journal of Information Technology Education, 6,* 499–514.
45. Haleem, A., Javaid, M., Vaishya, M. R., & Deshmukh, S. G. (2020). Areas of academic research with the impact of covid-19. *American Journal of Emergency Medicine, 38,* 1524–1526.

Dynamics of Inclusive and Lifelong Learning Prospects Through Massive Open Online Courses (MOOC): A Descriptive Study

Edwin Hernan Ramirez-Asis[ID], **Kolachina Srinivas**[ID], **K. Sivasubramanian**[ID], and **K. P. Jaheer Mukthar**[ID]

1 Introduction

The Massive Open Online Courses (MOOC) are available at no cost online course for all, irrespective of age, sex, religion and caste. The MOOC platform integrates community networking and online resource access and promotes learning in the education sector. It builds on the learner's engagement through self-learning to reach the learning goals, gain knowledge, and promote a common interest. This platform is easily accessible and convenient to take the new course. It will help the learners to obtain new knowledge and equip their skills for career advancement. The online platform provides better and enhanced additional knowledge to the existing and new learners. Many people worldwide started using the MOOC platform for various reasons such as career enhancement, career upliftment, preparing for higher studies, additional learning during graduation, continuous learning process, corporate development, training, etc. The MOOC online learning has dramatically altered the world of learning and its pedagogical system. The top ten MOOC platforms are Canvas network, iversity, Cognitive Class, Kadenze, Coursera, Khan Academy, edX, Udacity

E. H. Ramirez-Asis
Management and Tourism Faculty, Universidad Nacional Santiago Antúnez de Mayolo, Huaraz, Perú
e-mail: ehramireza@unasam.edu.pe

K. Srinivas (✉)
KL Business School, KL Deemed to Be University, KLEF, Vaddeswaram, Guntur District, Andhra Pradesh, India

K. Sivasubramanian · K. P. Jaheer Mukthar
Department of Economics, Kristu Jayanti College Autonomous, Bengaluru, India
e-mail: sivasubramanian@kristujayanti.com

K. P. Jaheer Mukthar
e-mail: jaheer@kristujayanti.com

etc. Every MOOC course provider has its technical infrastructure and business model. The MOOCs are the online classes or courses which allows the learners free access and free participation to all courses on their own choices. Also, the conservative mode of teaching like lectures, PowerPoint presentation and videos. MOOCs also offers a platform for discussion forums. The MOOC course is divided into two categories, namely, cMOOC and xMOOC. The cMOOC provides the dynamic system of development of course material. This type of MOOC course develops the course content through discussion rather than the pre-developed study material. The xMOOC course, on the other hand, works with the conventional method where the courses were well defined, pre-determined study materials and references. The MOOC was first invented in 2008 which was created and practised by George Siemen and Stephen Down. In the beginning time, it was termed Connectivism and Connective Knowledge (CCK08). The University of Manitoba was the first institution to introduce the MOOC as a credit course. The MOOC is become more familiar and took off in a faster manner from 2012 onwards with a starting course of Artificial Intelligence (AI). This particular course was enrolled by 1600,000 learners in approximately 190 countries [1].

During the pre-pandemic, the interesting courses are different, and it is fully changed during the post-pandemic. The major MOOCs course service providers such as Coursera, edX, Future Learn, and Class Central have a gross enrolment of 8 million, 5 million, 1.3 million, and 0.35 million, respectively, during the year 2019. It has a tremendous growth during 2020, mainly due to the covid-19 pandemic situation. It has grown about 150% of registration for Coursera courses, 60% growth for edX platform, 208.25% for Future Learn course and 100% growth for class central over the last year. The top ten MOOCs course offerings based on the languages is demonstrated in the Table 2. English is playing a key role in providing the course as a medium of instruction. It is evident from the data that 87.6% of the course through MOOCs platform is offered in the English language. The second is the Chinese language with 3.6%. The Malayalam is 0.1%, with the tenth place in the list. The maximum enrolment on MOOC platforms is happening in the United States of America, with a total enrolment of 242,279 with 42.29%. India is coming in second place among the ten places across the globe. The enrolment of 54,230 with 9.47% and Canada comes in the next place with 21,853 with 3.81%. Australia, Nigeria, Brazil, Spain, Philippines, Pakistan and UK are coming after Canada with 2.18, 2.11, 1.96, 1.85, 1.76, 1.66 and 1.41% of share, respectively.

2   Characteristics of MOOC Platform

The MOOC courses are highly flexible and self-directed open process through online mode, which designed for huge participation. It does not contain any form of fees or essential prerequisites for entry, and no formal educational credit can be secured. The completion rates of these courses were comparatively very less due to various reasons. Accessing the course material also costs less, and some courses may be charged during certificates for the respective courses. These MOOC platforms would also have the memorandum of understanding with educational institutions to provide cloud-based host courses. The MOOC technology consists of educators' involvement in many variables such as designing the courses, preparing study materials, and producing MOOC courses, which includes recording, editing and delivery, and running the course through online mode. It is possible to engage with a huge number of learners through the analysis and discussion slots. It has another vital advantage of re-watch the video lecture any time and any number of times. It is designed to attain the maximum number of learner's enrolment. The majority of the online MOOC courses include the video lecture series, assessment quizzes, feedback, and peer review of the content. The most advantageous aspect of the MOOC online platform course is the convenient and customized learning environment [2]. Scenery measures for decision-making around platforms and the assortment and value guarantee of courses are intense for various recognized personnel. The early upfront venture may be essential for some global online platforms. Usually, MOOCs necessitate a sustenance crew of learning creators, video making specialists and theoretical content supporters work on generating the course gratified—while there are a variety of options, from reusing the existing core units to scenery up the regional partnerships between institutes for communal capacity building, to outsource these roles and responsibility to private portal providers. The definite costs will hinge on the behaviour of specific course content—yet, it could be expected that one course would require the inlay of a particular team of three or four support staff members for many months. Dependent on the category of course and sustenance availability, academics or educational sectors typically report expenditure between hundred and four hundred hours in the production procedure. Disparate traditional sequences, the substance of the involvement is finished in production, whereas meaningfully less time is expended when the course is running. Effectually, instruction time is being constructed into the course throughout production. In some occurrences, academics will necessitate finance to pay for someone to impart in their residence for the period dedicated to the invention. MOOCs on the main global online platforms favour short duration recorded video contented lectures as one of the major media for e-content content. High-quality video content is relatively costly, and it requires access to equipment and technology. Those materials or technology might be hired or purchased for the respective electronic portals. The real cost for one minute of a particular video would depend on the specific style of the section, whether it may be an inexpensive screen or a high-level video generation. These costs might be ranged widely from approximately $200 to $1200 for a short video segment. While estimating the cost of production, the staff used and materials applied are included. In its place of complete investment in capital gear and employees, it is conceivable to subcontract to a marketable video making firm or hire contractors and sub-contractors for rigorous film shoot times. About study material, if the particular course material is under the copyright issue is required, then there may be a cost occurred in purchasing the right to apply it. In general, the open-source study material or the created study material will be favoured in production in the MOOC course [3].

The above Table 1 emphasizes the availability of various courses in the MOOC platforms. The major course enrolment has listed in the table, such as business management courses, economics courses, and statistics courses. Many courses in the business management category were listed as accounting, business law, business development, retail, finance, marketing, entrepreneurial development, public management and theory, and risk management. Many courses available in the economics category are microeconomics, macroeconomics, monetary economics, international economics, labour economics, managerial economics, fiscal economics, gender economics, international finance and trade. It is also covering all the statistical and software package courses for analysis. Ultimately, the MOOC course platform is a boon for learners and inclusive learning. It is not limited to only commerce, economics, management and statistics. All science courses, engineering, medicine, literature and other social sciences are also included.

Table 1 Course availability in MOOC platforms

Business management courses	Economics courses	Statistics courses
Accounting Science	Behavioural Economics	Algebra
Actuarial Science	Economic Policy	Calculus
Business Analytics	Game Theory	Algorithms
Business Development Sciences	International Trade	Geometry
Business Law and Practice	Econometrics	Inferential Statistics
Business Communication	Macro Economics	Probability
Business Review	Micro Economics	Linear Programming
Digital Marketing	Public Economics	Regression
Sales and Marketing Management	Fiscal Economics	Fundamentals of Statistics
Retail Management	Managerial Economics	Statistics & R
Financial Management	Economics for Competitive Examination	Basic Statistical Modelling
Entrepreneurial Development	Rural Development	Statistics & Python
Healthcare System Management	Gender Economics	Public Sector Debt Statistics
Human Resources	Monetary Economics	Statistical Learning
Innovation & Technology	International Economics	Data Analysis for Social Scientists
International Management	Financial Economics	Government Finance Statistics

(continued)

Table 1 (continued)

Business management courses	Economics courses	Statistics courses
Operation Management	Urban Development	Bio-Statistics
Quality Management	Agricultural Economics	Data Science: Probability
Supply Chain Management	Welfare Economics	Basic Statistics
Organizational Management	Health Economics	Statistics for Economists
Product Management	Banking and Taxation	Business Statistics
Managerial Economics	Insurance Policy	Statistical Methods
Public Management	Economics of Globalization	Statistical Models
Management Theory	International Finance	Statistics for Social Sciences
Project Management	Labour Economics	Statistics for Management
Risk Management	Economics of Trade and Business	Advanced Statistics

Source https://www.classcentral.com/report/mooc-stats-pandemic/

3 Advantages and Disadvantages of MOOC Courses

The MOOC online courses are ultimately a new pattern of the educational system for everyone and at any time. It came with many opportunities for both the teachers and students. The current learning process has been developed to enhance the knowledge and skill set of the learners. It has both advantages and disadvantages as follows.

3.1 Advantages

MOOC generates the opening for share ideas, knowledge, skills, and supports in improving inclusiveness and lifelong learning capacities by providing easy accessibility to all resources globally.

It helps to learn cross-cultural studies without any restrictions, leading to association with institutional educators and learners nationally and internationally.

It also helps to find an idea of where the students are standing in this competitive world because the registration occurs globally. It also helps the students to have group discussions with other learners internationally and share their thoughts.

Many studies have found that the learners could equip the knowledge through active learning rather than listening to the captured lectures. The learners listen to the video lectures more considerately while they have given an issue to solve before the lecture begins. In this situation, the MOOC course format has short lectures, no writing assignments, and certain quizzes based on the lectures would become the ideal methodology.

MOOC online course enhances the student-teacher connect time usually applied for lectures can be used inversely, for example, discussions, analysis, experiments, group work, project work, and joint work with peers. The learners will watch the lecture series online from their home and interact with teachers for their doubts and clarifications during the classes. It is strongly supported and demonstrated in the teaching. The faculty members are also got time to discuss with the learners individually [4].

The MOOC online platform helps through the discussion forum segment, reflecting active and progressive learners. Reflective students who could not distribute ideas and strategies inside the walls of physical classes could visualize their ideas on the discussion forum. The international learners who felt they missed it could share their diverse insights on the discussion segment. These activities will help all learner to bring out the solutions for various problems discussed in the segments [5].

It will also protect from creating examination fear among the learners. It encourages a very profound tactic of learning against the shallow and deliberate strategy of the learning system.

The evaluation process promotes the opportunity to gain new learning, which influences learning and teaching [6].

The MOOC platform arranges to learn from the best higher educational institutions across the globe based on the current situations and competitive requirements. It also provides highly recognized and verified international certificates on completing the MOOC course by the learners [7].

MOOC is a boon for the learning community in offering quality learning through online platform. These kind of educational service providers are arranging world-class faculties for video lecturing [8].

Perhaps the most significant benefit of MOOCs is their low cost. When compared with On campus courseware, anyone can receive a full-fledged graduate credential from a prestigious institute for almost half of the price. All this, in addition to the travel, lodging, and other expenses, thus making education more and more accessible.

3.2 Disadvantages

The MOOC course providing platform offers video contented lectures, tutorials, PowerPoint slides, and relevant study materials. It helps the learner's option for not moving through faculty lectures slowly but entire lectures in one day, which did not lead to a wider and deeper understanding of the subject content [9, 10].

In MOOC online courses, the instructor and learner's real-time interaction is not possible [11]. Many sciences, technical and practical courses required hands-on practical sessions, it is not possible in MOOC platforms [12].

Through MOOC platform online courses, effective assessment such as interactive question and answer segment in the classroom, surprise quizzes, and activities [13].

The system and procedure for evaluation are fully automatic, which is not as effective as the physical evaluation system. It leads to discontinuous of online course [14].

The online MOOC system of education slowly destroy empathy, teacher-student connect and respect towards the teacher by the learners. It encourages only virtual social friendship, but not real-time classroom connects [15].

MOOC course platform is limited to commerce, management, economics and statistics (Table 2). The platform is offering many courses, including engineering, languages and humanities as well. The engineering courses include aerospace engineering, biomedical engineering, chemical engineering, civil engineering, computer engineering, electrical engineering, mechanical engineering, software, structural and hardware engineering. Many foreign languages are also included; it helps many learners learn foreign languages. The other significant course such as art, child development psychology, human development and fashion design are also included in the MOOC online platform. It will help the overall learning community to enhance and enrich their knowledge and skills at the global level.

There is a major difference between the normal online classes that the course teacher handles online or virtual mode rather than physical classroom instruction (Table 3). On the other hand, the online MOOC courses, which is video recorded, online classes. The regular online classes are having course fees, whereas the MOOCs does not have any course fees. So, there is no cost for the MOOC courses. There are entrance requirement needs for the regular online courses, but there are no prerequisites for MOOC online courses. The resources and study materials are relatively limited in regular virtual courses. There is a limitation in resources on MOOC courses. Similarly, there are many advantages available such as flexible curriculum, open-source materials, universities are partnered with private organizations, providing

Table 2 Top ten subjects available on the MOOCs platform

Engineering courses	Language courses	Humanities courses
Aerospace	English	Art
Bio-Medical	Grammar	Child Development
Chemical	Spanish	Women Empowerment
Civil	Italian	Epidemics
Computer	Writing Skills	Fashion Design
Electrical	Professional Writing	Psychology
Industrial	Japanese	Philosophy
Mechanical	Russian	Public Speaking Course
Software	Latin	Child Rights
Structural	Composition	Gender Studies
Hardware	French	Human Rights

Source https://www.classcentral.com/report/mooc-stats-pandemic/

Table 3 Difference between Regular Online Course and the MOOCs

Particulars	Regular online course	MOOCs
Cost to user	Fees	No fees; possibly certificates and/or support
Entrance necessities	Yes, as per conventional courses	No
Rule	Limited. Capped by resources available for support and assessment	Thousands. Savings due to limited lecturer support
Teacher's role	Responsible for curriculum alignment, quality assurance (QA) and support	Flexible role regarding the curriculum. Limited individual support
Copyright	Largely proprietary. Some open	Content may be proprietary or open. User-generated content often © MOOC provider
Providers	Distance education providers	Traditional residential research universities partnered with private companies
Analytics	No, not usually	Yes, one of the promises
Certification	Conventional	Non-conventional
Quality assurance	Aligned with the usual formal courses, QA processes	As per non-formal offerings

Source https://www.classcentral.com/report/mooc-stats-pandemic/

analytics, a modern type of course, and non-formal education. It leads to open up an avenue for lifelong learning and inclusiveness.

4 Literature Review

MOOC course platforms are globally used for providing courses through online mode. There are many MOOC course online platforms used in many countries [16]. The institutes with the capabilities of the organizations and government officials aim to serve the improved educational requirement of the learners by providing the MOOC in the country [17]. The massive growth in enrolment is assuring the vast amount of online learners is started occurring in various countries, and surely it will move further in forthcoming years [18]. Information and Library Network Centre connects the university libraries and develops several online programmes [19]. MOOC is one of the major alternates for many developing nations accessing higher education and skill development [20]. Many studies state that the government and educational institutions have to provide financial support to enhance the MOOC courses [21]. The MOOC analyzed not just only as an effective instrument to provide quality education vastly and singly. It also provides the main raising control to comprehensive with the old procedure of regular education in educational institutions. Omnipresent education is another instrument that would also effort the same area in dispersing improvement and opening up the area to an enormous global content. Essentially, MOOCs would endorse Ubiquitous education, altering the enlightening epicentre from board to console. Consequently, the total instructive excellence would see an enormous uplift as an unabridged [22]. In creating MOOCs very popular to a wide extent, the learners' basic digital learning and popularity with the virtual world is an obvious and definite requirement [23]. In recent years, online conscription in MOOCs increased drastically. India after the United States of America is controlling the international growth in enrolment. Viewing the development of learning enrolment from the nation and satisfy the learner need for education, India started various projects for providing MOOC course [24]. It is to overwhelm the digital learning challenges and crisis due to accessibility [25]. Digital learning and upskilling are consistent and continuous procedure. It is applicable for all age of learners and professions [26]. MOOC course would appear as one of the most potent instruments in evidencing the quality education and massive learning to a large level of learners. It reaches all sets of peoples such as students, teachers and corporate professionals. Contempt intensified compensations and competence, learners retrieving MOOCs from the developing nations [27]. Massive Open Online Courses (MOOCs) are courses carried in an online setting with numerous diverse structures from preceding methods to online tutoring. The efficiency of MOOCs is an exposed query as accomplishment rates are considerably less than old online education sequences. The impartial of this study is to recognize influences that improve an individual' purpose to endure using MOOCs, which an incomplete amount of exploring has beforehand discovered. An investigation model based on the material systems extension expectation-confirmation model is planned and verified with data composed in a large-scale education. The investigation model clarified a substantial percentage of the alteration for the intent to endure using MOOCs, which is meaningfully prejudiced by apparent reputation, alleged openness, perceived practicality, perceived, and user gratification. Supposed standing and perceived directness were the stoutest predictors and have not formerly been inspected in the setting of MOOCs [28–30]. The various studies examine that the Lecturers in transcontinental higher learning professed that MOOCs were not appropriate for attributed preceding knowledge but would be valuable as an additional reserve for student learning and teaching. There was robust confidence that as global division confines obtainable a commodified creation, MOOCs were improbable to be accepted as an additional for outdated programme distribution methods, as learners strongly desire face-to-face teaching and learning support [31–33].

5 Research Methodology

The present study is based on the descriptive observational research design. Based on the secondary data sources, the objectives of the present study is framed along with the variables surveyed from various literature reviewed. The secondary data is collected from various published journals, government reports and unpublished PhD thesis. The collected data is tabulated and analyzed with interpretation.

6 Objectives

To find out the role, nature and significance of MOOC courses for the teachers and learners. To examine the impact of Covid-19 on MOOC online course enrolment.

Above Table 4 exhibits the growth of MOOC online course for the year 2019 and 2020. The major MOOCs course service providers such as Coursera, edX, Future Learn, and Class Central have a gross enrolment of 8 million, 5 million, 1.3 million, and 0.35 million, respectively, during the year 2019. It has a tremendous growth during 2020, mainly due to the covid-19 pandemic situation. It has grown about 150% of registration for Coursera courses, 60% growth for edX platform, 208.25% for Future Learn course and 100% growth for class central over the last year, as shown in Fig. 1.

The course interest by the learners was analyzed in above Table 5. The top ten-course interest is tabulated for pre-pandemic and post-pandemic situations. The interesting courses are different during the pre-pandemic, and it is fully changed during the post-pandemic, as shown in Fig. 2.

The top ten MOOCs course offerings based on the languages are demonstrated in Table 6 and Fig. 3. English is playing a key role in providing the course as a medium of instruction. It is evident from the data that 87.6% of the course through MOOCs platform is offered in the English language. The second is the Chinese language with 3.6%. The Malayalam is 0.1% with the tenth place in the list, as shown in Fig. 4.

The maximum enrolment on MOOC platforms is happening in the United States of America, with a total enrolment of 242279 with 42.29%. India is coming in second place among the ten places across the globe. The enrolment of 54230 with 9.47% and

Table 4 The familiar MOOC course providers 2019 & 2020

New registered	2019	2020	Total	Growth percentage
Coursera	8000000	20000000	65000000	150
edX	5000000	8000000	32000000	60
Future Learn	1300000	4000000	13500000	208.25
Class Central	350000	700000	2200000	100

Source https://www.classcentral.com/report/mooc-stats-pandemic/

MOOC Course Providers 2019 & 2020 by Enrolement

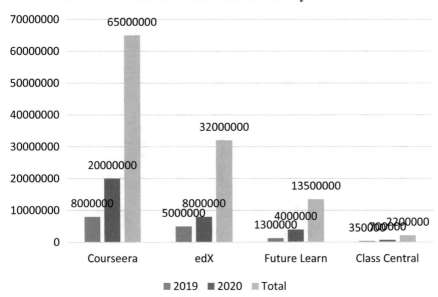

Fig. 1 MOOC course providers 2019 and 2020 by enrolment

Table 5 Course interest of the learners

S. No	Pre-pandemic	Course follows	Post-pandemic	Course follows
1	Computer Science	336	Personal Development	167
2	Programming	290	Business	**167**
3	Business	276	Art & Design	117
4	Personal Development	257	Management & Leadership	115
5	Management & Leadership	240	Self-Improvement	113
6	Data Science	235	Humanities	112
7	Artificial Intelligence	208	Computer Science	110
8	Information Technology	192	Communication Skills	107
9	Career Development	189	Health & Medicine	106
10	Entrepreneurship	188	Foreign Language	106

Source https://www.classcentral.com/report/mooc-stats-pandemic/

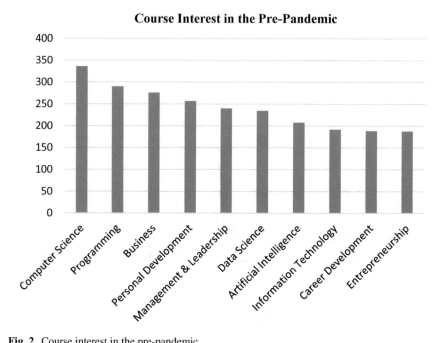

Fig. 2 Course interest in the pre-pandemic

Table 6 Top ten MOOC course languages

S. No	Languages	User percentage
1	English	87.6
2	Chinese	3.6
3	Spanish	2.8
4	Arabic	2.4
5	Portuguese	1.1
6	French	1.1
7	Japanese	0.6
8	Russian	0.1
9	Persian	0.1
10	Malayalam	0.1

Canada comes in the next place with 21853 with 3.81%. Australia, Nigeria, Brazil, Spain, Philippines, Pakistan and UK are coming after Canada with 2.18, 2.11, 1.96, 1.85, 1.76, 1.66 and 1.41% of share, respectively, as shown in Table 7, as shown in Fig. 5.

The MOOC platform integrates community networking and online resource access and promotes learning in the education sector. It builds on the learner's engagement through self-learning to reach the learning goals, gain knowledge, and promote

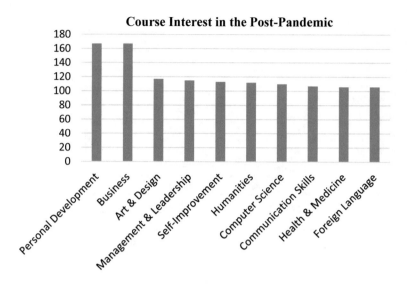

Fig. 3 Course interest in the post pandemic

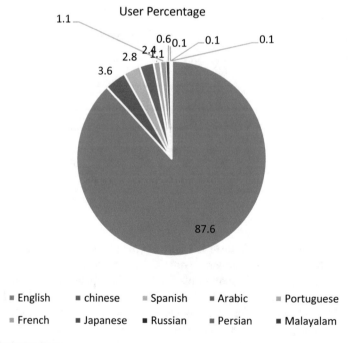

Fig. 4 User percentage

Table 7 Country-wise top ten enrolment in MOOC courses

Continent	Country	Total Enrolment	Percentage
North America	USA	242279	42.29
Asia	India	54230	9.47
North America	Canada	21853	3.81
Australia	Australia	12474	2.18
Africa	Nigeria	12067	2.11
South America	Brazil	11243	1.96
Europe	Spain	10582	1.85
Asia	Philippines	10099	1.76
Asia	Pakistan	9505	1.66
Europe	UK	8066	1.41

Source https://www.classcentral.com/report/mooc-stats-pandemic/

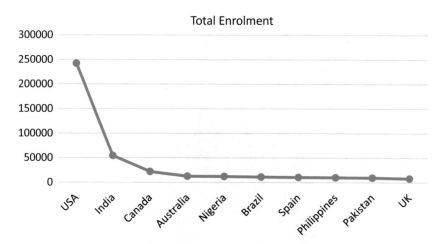

Fig. 5 Total enrolment

a common interest. This platform is easily accessible and convenient to take the new course. It will help the learners to obtain new knowledge and equip their skills for career advancement. The online platform provides better and enhanced additional knowledge to the existing and new learners. Many people worldwide started using the MOOC platform for a wide range of reasons such as career enhancement, career upliftment, preparing for higher studies, additional learning during graduation, continuous learning process, corporate development, training, etc. The MOOC online learning has dramatically altered the world of learning and its pedagogical system. MOOCs also offers a platform for discussion forums. The major course enrolment has listed in the tables, such as business management courses, economics courses, and statistics courses. Many courses in the business management category were listed as accounting, business law, business development, retail, finance, marketing, entrepreneurial development, public management and theory, and risk management. Many courses available in the economics category are microeconomics, macroeconomics, monetary economics, international economics, labour economics, managerial economics, fiscal economics, gender economics, international finance and trade. It is also covering all the statistical and software package courses for analysis. Ultimately, the MOOC course platform is a boon for learners and inclusive learning. The MOOC courses are highly flexible and self-directed open process through online mode, which designed for huge participation. It does not contain any form of fees or essential prerequisites for entry, and no formal educational credit can be secured. The completion rates of these courses were comparatively very less due to various reasons. Accessing the course material is also costless, and some courses may be charged during certificates for the respective courses. These MOOC platforms would also have the memorandum of understanding with educational institutions to provide cloud-based host courses. The MOOC technology consists of educators' involvement in many variables such as designing the courses, preparing study materials, and producing MOOC courses, which includes recording, editing and delivery, and running the course through online mode. It is possible to engage with a huge number of learners through the analysis and discussion slots. It has another vital advantage of re-watch the video lecture any time and any number of times. It is designed to attain the maximum number of learner's enrolment. The majority of the online MOOC courses include the video lecture series, assessment quizzes, feedback, and peer review of the content. The various studies examine that the Lecturers in transcontinental higher learning professed that MOOCs were not appropriate for attributed preceding knowledge but would be valuable as an additional reserve for student learning and teaching. There was robust confidence that as global division confines obtainable a commodified creation, MOOCs were improbable to be accepted as an additional for outdated programme distribution methods, as learners strongly desire face-to-face teaching and learning support [9, 31–33]. The regular online classes are having course fees, whereas the MOOCs does not have any course fees. Massive Open Online Courses (MOOCs) have arisen as one of the greatest possible gears in extending excellent learning and enormous exercise to a huge area of spectators internationally. On the one hand, it spreads an international field of beginners. Similarly, it propagates knowledge in a well-organized virtual platform, interweaving an enormous system of students, inventors, professors, researchers, teachers, and diverse stakeholders connected to education. Contempt strengthened advantages and competence; people retrieving MOOCs from emerging nations do not count to a noteworthy number [34–41].

7　Conclusion

India is a talented dwelling of learning that offers a potential area to MOOCs for big-scale application. But numerous issues and restraints like little digital literateness and the absence of immense digital substructure hamper this wide application of MOOCs. So, there is no cost for the MOOC courses. There are entrance requirement needs for the regular online courses, but there are no prerequisites for MOOC online courses. The resources and study materials are relatively limited in regular virtual courses. There is a limitation in resources on MOOC courses. You can learn from your friends through a social networking site. Besides various advantages like, access to the course is free of cost, material is advocated by top professors, the course material of MOOCs has been selected by top professors in universities. Access to learn different languages language-based learning is also available in MOOCs, they prepare students for entrance tests in various languages, there are certain limitations such as Attrition rates being high, learners with visual problems cannot be forced to take such classes. Students are unable to pay attention in such courses because they only have to look over the course materials. A MOOC is a game-changing concept that has revolutionized online learning. MOOCs removed the time and geographic barriers that previously existed for learners, allowing them to take classes from some of the world's most brilliant intellectual minds without having a need to travel or spend to attend a prestigious college. Besides various aspects dealt in the research article on the MOOCs platform, it can be concluded that MOOCs has been a reformatory change in the learning system and has been quire resourceful especially in this new normal of virtual world.

References

1. Suresh, P., & Rajest, S. S. (2019). An analysis of psychological aspects in student-centered learning activities and different methods. *Journal of International Pharmaceutical Research*, *46*(01), 165–172.
2. Cizerniewicz, L., Deacon, A., Fife, M.-A., Small, J., & Walji, S. (2015). CILT position paper: MOOCs. CILT, University of Cape Town.
3. Hollands, F. M., & Tirthali, D. (2014). MOOCs: Expectations and reality. Center for Benefit-Cost Studies of Education, Teachers College, Columbia University.
4. Bogdan, R., Holotescu, C., Andone, D., & Grosseck, G. (2017). How MOOCs are being used for corporate training? *eLearning & Software for Education, 2*.
5. Bragg, A. B. (2014). MOOCs: Where to from here? *Training & Development, 41*(1), 20.
6. Setiawan, R., Cavaliere, L. P. L., Hussaini, T., Villalba-Condori, K. O., Arias-Chavez, D., Gupta, M., Untari, D. T., & Regin, R. (2020). The impact of educational marketing on universities performance. *Productivity Management, 25*(5), 1278–1296.
7. Breslow, L., Pritchard, D. E., DeBoer, J., Stump, G. S., Ho, A. D., & Seaton, D. T. (2013). Studying learning in the worldwide classroom: Research into edX's first MOOC. *Research & Practice in Assessment, 8*, 13–25.
8. Byerly, A. (2012). Before you jump on the bandwagon. *The Chronicle of Higher Education, 59*(2).

9. Bandura, A. (2001). Social cognitive theory: An agentic perspective. *Annual Review of Psychology, 52*(1), 1–26.
10. Bandura, A. (2002). Growing importance of human agency in adaptation and change in the electronic era. *European Psychologist, 7*, 2–16.
11. Bates, T. (2014). A balanced research report on the hopes and realities of MOOCs. Retrieved May 15, 2014, from http://www.tonybates.ca/2014/05/15/abalanced-research-report-on-the-hopes-andrealities-of-moocs/.
12. Bates, T. (2012). What's right and what's wrong about Coursera-style MOOCs. Retrieved August 5, 2012, from http://www.tonybates.ca/2012/08/05/whatsright-andwhat-wrong-about-coursera-style-moocs.
13. Bayne, S., Gallagher, M. S., & Lamb, J. (2014). Being "at" university: The social topologies of distance students. *Higher Education, 67*(5), 569–583.
14. Belanger, Y., & Thornton, J. (2013). Bioelectricity: A quantitative approach. Duke University's first MOOC. Retrieved February 5, 2013, from http://dukespace.lib.duke.edu/dspace/bitstream/handle/10161/6216/Duke_Bioelectricity_MOOC_Fall2012.pdf.
15. Bogdan, R. (2016). Guidelines for developing educational environments in the automotive industry. *Interaction Design & Architecture*(s), *31*, 59–73.
16. Chauhan, J., & Goel, A. (2017). An overview of MOOC in India. *International Journal of Computer Trends and Technology, 49*(2), 111–1222.
17. Shah, D. (2016). By the numbers: MOOCS in 2016, how has the MOOC space grown this year? Get the facts, figures, and pie charts. https://www.class-central.com/report/mooc-stats-2016/.
18. Pandey, N. (2016). Open online course providers upbeat as enrolments jump. http://www.thehindubusinessline.com/news/education/open-online-course-providers-upbeat-as-enrolments-jump/article8850328.ece.
19. Damodharan, D. (2016). India represents the second highest enrolments in our courses': EdX CEO Anant Agarwal. http://www.huffingtonpost.in/dipin-damodharan/our-mission-is-to-dem ocra_b_8607538.html.
20. Aspiring Minds. (2016). National employability report 2016 - Aspiring minds. Annual report.
21. Miller, A. (2017). Sites powered by Open edX. Accessed from: https://github.com/edx/edx-pla tform/wiki/Sites-powered-by-Open-edX.
22. Chatterjee, P., & Nath, A. (2014). Massive open online courses (MOOCs) in education—A case study in Indian context and vision to ubiquitous learning. In *MOOC, 2014 IEEE International Conference on Innovation and Technology in Education (MITE)* (pp. 36–41). IEEE.
23. Garg, D. (2017). Major challenges in online higher educational environment. http://blogs.tim esofindia.indiatimes.com/breaking-shackles/major-challenges-in-online-higher-educational environment/.
24. Sethi, K. K., Bhanodia, P., Mishra, D. K., Badjatya, M., & Gujar, C. P. (2016). Challenges faced in deployment of e-learning models in India. In *Proceedings of the International Congress on Information and Communication Technology* (pp. 647–655). Springer Singapore.
25. Insights. (2015). Critically comment on the major challenges facing higher education in India and measures needed to overcome them. *Economic Times*. Accessed from: http://www.insightsonindia.com/2015/08/11/5-critically-comment-on-the-major-challe nges-facing-highereducation-in-India-and-measures-needed-to-overcome-them/.
26. Mukul, A. (2016). Microsoft to design app for Indian govt's open online courses. *Economic Times*. Accessed from: http://tech.economictimes.indiatimes.com/news/internet/microsoft-to-design-app-for-indian-govts-open-online-courses/52790346.
27. Srikanth, M. (2020). The advantages and disadvantages of MOOCs for learning. *Infopro Learning*. Accessed from: https://www.infoprolearning.com/blog/advantages-and-disadvant ages-of-moocs-massive-open-online-courses-for-learning/.
28. Hattangdi, A., & Atanu, G. (2008). Enhancing the quality and accessibility of higher education through the use of information and communication technologies. In *International Conference on Emergent Mission, Resources, and the Graphic Locus in Strategy as a Part of the 11th Annual Convention of the Strategic Management Forum (SMF). India 2008* (Vol. 2011).

29. Alraimi, K. M., Zo, H., & Ciganek, A. P. (2015). Understanding the MOOCs continuance: The role of openness and reputation. *Computers & Education, 80*, 28–38.
30. Anders, A. (2015). Theories and applications of massive online open courses (MOOCs): The case for hybrid design. *The International Review of Research in Open and Distributed Learning, 16*(6).
31. Anita, R., & Swamy, V. K. (2016). Skilling India-initiatives to create a global workforce. *International Journal of Research in Economics and Social Sciences, 6*(3), 365–371.
32. Annabi, C. A., & Wilkins, S. (2016). The use of MOOCs in transnational higher education for accreditation of prior learning, programme delivery, and professional development. *International Journal of Educational Management, 30*(6), 959–975.
33. Bandura, A. (1986). Fearful expectations and avoidant actions as coeffects of perceived self-inefficacy. *American Psychologist, 41*(12), 1389–1391.
34. Castellano, S. (2014). E-learning in higher education. *TD: Talent Development, 68*(11), 64–66.
35. Cepeda, N. J., Coburn, N., Rohrer, D., Wixted, J. T., Mozer, M. C., & Pashler, H. (2009). Optimizing distributed practise: Theoretical analysis and practical implications. *Experimental Psychology, 56*(4), 236–246.
36. Chang, H. M., Kuo, T. M. L., Chen, S. C., Li, C. A., Huang, Y. W., Cheng, Y. C., & Tzeng, J. W. (2016). Developing a data-driven learning interest recommendation system to promoting self-paced learning on MOOCs. In *Advanced Learning Technologies (ICALT)* (Vol. 5, pp. 23–25).
37. Chatterjee, P., & Nath, A. (2014). Massive open online courses (MOOCs) in education: A case study in Indian context and vision to ubiquitous learning. In *2014 IEEE International Conference on Innovation and Technology in Education (MITE)* (pp. 36–41).
38. Chatterjee, P., & Nath, A. (2014). Massive open online courses (MOOCs) in higher education-Unleashing the potential in India. In *2014 IEEE International Conference on MOOC, Innovation and Technology in Education (MITE)* (pp. 256–260).
39. Chauhan, J. (2017). An overview of MOOC in India. *International Journal of Computer Trends and Technology, 49*(2), 111–120.
40. Chin, C., & Brown, D. E. (2000). Learning in science: A comparison of deep and surface approaches. *Journal of Research in Science Teaching, 37*(2), 109–138.
41. Christensen, G., Steinmetz, A., Alcorn, B., Bennett, A., Woods, D., & Emanuel, E. J. (2013). The MOOC phenomenon: who takes massive open online courses and why? https://doi.org/10.2139/2350964.

The Impact of Artificial Intelligence on Instructional Leadership

Sabah Hejres

1 Introduction

This paper highlights the role of the school principals as instructional leaders and describe their behaviours focusing on improving the educational technology practice, following up the progress of the curriculum and technology. Schools have experienced an increase in responsibilities and tasks. Research on instructional leadership has experienced marked growth over the last three decades about the roles of school principals. In addition, Ezzat et al. [1] suggest a link between innovating instructional leadership and other styles of theories and practices. This paper describes the behaviour of the principal as an instructional leader and the impact with one of the big topics of discussion, which is artificial intelligence.

The artificial intelligence phenomenon has begun in many sectors, including the education system. Artificial intelligence enhances in education management and shown effectiveness in teaching. According to Sabuncuoglu [2], artificial intelligence has become a common ingredient nowadays as products and a part of education. Therefore, school leaders and educators take the responsibilities to teach the subject to inform students about their possible advantages and risks.

According to Timms [3], concludes that Artificial Intelligence in Education (AIED) is about the combination of computer and other devices such as robots and smart classrooms, which needs time to get the impact of the smart board and robot. It may need a set of another technological device. It could set courses that aim to embed technology in the classroom that directs to inevitable expansion for the using robotics and electrical engineering in educational areas include new creative systems. This development would provide experiences for both teachers and students in the application of technological resources to support education.

S. Hejres (✉)
Brunel University London, London, UK

Ahlia University, Manama, Bahrain

© The Author(s), under exclusive license to Springer Nature Switzerland AG 2022   697
A. Hamdan et al. (eds.), *Technologies, Artificial Intelligence and the Future of Learning Post-COVID-19*, Studies in Computational Intelligence 1019,
https://doi.org/10.1007/978-3-030-93921-2_36

2 Background

The role of the principal as instructional leadership is a type of leadership. It is one of the means to communicate high expectations of teachers and students, supervising teaching and learning methods, monitoring assessment and student progress, coordinating the school curriculum, promoting a climate for learning, and creating a supportive work environment [4, 5].

School principal behaviours as an instructional leader focus on improving the education process, following up the progress of the curriculum and technology, solving the educational problem and increasing student achievement and supportive school environment. Instructional leader helps the teacher to overcome the obstacles and difficulties of the learning process, expresses rewarding verbal terms about the teachers achievement, supports and follows up the application of technology, curriculum and environmental resource.

The school principal takes the responsibilities about the extent of activating technology in reforming school results because technology is progress and effects on living, work, and play. It means keep following of technological trends and watching the future because this is the demand for students need to improve their skills, knowledge for the rest of their education as the labour market know the types of jobs they want to do.

According to a computer scientist, John McCarthy coined the term artificial intelligence in 1956, "the science and engineering of making intelligent machines." ([6], p. 10). Studies confirmed that artificial intelligence aims to create functional smart like a human it may exceed the human ability in mental or physiology, this machine works by guider orders in a different process such as learning, reasoning, and self-correction and information analysis.

Artificial intelligence is an aspect of computer science that attempts to produce a new intelligence machine with setting a basic of intelligence order, which, respond as a creation similar to human. This creating (Robotics) can able to speak several languages same time, recognition images, doing different processing understanding. Artificial intelligence began with theory and principles; this application develops by the time and expands. It has similar activities in human. Artificial intelligence can think, stimulate, analysis information, artificial intelligence responds faster than a human does and it may exceed human intelligence.

Since the beginning of artificial intelligence, the theory and principles of technology have become more and more developed, and the application area has been expanding. The effective integration of technology into school's curriculum strongly influenced by the role played by a school principal. The principal as instructional leader focuses on the impact of technology integration and the characteristics needed by principals and teachers teaching and learning professional development and the challenges that they may probably face.

This paper indicates principals are crucial to the success of school initiatives, and that they and teachers need more training opportunities to have a positive impact desired for successful technology in a school system. According to Brockmeier et al.

[7] conclude, "Achieving successful learning outcomes with technology requires leadership with vision and expertise." (p. 55).

3 The Objective of the Study

1. To determine the factors impacted by artificial intelligence in the educational system.
2. To determine the artificial intelligence impact on instructional leadership
3. To emphasis the challenges facing the instructional leader in applying artificial intelligence.

4 Educational Factors Impacted by Artificial Intelligence

The principal can do several things to improve the leading of school by integrating the technology to keep up with development. This era of technology is not going to end any time soon. It will start with almost 100% of schools having Internet access [8]. Based on the available technology, which used, there is an expectation for demand to use technology by teachers and students, instructional leaders take the responsibilities to provide the technology atmosphere and encourage and increase the use of technology.

A study by Zhang [9] confirms in China and the United States in using technology as a tool in school improvement and raise the quality of education by experiences with technology, creativity, and dialogue for schools to generate better teaching and improve the principal preparation.

Technology, particularly computers, is now in extensive use within the educational system. Recent and current research within the field of artificial intelligence (AI) is having a positive impact on educational applications. According to Recalde [10], finds that principals who invested in the implementation of the artificial intelligence humanoid robot in the classroom fostered a positive learning community, and were directly supporting and encouraging their teachers led their team to a successful implementation of the new technology in their schools.

There are four fundamental factors of instructional leadership influenced by the technological application of artificial intelligence:

1. Curriculum

    As an instructional leader, the principal expected to understand the foundation of quality integration of technology as well as have enough knowledge of the school curriculum to make sure that appropriate content taught to all students in a quality way. Researchers recommend examining the impact of technology on pedagogy, curriculum, professional development, assessment, and school leadership [11]. According to Whitehead et al. [12] the role and importance of technology in the curriculum to give education of reality, this reality, facing educational leaders and technology coordinators to turn to the future direction of educational trends and better coincided with learning with modern, technological pedagogical and curriculum.

To promote the use of technology as an instructional tool in the education curriculum. According to Miller [13]. "Using technology as an influential instructional tool in the educational content must be on the curriculum and learning, not on the amount or type of technology used." ([5] p. 28). Therefore, the school principal has impacts on applying technology into a school curriculum by supportive supervision lessons of teachers.

According to Kumar [14] describes the integration and evaluation of technology into artificial intelligence curriculum by using the robot for students that increase in student excitement while this increased preparation time for teachers. Furthermore, according to Brockmeier et al. [7], confirm that what principals do to facilitate the integration of technology into the curriculum is a crucial variable. Principals need to model the use of technology, demonstrate to the staff to what extent the importance and use the tools are. Therefore, the school principal is one of the main factors to play a fundamental role in the facility the use of technology.

2. Teaching and learning

Technology-enhanced the teaching and learning practice. Artificial intelligence application exists in technologies that support all forms of teaching and learning activities. According to Johnson [15], there is extensive research and development effort in the pedagogical institution, autonomous software entities aimed at supporting student learning by interacting with students and teachers and by collaborating with other similar agents, in the context of interactive learning environments using smart device technology.

According to Timms [3], the important for school leaders to learn about how to model knowledge and what the student knows together with ways of teaching and learning strategies to provide feedback and pedagogy could have artificial intelligence when transported to robots for example. It would be probably challenging and would be very stimulating for artificial intelligence in education.

The principal supervision of teaching and learning and the embedding of artificial intelligence in education will support the production of data into smart classrooms would stream from lots of sensors in there. In addition, focusing on the students would require amounts of Educational Data Mining and would lead to new models of how students behave in the around environment of the classroom and not just with the instructional packages [3].

According to du Boulay (2020), artificial inelegance in education systems performance is likely for subjects such as Science, Technology, Engineering and Mathematics or (STEM). These kinds of learning subjects are easy to build and should be adopted and blending artificial inelegance in education technology with other forms of teaching.

3. Principal professional development

Principals as instructional leaders need more training opportunities to obtain a positive impact on their leadership styles need to the school system. Professional development program provides the practice of the artificial intelligence concept to build successful technology system and can make a change to support teacher's role in teaching and learning process. According to the National Education Technology Plan states that "the problem of technology, integration is not necessarily lacking funds, but lack of adequate training and understanding of how computers used to enrich the teaching and learning process." ([16], p. 25, [17]).

The role of the school leaders is changing fast pace with the constant changes happening in education sector development. The instructional leader being the main component in guiding the teaching-learning process is necessary for preparing students with relevant knowledge and skills technology. According to "The National Center for Education Statistics (2000) indicates that principal leadership described as one of the most factors affecting the effectiveness of the use of technology in classrooms. Additionally, principals who exhibit leadership are instrumental in modelling the use of technology in classrooms" ([18], p. 3). Therefore, school leaders attend a professional development program on the use of applications of new technologies and practices.

The training programs enable them to discuss with teachers through attendance for supervising teaching and learning methods at these sessions to come together and fund proper methods for integrating the teaching and learning with technology. According to Campolo et al. [19], the school principal plays a fundamental role in adopting the trend to apply artificial intelligence application. Principals need to consider the quality of data used to train artificial intelligence systems. This importance of roles and responsibilities demanded in helping teachers establish the proper learning environment for students.

School leaders can provide opportunities for teachers to use technology tools to track student achievements and attainment of learning goals. The school leader arranges school level meetings or plan training programs for demonstration of these tools for teacher using the teacher-to-teacher model for example. Give insights about how technology tools used to report student progress to parents and guardians (Priyanka, 2016, [20].

A study by Lennon [21] showed a significant difference in the integration of technology into the schools between the principals receiving the fewest hours of training and those receiving the most hours. The principals with more training had a higher integration score.

The principal needs professional development in assessing computer technology's influence on students achievement, using computer technology to collect and analyze data, integrating computer technology into the curriculum, using computer technology in their work as a principal, and using computer technology to facilitate her.

4. Educational policy

Artificial intelligence introduced into education policy areas. It has become a hot talk amongst the educationists, stakeholders due to excellent performance in recent years. Artificial intelligence present in education policy areas. According to the US Congress Office of Technology Assessment confirms, "Incorporating technology into the instructional process was one of the most important steps the nation can take to make the most of past and continuing investments in educational technology" ([22], p. 213, [23]).

However, artificial intelligence is an urgent topic of public concern. The legal regulatory require agreed-upon definitions. Conversations experts and review published policy documents to examine researcher and policy-maker conceptions of artificial intelligence prefer definitions of artificial intelligence that emphasize technical functionality, instead, using the definition that compares systems to human thinking and behaviour. Because of this gap, ethical and regulatory efforts may overemphasize concern about future technologies at the expense of pressing issues with existing deployed technologies [24, 25].

Educational decision-makers can use analytic tools to anticipate how artificial intelligence-enabled innovations might affect their school competencies. When principals investigate the processes or products could be affected by artificial intelligence and whether the outcomes would be competence enhancing or - problems, they can make the decision more accurately, schools exposure to artificial intelligence that based risks or opportunities and plan accordingly.

Managers can benefit from understanding the variability of artificial intelligence innovations, and they can build a competitive advantage even if they never adopt artificial intelligence but work with those who do in competence-enhancing ways [26].

For instructional leadership, there are specific technology requirements regarding licensure preparation for school leadership, which require the status of artificial intelligence courses, and current technology-professional administrators report about instructional leader preparation for promoting artificial intelligence.

Decision-makers will also need to design carefully, the privacy measures. Any artificial intelligence, system intelligence depends on ingesting as much training data as possible, and, to develop more sophisticated systems, the nature of the training data is becoming increasingly sensitive ([19], Powles & Hodson, 2017).

Paschen [26], the school principal must, therefore, be ready to answer questions such as how notice give, how consent attained, or what access to one data might imply. The proved challenging in practice since learning algorithms are constantly changing, and even the scientists who created them sometimes stumped by the results they produce.

5 Artificial Intelligence Impact on Instructional Leadership

Few studies address the influence of artificial intelligence on instructional leadership. According to Lynch's study [28], instructional leaders who use empathy to help their faculty and teachers further understand the technology and its possible role in our lives can get their teams ready to embrace artificial intelligence. Bourton et al. [29], describe the relationship between leadership and artificial intelligence as the potential to help leaders with clarity and specificity to make the decision. Robots tend to be non-functional without leaders. Therefore, instructional leaders need professional coaches to overcome challenges, threats that could facilitate and innovation, which is a need in schools. Artificial intelligence promotes new models of digital education and artificial intelligence workforce development for principal, teachers and students to have the skills needed in the 21st-century economy.

Artificial intelligence can provide chatbots with the same potential to help teachers and students. Reframe their thoughts and strategize to improve teaching and learning, for example, the assistants like Watson Assistant and Azure Bot Service, they support voice assistant platforms the teaching process [30].

According to [31], there are several applications to enable instructional leader using the Watson platform. For example, LEADx developed Executive Coach Amanda as the first chatbot taught to serve as a professional leadership coach. A robot such as (Amanda) can answer basic instructional leadership questions regarding staff management and teacher's inquiries. In addition, (Amanda) can coach leaders, meaning it will guide school principals in formulating their ideas, decisions, authorities, and plans.

The artificial intelligence applications platform used across teachers and students. Thus, instructional leaders must support the school's educational environment for an atmosphere where artificial intelligence will be ubiquitous. Furthermore, instructional leaders would provide teachers and students with different kind of training than they currently receive to adhere to their needs. The educational situation around the world may not be ready enough for many teachers and students. Therefore, instructional leaders ought to receive instruction in the kinds of skills and knowledge that will need in the artificial intelligence-dominated landscape. Instructional leaders may face challenges because of current shortages of data scientists, computer scientists, engineers, coders, and platform developers. These skills are in short supply; unless our educational system generates more specialists with these capabilities, it will limit artificial intelligence development in the school system, and this will consequently prevent the dynamic innovation needed in schools.

School leaders expect artificial intelligence to permeate every aspect of education that they may as well be ready to transform the educational space to achieve educational innovation. Moreover, the traditional physical space of the schools will likely change. The environment, in particular, the classroom also will transform into something more fluid, flexible and supported by artificial intelligence such as AR/VR and technology. Therefore, the instructional leader would guide teachers while students are to follow highly customized learning paths, and the teachers function as guides and facilitators of their progress.

School principals turn to the artificial intelligence software to guide the teacher's needs in teaching and learning process. There is similar software that can be developed for instructional leaders to support artificial intelligence's application, under the direction of Charu Sharma, created an Ellen application of intelligent mobile, which personalizes career development allocated for every employee widely. The artificial intelligence of this app incorporates what educational leaders need, mentorship, personalization, workflow synchronization, and data analytics. Although these applications are not the educational focus but are imperative to addresses many leadership issues that instructional leader needed to improve the school system. There will develop as long as there an artificial intelligence coach.

Summarizing what has reported about the effects of artificial intelligence on instructional leadership:

1. Artificial intelligence as the potential to help leaders with clarity and specificity to make the decision.
2. Several platform applications enable instructional leader using the developed Executive Coach to serve as a professional leadership coach.
3. Platform applications enable instructional leader in formulating their ideas, decisions, authorities, and plans.
4. Instructional leaders prepare to support the school's educational environment for an atmosphere where artificial intelligence will be ubiquitous.
5. The instructional leader provides teachers and students with different kind of training than they currently receive to adhere to their dynamic innovation needed in schools.
6. Instructional leader transform the traditional physical space educational space to achieve educational innovation into something more fluid, flexible and supported by artificial intelligence.
7. The artificial intelligence app incorporates what instructional leaders need, mentorship, personalization, workflow synchronization, and real-time data analytics.

6   The Challenges of Artificial Intelligence Applied in Schools

The instructional leader should take into account the disadvantage of artificial intelligence systems to education. There are barriers in applied artificial intelligence consider as a challenge for an instructional leader.

1. The school principal may need to consider the costs of artificial intelligence. Artificial intelligence is updating every day the hardware and software need to update with time to meet the latest requirements. Machines need repairing and

maintenance, which need plenty of costs. Its creation requires costs, as they are very complex machines [32].

2. It is the importance of instructional leadership to consider that artificial intelligence is encouraging the student to get the answers directly from the electronic application without efforts, which effect on their skills and practice of problem-solving, critical thinking, and experiential-based learning for example. In particular, with smartphones and other technology already, students can become too dependent on artificial intelligence; therefore, they will become lazy and lose their mental capacities in learning.

3. It is very important that the school principal put into consideration that artificial intelligence in a particular robot or other machines is not secure when putting it in the wrong hands. Teachers and student s should be aware of the ethical and legality about applying artificial intelligence.

4. Artificial intelligence can prevent sympathizing through student responding and contact, or give an impression about student performance in case teachers when expressing their feeling, attitude, an emotion about student performance. Furthermore, the electronic lessons delivered through digital resources lack the face-to-face interaction between teacher and student that provides a more personal experience.

5. For instructional leaders and teachers to stay aware of the technology of education; they may need to develop the training along with technology development. The teachers or principals may be teaching all their lives using traditional ways in teaching and learning and may not be very ready for the changes applied. They may even see it as a threat to their career security. Furthermore, some teachers think that the constant use of digital technology is affecting student attention and his ability to persevere and challenging. Although such belief is subjective, scholars, experts, and teachers all agree that technology has positively changed the way students teach [33].

6 Students get fun and entertainment by using computers or mobile devices, while students deal with textbooks as tools for learning. Moreover, students are likely to tend towards learning when reading a book, while they are likely to use a mobile device, to play games or spend their time in social media. Therefore, it is required to guide student's behaviour to use electronic tools for learning knowledge and science and not only for playing.

7. Employees interference is becoming less because artificial intelligence is replacing the majority of the tasks and other works with robots, which will cause the problem in employee jobs. Therefore, schools will look to replace the minimum qualified individuals with artificial intelligence which can do work with more efficiency. According to Ottoman [34], an American company found that 85% of people believed that in the era of artificial intelligence, artificial intelligence technology could replace the work of employees in some sectors.

7   Educational Technology and the Challenge of COVID-19 Pandemic

Each year the school principal celebrates the students' graduation in different academic levels, but 2019–2020 is a big challenge and different year as a crucial shortened period in annually learning which expected to lead to adversely affected by school outcomes. The COVID-19 pandemic caused to closed schools to contain the spread in educational institutions around the world. This situation affects student's achievement, which demands to shift the virtual lessons by educational technology rather than regular lessons. Therefore, many school leaders concerned about how educational technology could support a student's achievement and how to provide results based on transparency and integrity requirement.

According to Soland [35], due to COVID-19, the students of K-12 in the United States are currently missing teaching and learning instruction face-to-face. Many parents and educators thus share a common worry. Most students already spend at home without explicit face-to-face instruction from teachers. While the teacher adapts content for the online platform, the parents are work responsibilities caring for and educating their children.

Therefore, school leaders, teachers, and parents concerned about the extent of education technology by digital learning included artificial intelligence as an alternative method to face this challenge particularly, when governments reopen the school gradually. Instructional leader collaborates across teachers and educational policymakers for the importance and urgent task of supporting students' opportunity to learn during this challenging crisis.

There are some resources, which considered being the most educational influential sections in the educational field are the curricula department and educational supervision and guide. According to Tedesco [36], the curricula concern about teaching the values, competency-based approaches, soft and hard skills, and scientific and digital culture it reflects society needs and the culture, provides a variety of learning experiences to train competent and active citizenship and must ensure quality and equity in learning outcomes. According to Sullivan [37], supervision is about improving teaching and learning strategies and technique, it intends as a practical guide to the varied and alternative approaches to supervising that aims in improving the classroom.

Therefore, we assume, how can we benefit from educational resources or tributaries such as supervision and guidance and curricula to enhance instructional leadership in guiding this educational technology in the event of COVID-19 pandemic. There are several ways to assume to promote instructional leadership as follow:

1.   The School leaders should cooperate and with curricula department, aiming to allow teachers to coordinate with curriculum experts. For example, to transfer student's regular book to the E-book, in a C.D focusing on priority subjects contain the most fundamental knowledge and skills, for each them and level that students need and that parents can deal smoothly, including; The Humanities; The Natural Sciences; and The Social Sciences [38].

The E-Book is prepare to present by cooperation relation between experts in curricula and educational technology using flexible audiovisual model, which content detector of the skills, knowledge and information, that student's needs. According to Hu et al. [39], the audiovisual as the model for inferring the correlation between knowledge and skills contents. To ease the difficulty of classroom learning, audiovisual propose a novel curriculum learning strategy that trains the model shows that such ordered learning procedure rewards the model and practical to train and fast convergence. Meanwhile, the audiovisual model can also provide effective individual representation for each student performance.

2. Moreover, school leaders should cooperate with educational supervision and guidance department. The supervisor and guide play a determining role in setting the instructional guides to help teachers in modifying the educational status, aiming at the promotion of quality of teachers' performance and removal of their problems, especially after COVID-19 pandemic. Educational supervision and guidance improve teaching methodology of teachers, through modifying the educational plan, to teaching methodology, on-the-job training, to encourage using educational aids during teaching, to conduct an effectiveness evaluation, and to modify learning conditions for students. Therefore, instructional leaders must enhance the relationship between teachers and educational supervisor and guides to apply such teaching and learning process that Consistent with educational technology using proper competencies activities and suitable instructional application in the case of the influential of COVID-19 pandemic.

The influential by the COVID-19 pandemic imposed digital methods to teachers. This situation requires a supervisor and advisor role to modify the plan, which is suitable for teaching and learning methods (pedagogical), and assessment. It is impossible to leave teachers they work hard and persevere on their own, advisors provide supervision over the teacher's performance by skilled teachers as an example or help teachers to discover educational rules and find the best methods in teaching and learning by using several technological ways.

There are several duties of an educational supervisor to keep teachers meeting respective technological standards using various methods focusing on distance learning;

2.1. The school principal and the guides have to make teachers familiar with the details of their work and methods. In addition, using distant learning and focus only on the first-important fundamental academic skills and knowledge for each subject (The Humanities; (ii) The Natural Sciences; and (iii) The Social Sciences) with required means for the achievement of desired results.

2.2. The school principal and the guides encourage teachers in various forms for the best creative and innovative teaching method. The using of transform teaching and learning into an innovative efficient is required a practical grounding in artificial intelligence and its educational devices and applications, equipping teachers with the knowledge and confidence they need to transform distant learning into an innovative.

2.3.   Educational guidance and supervision must prepare different extra- activities for students with special needs according to No Child Lift Behind (NCLD) which, aim to provide equal educational opportunities for disadvantaged students.

Therefore, school principal, teachers and parents held accountable in how students learn and achieve in several ways. The educational guidance and supervision share school principal accountability for those with low performance need individual education programs, and different activates using both short-term and long-term planning that needed for special teaching and learning by various activities for those categories of students.

Therefore, artificial intelligence provides many types of different application that promote the disadvantaged of students. An application can find that how a student with disadvantaged predicts what is going to type and provide the suggestion to correct errors in spelling, reading and math. The students with disadvantaged individually utilise computer game, the machine itself plays the game like as an opponent based on educational activity in the game. Gaming like robot considered one of the most common uses of the advantages of artificial intelligence technology brings many benefits to students and teachers.

8   Methodology

The objectives of this paper did not include the methodology section. Researchers could replicate this study for further research using the method to gather data (survey, interviews, experiments, etc.), and describing in details to allow readers to evaluate the reliability and validity of this study.

9   Conclusion

This paper determines the principal's role as an instructional leader and technology integration, in particular artificial intelligence. The use of artificial intelligence is becoming one of the technology concepts exists more commonplace all around us— in our workplaces, or at homes, and especially in schools. Robot as an example of artificial intelligence that can be fantastic valuable teaching tools and the importance to implement them properly for the best performance results.

According to Vinuesa el at. [40], the development of artificial intelligence needs to support by the research and persistence, focuses on the main aspects, regulatory, ethical standard, and oversight for artificial intelligence-based technologies to avoid what could result in gaps in transparency and safety. School principal leadership is affected by the education technology trend of artificial intelligence which, becoming more commonplace all around us.

Artificial intelligence has some advantages. School leaders can facilitate the teaching and learning process and motivates teachers. In terms of setting their technological goals along with the disadvantages, robots as an example of artificial intelligence can be amazingly valuable teaching tools, but to implement them properly for the best possible results. Therefore, the school principal should be aware of the consequences that may become challenges affecting the school reform.

10 Limitation of the Study

Few studies addressed the impact of artificial intelligence on instructional leadership. Therefore, the author's personal experiences have taken into consideration in the paper discussion and the methodology of the study. Nonetheless, there is a need for further research in the educational field of instructional leadership and artificial intelligence in a particular educational stage. The same results should undertake several times, in different contexts, different researchers should treat the study before claiming its generalizability.

References

1. Ezzat, H., Agogué, M., Le Masson, P., & Weil, B. (2017). Solution-oriented versus novelty-oriented leadership instructions: Cognitive effect on creative ideation. In *Design Computing and Cognition'16* (pp. 99–114). Cham: Springer.
2. Sabuncuoglu, A. (2020). Designing one-year curriculum to teach artificial intelligence for middle school. In *Proceedings of the 2020 ACM Conference on Innovation and Technology in Computer Science Education, June* (pp. 96–102).
3. Timms, M. J. (2016). Letting artificial intelligence in education out of the box: Educational robots and smart classrooms. *International Journal of Artificial Intelligence in Education, 26*(2), 701–712.
4. Marks, H. M., & Printy, S. M. (2003). Principal leadership and school performance: An integration of transformational and instructional leadership. *Educational Administration Quarterly, 39*(3), 370–397.
5. Murphy, J. (1990). Principal instructional leadership. *Advances in Educational Administration, 1*(B: Changing Perspectives on the School), 163–200.
6. McCarthy, J. (1998). What is artificial intelligence? cogprints.org.
7. Brockmeier, L., Sermon, J., & Hope, W. (2005). Principals' relationship with computer technology. *NASSP Bulletin, 89*(643), 45–63.
8. NCREL. (2011). Critical issue: Providing effective schooling for students at risk retrieved on 15.07.2013 at URL: http://www.ncrel.org/sdrs/areas/issues/students/atrisk/at600.htm.
9. Zhang, W., & Koshmanova, T. (2020). A comparative study of school principal experiences: Recontextualization of best American school principals of using technology in China. In *Society for Information Technology & Teacher Education International Conference, April* (pp. 651–656). Association for the Advancement of Computing in Education (AACE).

50 Recalde, J. M., Palau, R., Galés, N. L, & Gallon, R. (2020). Developments for Smart Classrooms: School Perspectives and Needs. *International Journal of Mobile and Blended Learning (IJMBL), 12*(4), 34–50.

11. Roschelle, J., Pea, R., Hoadley, C., Gordin, D., & Means, B. (2000). Changing how and what children learn in school with computer-based technologies. *The Future of Children, 10*(2), 76–101.

12. Whitehead, B. M., Jensen, D. F., & Boschee, F. (2013). *Planning for technology: A guide for school administrators, technology coordinators, and curriculum leaders.* Thousand Oaks: Corwin Press.

13. Miller, M. (2008). A mixed-methods study to identify aspects of technology leadership in elementary schools. *Dissertation Abstracts International Section A: Humanities and Social Sciences, 69*(2-A), 579.

14. Kumar, A. N. (2004). Three years of using robots in an artificial intelligence course: Lessons learned. *Journal on Educational Resources in Computing (JERIC), Special Issue on Robotics in Undergraduate Education, part I, 4*(3_p).

15. Johnson, W. L., Rickel, J., & Lester, J. C. (2000). Animated pedagogical agents: Face-to-face interaction in interactive learning environments. *International Journal of Artificial Intelligence in Education, 11*, 47–78.

16. Kozloski, K. C. (2006). *Principal leadership for technology integration: A study of principal technology leadership.*

17. Aminova, M., Mareef, S., & Machado, C. (2020). Entrepreneurship ecosystem in Arab world: The status quo, impediments and the ways forward. *International Journal of Business Ethics and Governance, 3*(3), 1–13. https://doi.org/10.51325/ijbeg.v3i3.37

18. Kincaid, T., & Feldner, L. (2002). Leadership for technology integration: The role of principals and mentors. *Educational Technology & Society, 5*(1). Retrieved June 6, 2011, from http://ifets.ieee.org/periodical/vol_1_2002/kincaid.html.

19. Campolo, A., Sanfilippo, M., Whittaker, R., & Crawford, K. (2017). *AI Now 2017 report.* New York: AI Now. Retrieved from https://ainowinstitute.org/AI_Now_2017_Report.pdf.

20. M. Nassar, R. M., & Battour, M. (2020). The impact of marketing ethics on customer loyalty: A conceptual framework. *International Journal of Business Ethics and Governance, 3*(2), 1–12.

21. Lennon, L. (2012). The role of the school principal in technology integration: A literature review.

22. Leonard, L., & Leonard, P. (2006). Leadership for technology integration: Computing the reality. *The Alberta Journal of Educational Research, 52*(4), 212–224.

23. Salman, M., & Battour, M. (2020). Career excellence between leadership roles and achievement motivation for employees in the ministry of education in the United Arab Emirates. *International Journal of Business Ethics and Governance, 3*(1), 46–96. https://doi.org/10.51325/ijbeg.v3i1.33

24. Krafft, P. M., Young, M., Katell, M., Huang, K., & Bugingo, G. (2020). Defining artificial intelligence in policy versus practice. In *Proceedings of the 2020 AAAI/ACM Conference on AI, Ethics, and Society (AIES).*

25. Gupta, I., Mishra, N., & Tandon, D. (2020). Triple bottom line: Evidence from aviation sector. *International Journal of Business Ethics and Governance, 3*(1), 97–104. https://doi.org/10.51325/ijbeg.v3i1.32

26. Paschen, U., Pitt, C., & Kietzmann, J. (2020). Artificial intelligence: Building blocks and an innovation typology. *Business Horizons, 63*(2), 147–155.

51 Powles, J., & Hodson, H. (2017). Google DeepMind and healthcare in an age of algorithms. *Health and technology, 7*(4), 351–367.

28. Lynch, M. (2019). Will artificial intelligence take over educational leadership? https://www.thetechedvocate.org/will-ai-take-over-educational-leadership/.

29. M. Bourton, J. S., Lavoie, J., & Vogel, T. (2018). Will artificial intelligence make you a better leader? *The McKinsey Quarterly, 2*, 72–75. McKinsey Quietly Article.

30. Da Silva Oliveira, J., Espíndola, D. B., Barwaldt, R., Ribeiro, L. M., & Pias, M. (2019). IBM Watson Application as FAQ Assistant about Moodle. In *2019 IEEE Frontiers in Education Conference (FIE), October* (pp. 1–8). IEEE.

31. https://www.kevinkruse.com/7-things-i-learned-from-building-an-ai-chatbot-for-leadership-development/.
32 Suguna, S. K., & Kumar, S. N. (2019). Application of cloud computing and internet of things to improve supply chain processes, *In Edge Computing*, 145–170). Springer, Cham.
33 Akapame, R., Burroughs, E., & Arnold, E. (2019). A Clash Between Knowledge and Practice: A Case Study of TPACK in Three Pre-service Secondary Mathematics Teachers, *Journal of Technology and Teacher Education, 27*(3), 269–304.
34 Zuo Nina, & Li Dongqing. (May 2019) Research on the Cultivation of "Artificial Intelligence + Law Science" Composite Talents, *Journal of Management School of Guangxi University of Political Science and Law*, 125, (in Chinese).
35. Soland, J., Tarasawa, B., Johnson, A., Ruzek, E., & Liu, J. (2020). Projecting the potential impacts of COVID-19 school closures on academic achievement.
36. Tedesco, J. C., Opertti, R., & Amadio, M. (2014). The curriculum debate: Why it is important today. *Prospects, 44*(4), 527–546.
37. Sullivan, S., & Glanz, J. (2005). *Supervision that improves teaching: Strategies and techniques.* Thousand Oaks: Corwin Press.
38. Beauchamp, G. A. (1982). *Curriculum theory: Meaning, development and use.* New York: Routledge.
39. Hu, D., Wang, Z., Xiong, H., Wang, D., Nie, F., & Dou, D. (2020). Curriculum audiovisual learning. arXiv preprint arXiv:2001.09414.
40. Vinuesa, R., Azizpour, H., Leite, I., Balaam, M., Dignum, V., Domisch, S., & Nerini, F. F. (2020). The role of artificial intelligence in achieving the sustainable development goals. *Nature Communications, 11*(1), 1–10.
41. Du Boulay, B. (2016). Artificial intelligence as an effective classroom assistant. *IEEE Intelligence Systems, 31*(6), 76–81.
42. https://edtechreview.in/trends-insights/insights/2541-role-of-school-leader-principal-in-school-technology.

Printed in the United States
by Baker & Taylor Publisher Services